Spectroscopy Essentials

Volume I

Spectroscopy Essentials
Volume I

Edited by **Allegra Smith**

NY RESEARCH PRESS

New York

Published by NY Research Press,
23 West, 55th Street, Suite 816,
New York, NY 10019, USA
www.nyresearchpress.com

Spectroscopy Essentials: Volume I
Edited by Allegra Smith

International Standard Book Number: 978-1-63238-422-5 (Hardback)

Printed in China.

Contents

Preface

The phenomenon of interpretation of spectra of absorption and emission of electromagnetic radiation due to contact between an energy that depends on some particular wavelength of EM radiation and matter, is called Spectroscopy. Some of the prominent names in this field include Max Planck and his theory of blackbody radiation, Albert Einstein's photoelectric effect and Neil Bohr's theories related to atomic structure and spectra.

This phenomenon has established itself as a research tool in numerous disciplines of science such as; chemistry, physics, biology, medicine, and ecology. One such popular application of Spectroscopy is the use of lasers. This field is growing every day, as scientists are exploring new techniques. One of the central concepts which is crucial to everyone interested in spectroscopy is resonance and its corresponding resonant frequency. Spectroscopic studies have played a crucial role in the development of quantum mechanics. The measurement devices used to measure spectroscopic information include spectrometers, spectrophotometers, spectrographs or spectral analyzers.

The research field of Spectroscopy is quite large and the goal of this book is to cover the most advanced topics like atomic spectroscopy, laser spectroscopy, molecular spectroscopy, nanostructure spectroscopy, physical, and organic spectroscopy.

I wish to thank all the authors and scientists from all across the world for their contribution to this book. Without their timely submissions and dedication to the project, this publication wouldn't have been possible. Also, I'm grateful to the team of editors and the publishing house for providing technical assistance at each and every juncture of the project. Lastly, I wish to thank my family for their constant support throughout the publication process.

Editor

1

Vital Autofluorescence: Application to the Study of Plant Living Cells

Victoria V. Roshchina

Laboratory of Microspectral Analysis of Cells and Cellular Systems and Optical Department Russian Academy of Sciences, Institute of Cell Biophysics, RAS, Institutskaya Str. 3, Pushchino, Moscow Region 142290, Russia

Correspondence should be addressed to Victoria V. Roshchina, roshchinavic@mail.ru

Academic Editor: Jin Zhang

The application of various microscopy methods such as luminescence microscopy, microspectrofluorimetry and laser-scanning confocal microscopy has been considered as an approach to study the autofluorescence of plant living cells—from cell diagnostics up to modelling the cell-cell contacts and cell interactions with fluorescent biologically active substances. It bases on the direct observations of secretions released from allelopathic and medicinal species and the cell-donor interactions with cell-acceptors as biosensors (unicellular plant generative and vegetative microspores). Special attention was paid to the interactions with pigmented and fluorescing components of the secretions released by the cells-donors from plant species. Colored components of secretions are considered as histochemical dyes for the analysis of cellular mechanisms at the cell-cell contacts and modelling of cell-cell interactions. The fluorescence of plant biosensors was also recommended for the testing of natural plant excretions as medical drugs.

1. Introduction

The visible (400–700 nm) fluorescence of intact living cells (usually called autofluorescence) excited by ultraviolet, violet, or blue light is often observed under luminescence microscope [1, 2]. Due to special microspectrofluorimeters or confocal microscope the fluorescence spectra of objects studied and the emission intensity can be also measured. Autofluorescence is the natural emission of cellular components due to a contribution of fluorescent compounds located in different cellular compartments. In general blue and green emission reflects the participation of universal cellular fluorophores NAD(P)H (max. 460 nm), pterins (max. 450 nm), and flavins (max. 520 nm) [3], although in dry state weak blue fluorescence (420–450 nm) is also peculiar to isolated nucleotides, RNA, and DNA as well as some proteins, and seen in the films under luminescence microscope [4, 5], but water quenches the emission [4]. Besides, there are characteristic fluorescent compounds in various organisms [6]. In animal cells chitin (max. 450–460 and 520 nm), collagen and elastin (max. 400–430, 465, 495 and 520 nm) or cellulose (max. 420–430 nm) in plants, and fungi can also contribute in blue fluorescence [6]. Orange-red fluorescence depends on organism. In plant cells main known fluorophore used for environmental monitoring is chlorophyll with maximum 675–680 nm [7] and in photosynthesizing bacteria: bacteriochlorophyll [6]. Uroporphyrinogen and haemoglobin contribute in red lightening of mammalians, while green-red range emission may reflect a contribution of sea animal fluorescent proteins used for the genetic experiments as vector for the study of cellular components [6–8]. Terpenes, some alkaloids (colchicine) and flavonoids in plants, emit in blue or in blue-green (470–525 nm) region of the visible spectrum whereas polyacetylenes, isoquinoline, and acridone alkaloids emit in yellow and orange [8]. Red fluorescence of plant cells may be also due to anthocyanins and azulenes (600–630 nm) [8]. At ageing or, if the object being undergone ozone and relative reactive oxygen species, fluorescent pigments lipofuscins are formed that may emit in wider spectral range—from blue (450–480 nm) to red

(>600 nm) in a dependence on the nature of the organism and duration of the exposure under unfavourable factor [9, 10].

Autofluorescence of the cellular compounds may serve (1) as markers for cytodiagnostics and (2) considerations of as natural indicator of cellular state because the emission often shows changes in cellular metabolism and responses to the external and internal signals [8]. This is true for the plant secretory structures analyzed under luminescent microscope (the structures are not seen in usual microscope without special histochemical staining) or laser-scanning confocal microscope (LSCM images) [8, 10, 20–22, 37–39]. The intensity and spectral composition of the emission differ in a dependence on, namely, (1) nature of the organism or (2) cell analyzed, (3) a taxonomic position of the organism tested, and (4) environmental and experimental conditions (actinic light, temperature, humidity, physiological state of a cell as a whole and phase of development, influence of neighbour cells or organisms, including parasites). Autofluorescence could be used (i) in express-microanalysis of the accumulation of the secondary metabolites in secretory cells without long biochemical procedures; (ii) in diagnostics of cellular damage; (iii) in analysis of cell-cell interactions [8, 10, 20, 21, 39].

In any case the phenomenon of autofluorescence could be used as a resource for the nondestructive vital microscopic analysis of natural organisms [41]. It has already practical application in various diagnostic procedures for cell biology analysis, ecology, medicine, and pharmacology. This paper considers the application of various types of luminescence microscopy to the study of plant cells, mainly such as secretory systems.

2. Images and the Fluorescence Spectra of Fluorescing Cells in Cytodiagnostics

Today our knowledge in the problem of autofluorescence, peculiar to living cells, mainly include results of experiments made on big plant cells that are suitable for observation both *in vivo* (such as in nature) and *in vitro* (in cultivated systems). Among similar objects are generative cells of flowers: pollen (male gametophyte) and pistil (female gametophyte) and nongenerative cells, such as plant vegetative microspores and special secretory cells. Some examples dealt with the application of luminescence microscopy in various modifications to analysis of the cells are shown on Figure 1. In particular there are the images seen under usual luminescence or confocal microscopes and the fluorescence spectra measured by microspectrofluorimeter or some types of confocal microscopes. We shall also comment about Figure 1 in sections below.

The research of fluorescence maxima in secretory cells showed the correlations and differences in the emission spectra excited by UV-light among the plant species studied and their specialized structures [8]. For example, nectaries mainly fluoresce, in blue region, while sporangia of fungi, mosses, horsetails, and ferns are in green-yellow (500–550 nm). Except fungi, the structures have also maximum 680 nm in red, peculiar to chlorophyll located in chloroplasts. Crystals of secretions emit more brightness than liquids. Today the secretion peaks were compared with the maxima of individual components of the same secretions in solutions or films [5, 8, 10]. Interference of various fluorophores in total emission may take place, but diagnostics of prevailing component of the secretion is real. Changes in the fluorescing compounds depend on stage of plant development, the influence of various factors—ozonation, redox agents and excretions of other organisms in the environment.

2.1. Pollen. Pollen is the unicellular structure (served for plant breeding of seed-bearing plants) that germinates on pistil stigma of flower forming pollen tube that grows within pistil style and carries out two spermia—male cells. As soon as the pollen tube achieves ovule (seed bag), here one sperm interacts with egg cell, and the fertilization is over by the embryo arising. Pollen often show visible autofluorescence excited by ultra violet or violet light. Luminescence microscope or laser-scanning confocal microscope was used for both the receiving fluorescing cell images and (in some cases) fluorescence spectra. Some examples of the emission are on Figure 1(a) and Table 1. Full information about the earlier studies of pollen autofluorescence as well as apparatuses for the observation and measurement has been summarized in reviews [18, 19] and monograph [8]. Here there are the references related to pioneers who first have seen autofluorescence of pollen under a luminescence microscope [43–45] and later measured first fluorescence spectra of pollen from various species [46, 47] with a microspectrofluorimeter.

Figure 1 shows the LSCM images of pollens (from medicinal plants) released fluorescing secretions from apertures, through which pollen tube enters at pollen germination. We see different colors of autofluorescence from pollen surface and drops of secretions (marked with arrows) released from apertures (green emission of the secretions from *Solidago virgaurea*, orange-red lightening from one of *Calendula officinalis*, and colourless from *Oenothera biennis*). Today the phenomenon of autofluorescence is applied to the various diagnostics seen from Table 1. First papers in this line dealt with analysis of the exine autofluorescence by luminescence microscope. Exine, the outer layer of pollen cover, consists of sporopollenin and in various species may include various fluorescent pigments from phenols and carotenoids up to azulenes and proazulenes [8, 15]. Structure of exine was also analyzed with confocal microscopy [25], permitting to receive optical slices [8, 20, 21, 38, 39], and the fluorescence spectra were also recorded [8, 15–19]. Based on the autofluorescence the various steps in pollen development were studied [8, 13]. Moreover, the phenomenon is applied to pollen classification [30, 31], including the search pollen grains of fruitful varieties from certain plant species [33]. For understanding of sporopollenin synthesis and exine assembly a simple effective genetic screen based on their autofluorescence characteristics was represented for the *Arabidopsis thaliana* pollen (the recovery of mutants with a variety of defects in exine structure) to identify genes involved in pollen exine production [34].

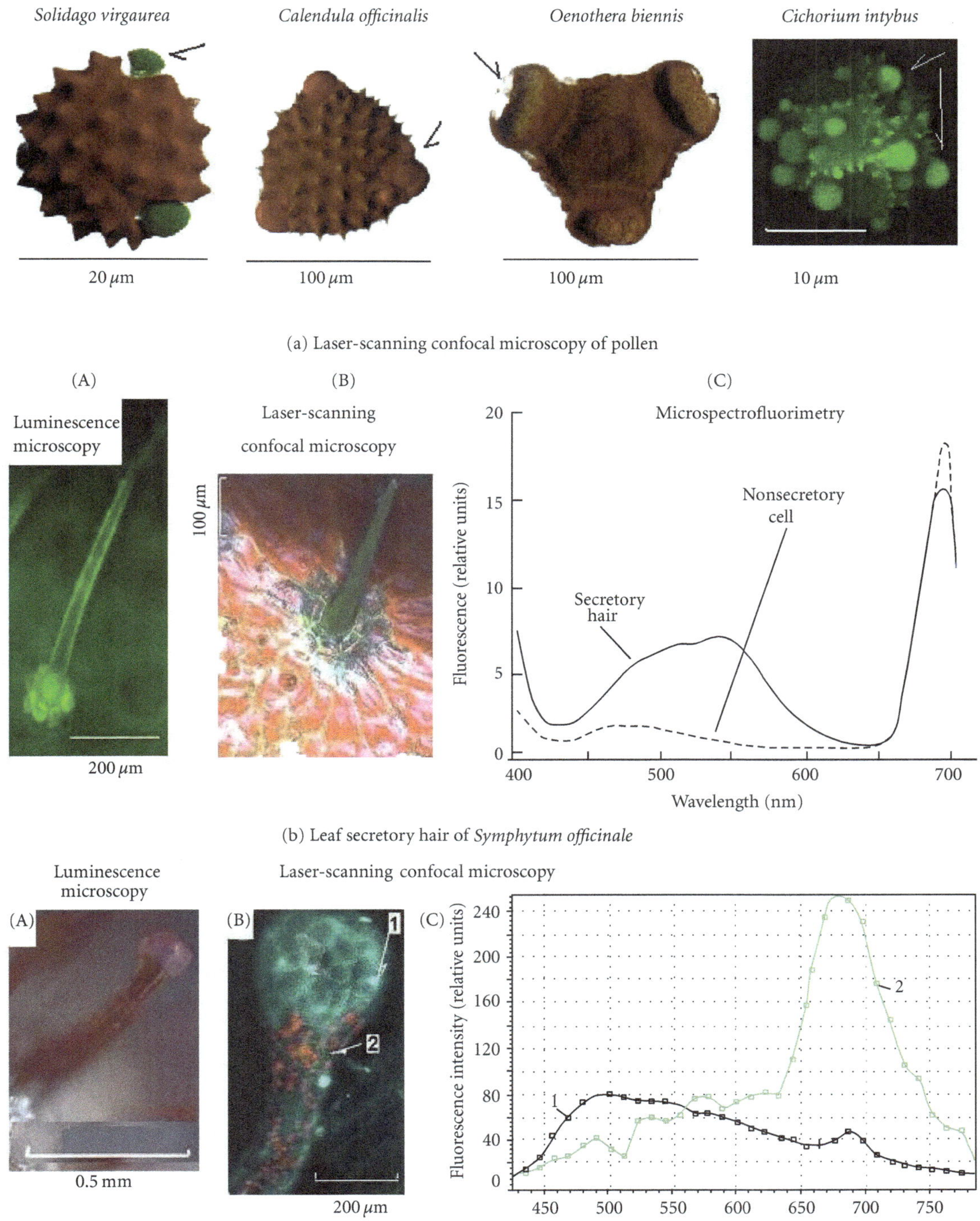

(a) Laser-scanning confocal microscopy of pollen

(b) Leaf secretory hair of *Symphytum officinale*

(c) Secretory hair on flower calyx from *Rubus odoratus*

FIGURE 1: Images and the fluorescence spectra of some intact fluorescing plant secreting cells. Source: adapted from [21, 22] and unpublished data of the review' author. (a) LSCM images of pollen grains from various species under confocal microscope LSM 510 NLO "Carl Zeiss" (Laser 488 nm). Drops of secretions from apertures are seen. (b) and (c) Views of secretory hairs on leaf of comfrey *Symphytum officinale* and on the flower calyx of flowering raspberry *Rubus odoratus* under (A) luminescence microscope Leica 6000 (Excitation 360–380 nm) and (B) laser-scanning confocal microscope Leica TSC SP 5 (laser 405 nm) and the (C) fluorescence spectra measured by registered microspectrofluorimeter MSF-1 (Excitation 360–380 nm) or laser-scanning confocal microscope Leica TSC SP 5 from secretory head with released secretion (1) and stalk (2) of secretory hair.

Table 1: Autofluorescence in pollen diagnostics (LM—Luminescence microscopy, MS—Microspectrofluorimetry, LSCM—Laser-scanning confocal microscope).

Diagnostics	Microscopic method	Fluorescence characteristics
Vital pollen characteristics. Changes in the autofluorescence of the pollen wall during microsporogenesis and chemical treatments [11, 12]	LM, MS	Color and fluorescence spectra
Wall development of sterile and fertile pollen [13]	LM	Color
Assessment of pollen retention using exine autofluorescence [14]	LM	Images
Matured and non-matured pollen [15–19]	MS	Fluorescence spectra (pigment composition)
Self-incompatibility and self-compatibility of pollen [17]	MS	Fluorescence spectra (pigment composition)
Pollen loads analysis as well as pollen-derived products of bees [8, 17]	MS	Fluorescence spectra (pigment composition—azulenes, carotenoids)
Pollen-pollen interactions in the mixtures [8, 20–23]	LSCM, MS	Changes in the fluorescence spectra of pollen from different species reflect changes in pigments' state
Pollen viability [8, 17]	MS	Fluorescence intensity after moistening
Pollen-pistil interaction at fertilization [8, 17, 24]	MS	Changes in the fluorescence spectra reflect the receptor state
Exine structure [25]	LSCM	Images
Quantifying dynamics of pollen shed [26]	LM	Images
Ozone damages [27–29]	MS	Changes in the fluorescence spectra reflect pigments' state and forming pigment of aging lipofuscin
Damages by reactive oxygen species [28, 29]	MS	Changes in the fluorescence spectra
Classification of pollen grain species using autofluorescence image analysis [30, 31]	LM	Images and blue/red autofluorescence ratio
Testing in nanotechnology: Commercially produced fullerene fine particles [32]	LM	Adhesion of nanoparticles estimated on the autofluorescence intensity changes
The search of fruitful varieties from certain plant species [33]	LSCM	Images
Genetic screen to identify genes involved in pollen exine production [34]	LSCM	Wild-type exine fluoresces with two peaks 460 nm (higher) and 505 nm (lower)
A novel screen identifies multiple pollen mutants that have defective exine [34]	LSCM	Defective mutants differ in the emission intensity in the above-mentioned peaks
Taxonomic classification of pollen [35]	LSCM	Images
Pollen concentration in air [6, 36]	LM, LSCM	Images, emission intensity

The fluorescence spectra of pollen from 20 plant species, both wind—and insect-pollinated, were analyzed [8]. Pollen grains from wind-pollinated species (birch, pine and others) have mainly blue fluorescence whereas insect-pollinated ones belonged to different species—in blue, yellow, green, or blue-green. Nonmatured pollen demonstrated red fluorescence due to the presence of chlorophyll, while matured grains changed the color of the emission. The autofluorescence spectra of many pigmented and colorless pollens treated with ozone in various concentrations was also studied, and some characteristic damages were shown [27–29]. The fluorescence maximum at 530–550 nm disappeared in carotenoid-containing pollen grains of *Philadelphus grandiflorus* and *Epiphyllum hybridum*, and a new maximum at 475–480 nm arose (unlike carotenoid-less pollen of *Plantago major*) that correlated with lipofuscin-like substances observed in the extracts from the pollen grains [27–29].

In pollen mixtures changes in the fluorescence of pollens from different species may reflect their chemical (allelopathic) relations [8, 21]. LSCM images directly demonstrated the alterations in the emission and the state of pollen grains from Field Scabious *Knautia arvensis* (L.) Coult. (Dipsacaceae) contacting with pollen from wormwood *Artemisia absinthium* L. (Linn.) (Asteraceae) [20].

The autofluorescence imaging alone or in a combination with other morphological parameters [33] can be considered as nondisruptive method for pollen classification since it does not require any treatment of fixing or staining of the specimens. Therefore, this simple and fast method can be used for pollen counting and taxonomical classification in multiple applications that concern disciplines like aerobiology for identification of pollens in bioaerosols [6, 36, 48, 49]. Today the pollen study is very important, because air-occurring pollen from certain plant species may induce allergy, although on the other side pollen grains of medicinal plants are used for drug preparation. Knowledge of pollen state is necessary for analysis of plant fertility in genetics and biotechnology or of the meteorological situation in ecological monitoring as well as in pharmacy.

2.2. Pistil. Pistil of flower is female gametophyte which stigma receives chemosignal from pollen as male gametophyte and recognizes pollen grains from the same plant species (as well as self-compatible or self-incompatible pollen of the species) or from foreign species. Autofluorescence reflects the pistil state before pollen application and after the pollen grains addition on pistil stigma [8, 17, 24]. If self-pollination occurs one can see changes in the peaks positions and the emission intensity in a comparison with the pistil fluorescence reply on pollen that originates from another plant of the same species like was seen on the example with *Hippeastrum hybridum* [16, 17]. When foreign pollen was applied to the pistil stigma often there are no changes in the pistil fluorescence [8]. Moreover pistils from self-compatible and self-incompatible lines of the one and same plant species, in particular as in *Petunia hybrida* clones, have different maxima in the fluorescence spectra [8].

2.3. Secretory Cells—Donors of Biologically Active Fluorescing Compounds. Any living cells may form and release secretions out, and it is better seen in specialized secretory plant cells with fluorescent products [8, 10]. Images of the secreting cell fluorescence excited with ultraviolet or violet light are especially valuable in allelopathic and medicinal plants which contain compounds named allelochemicals (simultaneously being valuable pharmaceuticals) in secretory cells [50]. Many of the chemicals excreted act on the other organisms in biocenosis. The secretory cells often show bright fluorescence at the excitation of ultraviolet (360–390 nm) or violet (400–430 nm) light [8, 10, 51, 52] due to the presence of phenols, alkaloids, terpenoids, etc.). The fluorescing images of similar cells—glandular hairs of leaf and flower—are shown under usual luminescence and confocal microscopes as seen on Figures 1(b) and 1(c). Leaf secretory hair of comfrey *Symphytum officinale* from family Boraginaceae under the violet or blue-green excitation (the structures weakly lighten at the ultraviolet) has the green-yellow emission maximum 550 nm and shoulder 460–500 nm in the fluorescence spectrum. Another plant species—purple flowering raspberry *Rubus odoratus*—shows the emission of calycinal secretory hair on flower in blue-green with maximum 475 nm unlike nonsecretory stalk cells that have maximum 675–680 nm, peculiar to chlorophyll.

The images and the emission intensity of secretory cells included fluorescent drugs is important for medicinal plant species because this may characterise the state of pharmaceutically valuable material [40] as shown in Table 2 for fresh and dried leaves of warmwood *Artemisia* species (in pharmacy used as tonic, stomachic, febrifuge, and antihelmintic and a good remedy for enfeebled digestion and debility). The species contained fluorescent terpenes, sesquiterpenes and phenols [8]. The green emission of secretory cells on fresh leaves of wormwood *A. absinthium* L. was about 4 times higher, than in non-secretory cells. In dried material the difference increased significantly up to 20 times. This demonstrates a difference in the fluorescence intensity between secretory cells of fresh leaves than in dried leaves. In mugwort *A. vulgaris* L. dried leaf secretory cells emitted 2 times brightly in a comparison with secretory cells of fresh material. The visible difference in the fluorescence intensity between two species reflects the difference in an accumulation of fluorescent pharmaceuticals (azulenes and flavonoids) [40]. Flavonoids and coumarins as fluorophores may be also included in the leaf fluorescence of *A. vulgaris*.

3. Images of Fluorescent Plant Secretory Cells in Cell-Cell Interactions

Cytodiagnostics of cell-cell interactions has perspectives also in the analysis of autofluorescence in the studies of chemical relations (allelopathy) between species in biocenosis relations). The examples may be a visualization of the insect attack [53] or fungal invasion [54–57] and relationships between pollen grains of different species [20, 21]. Fungi or insects such as ticks, spiders, or human skin fluoresce, mainly, in blue (460–470 nm) or blue-green (<520–530 nm) whereas muscle—in blue with maxima 410 and 480 nm. Insects fluoresce by all parts of their body, and it may observe separately on the plant cell surface. The blue emission intensity of fungal hypha and mycelium are much higher than in fluorescing secretory plant cells that makes possible to recognize fungal infection.

3.1. The Parasitive Invasion. Figure 2 demonstrates the example for leaf of medical plant mint *Mentha piperita* used by aphid for nutrition. On the leaf colorless microscopic larvae are plant-louse not seen well up to the excitation by actinic light of luminescence microscope. Blue-fluorescing Mentha gland (if excited by UV-light) with maxima 475, 540, 675–680 nm in the fluorescence spectrum is covered by

Table 2: The fluorescence intensity (I) of leaf secretory cells of medicinal plants *Artemisia absinthium* and *Artemisia vulgaris* measured by double-beam microspectrofluorimeter MSF −2 at 520–540 nm or at 640–680 nm. Source: [40]. Excitation 420 nm.

Cell	Plant material	$I_{520-540}$	$I_{640-680}$
	Artemisia absinthium		
Glandular cell and hair	Fresh leaves	0.84 ± 0.07	5.76 ± 0.9
Nonsecretory cells	Fresh leaves	0.21 ± 0.01	7.99 ± 1.0
Glandular cell and hair	Dried leaves	0.28 ± 0.01	3.15 ± 0.08
Nonsecretory cells	Dried leaves	0.01	2.28 ± 0.03
	Artemisia vulgaris		
Glandular cell and hair	Fresh leaves	0.12 ± 0.01	2.26 ± 0.09
Nonsecretory cells	Fresh leaves	0.04 ± 0.01	4.89 ± 0.20
Glandular cell and hair	Dried leaves	0.26 ± 0.02	1.83 ± 0.08
Nonsecretory cells	Dried leaves	0.14 ± 0.01	2.40 ± 0.5

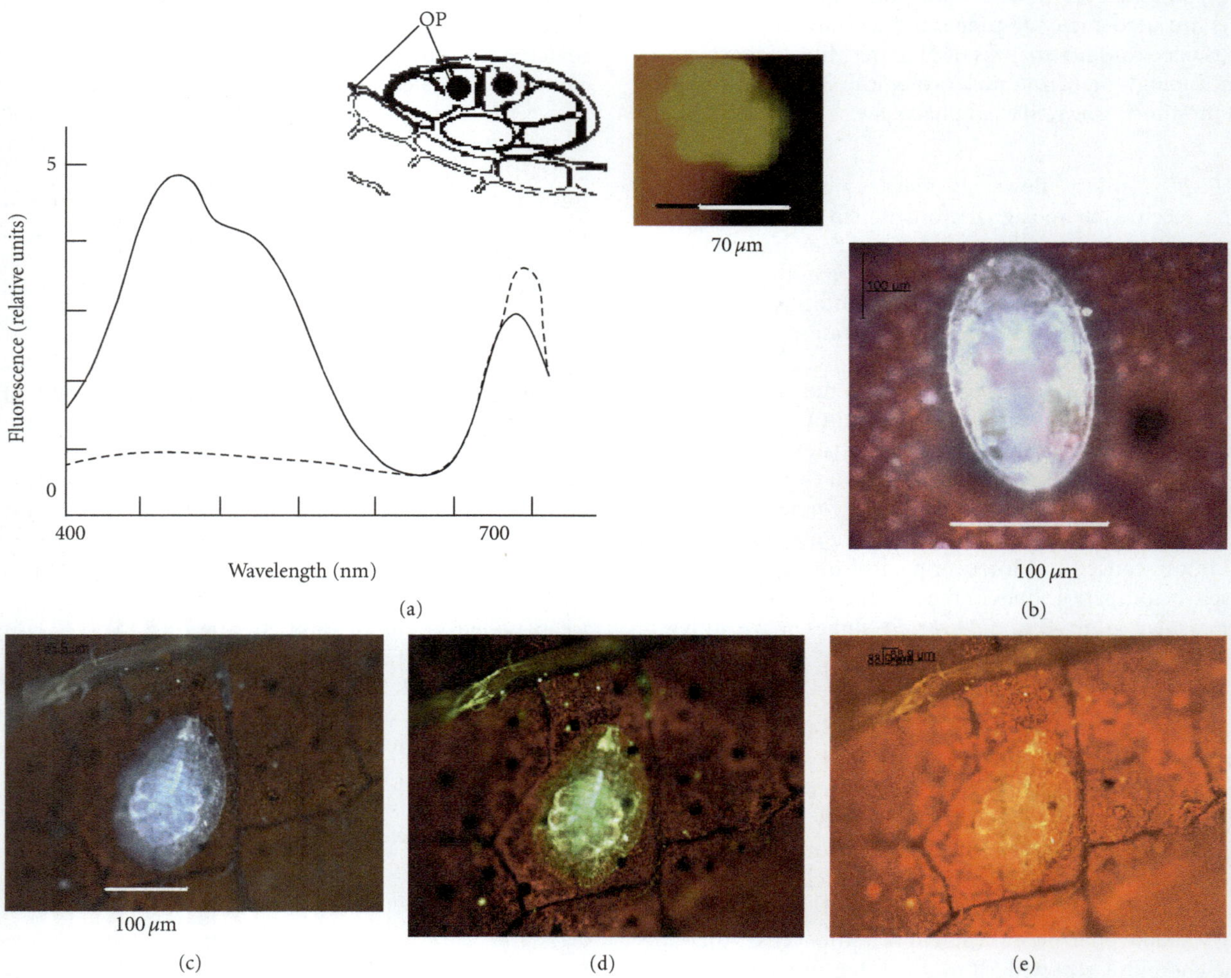

Figure 2: Fluorescent larva of aphis interacts with cellular complex of leaf gland of mint *Mentha piperita*. (a) Left: the fluorescence spectra of secretory cells measured by microspectrofluorimeter MSF-1. Excitation: 360–380 nm. Unbroken line—secretory cells; broken line nonsecretory cells. The position of optical probe (OP) is shown. Right: the common view of gland under luminescence microscope "Carl Zeiss CD-28". ((b)–(e)) Images under luminescence microscope *Leica* DM 6000. (b) The common view of aphis larva out of the gland, ((c)–(e)) views of larva on the mint gland. Excitation by light 360–410 nm (b), 420–450 nm (c), 480 nm (d), and 560 nm (e).

weak blue-fluorescing aphis larva (chitin emission, mainly at 450 nm). At excitation by violet or green-yellow light yellow-fluoresced mint gland (surrounded by red fluorescing non-secretory cells of leaf mesophyll due to chlorophyll presence) mainly was seen through nonvisible (transparent) larva body.

Many pathogenic and saprophytic fungi fluoresce in blue green (in 390–490 nm spectral region) under actinic ultraviolet light [54–57]. Moreover the phenomenon was better seen under stress (desiccation, heat, chemicals) [56]. Recently blue-green autofluorescence of fungi of *Ustilago maydis* on maize and cell wall of this plant is an indicator of the invasion [57]. At the fungal infection the value of autofluorescence as a screening method for detecting fungi in tissues was compared with conventional histochemical stains. Autofluorescence can be used as a rapid screening method for identification of fungi in tissue sections as it does not require any other specialized staining procedure. Confocal microscopy appears to be more suitable for plant-pathogen interaction in details [58]. Fungal fluorophores may depend on species and are not well studied yet. Today fungal infection is better registered observing the enhanced blue-green autofluorescence of plant cell walls related to their phenols and sometimes auxin exchange [59].

3.2. Pollen-Pollen Interactions. Special interest is the visualization of the interaction between pollen grains of various species that named as pollen allelopathy. At the contact, excretion from pollen-donor acts as chemosignal on pollen-acceptor, inducing the changes in the fluorescence of the microspores and their germination. The interaction of big pollen from *Knautia arvensis* (fluorescing in green) and small pollen grains from *Artemisia absinthium* or from *Solidago virgaurea*, which have released drops of red-fluorescing secretion as seen under laser-scanning confocal microscope [8, 20, 21]. This secretion was seen on the surface of non-germinated pollen from *Knautia arvensis* and also on the surface of pollen tubes. The interaction of green fluorescent pollen of *K. arvensis* and red-fluorescing secretion of pollen from *Artemisia absinthium* decreased the germination in both contacting types of microspores. In this case, perhaps, the enhancement in fluorescence is related to the energy loss peculiar to nonviable pollens. On LSCM images of fluorescent pollens in their mixtures red-fluorescing secretion for large pollen of *Oenothera biennis* contacted with small pollens of *Matricaria chamomilla* emitted in green and liberated green-fluorescing secretions were seen [21]. Optical slices showed the interior of contacted pollen grains, green fluorescence was seen only on the surface of *M. chamolilla* microspore, while red emission was also seen inside the pollen grain of *O. biennis*. LSCM images of fluorescent pollen made it possible to observe directly (i) the contact of intact cell, (ii) changes in their emission, and (iii) location of contacting secretions.

Intensity of autofluorescence from interacting cells in pollen mixtures also reflects the mutual influence of contacted secretions as shown for microspores from *Hippeastrum hybridum*, *Philadelphus grandiflorus*, *Knautia arvensis*, *Plantago major*, and *Oenothera biennis* used as pollen-acceptors, while other pollen grains from 42 plant species—as donor pollens [21]. Change of the light emission (excited by UV 360–380 nm or violet 430 nm light) is a visual chemical response to the pollen excretion from foreign pollens. To develop models for pollen-pollen interactions, the effects of allelochemicals secreted by foreign pollen (i.e., donor pollen) were studied *in vitro* or *in vivo* on autofluorescence concerned with the germination of acceptor-pollen [21]. Autofluorescence of acceptor and donor pollens in mixtures was changed to blue-green and red spectral regions that may serve as the indicator reply on foreign microspores. Generally increased blue-green and occasionally red emission indicated the decrease in the rate of pollen germination. However, red fluorescence was not related to germination and depended, mainly, on species used. It may be explained by redox changes in pollen components (often cover constituents) or by the synthesis of new fluorescing metabolite. Perhaps, the enhancement in blue-green emission of pollen could be related to the formation of fluoresced Schiff bases or to the formation of NAD(P)H emitted with maxima 460 nm. Red fluorescence of matured pollen is often correlated with the presence of anthocyanins or azulenes like occurs in overgrown parts of plants [8, 21].

4. Autofluorescence as Biosignal or Marker to Study Localization of Secondary Metabolites

Even 50 years ago Frey-Wyssling and Agthe have discovered fluorescence of flower nectar [8], as possible signal for insects-pollinators [60], and today the phenomenon has been demonstrated for various floral and extrafloral nectaries [8]. Moreover, some secondary metabolites served as the defense agents against parasitive invasion, for example, phenols emit under the excitation of UV-light [3], and their autofluorescence should mark their location in cells and tissues [61–63].

4.1. Biosignal for Insects. New look on autofluorescence appears with several works dealt with the search of signal for pollinators which prefer flowers of the certain plant species [64, 65]. A quantitative evaluation of the light emerging from intact petals of *Rhododendron indicum* flowers of different colors was performed based on the measurement of reflectance and fluorescence emission. Emitted photons as fluorescence were compared with reflected photons. The fluorescence quantum yield values (ϕ_f) varied from 7.6×10^{-5} to 6.3×10^{-4} for the emission in the blue region of the electromagnetic spectrum and from 2.4×10^{-5} to 1.9×10^{-4} for the emission in the red one [64]. (ϕ_f) was calculated for flowers of *Bellis perennis* (white, yellow, pink, and purple), *Ornithogalum thyrsoides* (petals and ovaries), *Limonium sinuatum* (white and yellow), *Lampranthus productus* (yellow), *Petunia nyctaginiflora* (white), *Bougainvillea spectabilis* (white and yellow), *Antirrhinum majus* (white and yellow), *Eustoma grandiflorum* (white and blue), *Citrus aurantium* (petals and stigma), and *Portulaca grandiflora* (yellow) [65]. The highest values were obtained for the ovaries of

O. thyrsoides (ϕ_f = 0.030) and for *Citrus aurantium* petals (ϕ_f = 0.014) and stigma (ϕ_f = 0.013). The fluorescence emission as an optical signal in biocommunication was negligible if to compare to the light reflected by the petals. Nevertheless, the calculation of quantum catches for each studied flower species described the visual sensitization of insect eye photoreceptors.

4.2. Localization of Secondary Metabolites in Cells and Tissues. Location of secondary metabolites fluoresced under UV or violet excitation is a new approach to visualize secondary metabolites in plants [61–63, 66–70]. Among the compounds are fluorescent phenol compounds that play a major role in the interaction of organisms in biocenosis attracting insects and functioning as signals between plants (allelopathy), as signals between plants and phenols in cell walls, vacuoles, and associated with N_2-fixing bacteria or pathogenic organisms [61–63]. The distribution of flavonoids such as anthocyanins in grape (*Vitis vinifera* L.) was measured nondestructively by chlorophyll fluorescence imaging using two excitation light bands at 550 and 650 nm [62]. Strong blue autofluorescence of living young leaf of *Coffea canephora*, species which contains very high level of chlorogenic acid, relates to vesicles filled with the compound and where the similar emission was also observed [63]. The phenol localization in the vacuole of epidermis cells, in vesicles and chloroplasts of mesophyll cells strengthen the hypothesis of its involvement in leaf protection against UV damages or pathogens. Besides phenols, yellow pigments betaxanthins also fluoresce in flowers in green region [66] and according the emission it is possible to determine dopamine-pigment complex in living *Portulaca oleracea* cells [67].

Especial attention was paid to localization in cells and tissues such flavonoids as anthocyanins in grape and apples [68, 69]. Practical tendency in the investigations is the use of multiparametric fluorescence sensor in apple quality [69]. The pixel intensity in the derived logarithm of the fluorescence excitation ratio image was directly related to the anthocyanin concentration of berry extracts. The method will be useful for the assessment of the heterogeneity of anthocyanin accumulation in berries depending on physiologic and climatic factors. It can also represent a new, rapid and noninvasive technique for the assessment of fruit ripening and the appropriate time of harvest. Moreover, well-seen autofluorescence of the anthocyanins is recommended for their traffic within cell [70]. The detection of pigments and colorless flavonoids can provide a useful indication of fruit quality [68, 69] or to localize some phenols such as chlorogenic acid in leaf tissues, demonstrating the great potential of multiphoton microscopy combined with spectral deconvolution [63].

5. Autofluorescence of Biosensors

Due to their autofluorescence plant intact cells may serve as biosensors and bioindicators because they have natural structure that determinates their high activity and stability [71]. Criteria in the screening of the plant cells as biosensors for biologically active compounds should be as follows: (1) the object use is easy in laboratory and in the field and (2) autofluorescence should be one of the physiological responses on the external chemicals because reflects the fast biochemical changes. They could serve as the biosensors of some biologically active excretions (having drugs' characteristics) from plants known as weeds, medicinal, or poisonous plant species [37, 42, 71]. Plant biosensors are more suitable for similar testing, than animals, which use needs procedures of vivisection [40, 71].

5.1. Characteristics of Biosensors. As unicellular biosensors vegetative microspores of spore-bearing plants and pollens of seed-bearing plants [8, 42, 71] may be used as well as multicellular biosensor—pistil stigma in flowers [8, 24].

Figure 3 shows examples of fluorescent analysis of vegetative microspores of horsetail *Equisetum arvense* using various technique. Under luminescence microscope, green fluorescence of cell wall and elaters (shell employees to attach microspore to the substrate) is visible. Through the rather transparent cover bright red fluorescence of chloroplasts, located inside the cell among nonlightening cytoplasm, is also observed. The total spectra of fluorescence from cells measured by microspectrofluorimeter depended on the presence of appropriate pigments and had the three peaks—in blue with maximum 450–460 nm (azulenes, phenols), in green with maximum 540–550 nm (carotenoids), as well as the most intense red emission with maximum 675–680 nm (chlorophyll). Cell cover liberated during the microspore development does not have a maximum in red spectral region [72]. Average fluorescence intensity of 100 cells measured on the one subject slide by dual-beam microspectrofluorimeter in green (520 nm) and red (640–680 nm) was differed in many times as seen from numerical data of summary emission shown in the histogram area (3.15 relative units in red against 0.13 relative units in green region). Red emission prevailed. Its intensity increases during the microspore development in comparison with dry rest (nondeveloped) cells [72, 73].

If the division of the spore cell occurs (and duplication of chloroplasts takes place) the 2 times increase in red fluorescence is the indicator of cellular division because amounts of chloroplasts and fluorescing chlorophyll are doubled. Unlike the microspectrofluorimetry laser-scanning confocal microscopy permits to see images of optical slices of intact microspore cells at different depths and register the spectra of fluorescence from individual parts (marked as numerals on Figure 3), namely, cell wall, including its outer layer exine (1), its inner layer intine with plasmalemma (2), and chloroplasts (3).

The natural fluorescence of pollen and pistils as biosensors is served to study of various processes [10, 21, 22, 24, 27–29] from the damages by ozone and other reactive oxygen species to chemosignaling. Indicators of the emission (Table 3) based on color and intensity of fluorescence altered under ozone and reactive oxygen species or the addition

Table 3: Pollen biosensors.

Plant species	Colour of pollen autofluorescence	Prevailing fluorophore	Indicator reaction
Hippeastrum hybridum L. (Herb), knight's star (Amaryllidaceae)	Blue-green	Flavins, phenols	Shifts in the fluorescence maxima and intensity at pollen-pistil or pollen-pollen interactions and neurotransmitters additions [8, 16, 17, 19, 21, 23]
Philadelphus grandiflorus, wild mock-orange (Hydrangeaceae)	Yellow-orange	Carotenoids, azulenes, anthocyanins	Shifts in the fluorescence maxima and intensity after ozone or other reactive oxygen species treatment [26–28]
Knautia arvensis (L.) Coult., field scabious (Dipsacaceae)	Green	Anthocyanins	Shifts in the fluorescence intensity at pollen-pollen interactions [20, 21]

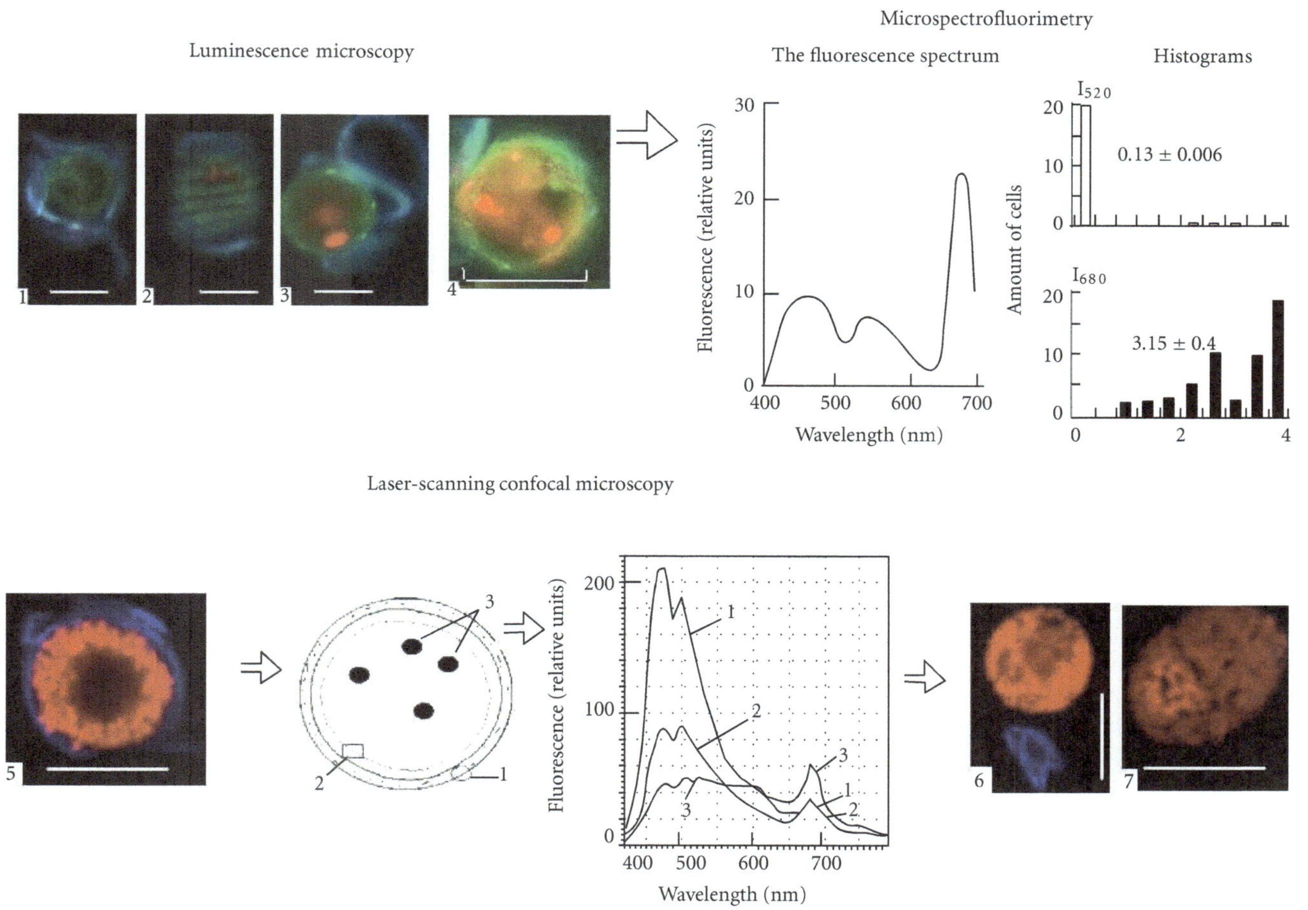

Figure 3: The images and fluorescence spectra of the vegetative microspores of horsetail *Equisetum arvense*. Sources: [8, 42] and non-published data of author. Upper side: Luminescence microscopy (left). Views (1–4) under the luminescence microscope Leica 6000. Micro-spectrofluorimetry. The fluorescence spectrum (middle) and the emission intensity distribution in form of numerical data or histogram (right). The fluorescence intensity (I) on histograms measured by dual-wavelength microspectrofluorimeter MSF-2 shows as I_{520} and I_{680}, respectively, the emission intensity in the green region with maximum 520 nm and red region with a maximum 680 nm. Lower side: Laser-scanning confocal microscopy. Images 5–7 under laser-scanning confocal microscope LSM 510 NLO (5–moistured spore) and Carl Zeiss (6 and 7 developing spore that put off the blue-fluoresced rigid cover-envelope and then divided, resp.). On the scheme of optical image slice under laser-scanning confocal microscope (at the middle of the picture numbers mean cellular parts, from which the appropriate fluorescence spectra have been received with laser-scanning confocal microscope *Leica* TCS SP-5. By numerals exine (1), its inner layer intine with plasmic membrane (2) and chloroplast (3) are marked. Excitation for the luminescence microscope or microspectrofluorimeter was 420 nm and for laser-scanning confocal microscopy—405 nm laser.

TABLE 4: Fluorescence of biologically active compounds (medical drugs) out and within cells of microspores—biosensors. Source: [38, 40].

Component of plant secretions with pharmaceutical features	Fluorescence (color/maxima in nm)			Cellular component—target of a fluorescent staining
	Pure compound	At the interaction with microspores		
		E. arvense	*H. hybridum*	
Azulene	Blue/425	Blue/425, 550	Blue/425, 550	Nucleus, chloroplasts
Atropine	Blue/415	Blue/465–470	Blue/465–470	ChR in Plm
d-Tubocurarine	Blue/415	Blue/425	Blue/465–475	ChR, Plm
Yohimbine	Blue/455	Blue/4305	Blue/475	AdR, Plm

* Plm: plasmalemma, CW: cell wall, ChE: cholinesterase, ChR: cholinoreceptor, AdR: adrenoreceptor, and Glc: glucosidase.

of some natural compounds (including complex secretions) and artificial chemicals. Pollen grains of *Hippeastrum hybridum* were often used as biosensors [37, 42, 71] as well as a pistil stigma of the same species [8, 24]. Multicolor pictures received looked as fantastic art photo that makes them fine illustrations [20, 21]. Besides fluorescent pollens, weak-emitted pollen such as from plantain *Plantago major* is also used for the analysis of interactions with fluorescent plant secretions, in particular from colored pollen grains) and the action mechanisms of drugs [8, 20–22, 40].

5.2. Biosensors in the Study of Chemosignal Processes. Chemosignaling is a first process occurred at contacts of sexual and somatic cells. For the studies it is necessary to distinguish two groups of fluorescing objects—cells which serve as donors of fluorescing compounds excreted and cells-acceptors of the chemical signal that are biosensors. Analyzing both cell types the researcher can observe, namely: (1) the changes in natural fluorescence of contacting cells, (2) the penetration of emitted fluorophore into the cell acceptor, (3) binding of the lightening compounds within cells, and (4) quenching or strengthening the total cell-acceptor fluorescence.

Chemosignaling at direct cell-cell contacts is possible to observe, using some sensitive pollen as biosensor (like to study of pollen allelopathy in Section 2.1). Application of LSCM technique permits to observe direct interaction of the contacted pollens and the alterations in their state, basing on the fluorescence changes. In example with LSCM images the secretion released from red-fluorescing pollen of *Hypericum perforatum* may stain the surface of the color-less pollen from *Plantago major* [8, 20]. The fluorescence of slices showed that the red-emitting pigment hypericin concentrated only on the surface without penetration into a cell.

Besides the direct observation of chemosignaling at cellular communications on LSCM-images, there is modelling of the interaction in the system biosensor–contacting chemical. Use of cellular autofluorescence as a response to the chemosignal can be one of the approaches to study mechanisms of action of biologically active substances on sensitive cells [37, 59]. Cellular mechanisms of the effects of many natural compounds as chemosignals often are not suitable to observe on the tissue slices of animals after vivisection or special conditions of cell storing and cultivation; plant cells have advantages as biosensors. For example, neurotransmitters acetylcholine, catecholamines, serotonin and histamine that also found in excretions of plant and microorganisms [74, 75] changed the position of maxima in the emission spectra and intensity of pollen and pistil autofluorescence in *Hippeastrum hybridum* [8, 24] or vegetative microspores of horsetail *Equisetum* arvense [42]. The neurotransmitters in the presence of other components of secretions—lipids and proteins or membrane-active agents of fungal infection—decreased or enhanced the fluorescence intensity at 680 nm of biosensors-vegetative microspores from horsetail *Equisetum arvense* [42]. Some of the contacted secretory products also fluoresce that allows observing how they interact with unicellular biosensors [8, 21, 22, 24]. Fluorescent components of plant secretions are used to study of chemosignal processes at cellular level, stain living cells and provide valuable information about the level of their entering into a cell and binding sites [21, 22] as well as the mechanisms of chemoreception and action of some drugs [40, 42, 71]. Blue-fluorescent antagonists (blockers of appropriate receptors), in particular antagonist of acetylcholine atropine or d-Tubocurarine and antagonist of catecholamines yohimbine, were binding with plasmatic membrane while histaminic antagonist azulene penetrated into a cell that was seen on LSCM images [42]. Table 4 demonstrates the targets of the compounds' bindings. Antagonists of neurotransmitters are included in secretions of medicinal (allelopathically active) plants as pharmaceuticals with own fluorescence. All compounds studied decreased number of the cells of *E. arvense* fluorescing in red (Figure 4). d-Tubocurarine and yohimbine did not penetrate the cells binding on the cellular surface, but azulene was linked at both the cell surface and inside the cell, demonstrating blue or blue-green emission. Tubocurarine bound with cholinoreceptor in plasmatic membrane stimulated blue emission and decreased red chlorophyll fluorescence masking plastidal emission. The evidence that tubocurarine and acetylcholine are linked in one and the same site of the membrane is the absence or small blue fluorescence and enhanced red fluorescence in the variant with preliminary treatment of the biosensor cells with acetylcholine.

Besides unicellular microspores, the long-lengthed pistil of *Hippeastrum hybridum* which the stigma autofluorescence (maxima in the fluorescence spectra and the emission intensity) is sensitive to acetylcholine and histamine as well as their antagonists d-Tubocurarine and azulene [8, 24] may be suitable biosensor for the study of chemosignalling.

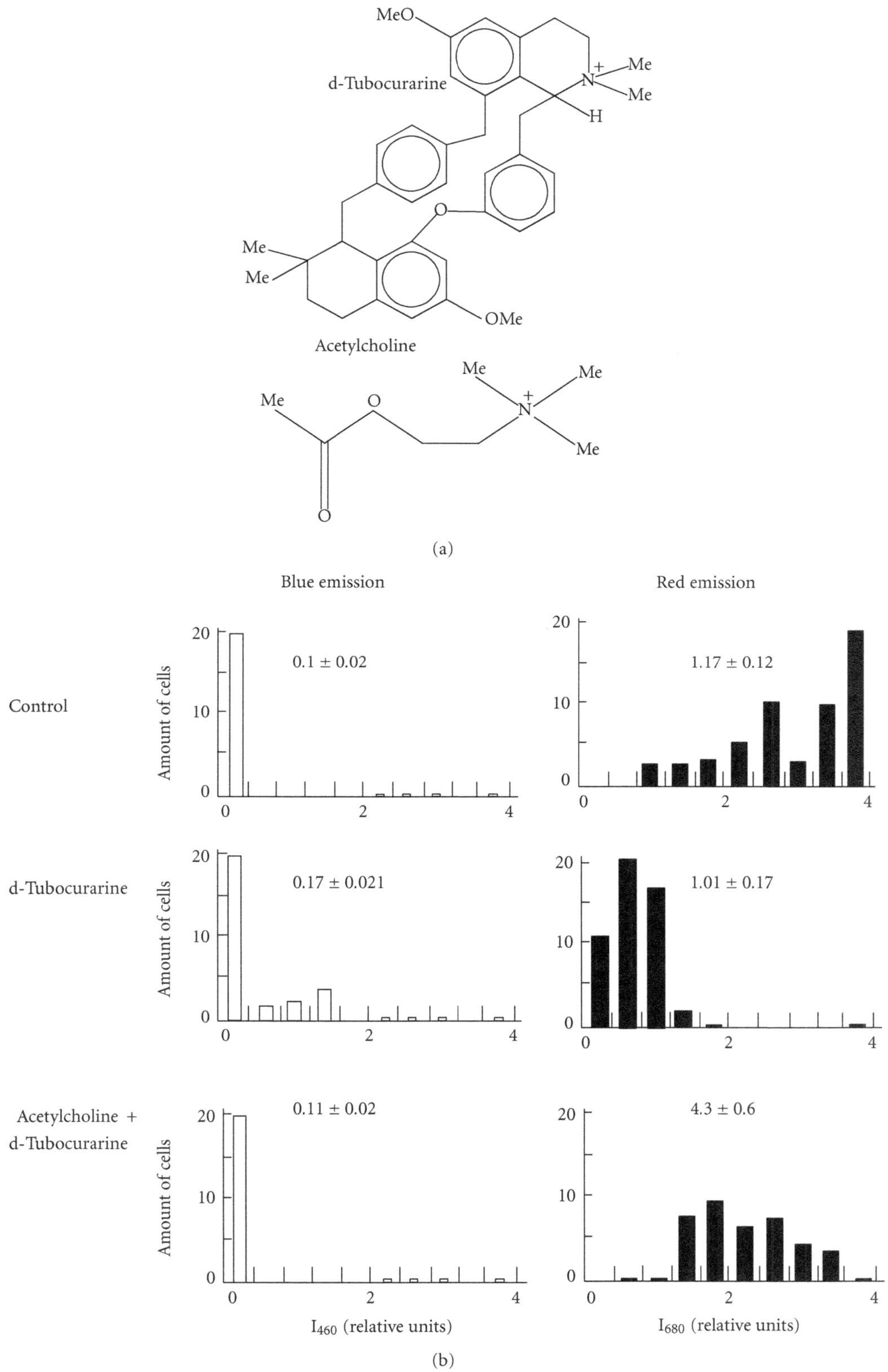

FIGURE 4: Summed distribution histograms of blue (maximum 460 nm) fluorescence intensity of vegetative microspores of horsetail *Equisetum arvense* before and after the treatment with components of plant secretions during 24 h. Excitation 360–380 nm. Total amount of studied cells in a variant = 100. Summed values for 100 analyzed cells are also given.

5.3. Potentials for a Search of New Biosensors. Among potential biosensors may be multicellular systems having autofluorescence. Similar biosensor based on petal fluorescence in flowers such as *Mirabilis jalapa* may be also applied to analysis of chemosignalling dealt with certain pigments-fluorophores [60]. The visible fluorescence emitted by one pigment, a yellow betaxanthin, is absorbed by another, a violet betacyanin, to create a contrasting fluorescent pattern on the flower's petals for pollinators. The analysis of free radical processes with the participation of the pigments [76] opened new horizons to search biosensors among pigmented plant cells.

Perspective biosensors today appears to be mycorrhizal roots of plants sensitive to a colonization of arbuscular mycorrhizal fungal structures (such as hyphae, vesicles and spores) that brightly autofluorescence among root tissues under ultraviolet, blue or green light excitation [76–78]. Whole roots of lucerne *Medicago sativa* may be suitable for the same aim.

6. Conclusion

Autofluorescence of living cells may serve as biosensor and bioindicator reaction for various studies of plants–from taxonomic position of plant or pollen viability to the chemosignalling processes. This reaction is analyzed by following ways, namely: (1) on undamaged structures, if their surface composed of fluorescent components or interior of the structures filled of fluorescent secretion; (2) modeling with individual cellular components showing the picture of the fluorescing cellular structure observed. The autofluorescent structure images are studied in a dependence on a purpose either for anatomic characteristics applying to taxonomic sign in plant systematics or for a comparison of samples in the control variants and those that exposed under someone factor. Moreover, optical slices made with laser-scanning confocal microscopy permit to show a location of fluorescing secondary metabolite naturally synthesized within the cell or exometabolite interacted with the cell at cell-cell communications or at experimental addition of the chemical agent. Besides images, the fluorescence spectra of emitted structures may be recorded with microspectrofluorimetry or certain laser-scanning confocal microscopes. Measuring the absorbance and fluorescence spectra there is a possibility to estimate quantitative yield of fluorescence not only for individual fluorophore, but also for certain cells, such as flower petals.

References

[1] D. L. Taylor and E. D. Salmon, "Basic fluorescence microscopy," in *Methods in Cell Biology: Living Cell in Culture*, J. L. Wang and D. L. Taylor, Eds., pp. 207–237, Academic Press, San Diego, Calif, USA, 1989.

[2] H. Andersson, T. Baechi, M. Hoechl, and C. Richter, "Autofluorescence of living cells," *Journal of Microscopy*, vol. 191, no. 1, pp. 1–7, 1998.

[3] O. S. Wolfbeis, "The fluorescence of organic natural products," in *Molecular Luminescence Spectroscopy: Methods and Applications*, S. G. Schulman, Ed., pp. 167–370, John Wiley & Sons, New York, NY, USA, 1985.

[4] V. V. Roshchina, "Cellular models to study the allelopathic mechanisms," *Allelopathy Journal*, vol. 13, no. 1, pp. 3–16, 2004.

[5] V. V. Roshchina, V. A. Yashin, A. V. Yashina, and M. V. Gol'tyaev, "Colored allelochemicals in modelling of cell-cell allelopathic interactions," *Allelopathy Journal*, vol. 28, no. 1, pp. 1–12, 2011.

[6] C. Pöhlker, J. A. Huffman, and U. Pöschl, "Autofluorescence of atmospheric bioaerosols—fluorescent biomoleculesand potential interferences," *Atmospheric Measurement Techniques*, vol. 4, no. 5, pp. 5857–5933, 2011.

[7] G. Agati, "Response of the in vivo chlorophyll fluorescence spectrum to environmental factors and laser excitation wavelength," *Pure and Applied Optics*, vol. 7, no. 4, pp. 797–807, 1998.

[8] V. V. Roshchina, *Fluorescing World of Plant Secreting Cells*, Science, Enfield, UK, 2008.

[9] M. N. Merzlyak, "Liposoluble fluorescent "aging pigments" in plants," in *Lipofuscin-1987: State of the Art*, I. Nagy, Ed., pp. 451–452, Akademiai Kiado, Elsevier, Budapest, Amsterdam, The Netherlands, 1988.

[10] V. V. Roshchina, "Autofluorescence of plant secreting cells as a biosensor and bioindicator reaction," *Journal of Fluorescence*, vol. 13, no. 5, pp. 403–420, 2003.

[11] M. T. M. Willemse, "Autofluorescence of pollen wall of Lilium and changes in pollen wall of *Gasteria* and *Lilium* anther," *Acta Societatis Botanicorum Poloniae*, vol. 50, pp. 103–110, 1981.

[12] M. N. B. M. Driessen, M. T. M. Willemse, and J. A. G. van Luijin, "Grass pollen grain determination by light- and UV-microscopy," *Grana*, vol. 28, no. 2, pp. 115–122, 1989.

[13] J. C. Audran and M. T. M. Willemse, "Wall development and its autofluorescence of sterile and fertile *Vicia faba* L. Pollen," *Protoplasma*, vol. 110, no. 2, pp. 106–111, 1982.

[14] S. A. Weinbaum and V. S. Polito, "Assessment of pollen retention using exine autofluorescence," *Mikroskopie*, vol. 42, pp. 278–280, 1985.

[15] V. V. Roshchina, E. V. Melnikova, N. A. Spiridonov, and L. V. Kovaleva, "Azulenes, the blue pigments of pollen," *Doklady Biological Sciences*, vol. 340, no. 1, pp. 93–96, 1995.

[16] V. V. Roshchina, E. V. Mel'nikova, and L. V. Kovaleva, "Autoflourescence in pollen-pistil system in *Hippeastrum hybridum*," *Doklady Akademii Nauk*, vol. 349, no. 1, pp. 118–120, 1996.

[17] V. V. Roshchina, E. V. Mel'nikova, and L. V. Kovaleva, "Changes in fluorescence during development of the male gametophyte," *Russian Journal of Plant Physiology*, vol. 44, no. 1, pp. 36–44, 1997.

[18] V. V. Roshchina, E. V. Melnikova, L. V. Mit'kovskaya, and V. N. Karnaukhov, "Microspectrofluorimetry for the study of intact plant secreting cells," *Zhurnal Obshchei Biologii*, vol. 59, no. 5, pp. 531–554, 1998.

[19] V. V. Roshchina and E. V. Melnikova, "Microspectrofluorimetry of intact secreting cells, with applications to the study of allelopathy," in *Principles and Practices in Plant Ecology: Allelochemical Interactions*, Inderjit, K. M. M. Dakshini, and C. L. Foy, Eds., pp. 99–126, CRC Press, Boca Raton, Fla, USA, 1999.

[20] V. V. Roshchina, A. V. Yashina, and V. A. Yashin, "Cell communication in pollen allelopathy analyzed with laser-scanning confocal microscopy," *Allelopathy Journal*, vol. 21, no. 2, pp. 219–226, 2008.

[21] V. V. Roshchina, A. V. Yashina, V. A. Yashin, and N. K. Prizova, "Models to study pollen allelopathy," *Allelopathy Journal*, vol. 23, no. 1, pp. 3–24, 2009.

[22] V. V. Roshchina, A. V. Yashina, V. A. Yashin, and M. V. Gol'tyaev, "Fluorescence of biologically active compounds in plant secretory cells," in *Research Methods in Plant Science, Vol. 2. Forestry and Agroforestry*, S. S. Narwal, P. Pavlovic, and J. John, Eds., pp. 3–25, Studium Press, Houston, Tex, USA, 2011.

[23] V. V. Roshchina and E. V. Melnikova, "Microspectrofluorometery: a new technique to study pollen allelopathy," *Allelopathy Journal*, vol. 3, no. 1, pp. 51–58, 1996.

[24] V. V. Roshchina and E. V. Melnikova, "Chemosensory reactions at the interaction pollen-pistil," *Biology Bulletin*, no. 6, pp. 678–685, 1998.

[25] A. Salih, A. S. Jones, D. Bass, and G. Cox, "Confocal imaging of exine as a tool for grass pollen analysis," *Grana*, vol. 36, no. 4, pp. 215–224, 1997.

[26] A. E. Fonseca, M. E. Westgate, and R. T. Doyle, "Application of fluorescence microscopy and image analysis for quantifying dynamics of maize pollen shed," *Crop Science*, vol. 42, no. 6, pp. 2201–2206, 2002.

[27] V. V. Roshchina and V. N. Karnaukhov, "Changes in pollen autofluorescence induced by ozone," *Biologia Plantarum*, vol. 42, no. 2, pp. 273–278, 1999.

[28] V. V. Roshchina and E. V. Mel'nikova, "Pollen chemosensitivity to ozone and peroxides," *Russian Journal of Plant Physiology*, vol. 48, no. 1, pp. 74–83, 2001.

[29] V. V. Roshchina and V. D. Roshchina, *Ozone and Plant Cell*, Kluwer Academic, Dordrecht, The Netherlands, 2003.

[30] K. Mitsumoto, K. Yabusaki, and H. Aoyagi, "Classification of pollen species using autofluorescence image analysis," *Journal of Bioscience and Bioengineering*, vol. 107, no. 1, pp. 90–94, 2009.

[31] K. Mitsumoto, K. Yabusaki, K. Kobayashi, and H. Aoyagi, "Development of a novel real-time pollen-sorting counter using species-specific pollen autofluorescence," *Aerobiologia*, vol. 26, no. 2, pp. 99–111, 2010.

[32] H. Aoyagi and C. U. Ugwu, "Fulerene fine particles adhere to pollen grains and affect their autofluorescence and germination," *Nanotechnology, Science and Applications*, vol. 4, pp. 67–71, 2011.

[33] A. Speranza, R. Crinelli, V. Scoccianti, and A. Geitmann, "Reactive oxygen species are involved in pollen tube initiation in kiwifruit," *Plant Biology*, vol. 14, no. 1, pp. 64–76, 2011.

[34] A. A. Dobritsa, A. Geanconteri, J. Shrestha et al., "A large-scale genetic screen in Arabidopsis to identify genes involved in pollen exine production," *Plant Physiology*, vol. 157, no. 2, pp. 947–970, 2011.

[35] A. J. Castro, J. D. Rejon, M. Fendri et al., "Taxonomical discrimination of pollen grains by using confocal laser scanning microscopy (CLSM) imaging of autofluorescence," in *Microscopy: Science, Technology, Application and Education*, A. Mendez-Vilas and J. Diaz, Eds., pp. 607–613, Formatex, Badajoz, Spain, 2010.

[36] M. Suzuzi, M. Tonouchi, K. Muyarama, and K. Mitsumoto, "An analysis of high concentration of pollen at early morning in 2005 spring and a measurement of Japanese cedar and cypress pollen using autofluorescence characteristics," *Journal of Aerosol Research*, vol. 20, no. 4, pp. 281–289, 2005.

[37] V. V. Roshchina, "Cellular models as biosensors," in *Cell Diagnostics: Images, Biophysical and Biochemical Processes in Allelopathy*, V. V. Roshchina and S. S. Narwal, Eds., pp. 5–22, Science, Enfield, UK, 2007.

[38] V. V. Roshchina, "Luminescent cell analysis in allelopathy," in *Cell Diagnostics: Images, Biophysical and Biochemical Processes in Allelopathy*, V. V. Roshchina and S. S. Narwal, Eds., pp. 103–115, Science, Enfield, UK, 2007.

[39] V. V. Roshchina, V. A. Yashin, A. V. Kononov, and A. V. Yashina, "Laser-scanning confocal microscopy (LSCM): study of plant secretory cell," in *Cell Diagnostics: Images, Biophysical and Biochemical Processes in Allelopathy*, V. V. Roshchina and S. S. Narwal, Eds., pp. 93–102, Science, Enfield, UK, 2007.

[40] V. V. Roshchina and V. N. Karnaukhov, "The fluorescence analysis of medicinal drugs' interaction with unicellular biosensors," *Farmatsiya*, vol. 7, no. 3, pp. 43–46, 2010.

[41] A. G. Valdecasas and A. Abad, "Morphological confocal microscopy in arthropods and the enhancement of autofluorescence after proteinase K extraction," *Microscopy and Microanalysis*, vol. 17, no. 1, pp. 109–113, 2011.

[42] V. V. Roshchina, V. A. Yashin, and I. M. Vikhlyantsev, "Fluorescence of plant microspores as biosensors," *Biochemistry Supplemental Series A*, vol. 6, no. 1, pp. 105–112, 2012.

[43] W. Rühland and K. Wetzel, "Der Nachweis von Chloroplasten in den generativen Zellen von Pollenschläuchen," *Berirhte Der Deutarhen Botanischen Gesellschaft*, vol. 42, pp. 3–14, 1924.

[44] F. Berger, "Das Verhalten der heufieber-erregenden Pollen in filtrien ultravioleten Licht," *Beiträge zur Biologie der Pflanzen*, vol. 22, no. 1, pp. 1–12, 1934.

[45] F. Asbeck, "Fluorescezieren der Blutenstaub," *Naturwissenschaften*, vol. 42, no. 5, p. 632, 1955.

[46] P. van Gijzel, "Autofluorescence and age of some fossil pollen and spores," *Proceedings of the Koninklijke Nederlandse Akademie Van Wetenschappen—Series B*, vol. 64, no. 1, pp. 56–63, 1961.

[47] M. T. M. Willemse, "Morphological and fluorescence microscopical investigation on sporopollenin formation at *Pinus sylvestris* and *Gasteria verrucosa*," in *Sporopollenin. Proceeding Symposim at Geology Department, Imperial College, London, 1970. 23–25 September*, J. Brooks, P. R. Grant, M. Muir, and P. R. van Gijzel, Eds., pp. 68–91, Academic Press, New York, NY, USA, 1971.

[48] D. J. O. 'Connor, D. Iacopino, D. A. Healy, D. O'Sullivan, and J. R. Sodeau, "The intrinsic fluorescence spectra of selected pollen and fungal spores," *Atmospheric Environment*, vol. 45, no. 35, pp. 6451–6458, 2011.

[49] Y. L. Pan, S. C. Hill, R. G. Pinnick, J. M. House, R. C. Flagan, and R. K. Chang, "Dual-excitation-wavelength fluorescence spectra and elastic scattering for differentiation of single airborne pollen and fungal particles," *Atmospheric Environment*, vol. 45, no. 8, pp. 1555–1563, 2011.

[50] V. V. Roshchina and V. D. Roshchina, *The Excretory Function of Higher Plants*, Springer, Berlin, Germany, 1993.

[51] V. V. Roshchina, E. V. Melnikova, V. N. Karnaukhov, and B. N. Golovkin, "Application of microspectrofluorimetry in spectral analysis of plant secretory cells," *Biology Bulletin*, no. 2, pp. 167–171, 1997.

[52] V. V. Roshchina and E. V. Melnikova, "Spectral analysis of intact secretory cells and excretions of plants," *Allelopathy Journal*, vol. 2, no. 2, pp. 179–188, 1995.

[53] S. Lee, R. L. Brown, and W. Monroe, "Use of confocal laser scanning microscopy in systematics of insects with a comparison of fluorescence from different stains," *Systematic Entomology*, vol. 34, no. 1, pp. 10–14, 2009.

[54] Z. Žižka and J. Gabriel, "Primary fluorescence (autofluorescence) of fruiting bodies of the wood-rotting fungus *Fomes fomentarius*," *Folia Microbiologica*, vol. 51, no. 2, pp. 109–113, 2006.

[55] C. H. Wu and H. L. Warren, "Natural autofluorescence in fungi and its correlation with viability," *Mycologia*, vol. 76, no. 6, pp. 1049–1058, 1984.

[56] C. H. Wu and H. L. Warren, "Induced autofluorescence in fungi, and its correlation with viability: potential application of fluorescence microscopy," *Phytopathlogy*, vol. 74, no. 6, pp. 1353–1358, 1984.

[57] G. Doehlemann, K. van der Linde, D. Aßmann et al., "Pep1, a secreted effector protein of *Ustilago maydis*, is required for successful invasion of plant cells," *PLoS Pathogens*, vol. 5, no. 2, Article ID e1000290, 2009.

[58] A. R. Hardham, "Confocal microscopy in plant-pathogen interactions," in *Plant Fungal Pathogens: Methods and Protocols*, M. D. D. Bolton, P. H. J. Bart, and P. Thomma, Eds., vol. 835 of *Methods in Molecular Biology*, pp. 295–309, Springer, Berlin, Germany, 2012.

[59] F. Perrine-Walker, P. Doumas, M. Lucas et al., "Auxin carriers localization drives auxin accumulation in plant cells infected by *Frankia* in *Casuarina glauca* actinorhizal nodules," *Plant Physiology*, vol. 154, no. 3, pp. 1372–1380, 2010.

[60] F. Gandía-Herrero, F. García-Carmona, and J. Escribano, "Botany: floral fluorescence effect," *Nature*, vol. 437, no. 7057, p. 334, 2005.

[61] P. Hutzler, R. Fischbach, W. Heller et al., "Tissue localization of phenolic compounds in plants by confocal laser scanning microscopy," *Journal of Experimental Botany*, vol. 49, no. 323, pp. 953–965, 1998.

[62] G. Agati, Z. G. Cerovic, P. Pinelli, and M. Tattini, "Light-induced accumulation of ortho-dihydroxylated flavonoids as non-destructively monitored by chlorophyll fluorescence excitation techniques," *Environmental and Experimental Botany*, vol. 73, pp. 3–9, 2011.

[63] G. Conejero, P. Talamond, and J. L. Verdeil, "A new approach to visualize secondary metabolites in plants. Histocytology and plant cell imaging control platform," La plate-forme regionale d'imagerie "Montpellier Rio Imaging" (MRI) organise une journee "Imagerie", Lafayette, La, USA, Octobre 2009, http://phiv.cirad.fr/download/poster.pdf.

[64] A. Iriel and M. G. Lagorio, "Implications of reflectance and fluorescence of *Rhododendron indicum* flowers in biosignaling," *Photochemical and Photobiological Sciences*, vol. 9, no. 3, pp. 342–348, 2010.

[65] A. Iriel and M. G. Lagorio, "Is the flower fluorescence relevant in biocommunication?" *Naturwissenschaften*, vol. 97, no. 10, pp. 915–924, 2010.

[66] F. Gandía-Herrero, J. Escribano, and F. García-Carmona, "Betaxanthins as pigments responsible for visible fluorescence in flowers," *Planta*, vol. 222, no. 4, pp. 586–593, 2005.

[67] F. Gandía-Herrero, M. Jiménez-Atiénzar, J. Cabanes, J. Escribano, and F. García-Carmona, "Fluorescence detection of tyrosinase activity on dopamine-betaxanthin purified from *Portulaca oleracea* (common purslane) flowers," *Journal of Agricultural and Food Chemistry*, vol. 57, no. 6, pp. 2523–2528, 2009.

[68] G. Agati, M. L. Traversi, and Z. G. Cerovic, "Chlorophyll fluorescence imaging for the noninvasive assessment of anthocyanins in whole grape (*Vitis Vinifera* L.) bunches," *Photochemistry and Photobiology*, vol. 84, no. 6, pp. 1431–1434, 2008.

[69] D. L. Betemps, J. C. Fachinello, S. P. Galarca et al., "Non-destructive evaluation of ripening and quality traits in apples using a multiparametric fluorescence sensor," *Journal of the Science of Food and Agriculture*, vol. 92, no. 9, pp. 1855–1864, 2012.

[70] C. Gomez, G. Conejero, L. Torregrosa, V. Cheynier, N. Terrier, and A. Ageorges, "In vivo grapevine anthocyanin transport involves vesicle-mediated trafficking and the contribution of anthoMATE transporters and GST," *Plant Journal*, vol. 67, no. 6, pp. 960–970, 2011.

[71] V. V. Roshchina, "Plant microspores as biosensors," *Trends in Modern Biology*, vol. 126, no. 3, pp. 262–274, 2006.

[72] V. V. Roshchina, E. V. Mel'nikova, V. A. Yashin, and V. N. Karnaukhov, "Autofluorescence of intact *Equisetum arvense* L. spores during their development," *Biofizika*, vol. 47, no. 2, pp. 318–324, 2002.

[73] V. V. Roshchina, V. A. Yashin, and A. V. Kononov, "Autofluorescence of developing plant vegetative microspores studied by confocal microscopy and microspectrofluorimetry," *Journal of Fluorescence*, vol. 14, no. 6, pp. 745–750, 2004.

[74] V. V. Roshchina, *Neurotransmitters in Plant Life*, Science, Enfield, UK, 2001.

[75] V. V. Roshchina, "Evolutionary considerations of neurotransmitters in microbial, plant and animal cells," in *Microbial Endocrinology: Interkingdom Signaling in Infectious Disease and Health*, M. Lyte and P. P. E. Freestone, Eds., pp. 17–52, Springer, Berlin, Germany, 2010.

[76] R. N. Ames, E. R. Ingham, and C. P. P. Reid, "Ultraviolet-induced autofluorescence of a mycorrhizal root infestions and alternative to clearing and staining methods for infections," *Canadian Journal of Microbiology*, vol. 28, pp. 351–355, 1982.

[77] B. Dreyer, A. Morte, M. Pérez-Gilabert, and M. Honrubia, "Autofluorescence detection of arbuscular mycorrhizal fungal structures in palm roots: an underestimated experimental method," *Mycological Research*, vol. 110, no. 8, pp. 887–897, 2006.

[78] B. Dreyer, "Use of the autofluorescence properties of AM fungi for AM assessment and handling," *Soil Biology*, vol. 18, pp. 123–140, 2009.

2

Resonance-Enhanced Raman Spectroscopy on Explosives Vapor at Standoff Distances

Anneli Ehlerding, Ida Johansson, Sara Wallin, and Henric Östmark

Department of Energetic Materials, Swedish Defense Research Agency (FOI), 147 25 Tumba, Sweden

Correspondence should be addressed to Anneli Ehlerding, anneli.ehlerding@foi.se

Academic Editor: S. Michael Angel

Resonance-enhanced Raman spectroscopy has been used to perform standoff measurements on nitromethane (NM), 2,4-DNT, and 2,4,6-TNT in vapor phase. The Raman cross sections for NM, DNT, and TNT in vapor phase have been measured in the wavelength range 210–300 nm under laboratory conditions, in order to estimate how large resonance enhancement factors can be achieved for these explosives. The results show that the signal is enhanced up to 250,000 times for 2,4-DNT and up to 60,000 times for 2,4,6-TNT compared to the nonresonant signal at 532 nm. Realistic outdoor measurements on NM in vapor phase at 13 m distance were also performed, which indicate a potential for resonance Raman spectroscopy as a standoff technique for detection of vapor phase explosives. In addition, the Raman spectra of acetone, ethanol, and methanol were measured at the same wavelengths, and their influence on the spectrum from NM was investigated.

1. Introduction

Improvised explosive devices (IEDs) are a common and a growing threat to the civilian society as well as to peace-keeping operations. Today, detection and identification of explosive charges and IEDs in real environments (indoors or outdoors) is a very challenging task and can in many cases also be a very hazardous operation. IEDs are in a way unique compared to regular ordnances, since the IED manufacturer has to improvise the design and use whatever materials are available. For this reason, the threat varies over time and in different parts of the world; it also evolves as countermeasures are taken. The diverse and varying nature of IEDs poses great challenges to the methods for their detection since they can contain any explosive which is available; this means commercial, military, or homemade explosives (HMEs). An overview of HME properties is given by Menning and Ostmark [1]. From the perspective of the user of the detection equipment, much would be gained if detection and identification could be reliably performed from a safe distance. This drives the research of possible standoff detection methods, where the developed methods will need to be both sensitive and selective in order to detect an explosive material in vapor phase and at low concentrations, or as trace amounts of particles on a surface, among an unknown number and composition of possible interfering substances. In addition, the method should be applicable at a distance of some tenths to a few hundreds of meters.

Techniques that are anticipated to fulfill these requirements are laser-based spectroscopy methods. A number of different laser-based methods for stand-off detection and identification of explosives have been proposed and tested, for example, Raman, LIBS, and PF/LIF. For reviews see Wallin et al. [2] and Gottfried et al. [3]. Raman scattering is an instantaneous inelastic scattering of photons where some energy is lost to (or gained from) the target molecule, returning scattered light with a different wavelength. The difference in energy corresponds to a vibrational mode of the target molecules, and these vibrational modes can be regarded as a fingerprint that uniquely identifies the substance or substances in a sample. Also complex mixtures can often be analyzed using algorithms for pattern recognition [4].

Standoff Raman explosives detection systems have been developed and tested by a number of groups; Sharma et al. demonstrated explosives detection (TATB and HMX) at 10 m distance using a frequency-doubled Nd:YAG laser

[5]. Detection of diluted explosives in a silicon matrix at longer distances, 27 m and 50 m, was demonstrated by Carter et al. in a laboratory environment using a frequency doubled Nd:YAG and gated detection [6]. The usefulness of spontaneous Raman spectroscopy for stand-off detection in more realistic scenarios has also been demonstrated [7, 8]. The instrumentation used was based on a frequency doubled pulsed Nd:YAG laser, a Schmidt-Cassegrain telescope, a spectrometer for wavelength dispersion, and an ICCD for gated detection. These measurements showed the applicability of 532 nm radiation for examination through different container materials (transparent or semitransparent) and during varying weather conditions at distances of up to 55 m. Recent results using this equipment have shown that the detection limit for, for example, TNT at 30 m and a detection time of 5 s is approximately 500 μg [4, 9]. This indicates the difficulties in using conventional Raman for stand-off trace and vapor detection.

Several groups have pursued Raman spectroscopy at UV wavelengths. The reasons for doing so are multiple; the obvious one is that the Raman signal intensity scales as $1/\lambda^4$, thus leading to an improvement of the inherently weak Raman response upon examination. Another advantage is the much reduced interference of fluorescence [10]. At wavelengths shorter than 250 nm, the full Raman range, up to 4,000 cm^{-1}, is normally free from luminescence. This can be used for identifying species in a fluorescent matrix [11]. If a laser wavelength outside the visible spectral region is used, the problem of eye hazards is significantly decreased since the maximum permissible exposure is markedly higher for wavelengths that are not focused by the lens of the human eye onto the retina. Of all the advantages of turning to UV wavelengths, perhaps the most intriguing, is the possible resonance enhancement that can occur when the probe laser wavelength is nearing an electronic transition within the molecule. This can give rise to enhancements of certain Raman lines (those coupled to the electronic transition) for which the Raman cross section can be significantly enhanced by a factor of 10^2–10^6 [12]. A number of Raman publications have had a focus on resonance-enhanced UV Raman spectroscopy of explosives [11, 13, 14]. Absolute Raman cross sections for UN, TATP, RDX, TNT, and PETN at 248 nm excitation have been published by Nagli et al. [13], and for TNT, PETN, RDX, and HMX at 229 nm excitation by Tuschel et al. [14].

The ability to detect explosives in vapor phase depends on the vapor pressure of the explosive. Some explosives, such as triacetone triperoxide (TATP), have a relatively high vapor pressure (4.65×10^{-2} Torr at 25°C [15]) and can be expected to be found in concentrations detectable also with methods with high detection limits (4.65×10^{-2} Torr corresponds to 60 ppm at atmospheric pressure) whereas other explosives, such as HMX, have very low vapor pressure [16] and can probably not be expected to be detectable with any gas phase detection methods. In this paper we have measured the Raman signal and calculated the apparent Raman cross section in the UV region, which is the cross section based on the measured signal without taking into account the effects of absorption and self-absorption. The focus is on investigating if resonance-enhanced Raman spectroscopy could be used as a stand-off explosives detection method, and absorption and self-absorption do not have a large influence on that application. We have focused on nitromethane (NM), 2,4-DNT and 2,4,6-TNT. We have also studied the Raman spectra of a few solvents: acetone, ethanol, and methanol, to investigate their influence on the Raman spectrum of nitromethane.

2. Experiment

Measurements were performed both in laboratory settings with controlled ambient conditions and outdoors; the measurement setups are schematically described in Figure 1. The investigated substances were nitromethane (NM) (>98%), 2,4-DNT (>97%), and 2,4,6-TNT. Pure 2,4,6-TNT was obtained by dissolving military grade TNT in ethylacetate, washing three times with a 5% aqueous solution of $NaHSO_3$, drying the remaining organic phase over Na_2SO_4 and recrystallizing the solid residue once from 99.5% ethanol. The white crystals were dried at 40°C overnight. In addition, the solvents acetone (>99%), ethanol (99.7%), and methanol (99.8%) were studied.

The explosives vapor was generated from a liquid or solid sample placed on a Petri dish. For the measurements using the vapor chamber, the Petri dish was placed inside the chamber, nitrogen (AGA, OTC-50) was added to reach atmospheric pressure, and equilibrium was allowed to be established. Equilibrium was decided as being when the signal from the sample did not increase with time; this process required different amount of time for different samples. When measuring on a combination of substances (NM and solvent), they were placed on two separate Petri dishes. The concentration $c(T)$ of the vaporized explosive in the chamber was determined by the ratio of the vapor pressure of the sample, $P_{v,\mathrm{HE}}(T)$, and the total pressure in the chamber, P_{total} as

$$c(T) = \frac{N_{\mathrm{HE}}(T)}{N_{\mathrm{total}}} = \frac{P_{v,\mathrm{HE}}(T)}{P_{\mathrm{total}}}, \qquad (1)$$

where $N_{\mathrm{HE}}(T)$ is the number of molecules of the sample in gas phase and N_{total} the total number of gas molecules in the chamber. The vapor pressure of NM has been taken from McCullough et al. [17], the vapor pressure of 2,4-DNT from Pella [18] and of 2,4,6-TNT from Edwards [19]. The vapor chamber (Al, 340 mm long, ∅ 200 mm) has a high UV-transmitting window on each side (10 mm uncoated UV-FS). In the measurements on DNT and TNT, the temperature of the chamber was elevated to above the melting point of the samples in order to reach a high enough concentration while the temperature and pressure were continuously monitored. To avoid condensation of explosives on the windows, they were held at a higher temperature than the rest of the chamber. When changing the sample, the chamber was cleaned and heated to 80°C overnight, while evacuating the system. In the measurements without the vapor chamber the sample was put outside the laboratory on a Petri dish and

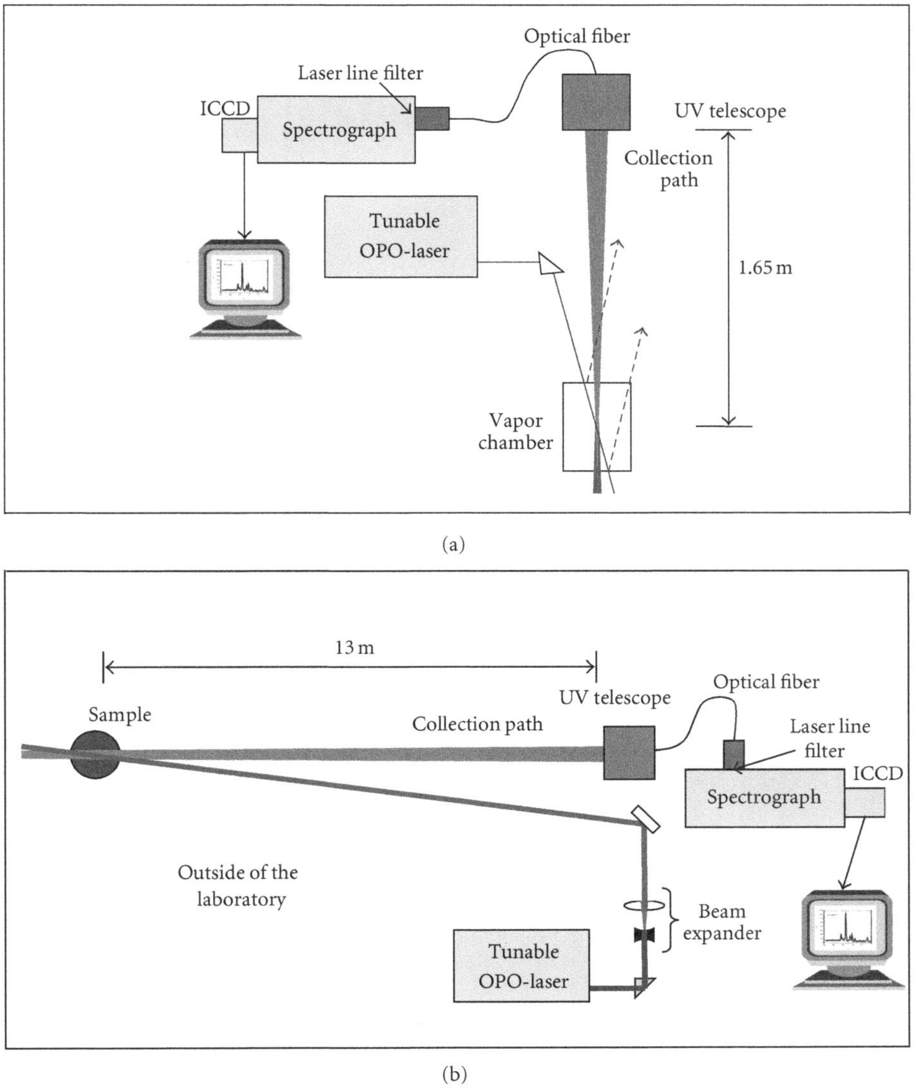

FIGURE 1: Schematic picture of the resonant Raman spectroscopy setup (a) using the vapor chamber and (b) outdoors in open air.

heated to about 328 K while the surrounding air temperature was around 274–278 K.

A tunable Optical Parametric Oscillator (OPO), pumped by the 3rd harmonic of a Nd:YAG laser (pulse length 6 ns, repetition rate 10 Hz, pulse energy 2–5 mJ), was used in the measurements, and the beam was directed towards the chamber via a right-angle prism (UV-FS). The signal was measured in a back scatter configuration, where the scattered light was collected by a 5″ UV telescope at a stand-off distance of 1.65 meters. The telescope was focused at the centre of the chamber where it collected light from a path with a width of 8 mm. The diameter of the laser beam was approximately 5–8 mm, depending on the wavelength. No expansion of the laser beam was used. The laser beam and the collection path of the telescope overlapped only inside the chamber, and thus the Raman signal was only collected from a volume (approximately 4 cm^3) where the explosive vapor concentration in relation to the nitrogen concentration was known. Nitrogen has a known Raman cross section [20] and is known not to undergo resonance enhancement in the spectral region of interest in these measurements. It was therefore used as an internal reference. Measurements performed without nitrogen in the chamber showed that the possible contribution to the nitrogen peak from air outside the chamber was negligible.

For the measurements outdoors on NM without the vapor chamber, the same OPO laser and equipment was used, but the laser beam was expanded 4-5 times and the stand-off distance was 13 meters. The beam was directed out from the laboratory by a mirror to coincide with the collection path of the telescope about 5 cm above the sample. Hence, the setup had an oblique configuration where the path of the laser beam and the collection path of the telescope were crossed only for a limited spatial section. The diameter of the collection path of the telescope was approximately 55 mm at the sample, which was slightly larger than the laser beam diameter.

The light was introduced into a spectrometer via an optical round-to-slit fiber bundle (450 μm diameter at the round end). The spectrometer was a USB-connected, computer controlled 500 mm Czerny-Turner spectrometer with a 2400 groves/mm grating, resulting in a covered spectral range of around 1000–3000 cm^{-1}, which is sufficient to see first order resonant peaks of the investigated substances and the nitrogen peaks for UV wavelengths. The Raman spectra were registered by an ICCD camera with nanosecond gating time. The camera was held at −20°C. All measurements shown in this paper are accumulated during 1000 laser pulses, that is, 100 seconds.

Laser line long pass filters were used in the 2,4,6-TNT measurements and in the 2,4-DNT measurements in order to reduce the background from stray laser light. The Raman signal from TNT was measured for incident wavelengths between 243 nm and 268 nm. Since a Raman shift of 1350 cm^{-1} (around the lowest Raman shift observed for TNT in these measurements) corresponds to a shift of around 8-9 nm for these wavelengths, the filters were used also for incident laser wavelengths below the filter cutoff, and three different laser line filters were used to cover the whole range of Raman scattered signal. A 248 nm filter was used for incident wavelengths in the range 243–250 nm, a 257 nm filter in the range 251–258 nm, and a 266 nm filter in the range 260–268 nm. The transmission of the filters drops somewhat at the lower wavelengths in the range, and corrections were performed to account for this using the filter data from the manufacturer.

The response to UV light of the setup was investigated by measuring the spectrum from a calibrated UV-light source. The response of the system decreases rapidly at smaller wavelengths. However, measuring the resonant Raman signal for pure N_2 for an excitation wavelength of 210 nm results in a clearly detectable Raman shifted signal at 215 nm. Thus, despite the low response to wavelengths below 230 nm, the sensitivity of the system still enables detection in this range.

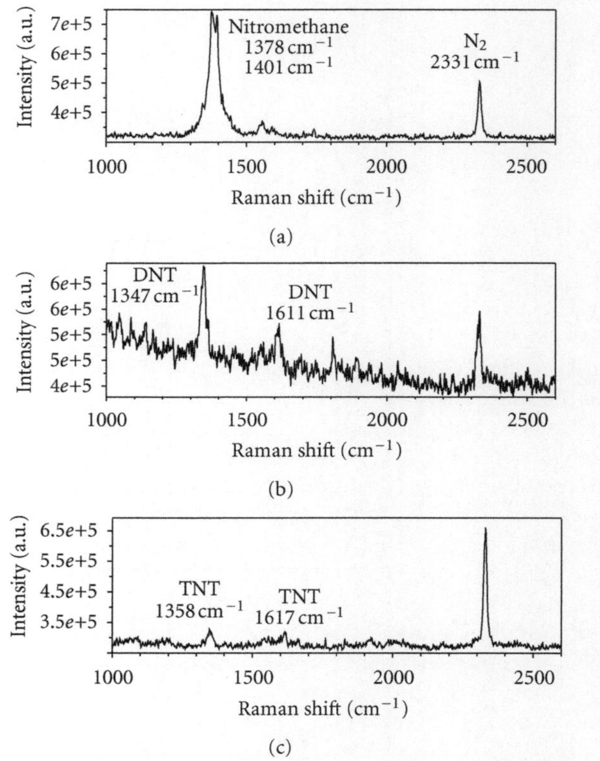

FIGURE 2: Resonant Raman signal from (a) NM at 232 nm excitation wavelength, concentration 32,500 ppm, (b) 2,4-DNT at 263 nm, concentration 1445 ppm, and (c) 2,4,6-TNT at 248 nm, concentration 341 ppm. All spectra are recorded during 1000 laser pulses, that is, 100 s, at a distance of 1.65 m. The TNT spectrum was recorded using a 248 nm long-pass filter, hence reducing the noise level.

3. Data Analysis

The areas of the Raman peaks from the NM, DNT, and TNT, $I_{HE}(\lambda)$, and from N_2, $I_{N_2}(\lambda)$, were extracted from the measured spectra (examples of spectra are shown in Figure 2) after subtracting the background noise level. From this measured signal and the pressure and temperature of the vapor chamber, the enhancement factors and measured cross sections were calculated. Nitrogen is used as an internal reference, and the Raman cross section of N_2, $\sigma_{N_2}(\lambda)$, is used for normalization. The Raman cross section for nitrogen behaves nonresonant in the wavelength range 210–300 nm and can thus be extrapolated from the value of $3.5 \times 10^{-30}\ cm^2\ str^{-1}$ at 337 nm [20]. The Raman cross sections of the studied substance, $\sigma_{HE}(\lambda)$, can then be expressed as

$$\sigma_{HE}(\lambda) = \sigma_{N_2}(\lambda)\frac{I_{HE}(\lambda)}{I_{N_2}(\lambda)}\frac{N_{N_2}}{N_{HE}}A_{abs}(\lambda, N_{HE}), \quad (2)$$

where N_{N_2} and N_{HE} are extracted from (1), and $A_{abs}(\lambda, N_{HE})$ takes into account the change in signal due to absorption and self-absorption. The effect of this will be discussed later however, in this paper we have not been able to assess these parameters, but focus has been on demonstrating the resonance enhancement. Therefore, (2) has been simplified to calculate the apparent cross section, $\sigma_{HE,app}$, which is the Raman cross section where the effect of absorption and self-absorption is not taken into account:

$$\sigma_{HE,app}(\lambda) = \sigma_{N_2}(\lambda)\frac{I_{HE}(\lambda)}{I_{N_2}(\lambda)}\frac{N_{N_2}}{N_{HE}}. \quad (3)$$

The apparent Raman enhancement factor compared to the Raman signal at 532 nm, $F_{532}(\lambda)$, can then be calculated as

$$F_{532}(\lambda) = \frac{\sigma_{HE,app}(\lambda)}{\sigma_{HE}(532)}. \quad (4)$$

The cross section at 532 nm for NM could not be found in the literature. Therefore, the Raman activities of NM and N_2 were calculated using Gaussian [21] (HF 6-31G(d,p)) and used to scale the Raman cross section for N_2 to give the cross section for NM. The Raman cross section for NM at 532 nm was found to be $3.65 \times 10^{-31}\ cm^2\ str^{-1}$ and $3.79\times10^{-31}\ cm^2\ str^{-1}$ for the 1378 cm^{-1} and 1401 cm^{-1} lines, respectively. The Raman cross section for the 1358 cm^{-1} line in TNT is 4.9×10^{-31} cm^2 str^{-1} [22]. The Raman cross section for DNT has not to our knowledge been published, and we estimate the cross section of the 1347 cm^{-1} line in DNT to be the same as the cross section for the 1358 cm^{-1} line in TNT at 532 nm.

4. Results and Discussions

4.1. Resonance-Enhanced Raman Cross Sections. Spectra from vapor phase resonance Raman measurements on NM, 2,4-DNT and 2,4,6-TNT in the vapor chamber were recorded at several wavelengths in the range 210–290 nm. Examples of the spectra are shown in Figure 2, showing measurements performed at an incident wavelength of 232 nm, 263 nm, and 248 nm for NM, DNT, and TNT, respectively, which is close to the observed maximum in the resonant Raman enhancement for each substance. However, as will be discussed below, the enhancement might be higher at lower wavelengths. In the displayed spectra, the background level has not been subtracted. The NM and DNT spectra in Figure 2 were recorded without the use of laser line filters, whereas for TNT a 248 nm long-pass filter was used.

The NM spectrum, Figure 2(a), shows a clear enhancement of the 1378 cm^{-1} and 1401 cm^{-1} lines. The resolution of the system does not allow them to be completely resolved and they are seen as a double peak. The signal at 2331 cm^{-1} is the signal from N_2. The measurement on NM was performed at room temperature, 291 K, and the vapor pressure was 24.35 Torr [17]. The atmospheric pressure varied between 744 Torr and 750 Torr during the measurements, resulting in a NM concentration between 32,500 and 32,700 ppm. The 2,4-DNT spectrum shown in Figure 2(b) exhibits an enhancement of the 1347 cm^{-1} and 1611 cm^{-1} lines, and for 2,4,6-TNT, correspondingly, the 1358 cm^{-1} and 1617 cm^{-1} lines were enhanced, as seen in Figure 2(c). DNT and TNT were heated in order to reach measurable concentrations; in the measurements presented in Figure 2, the temperature was 110°C and 128°C for DNT and TNT, respectively, resulting in concentrations of 1445 ppm DNT and 341 ppm TNT.

For each spectrum, the integrated signal of the 1378 cm^{-1} and 1401 cm^{-1} lines from NM, the 1347 cm^{-1} line from DNT and the 1358 cm^{-1} line from TNT was used to calculate the apparent Raman cross section, as described in (3). The apparent Raman cross sections as a function of incident wavelength are shown in Figures 3–5. The wavelength of maximum resonance enhancement should coincide approximately with an absorption line in the sample. In order to correlate the cross sections at different wavelengths with the absorption lines, the UV-absorption spectra of NM, DNT, and TNT were measured in vapor phase using a UV/visible spectrometer, and the results are shown in Figure 6. It should be noted that the absorption peaks could contain multiple absorption lines.

The arrows in Figure 6 indicate where the maxima in apparent Raman cross section are found in vapor phase: for NM at 232 nm, for DNT between 252 nm, and 258 nm and for TNT around 250 nm. Approaching an absorption

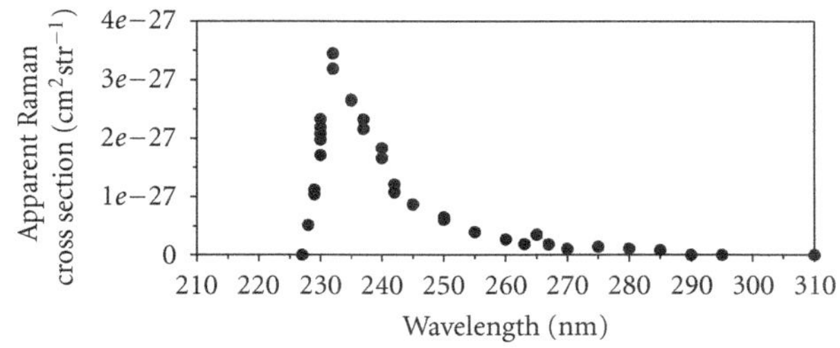

FIGURE 3: The apparent Raman cross section for the Raman shifted signal at 1378 cm^{-1} and 1401 cm^{-1} in NM as a function of incident wavelength. The concentration of NM was between 32,500 and 32,700 ppm at room temperature, 291 K.

peak, absorption and self-absorption are likely to occur in the measured Raman signal, resulting in a reduced signal as explained in Section 3. A large part of the photons from the laser will be absorbed in the vapor before reaching the center of the chamber where the Raman signal is detected. This is believed to be mainly what happens, for example below 230 nm in the NM measurements, where the Raman signal, also from N_2, entirely disappears. Transmission measurements of the laser light through the vapor chamber also show a sharp drop in transmittance between 230 nm and 240 nm, which support this interpretation. Continuing the example of NM, at wavelengths around 230 nm and above, the NM molecules will be illuminated by laser light, but they will also have a high probability to self-absorb the scattered light. This probability increases with the sample concentration in the chamber. The apparent cross section therefore gives a minimum value, and the absolute cross section is in fact higher than this. A third aspect which might contribute to the drop in signal is the weaker system response at low wavelengths. This is, however, expected to be a smaller issue than absorption and self-absorption.

For 2,4-DNT, the Raman signal was measured at four different temperatures, that is, concentrations, in the wavelength range from 220 nm to 280 nm. The apparent cross sections for the Raman shifted signal at 1347 cm^{-1} are shown in Figure 4. In the measurements of the wavelength-dependent signal of DNT, no laser-line filters were used. As seen, the apparent cross sections measured at different concentrations differ in both magnitude and wavelength for maximum cross section; at lower concentrations it is higher and the maximum tends to be shifted to shorter wavelengths, that is, closer to the absorption peak. The absolute cross section is of course the same independent of the concentration, and the observed change in magnitude is due to absorption and self-absorption in DNT in the same way as for NM. Similarly, the fact that the maximum in the apparent Raman cross section is slightly shifted towards lower wavelengths and increased in magnitude at lower concentrations indicates lower absorption at lower concentrations. Transmission measurements of the laser light through the vapor chamber shows a clear drop in transmission at around 230–260 nm, indicating that the absorption is strong enough to be important.

The wavelength dependence of the Raman signal of 2,4,6-TNT was only investigated at a concentration of 341 ppm,

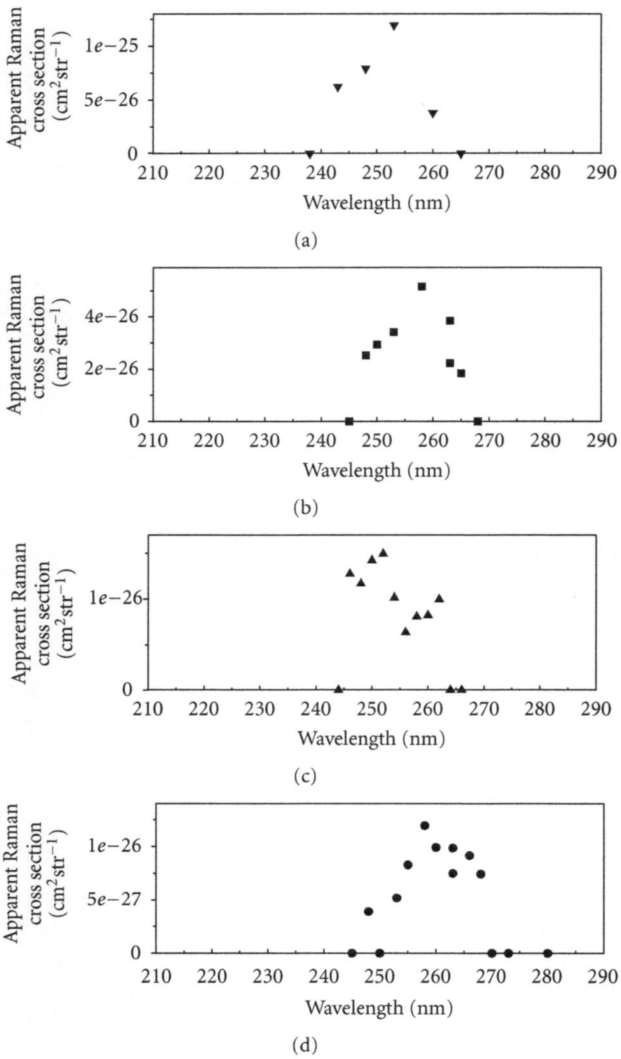

FIGURE 4: The apparent Raman cross section for the Raman shifted signal at 1347 cm^{-1} in DNT as a function of incident wavelength at (a) 354 K, concentration 126 ppm, (b) 368 K, concentration 432 ppm, (c) 371 K, concentration 550 ppm, and (d) 383 K, concentration 1445 ppm.

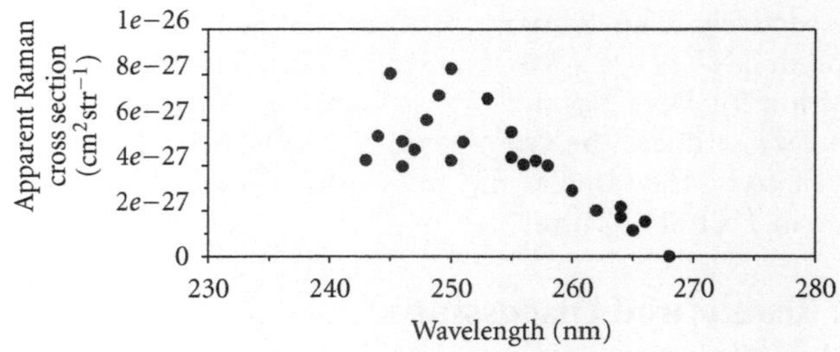

FIGURE 5: The apparent Raman cross section of the Raman shifted signal at 1358 cm^{-1} in TNT as a function of incident wavelength at 401 K, TNT concentration 341 ppm.

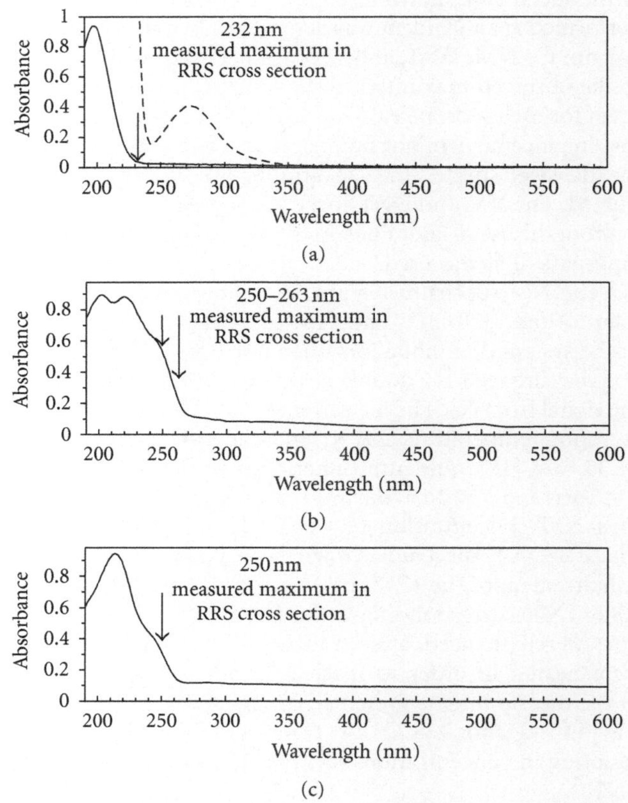

FIGURE 6: UV-absorption spectra of (a) NM, (b) 2,4-DNT, and (c) 2,4,6-TNT in vapor phase, measured with a UV/visible absorption spectrometer. The dashed line in (a) is a measurement with higher NM concentration, revealing an absorption maximum at 273 nm. The absolute sample concentrations were not measured. The arrows show the wavelengths where the maximum in apparent resonant Raman cross section is found in the vapor phase measurements of each substance. The UV spectra were taken at 291 K, 353 K, and 393 K for NM, DNT, and TNT, respectively.

and the apparent cross section was not measured in the wavelength region below 243 nm.

4.2. Enhancement Compared to 532 nm and the Dependence on Concentration. The apparent Raman cross sections for NM, DNT, and TNT presented in Figures 3–5 have been used in (4) to estimate the resonance enhancement compared to the Raman cross section at 532 nm for each substance. The Raman cross section for DNT was assumed to be the same as the cross section for TNT at 532 nm, that is, $4.9 \times 10^{-31}\,\text{cm}^2\,\text{str}^{-1}$. For all three substances a strong enhancement is seen, and the apparent enhancement factors F_{max} are shown in Table 1.

The DNT resonance enhancement increases with one order of magnitude as the concentration is reduced about the same factor, and although the signal at 354 K was very weak and embedded in the noise, which is why the temperature was raised to higher values, it is still clear that a stronger enhancement is measured at lower concentrations where the absorption is less.

The apparent Raman cross section of TNT at 248 nm excitation is one order of magnitude lower than the absolute

TABLE 1: Apparent maximum enhancements, F_{max}, compared to the Raman cross section at 532 nm for NM, DNT, and TNT at the given concentrations. λ_{max} is the incident wavelength resulting in the highest cross section, ω is the Raman wave number of the observed Raman peak, and $\sigma_{HE,app}$ is the apparent Raman cross section. The Raman signal for TNT at a concentration of 26 ppm is only measured at 248 nm; a higher value might be found at another wavelength. The apparent cross section is included in the table to facilitate comparison with other measurements and should not be mistaken for an absolute Raman cross section.

Substance	λ_{max} (nm)	ω (cm^{-1})	$\sigma_{HE,app}$ ($cm^2\,str^{-1}$)	Temp. (K)	Conc. (ppm)	F_{max}
NM	232	1378/1401	3.4×10^{-27}	291	32,500	9,300
2,4-DNT	262	1347	1×10^{-26}	383	1,445	24,000
2,4-DNT	252	1347	1×10^{-25}	354	126	240,000
2,4,6-TNT	250	1358	8×10^{-27}	401	341	17,000
2,4,6-TNT	248	1358	3×10^{-26}	369	26	61,000

cross section published for TNT in solution at an incident wavelength of 229 nm by Tuschel et al. [14]. Since the apparent cross sections presented here are believed to be reduced by absorption and to in fact increase at lower wavelength, this could account for the discrepancy. However, it cannot be assumed that the Raman cross section should be identical for solution and vapor phase samples. Absolute Raman solid-state cross sections of TNT for an incident wavelength of 248 nm reported by Nagli et al. [13], however, are around two orders of magnitude lower than the present apparent cross section. This is a surprisingly large difference for which we have no clear explanation; possible reasons could be absorption issues also in the solid-state experiments or discrepancies between the solid-state and vapor Raman cross sections.

The dependence on the concentration for the resonance Raman signal was investigated for 2,4-DNT and 2,4,6-TNT in the concentration range 10–1445 ppm and 26–364 ppm, respectively. This measurement was performed to get an estimate of the influence of self-absorption and investigate how the resonant behavior could be explored in an application. Without the absorption, factor the resonance enhancement is independent of the sample concentration. The resonance Raman signal for incident light of 248 nm was systematically monitored using a 248 nm long-pass filter while heating the chamber, that is, increasing the concentration. The apparent enhancement of the resonance Raman signal compared to the Raman cross section at 532 nm was then calculated for different concentrations and is shown in Figure 7.

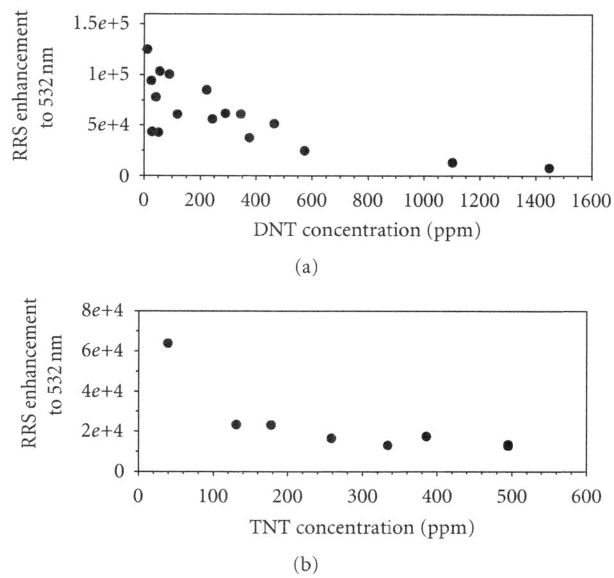

FIGURE 7: Apparent resonance enhancement of the Raman signal compared to the Raman cross section at 532 nm for (a) the 1347 cm^{-1} line in 2,4-DNT at different DNT concentrations and (b) the 1358 cm^{-1} line in 2,4,6-TNT at different TNT concentrations. The wavelength of the incident light was 248 nm, and a 248 nm long-pass filter was used in both measurements. The enhancement varies with concentration due to the absorption effects. Note that the maximum enhancement for the higher concentrations of DNT is found at longer wavelengths than 248 nm, whereas it is close to the wavelength giving maximum enhancement for TNT at 341 ppm.

4.3. Measurements on Solvents at 232 nm. The results from measurements on acetone, ethanol and methanol are presented in Figures 8, 9, and 10. The upper spectra were recorded for the pure solvents and the lower spectra were recorded for the solvent together with NM. The spectra were taken using a laser wavelength of 232 nm, which is where the maximum signal from NM was observed at room temperature. All solvent measurements were performed at room temperature. During the pure acetone measurements, the temperature was 294 K, resulting in a vapor pressure of acetone of 192 Torr [23], and the atmospheric pressure was 761 Torr giving an acetone concentration of approximately 201,000 ppm. For ethanol the corresponding temperature and atmospheric pressure were 297 K and 747 Torr, resulting in a vapor pressure of ethanol of 55.3 Torr [24] and an ethanol concentration of 69,000 ppm. During the methanol measurements, the temperature was 297 K, and the atmospheric pressure 747 Torr, resulting in a methanol vapor pressure of 120 Torr [25] and a methanol concentration of 138,000 ppm. For the measurements on a combination of solvent and NM, the concentrations were similarly determined: acetone and NM to 218,000 ppm and 33,700 ppm, respectively, ethanol and NM to 66,200 ppm and 40,400 ppm, respectively, and methanol and NM to 133,000 ppm and 37,500 ppm, respectively.

Acetone, ethanol, and methanol all show a clear Raman spectrum at 232 nm at the given vapor concentrations. The lines are not overlapping the lines of NM, and there is no strong fluorescence from the solvents.

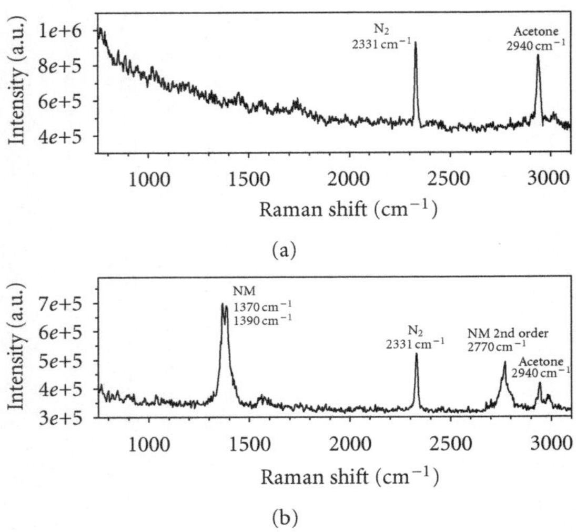

Figure 8: Raman signal from vapor phase measurements at a laser wavelength of 232 nm on (a) acetone at a concentration of 201,000 ppm and (b) acetone and NM at concentrations of 218,000 ppm and 33,700 ppm, respectively.

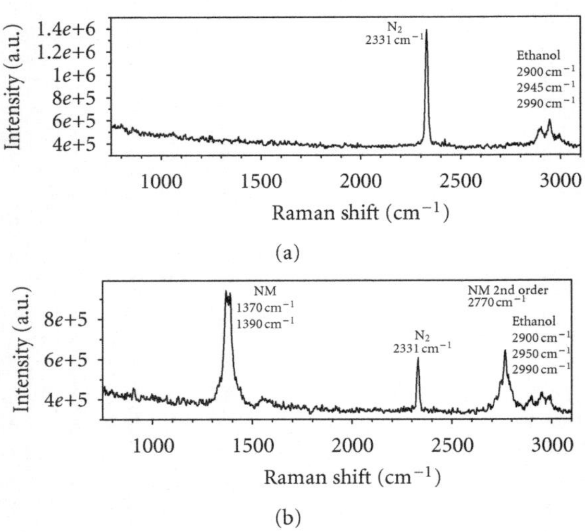

Figure 9: Raman signal from vapor phase measurements at a laser wavelength of 232 nm on (a) ethanol at a concentration of 69,000 ppm and (b) ethanol and NM at concentrations of 66,200 ppm and 40,400 ppm, respectively.

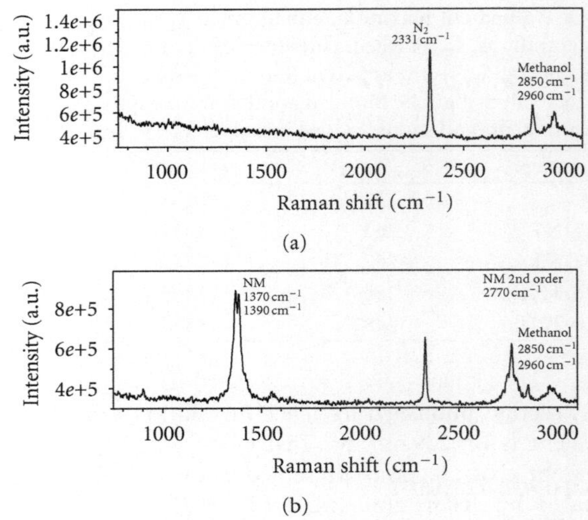

Figure 10: Raman signal from vapor phase measurements at a laser wavelength of 232 nm on (a) methanol at a concentration of 138,000 ppm and (b) methanol and NM at concentrations of 133,000 ppm and 37,500 ppm, respectively.

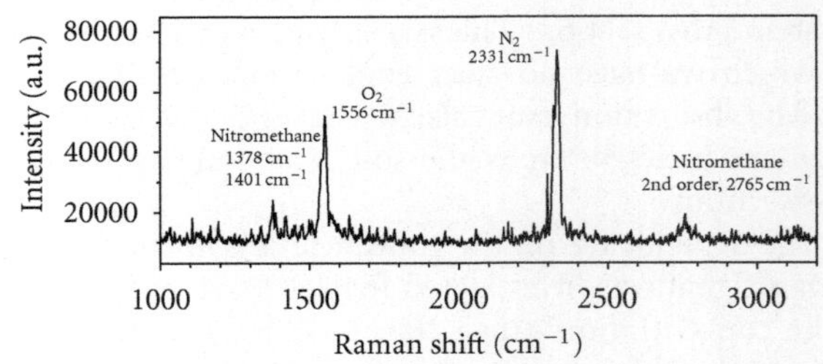

Figure 11: Resonance Raman spectrum from outdoor measurement on nitro methane in vapor phase at an irradiation wavelength of 220 nm. The sample temperature was approximately 328 K, the outdoor temperature was 274 K, and the atmospheric pressure was about 755 Torr. The spectrum was accumulated during 1000 laser pulses.

4.4. Outdoor Standoff Measurements on NM. Resonance Raman measurements were also made on NM in vapor phase outdoors, with the sample slightly heated and completely exposed to the wind. Figure 11 shows the NM spectrum measured outdoors at an irradiation wavelength of 220 nm. The concentration of NM was not determined and therefore the cross section could not be determined. At 2331 cm^{-1}, the N_2 peak is seen, and the peak at about 1550 cm^{-1} originates from O_2. The first order NM double peak is seen at approximately 1380 cm^{-1} and the second order Raman peak can be seen at about 2770 cm^{-1}. In the outdoor measurements, it was possible to see the signal from NM vapor at shorter wavelengths than was possible in the vapor chamber. The best signal was obtained at about 220 nm outdoors instead of at 232 nm as in the laboratory measurements. This is probably a result of a much lower NM concentration in the outdoor measurements, resulting in less absorption and self-absorption and making it possible to measure the signal closer to the absorption peak.

5. Conclusions

Resonance Raman spectroscopy measurements have been performed on NM, DNT, and TNT in vapor phase at standoff distances. In addition, the Raman spectra of acetone, ethanol, and methanol in vapor phase were measured at 232 nm. It was demonstrated that there is a clear enhancement of specific Raman lines in the explosive substances at UV wavelengths and that they can be used to identify vapor at a distance. The resonance enhancement for the explosive substances at different concentrations and wavelengths has been estimated to be between 10,000 and 250,000 compared to the Raman cross section at 532 nm. The apparent cross sections and enhancements are reduced by absorption and self-absorption in the sample vapor, and the sensitivity is affected by the small size of the illuminated volume. It is believed that an increased volume and lower concentration in open air will have a positive effect on the sensitivity. Measurements performed on NM outdoors at 13 m distance under more realistic conditions showed good results, supporting this idea, and together with the laboratory results, they demonstrate the capability of resonance-enhanced Raman spectroscopy as a method to detect small amounts of explosives in vapor phase and its potential for detection of hidden explosives at stand-off distances.

Acknowledgment

The presented work has been funded by the Defense Advanced Research Projects Agency (DARPA).

References

[1] D. Menning and H. Ostmark, "Detection of liquid and homemade explosives: what do we need to know about their properties?" in *Proceedings of the NATO Advanced Research Workshop on Detection of Liquid Explosives and Flammable Agents in Connection with Terrorism*, pp. 55–70, St Petersburg, Russia, 2007.

[2] S. Wallin, A. Pettersson, H. Östmark, and A. Hobro, "Laser-based standoff detection of explosives: a critical review," *Analytical and Bioanalytical Chemistry*, vol. 395, no. 2, pp. 259–274, 2009.

[3] J. L. Gottfried, F. C. de Lucia, C. A. Munson, and A. W. Miziolek, "Laser-induced breakdown spectroscopy for detection of explosives residues: a review of recent advances, challenges, and future prospects," *Analytical and Bioanalytical Chemistry*, vol. 395, no. 2, pp. 283–300, 2009.

[4] N. R. Butt, M. Nilsson, A. Jakobsson et al., "Classification of raman spectra to detect hidden explosives," *IEEE Geoscience and Remote Sensing Letters*, vol. 8, pp. 517–521, 2010.

[5] S. K. Sharma, A. K. Misra, and B. Sharma, "Portable remote Raman system for monitoring hydrocarbon, gas hydrates and explosives in the environment," *Spectrochimica Acta A*, vol. 61, no. 10, pp. 2404–2412, 2005.

[6] J. C. Carter, S. M. Angel, M. Lawrence-Snyder, J. Scaffidi, R. E. Whipple, and J. G. Reynolds, "Standoff detection of high explosive materials at 50 meters in ambient light conditions using a small Raman instrument," *Applied Spectroscopy*, vol. 59, no. 6, pp. 769–775, 2005.

[7] A. Pettersson, I. Johansson, S. Wallin, M. Nordberg, and H. Östmark, "Near real-time standoff detection of explosives in a realistic outdoor environment at 55 m distance," *Propellants, Explosives, Pyrotechnics*, vol. 34, no. 4, pp. 297–306, 2009.

[8] A. Pettersson, S. Wallin, H. Östmark et al., "Explosives standoff detection using Raman spectroscopy: from bulk towards trace detection," in *Proceedings of the Detection and Sensing of Mines, Explosive Objects, and Obscured Targets XV*, R. S. Harmon, J. H. Holloway Jr., and J. T. Broach, Eds., vol. 7664, April 2010.

[9] M. Nordberg, F. Akke, and A. Pettersson, "Detection limits of stand-off spontaneous Raman scattering," Tech. Rep., Swedish Defence Research Agency, Stockholm, Sweden, 2009.

[10] H. S. Sands, I. P. Hayward, T. E. Kirkbride, R. Bennett, R. J. Lacey, and D. N. Batchelder, "UV-excited resonance Raman spectroscopy of narcotics and explosives," *Journal of Forensic Sciences*, vol. 43, no. 3, pp. 509–513, 1998.

[11] M. Gaft and L. Nagli, "UV gated Raman spectroscopy for standoff detection of explosives," *Optical Materials*, vol. 30, no. 11, pp. 1739–1746, 2008.

[12] N. A. Szymanski, *Raman Spectroscopy: Theory and Practice*, Plenum, New York, NY, USA, 1967.

[13] L. Nagli, M. Gaft, Y. Fleger, and M. Rosenbluh, "Absolute Raman cross-sections of some explosives: trend to UV," *Optical Materials*, vol. 30, no. 11, pp. 1747–1754, 2008.

[14] D. D. Tuschel, A. V. Mikhonin, B. E. Lemoff, and S. A. Asher, "Deep ultraviolet resonance raman excitation enables explosives detection," *Applied Spectroscopy*, vol. 64, no. 4, pp. 425–432, 2010.

[15] H. Östmark et al., "Vapor pressure of explosives: a critical review," *Propellants, Explosives, Pyrotechnics*. In press.

[16] R. B. Cundall, T. F. Palmer, and C. E. C. Wood, "Vapour pressure measurements on some organic high explosives," *Journal of the Chemical Society, Faraday Transactions I*, vol. 74, pp. 1339–1345, 1978.

[17] J. P. McCullough, D. W. Scott, R. E. Pennington, I. A. Hossenlopp, and G. Waddington, "Nitromethane: the vapor heat capacity, heat of vaporization, vapor pressure and gas imperfection; the chemical thermodynamic properties from 0 to 1500° K," *Journal of the American Chemical Society*, vol. 76, no. 19, pp. 4791–4796, 1954.

[18] P. A. Pella, "Measurement of vapor pressures of TNT, 2,4-DNT 2,6-DNT, and EGDN," *The Journal of Chemical Thermodynamics*, vol. 9, no. 4, pp. 301–305, 1977.

[19] G. Edwards, "The vapor pressure of 2-4-6-trinitrotoluene," *Transactions of the Faraday Society*, vol. 46, pp. 423–427, 1950.

[20] H. Inaba and T. Kobayasi, "Laser-Raman radar—Laser-Raman scattering methods for remote detection and analysis of atmospheric pollution," *Optical and Quantum Electronics*, vol. 4, no. 2, pp. 101–123, 1972.

[21] M. J. Frisch, G. W. Trucks, and H. B. Schlegel, *Gaussian 03, Revision C.02*, Gaussian, Wallingford, Conn, USA, 2004.

[22] M. Gaft and L. Nagli, "Standoff laser based spectroscopy for explosives detection," in *Electro-Optical Remote Sensing, Detection, and Photonic Technologies and Their Applications*, vol. 6739 of *Proceedings of SPIE*, p. 673903, Florence, Italy, September 2007.

[23] D. Ambrose, C. H. S. Sprake, and R. Townsend, "Thermodynamic properties of organic oxygen compounds XXXIII. The vapour pressure of acetone," *The Journal of Chemical Thermodynamics*, vol. 6, no. 7, pp. 693–700, 1974.

[24] C. B. Kretschmer and R. Wiebe, "Liquid-vapor equilibrium of ethanol-toluene solutions," *Journal of the American Chemical Society*, vol. 71, no. 5, pp. 1793–1797, 1949.

[25] D. Ambrose and C. H. S. Sprake, "Thermodynamic properties of organic oxygen compounds XXV. Vapour pressures and normal boiling temperatures of aliphatic alcohols," *The Journal of Chemical Thermodynamics*, vol. 2, no. 5, pp. 631–645, 1970.

3

New Possibility for REE Determination in Oil

Alexander Soin,[1] Tatiana Maryutina,[1,2] Natalya Musina,[1] and Andrey Soin[1,2]

[1] *United Research and Development Centre, Leninsky prospect 55/1 Building 2, Moscow 119333, Russia*
[2] *Vernadsky Institute of Geochemistry and Analytical Chemistry, Russian Academy of Sciences, Kosygin Street 19, Moscow 119991, Russia*

Correspondence should be addressed to Tatiana Maryutina, tatiana@maryutina.com

Academic Editor: S. Michael Angel

A novel approach of rare earth elements (REE) determination in crude oil is suggested. Special application of countercurrent chromatography (CCC) is used as a sample pretreatment tool. An oil sample is continuously pumped through the rotating coil column (RCC) as a mobile phase, while an aqueous phase (nitric acid solution) is retained as a stationary phase. Two phases are kept well mixed and agitated, but there is no emulsion at the interface under the chosen conditions. Special features of CCC give an opportunity to vary the volume of oil samples to be analyzed from 10 mL to 1 L or more. Trace metals are preconcentrated into 10 mL of stationary phase (acidic solutions) pumped out of the column so that analysis can be easily determined with inductively coupled plasma mass spectrometry (ICP-MS) without additional sample preparation procedures. Optimal concentration of nitric acid in the stationary phase for preconcentration of REE from oil by CCC has been investigated. The combination of CCC with ICP-MS gives the possibility to develop a rapid, reliable, and accurate method of trace metal including rare earth elements (REE) determination in crude oils and oil products. Such method could be an alternative for unexpanded and expensive neutron-activation analysis (NAA).

1. Introduction

Microelemental and REE contents of oil are very important for estimating oil's age and for developing theories of oil's origin. It is known that information about hydrocarbons genesis could be obtained using element ratios. Such ratios also could be determinative as the georeconnaissance data. From this point of view, REE content/ratios are of the prime interest. REE ratios could be used as the reference points for different oils (from different deposits/fields) like for most geological samples as rocks, minerals, and ores [1, 2].

In the meantime, REE analysis of oils is still a very complicated analytical task. There are no methods of preconcentration of metals from oil to aqueous phases, and REE contents are usually at ppt level and lower. Also there are no standard methods (ASTM D, IP, EN, or UOP) or standard/reference samples for REE analysis of oils. It should be mentioned that articles and studies dedicated to REE content in oils could be found really seldom. Thus, ICP-MS determination of REE in oils was published [3] when acid digestion was used for sample preparation. As a result, most of REE, measured (including Pr, Sm, Eu, Gd, Dy, Er and Yb) were "not detected."

Suggesting hybrid method (CCC and ICP-MS) enables lessening detection limit (DL) and detecting most of metals in oil including REE.

Abilities of the liquid stationary phase retention during pumping of oil samples through column enable isolation and preconcentration of the inorganic impurities into acidic aqueous solutions. In this case, CCC can be used as a tool for the sample pretreatment and preparation. Preliminary investigations on retention features of oil/oil products—aqueous phase systems in CCC, were published [4, 5].

The main characteristic of sample preparation methods is recovery values. It is preferable to compare recoveries of inorganic impurities from oil by extraction method and other instrumentation techniques. If that instrumentation method enables direct oil analysis, then comparison will be more reliable. Thus, NAA was used in this work. Aforesaid issue of standard/reference samples is a real problem for extraction/leaching methods as well. Therefore, relative contents of REE were mainly used.

Thus, regardless of the geological and tectonic structure of the source regions, oil composition, depth of deposit occurrence, and lithology of the enclosing rocks, the distribution curves of REE are usually highly differentiated. It has been shown also that total quantity of lanthanides (ΣLn_N) has a very wide range from 0.5 to 5.8 mg/t (3.0 mg/t on average) in West Siberian and Kazakhstani oils. As an example, fluids of cooling magma pockets of increased alkalinity are the major source of lanthanides in oil. They are involved in oil enrichment with light lanthanides and formation of the positive Eu anomaly [6].

Main objective of this work was to estimate CCC-ICP-MS approach to REE determination of oils from different sources and to compare with the results obtained by NAA.

2. Experimental

2.1. Reagents and Materials. High-purity deionized water (resistivity 18.2 mΩ) obtained from a Milli-Q water purification system (Millipore, Canada) and ultrapure nitric acid HNO_3 (Merck, Germany) for aqueous solution preparing and decomposition of oil samples and as a stationary phase has been used.

In the quantitative elemental analysis by ICP-MC, we prepared standard solutions by diluting multielement agilent standard A (Al, As, Ba, Be, Bi, B, Cd, Ca, Ce, Cs, Cr, Co, Cu, Dy, Er, Eu, Gd, Ga, Ho, In, K, Fe, La, Li, Lu, Mg, Mn, Na, Nd, Ni, P, Pb, Pr, Rb, Re, Sm, Sc, Se, Sr, Tb, Th, Tl, Tm, U, Yb, Y, V, and Zn 10 μg/mL ±0.5% in 2% HNO_3). Elements concentration range for all the elements was 1–100 μg/L.

Calibration coefficients of determination for all elements were 0.9995–0.9999. Recovery tests were performed using "Trace metals in drinking water" solution (with 2–250 μg/L concentrations of Al, As, Ba, Be, Bi, Cd, Cr, Co, Cu, Fe, Pb, Li, Mn, Mo, Ni, Rb, Sb, Se, Sr, Te, Tl, U, V, Zn, and Ag), and recoveries were in the range of 91–112%.

Experiments on trace and rare earth elements preconcentration were performed with light Kazakhstan (Tengizskaya) oil with density ρ = 0.795 g/mL and viscosity η = 2.270 mm^2/s, West Siberian oil (region) with ρ = 0.870 g/mL and η = 17.90 mm^2/s and composite oil with ρ = 0.828 g/mL and η = 4.745 mm^2/s. Tengizskaya oil is a medium-sulfur oil (S content ~0.9 mass.%), West Siberian oil is a middle-sulfur oil (S content ~1.6 mass.%), and composite oil is a medium-sulfur oil (S content ~0.8 mass.%).

2.2. Instruments

2.2.1. RCC. Figure 1 schematically illustrates type J single-layer synchronous planetary centrifuge. Typical column is formed by winding Teflon tubing around a cylindrical holder. Column holder rotates around its own axis and revolves around the centrifuge axis at the same speed (ω) and in the same direction. An aqueous solution is retained in column due to the complex gravitational (centrifugal) field, while the mobile phase (oil) is pumped through the column.

RCC experiments were performed with "Spring-3 M" J-type single-layer synchronous horizontal planetary centrifuge (made in the Institute for Analytical Instrumentation,

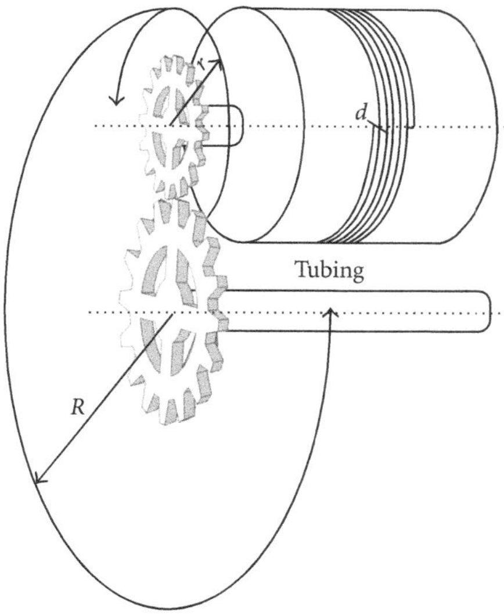

Figure 1: Scheme of planetary centrifuge.

St. Petersburg, Russia) with the total column volume of 19 mL at constant T = 20°C. Scheme of planetary centrifuge is in Figure 1. RCC coiling diameter was 120 mm, revolution radius was 93 mm, and length of the tube was 1.15 m. Phases were pumped through an RCC using a peristaltic pump (Masterflex, UK).

The column was filled with aqueous HNO_3 solution while stationary. Then the planetary centrifuge is set to rotate at constant speed, and the mobile phase (oil or diesel) was pumped through the column from the top of the column to the bottom or "Head to Tail" [4, 5, 7]. Some volume of the aqueous acid solution is retained in the column as the stationary phase. All working parameters for the planetary centrifuge are shown in Table 1. After the experimental run, the acid solution was pumped out of the nonrotating column and analyzed by ICP-MS.

Toluene, iso-octane, and acetone of chemically pure grade were used for column rinsing. After every CCC experiment, 70–100 of each solvent were pumped through the stationary column.

2.2.2. Autoclave Decomposition. To compare results of CCC elements isolation from oil with other techniques, autoclave decomposition of oil samples was used. Module "MKP-04" with autoclave NPVF "ANKON-AT-2" (Russia) was used for decomposition in a closed system. Standard conditions of autoclave decomposition are provided in Table 2. Oil samples were digested by concentrated ultrapure HNO_3 in sealed Teflon containers and than cooled and diluted to 50 mL by deionized water (washing of containers sides). Total heating time of oil sample at different temperature conditions takes 4 h. These diluted samples were analyzed by ICP-MS.

2.2.3. ICP-MS Determination. ICP-MS measures most of the elements with very low (ppb or even ppt) DLs. Minimization of large interferences was provided with a simple matrix of analyzed samples (diluted HNO_3 solutions), maximal comp-

Table 1: Working parameters of planetary centrifuge.

Mobile-phase volume	Stationary-phase volume	Centrifuge rotation speed	Beta ratio (spool radius/rotor radius)	Flow rate of mobile phase	Column tubing bore	Temperature
V_M, mL	V_S, mL	ω, rpm	$\beta = r/R$	F, mL/min	d, mm	T, °C
19–110	10	750	0.65	0.5; 1.2	0.8	20

Table 2: Conditions and working parameters of autoclaves decomposition of oil sample.

Sample mass	Acid volume	Time and temperature		
m_s, g	V_{HNO_3}, mL	1	2	3
0.5	10	1 h—160°C	1 h—180°C	2 h—200°C

Table 3: Instrumental parameters for determination of metals by ICP-MS.

Forwarded power	1200 W
Analog stage voltage	−1800 V
Pulse stage voltage	1100 V
Argon consumption	
Plasma generating flow	15 L/min
Auxiliary flow	0.8 L/min
Carrier flow	1.4 L/min
Sample consumption	0.5 mL/min
Peak resolution	∼0.6 amu
Detector operating modes	Pulse count and analog
Speed of peristaltic pump	24 rpm
Spray chamber	Cooling (4°C) quartz
Mass range	27–238 amu
Sweeps/reading	40
Replicates	3
Dwell time	50 ms
Scan mode	Peak hopping
Abundance sensitivity	
Low mass	$<5\cdot10^{-7}$
High mass	$<1\cdot10^{-7}$

liance with the standards matrix, and optimization of the instrument parameters.

Performance of the ICP-MS instrument strongly depends on the operating conditions. Tuning solution containing Li, Y, Ce, Tl, Co (10 μg/L) in 2 wt% aqueous HNO_3 was used to optimize the instrument (Agilent 7500c, USA) in terms of sensitivity, resolution, and mass calibration. $^{140}Ce^{16}O^+/^{140}Ce^+$ and $^{137}Ba^{2+}/^{137}Ba^+$ ratios were used to minimize oxide ions and double-charged ions levels in the plasma that could interfere with the determination of some elements.

Measurement parameters such as carrier gas flow, torch position, speed of peristaltic pump, and dwell time were optimized with the aim of high sensitivity and low isobaric interferences. ICP-MS analysis was performed following the operating program and parameters shown in Table 3.

All ICP-MS data (final concentrations in analyzed solutions, calibrations, tunes, statistics, etc.) were obtained from ICP-MS Top Agilent program pack.

Final REE concentrations (C, μg/g) in oils were calculated using the following standard formula:

$$C = \frac{\left(A_{\text{sample}} - \overline{A}_{\text{blanks}}\right) \cdot V}{m}, \quad (1)$$

where A_{sample} is the concentration of each element in sample solution (μg/L) obtained from ICP-MS (counted by ICP-MS Top Agilent software), A_{blanks} is the mean concentration of each element in blank solutions (nitric acid was pumped through the column in 3–5 repetitions), V is the volume of oil (mobile phase) pumped through the column in CCC experiments, and m is the mass of analyzed oil.

2.2.4. NAA. Measurements were conducted at Central Laboratory of Substance Analysis of Vernadsky Institute of Geochemistry and Analytical Chemistry RAS. Analyzed oil samples of 10–50 mg were placed in a bag of aluminum foil 10 × 10 mm in size together with certified reference materials (ST-1A, effusive basalt; SGD-1A, intrusive basic rock; AN-G, intrusive basic rock). The bag is placed in an aluminum case and irradiated with a neutron flux of 1.2×10^{13} neutron $cm^{-2}\ s^{-1}$ for 15 h. The samples are cooled for 6 days to reduce activity of radionuclides with $T_{1/2} < 15$ h. Then gamma spectra of samples are measured using a Ge or a Ge(Li) detector with a multichannel pulse analyzer for activity levels of 100 and 2000 s, respectively. Measurements are repeated after 6 to 10 days to distinguish gamma lines of relatively long-lived radionuclides. The gamma spectra obtained are processed using the program package described in [8, 9] and by comparing the normalized peak areas of the studied and reference samples. Element concentrations are calculated with regard to their concentrations in the reference samples.

2.3. Results and Discussion. Application of CCC for REE recovery was tested on Tengizskaya oil. Table 4 provides measured concentrations of 12 REE from Tengizskaya oil. Various volumes of analyzed oil (52, 104, and 624 mL) were pumped at the constant flow rate of 1.2 mL/min through the CCC column rotating at 750 rpm. Minimal HNO_3 concentration (0.5 M) was used for all CCC experiments because of no emulsion at the oil-water interface [10]. Nitric acid with higher concentrations can form some emulsion on the oil-water interface and that means bleed of the stationary phase. It is obvious that concentration factor (5, 10, and 60, resp.) did not affect obtained REE concentrations. The choice of the concentration factor is upon sample quantity and sample preparation time. It is also clearly seen that concentrations

Table 4: Measured REE concentration in Tengizskaya oil by ICP-MS after CCC preconcentration from different sample volumes of Tengizskaya oil.

Element	Concentration, μg/kg		
	V_M = 52 mL	V_M = 104 mL	V_M = 624 mL
Y	**0.019** ± 0.004	**0.021** ± 0.004	**0.023** ± 0.004
La	**0.133** ± 0.02	**0.075** ± 0.02	**0.091** ± 0.02
Ce	**0.06** ± 0.01	**0.06** ± 0.01	**0.05** ± 0.01
Pr	**0.006** ± 0.001	**0.006** ± 0.001	**0.007** ± 0.001
Nd	<0.001	<0.001	<0.001
Sm	**0.006** ± 0.001	**0.004** ± 0.001	**0.005** ± 0.001
Eu	**0.018** ± 0.004	**0.021** ± 0.004	**0.017** ± 0.004
Gd	**0.011** ± 0.003	**0.014** ± 0.003	**0.013** ± 0.003
Tb	**0.0020** ± $4E-4$	**0.0020** ± $4E-4$	**0.0020** ± $4E-4$
Dy	**0.0030** ± $6E-4$	**0.0030** ± $6E-4$	**0.0030** ± $4E-4$
Ho	**0.0010** ± $2E-4$	**0.0010** ± $2E-4$	**0.0010** ± $4E-4$
Er	**0.0020** ± $4E-4$	**0.0020** ± $4E-4$	**0.0030** ± $4E-4$
Yb	**0.005** ± 0.001	**0.005** ± 0.001	**0.005** ± 0.001
Lu	<0.0005	<0.0005	<0.0005

Table 5: Measured REE concentration in Tengizskaya, West Siberian, and composite oils by ICP-MS after CCC preconcentration.

Element	Concentration in oils, μg/kg		
	Tengizskaya	West Siberian	Composite
Y	**0.019** ± 0.004	**0.007** ± 0.001	**0.012** ± 0.002
La	**0.133** ± 0.023	**0.208** ± 0.036	**0.037** ± 0.004
Ce	**0.06** ± 0.01	**0.012** ± 0.002	**0.0040** ± $8E-4$
Pr	**0.006** ± 0.001	**0.0020** ± $4E-4$	**0.0020** ± $4E-4$
Nd	<0.001	**0.007** ± 0.002	**0.0030** ± $6E-4$
Sm	**0.006** ± 0.001	**0.0020** ± $4E-4$	**0.0020** ± $4E-4$
Eu	**0.018** ± 0.004	**0.0020** ± $4E-4$	**0.0004** ± $1E-4$
Gd	**0.011** ± 0.003	<0.0002	**0.0004** ± $1E-4$
Tb	**0.0020** ± $4E-4$	**0.0030** ± $6E-4$	**0.0002** ± $5E-5$
Dy	**0.0030** ± $6E-4$	**0.0010** ± $2E-4$	**0.0005** ± $1E-4$
Ho	**0.0010** ± $2E$-4	**0.0010** ± $2E-4$	**0.0003** ± $1E-5$
Er	**0.0020** ± $4E-4$	**0.0010** ± $2E-4$	**0.0002** ± $1E-5$
Yb	**0.005** ± 0.001	**0.0010** ± $2E-4$	**0.0005** ± $1E-4$
Lu	<0.0005	<0.0005	<0.0005

of REE are very low with ΣLn_N ~0.2 mg/t. In case of low concentration of REE in oil, the highest concentration factor (60 or even higher) should be used especially with the other instrumentation techniques (ICP-AES, AAS).

It should be noted that concentrations of REE in oils are at very low level that cause difficulties of their direct detection.

There are comparative data of REE determination in 3 different oils: Tengizskaya, West Siberian, and composite in Table 5. REE distributions and ratios are given in Figure 2 and Table 6, respectivelly. For both oils, ΣLn_N was about 0.2 mg/t, while composite oil was diluted by lighter kerosene fraction (ΣLn_N ~0.1 mg/t). It is also clear that REE ratios are differentiated.

REE contents and rations of oils from 3 source regions are different (Figure 2, Table 6). Such REE ratios are usually used as geochemical certificates for oils. In the meantime, we obtained "less DL" results for REE in all oils after autoclaves decomposition. DLs were in the range of 0.3–0.04 μg/kg and it was impossible to compare results of the two sample preparation techniques.

Therefore, NAA was used in this work as an instrumentation method that enables direct oil analysis (REE detection in the native forms). Only comparison with NAA could be applied as correctness criteria for combined CCC-ICP-MS method. Comparative REE ratios obtained via NNA and CCC-ICP-MS are presented in Table 7. These results prove correctness of both the methods of REE ratios

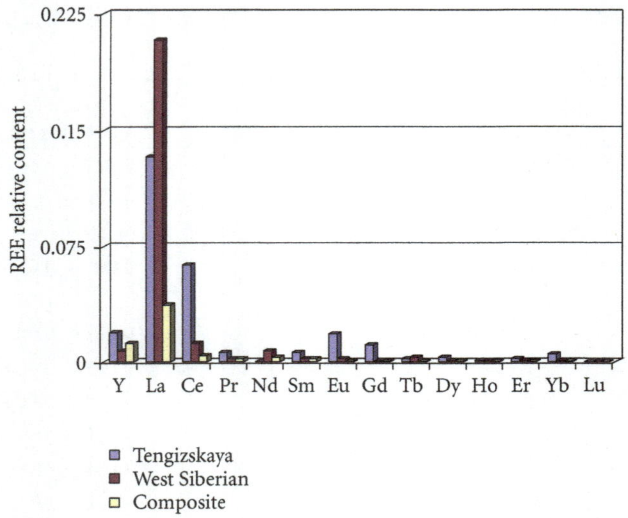

FIGURE 2: Relative REE distributions in Tengizskaya, West Siberian, and composite oil samples.

TABLE 6: Comparative REE ratios of Tengizskaya, West Siberian, and composite oils obtained by ICP-MS after CCC preconcentration.

Oil	Elements ratios			
	La/Ce	Nd/Sm	Eu/Gd	Dy/Ho
Tengizskaya	**2.1**	—	**1.8**	**3.0**
West Siberian	**17**	**3.5**	—	**1.0**
composite	**9.3**	**1.5**	**1.0**	**1.7**

TABLE 7: Comparative REE ratios of Tengizskaya oil obtained by ICP-MS after CCC preconcentration and by NAA.

Method	Elements ratios			
	La/Ce	Nd/Sm	Eu/Gd	Dy/Ho
NAA	**1.8**	**4.2**	**1.6**	**3.0**
ICP-MS	**2.1**	—	**1.8**	**3.0**

determination. Both results are not distorted, that is, very important for studying base geochemical processes of the oil genesis.

3. Conclusions

It has been shown that REE contents and rations of oils from 3 source regions are different. Concentrations of REE are very low with ΣLn_N ~0.2 mg/t in 3 oil samples studied. Composite oil has ΣLn_N ~0.1 mg/t as diluted one with light fraction. Such ratios are usually used as geochemical certificates for oils. Comparative REE ratios obtained via NNA and suggested combined CCC-ICP-MS method are in close agreement. Correct REE ratios could be used for geochemical studies dedicated to oil genesis research.

Acknowledgment

The authors are grateful to Russian Foundation for Basic Research (RFBR) Project no. 09-03-00757.

References

[1] A. F. Oluwole, O. I. Asubiojo, J. I. Nwachukwu et al., "Neutron activation analysis of nigerian crude oils," *Journal of Radioanalytical and Nuclear Chemistry*, vol. 168, no. 1, pp. 145–152, 1993.

[2] S. M. Eggins, J. D. Woodhead, L. P. J. Kinsley et al., "A simple method for the precise determination of ≥ 40 trace elements in geological samples by icpms using enriched isotope internal standardisation," *Chemical Geology*, vol. 134, no. 4, pp. 311–326, 1997.

[3] A. Akinlua, N. Torto, and T. R. Ajayi, "Determination of rare earth elements in niger delta crude oils by inductively coupled plasma-mass spectrometry," *Fuel*, vol. 87, no. 8-9, pp. 1469–1477, 2008.

[4] T. A. Maryutina, A. V. Soin, and O. N. Katasonova, "Countercurrent chromatography for oil analysis: retention features and kinetic effects," *Journal of Chromatography*, vol. 1216, no. 19, pp. 4232–4236, 2009.

[5] A. V. Soin, T. A. Maryutina, and N. S. Musina, "Behavior of oil—aqueous solution systems in rotating coil column," *Journal of Chromatography and Separation Techniques*, vol. 2, no. 1, pp. 106–110, 2011.

[6] R. P. Gottikh, S. F. Vinokurov, and B. I. Pisotskii, "Rare-earth elements as geochemical criteria of endogenous sources of microelements contained in oil," *Doklady Earth Sciences*, vol. 425, no. 2, pp. 325–329, 2009.

[7] W. D. Conway, *Countercurrent Chromatography. Apparatus, Theory and Applications*, VCH, New York, NY, USA, 1990.

[8] G. M. Kolesov, N. A. Shubina, and A. Y. Lyul, "Optimizing instrumental neutron activation analysis of extraterrestrial materials: fragments of lunar rocks, meteorites, chondrules, and ultrarefractory inclusions," *Journal of Analytical Chemistry*, vol. 56, no. 11, pp. 1022–1028, 2001.

[9] N. A. Shubina and G. M. Kolesov, "Advantage factors in optimized instrumental neutron activation analysis of some biosphere samples," *Journal of Analytical Chemistry*, vol. 55, no. 3, pp. 231–238, 2000.

[10] T. A. Maryutina and A. V. Soin, "Novel approach to the elemental analysis of crude and diesel oil," *Analytical Chemistry*, vol. 81, no. 14, pp. 5896–5901, 2009.

4

Detection of *Bacillus anthracis* Spores Using Peptide Functionalized SERS-Active Substrates

Atanu Sengupta, Chetan Shende, Stuart Farquharson, and Frank Inscore

Real-Time Analyzers Inc., Unit 8, 362 Industrial Park Road, Middletown, CT 06457, USA

Correspondence should be addressed to Atanu Sengupta, atanu@rta.biz

Academic Editor: Kevin Spencer

The need for portable technologies that can rapidly identify biological warfare agents (BWAs) in the field remains an international priority as expressed at the 2011 Biological Weapons Convention. In recent years, the ability of surface-enhanced Raman spectroscopy (SERS) to rapidly detect various BWAs at very low concentrations has been demonstrated. However, in the specific case of *Bacillus anthracis*, differentiation at the species level is required since other bacilli are common in the environment, representing potential false-positive responses. To overcome this limitation, we describe the use of a peptide attached to the SERS-active metal that selectively binds *Bacillus anthracis*-Sterne as the target analyte. Using this approach, 10^9 *B. anthracis*-Sterne spores/mL produced an intense dipicolinic acid spectrum upon the addition of acetic acid, while the same concentration and treatment of *B. cereus* and *B. subtilis* did not.

1. Introduction

Since the distribution of *Bacillus anthracis* spores through the US Postal System, there has been a persistent fear that biological warfare agents (BWAs) will be used by terrorist against our warfighters abroad and our civilians at home. The extreme lethality, speed of infection, low cost, and ability to aerosolize bioagents, such as *B. anthracis*, make this scenario attractive to terrorists and a serious concern. This concern was emphasized by the Secretary of State Hillary Rodham Clinton at the December 2011 Biological Weapons Convention held in Geneva, who stated "unfortunately, the ability of terrorists to develop and use these weapons is growing. Terrorist groups have made it known they want to acquire these weapons [1]."

Our inability to rapidly detect BWAs was revealed during the Fall of 2001 as the nation became anxious not knowing the extent of the anthrax attack. This anxiety was largely due to the several days required to grow spores in culture to sufficient numbers for detection. It became apparent that emergency and medical responders need portable technologies that can determine what has been contaminated by BWAs and who has been infected with sufficient speed and sensitivity to minimize fatalities. Since that time, there has been considerable effort to develop such technologies, such as polymerase chain reaction (PCR) systems [2] and immunoassay kits [3, 4]. While PCR systems are capable of identifying *B. anthracis* spores in ca. 2 hours and have been installed in over 300 regional US Postal Offices [5], each system is limited in the number of samples that it can process, and the systems are not portable. In contrast, immunoassays are very portable, but are unable to detect a lethal dose which has been estimated at ca. 10,000 spores (100 ng) by the Center for Disease Control [6], and, consequently, the false-positive identification rate remains unacceptably high [7]. In addition to PCR and immunoassays, other technologies are being developed that focus on speed, sensitivity, and portability. These are largely spectroscopic technologies, which include fluorescence [8], luminescence [9], mass spectrometry [10], infrared [11], and Raman spectroscopy (RS) [12, 13].

Prior to the 2001 attacks, we began investigating the capabilities of Raman and surface-enhanced Raman spectroscopy (SERS) to detect BWAs, specifically *B. anthracis*. RS has many of the desired attributes for this application. First, the rich molecular information of RS can be used to differentiate bacterial spores from common substances, such as hoax materials [14]. This can be accomplished in microseconds using spectral search and match routines. Second, RS, combined with chemometrics, allows bacterial identification at the genus level [11, 15]. Third, RS does not require sample preparation, as the excitation laser is simply focused on the sample. Fourth, spectral acquisition is fast, requiring a minute or less. Lastly, portable RS systems have been developed in the past decade making it suitable for field measurements. In fact the ability of Raman spectroscopy to measure *Bacillus* spores on mail sorting equipment has been demonstrated [12]. While sensitivity of the analyzer was sufficient to detect visible quantities of spores, it was unable to detect invisible quantities on contaminated surfaces that could result in infection. Recognizing this limitation we also have been investigating surface-enhanced Raman spectroscopy to detect trace quantities of spores, since it is capable of increasing Raman signal intensities by as much as 6 orders of magnitude [16]. These efforts focused on the detection of dipicolinic acid (DPA) as an anthrax biomarker [17] as it represents approximately 10% of the spore weight as calcium dipicolinate (CaDPA) [18, 19]. In the past few years we developed a method capable of detecting a few hundred spores on a surface, well below the infectious dose, within 5 minutes using a portable Raman analyzer and a SERS-active capillary to collect the sample [20]. Key to successful measurements was the use of acetic acid to digest the spores and release DPA, which occurs within 1 minute at room temperature [21, 22]. The value of this method is predicated on the fact that only two common spore-forming bacteria, *Bacillus* and *Clostridium*, contain CaDPA. Nevertheless, there still exists a chance that this method could result in false alarms since several common outdoor bacilli, such as *B. cereus*, *B. subtilis*, and *B. megaterium*, could also produce a DPA SERS signal. In an effort to provide differentiation at the species level we have been investigating the ability of functionalizing silver particles with species-specific molecular recognition elements, such as antibodies, aptamers, and small peptides, to bind target BWAs. Here we present, to our knowledge, the first report of a bacterium, *B. anthracis*-Sterne, selectively bound to a peptide and detected by surface-enhanced Raman spectroscopy.

2. Experimental

2.1. Materials. Acetic acid (glacial, 99.7%), cysteine, dipicolinic acid (2,6-pyridinedicarboxylic acid), and all chemicals used to produce the silver-doped sol-gels were obtained at their purest commercially available grade from Sigma-Aldrich (Milwaukee, WI) and used as received. Calcium dipicolinate was prepared from disodium dipicolinate (Na_2DPA), which was prepared from DPA according to previous publications [23]. *B. anthracis*-Sterne, *B. cereus*, and *B. subtilis* were obtained from American Type Culture Collection (Manassas, Virginia) and grown on nutrient agar plates at 30°C for 24 hours and then stored until used. For each spore sample, 10 nutrient agar plates were inoculated and incubated at 30°C for 4-5 days. Following incubation, growths on the plates were harvested with 50 mL of sterile distilled water using a sterile hockey stick. The harvested sample was centrifuged for 5 min at 10,000 rpm in a Sorval SS-34 Rotor (Thermo Fisher, Waltham, MA). The pellet was washed four times by resuspending it in 25 mL of sterile distilled water and repeating the centrifugation. The final pellet was resuspended to a total volume of 10 mL in sterile distilled water. This sample was diluted by a factor of 100 in distilled water. A portion of this sample was placed in a hemocytometer to perform a direct spore count. The counts for *B. anthracis*-Sterne, *B. cereus*, and *B. subtilis* were 5.7×10^9, 1.7×10^{10}, and 6.1×10^9 spores/mL, respectively. In all cases, the 10^{-2} diluted samples were further diluted to 10^{-7} in sterile distilled water, and in each case a 0.1 mL sample was plated in triplicate on nutrient agar plates that were incubated overnight at 30°C. The viability counts for *B. anthracis*-Sterne, *B. cereus*, and *B. subtilis* were 2.1×10^{10}, 9.3×10^9, and 6×10^9 CFUs/mL, respectively. Three mL aliquots of the above endospore suspension were placed into three lyophilization vials and stored overnight at −80°C. The frozen samples were dried under vacuum for 12–15 hr, sealed with butyl rubber stoppers, and screw-capped.

2.2. SERS-Active Capillaries. The SERS-active capillaries (Simple SERS Sample Capillary, RTA) were prepared according to published procedures [24, 25], by mixing a silver amine precursor and an alkoxide precursor at 1 : 1 v/v. The silver amine precursor consisted of a 1 : 1 : 2 v/v/v ratio of 1 N $AgNO_3$/28% NH_3OH/CH_3OH, while the alkoxide precursor consisted of methyltrimethoxysilane. The SERS capillaries were prepared by drawing 20 μL of the silver-doped sol-gels into 10 cm long, 1-mm diameter glass capillaries to produce ca. 1 cm plugs. The plugs were allowed to gel and cure, after which the incorporated silver ions were reduced with dilute $NaBH_4$.

The SERS capillaries were functionalized with a peptide synthesized to specifically bind *B. anthracis*-Sterne spores (not vegetative cells) by drawing a 5 mg/mL peptide/water solution into the SERS capillaries and allowing the peptide to attach to the silver overnight. The peptide was synthesized with a cysteine residue at the C-terminus so that it could form a covalent bond through the thiol group to the silver. Cysteine also serves as a spacer, in that the peptide is displaced from the silver surface by about 2–10 angstroms, providing space for the spores to bind. The functionalized sol-gel was then washed with a solvent to remove unbound peptide.

2.3. Sample Preparation for Measurement. Initially, 20 μL of each bacilli sample was drawn into a functionalized SERS capillary and allowed to bind for 5, 10, and 15 minutes. After this period, 5 mL of water was drawn through the capillary to remove any unbound spores. In subsequent experiments, the samples were diluted by approximately a factor of 10 to produce 10^9 spores/mL samples in HPLC grade water, and 20 μL of acetic acid was drawn into the SERS capillaries after the water wash to effect release of DPA.

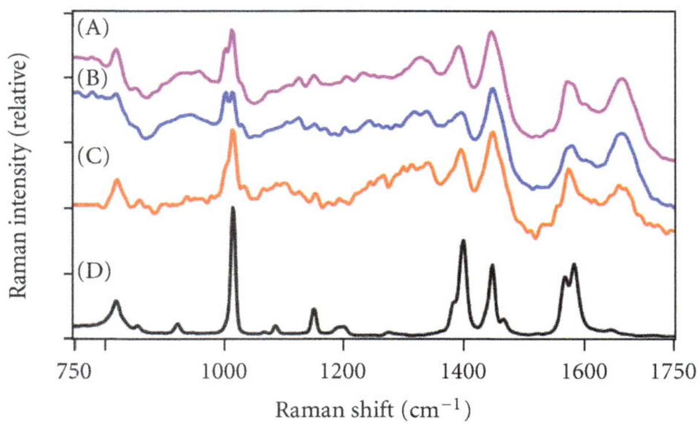

FIGURE 1: RS of (A) *B. cereus*, (B) *B. subtilis*, and (C) *B. anthracis*-Sterne spores, compared to (D) calcium dipicolinate. Spectral conditions: 500 mW of 1064 nm irradiation at the sample, (A–C) 20 min, (D) 5 min acquisition time, 8 cm^{-1} resolution.

2.4. Raman and Surface-Enhanced Raman Spectroscopy. RS measurements were typically performed using 500 mW of 1064 nm laser excitation at the sample, while SERS measurements were performed using 85 mW of 785 nm excitation at the sample (see figure captions for details). In both cases Fourier transform Raman spectrometers were used with a resolution of 8 cm^{-1} (RTA, model RamanID-785 and -1064). Fiber optics were used to deliver the excitation beam to the sample probe and the scattered radiation to the interferometer (2 m lengths of 200 and 365 μm core diameter, resp., Spectran, Avon, CT). For the RS measurements the samples were placed in sealed glass vials, mounted horizontally above the probe, and measured. For SERS measurements the capillaries were mounted horizontally on an *XY* plate reader (Conix Research, Springfield, OR), and in-house software was used to measure five points spaced along the length of the metal-doped sol-gel plug. The five-point relative standard deviation was typically 20% [22], and the reported spectra are the average of these five 1 min measurements. The spot size for both RS and SERS was approximately 300 microns in diameter.

3. Results and Discussion

The fingerprint region of the Raman spectra of bacilli is dominated by calcium dipicolinate peaks at 821, 1014, 1391, 1446, and 1573 cm^{-1}, which have been assigned to a ring CC bend, an out-of-plane CH bend, the symmetric pyridine ring stretch, a symmetric OCO stretch, a symmetric ring CH bend, and an asymmetric OCO stretch, respectively, (Figure 1) [12]. The remaining dominant spore peaks have been assigned to protein modes associated with the peptidoglycan cell wall, such as amino acids and peptide linkages [26–29]. The 1003 and 1598 cm^{-1} peaks are assigned to phenylalanine modes, while the 1318 and 1666 cm^{-1} peaks are assigned to amide III combination modes and the amide I mode, which is primarily a C=O stretch. In several cases, protein and CaDPA vibrational modes occur at or close to the same frequency, such as the 821 and 1446 cm^{-1} peaks. In fact the only obvious difference between the RS of the three bacilli is the relative intensity of the CaDPA pyridine ring stretching mode at 1014 cm^{-1}, especially when compared to the intensity of the phenylalanine ring stretch at 1003 cm^{-1}. While it is tempting to use this difference to identify the *Bacillus* species, it more likely reflects variations in the growth media and growth conditions. Indeed others have shown that dramatic spectra differences occur when different growth media and growth conditions are used [30], and only if identical growth conditions are used would identification of the bacilli at the species level be possible [15].

The greater limitation of Raman spectroscopy is its sensitivity. High quality spectra, such as that shown in Figure 1, require high laser powers and long acquisition times. As previously noted this can be overcome using surface-enhanced Raman spectroscopy, and several researchers have demonstrated that bacteria can be identified at the genus level [31–34]. Yet, in several cases spectra of different genera, such as *B. subtilis* and *E. coli* collected by the same research group, are virtually identical [31–33], while they are completely different between groups. Although this could be attributed to the use of gold versus silver to generate SERS, different spectra are still observed when different groups use silver [31, 32, 34]. At present there is not a consensus regarding the ability of RS or SERS to differentiate bacteria at the species level.

This uncertainty can be eliminated by functionalizing the SERS-active metal with molecular recognition elements that only bind one type of bacteria and preferably one bacterial species. Here, a peptide was synthesized to bind to the surface of *B. anthracis*-Sterne, but not other bacilli, such as *B. cereus* and *B. subtilis*. Cysteine was added to the C-terminus of the peptide so that it could form a chemical bond with the silver particles contained within the sol-gel. The SERS of the peptide functionalized silver-doped sol-gel is virtually identical to the SERS of cysteine (Figures 2(B) and 2(C)), dominated by a S-Ag stretching mode at 660 cm^{-1}, and cysteine peaks at 810, 875, 1060, 1300, 1355, and 1410 cm^{-1}. The first set of experiments involved adding various amounts of *B. anthracis*-Sterne spores to the peptide functionalized silver-doped sol-gel to see if it produced a unique spectral signature indicative of spore binding. Even after allowing the spores to bind for 15 minutes, no signature appeared, and, at best, the spores decreased the signal intensity of the peptide to a modest degree (Figure 2(A)).

Prior to the addition of acetic acid to the presumably bound *B. anthracis*-Sterne spores, the SERS for all three bacilli added to unfunctionalized silver particles was measured with the addition of acetic acid to ensure that they all produced equivalent spectra. This was indeed the case, as each *Bacillus* was dominated by the DPA peaks at 812, 1006, 1381, 1426, and 1567 cm^{-1} (Figure 3), which have the same assignments as the CaDPA peaks at similar wavenumbers (Figure 1).

In the last set of experiments, *B. anthracis*-Sterne, *B. cereus*, and *B. subtilis* samples were separately added to 3 SERS-active capillaries functionalized with the *B. anthracis*-Sterne-specific peptide, allowed to bind for 15 minutes, then washed with water to remove unbound spores, followed by

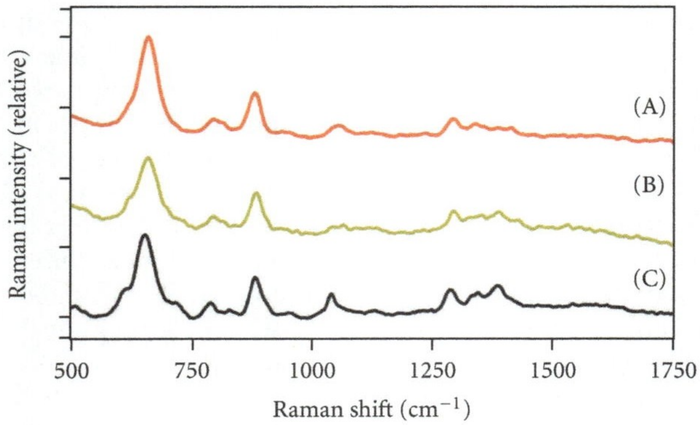

FIGURE 2: SERS of (A) 10^9 spores/mL *B. anthracis*-Sterne bound to the *B. anthracis*-Sterne peptide functionalized SERS-active substrate, (B) 5 mg/mL *B. anthracis*-Sterne peptide functionalized SERS-active substrate, and (C) 1 mg/mL cysteine. Spectral conditions: 80 mW of 785 nm irradiation at the sample, 1 min acquisition time, 8 cm^{-1} resolution.

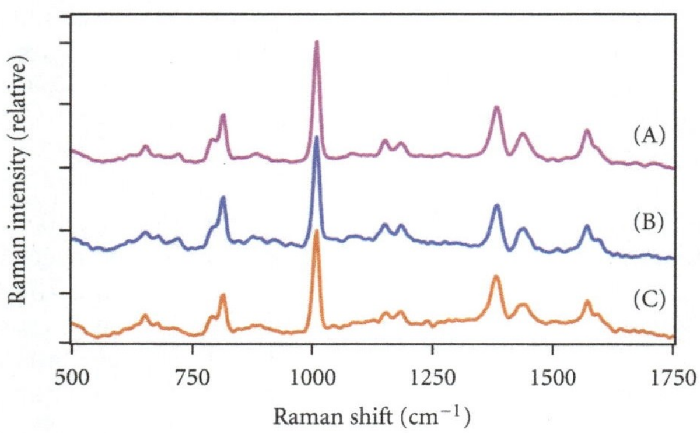

FIGURE 3: SERS of (A) *B. cereus*, (B) *B. subtilis*, and (C) *B. anthracis*-Sterne all at a concentration of 10^9 spores/mL treated with acetic acid. Spectral conditions as in Figure 2.

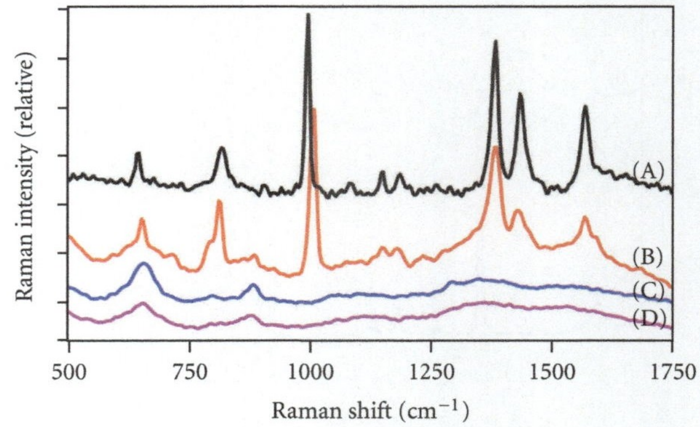

FIGURE 4: (A) RS of 0.55 mg/mL DPA in 1 N KOH in a plain glass capillary (intensity multiplied by 3.5 to match SERS intensity of 1000 cm^{-1} peak of (B)). Spectral conditions: 300 mW of 785 nm irradiation at the sample, 25 min acquisition time, 8 cm^{-1} resolution. SERS of (B) *B. anthracis*-Sterne, (C) *B. cereus*, and (D) *B. subtilis* all at a concentration of 10^9 spores/mL on the *B. anthracis*-Sterne peptide functionalized SERS-active substrate after a 15-min incubation period, a wash step, and the addition of acetic acid. Spectral conditions for (B)–(D) as in Figure 2.

the addition of acetic acid. Only the *B. anthracis*-Sterne sample produced a DPA spectrum, while the two negative control samples produced spectra only of the peptide at reduced intensity (Figure 4), clearly demonstrating selectivity.

While high quality spectra of 10^9 spores/mL *B. anthracis*-Sterne were repeatedly obtained, sensitivity quickly decreased at lower concentrations, and no signal was obtained at 10^7 spores/mL. While this sensitivity is similar to other work [34], it is considerably less than previous work that included the use of acetic acid, but not the use of a binding peptide, in which 2.2×10^5 spores/mL could be measured [22]. At present this loss is partially attributed to the peptide reducing the amount of silver surface available to the DPA molecules.

In an effort to estimate an enhancement factor, a 0.55 mg/mL DPA in 1 N KOH was measured by RS (Figure 4(A)). This concentration corresponds to 55×10^9 spores assuming each spore contains 10% DPA by weight. Taking into account the relative concentrations, the signal intensity of the symmetric stretching mode at ~1000 cm^{-1}, and the laser powers used for this measurement and the acetic acid treated peptide bound *B. anthracis*-Sterne measurement (Figure 4(B)), a modest enhancement factor of 722 is calculated.

Finally, it is worth stating that the *XY* plate reader, which held the SERS-active capillaries, was used as a matter of convenience and could easily have been replaced by a simple fixed sample holder coupled to a portable Raman analyzer so that measurements could be performed in the field. This is in contrast to previous work [31–34], in which the samples were measured using microscope systems, limiting portability.

4. Conclusion

In this work we have successfully demonstrated the use of peptide-functionalized silver particles to selectively bind a target analyte and generate surface-enhanced Raman scattering. Furthermore, it was shown that discriminate detection of spores at the species level, specifically *B. anthracis*-Sterne and not *B. cereus* and *B. subtilis*, could be performed. However, this improvement in specificity comes with a loss in sensitivity, as the inclusion of the peptide appears to reduce the available silver surface. It is also worth noting that at present the method is limited to bacteria with an extractable SERS-active signature, such as DPA. Current work is aimed at improving sensitivity by modifying the chemistry to increase the available silver surface as well as extending this analysis to other BWAs.

Authors' Contribution

All authors contributed equally to this work.

Acknowledgments

The authors are grateful to Professor Jay Sperry of the University of Rhode Island for growing and supplying the bacilli samples for the measurements presented here. They are also appreciative of funding from the National Science Foundation (DMI-0349687, IIP-0810335, and II-0956170).

References

[1] G. Hess, "Biosecurity: an evolving challenge," *Chemical & Engineering News*, vol. 90, no. 7, pp. 30–32, 2012.

[2] C. A. Bell, J. R. Uhl, T. L. Hadfield et al., "Detection of *Bacillus anthracis* DNA by light cycler PCR," *Journal of Clinical Microbiology*, vol. 40, no. 8, pp. 2897–2902, 2002.

[3] R. Mabry, K. Brasky, R. Geiger et al., "Detection of anthrax toxin in the serum of animals infected with *Bacillus anthracis* by using engineered immunoassays," *Clinical and Vaccine Immunology*, vol. 13, no. 6, pp. 671–677, 2006.

[4] S. Tang, M. Moayeri, Z. Chen et al., "Detection of anthrax toxin by an ultrasensitive immunoassay using europium nanoparticles," *Clinical and Vaccine Immunology*, vol. 16, no. 3, pp. 408–413, 2009.

[5] A. Thayer, "Homeland security: postal service readies defense—team will install PCR-based systems to detect biohazards in mail facilities," *Chemical & Engineering News*, vol. 81, no. 21, article 7, 2003.

[6] T. V. Inglesby, T. O'Toole, D.A. Henderson et al., "Anthrax as a biological weapon: updated recommendations for management," *The Journal of the American Medical Association*, vol. 287, no. 17, pp. 2236–2252, 2002.

[7] M. Kellogg, M. Hollis, C. Gauthier, and T. O'Brien, "Detection of biological agents used for terrorism: are we ready?" *Clinical Chemistry*, vol. 56, no. 1, pp. 10–15, 2010.

[8] R. Nudelman, B. V. Bronk, and S. Efrima, "Fluorescence emission derived from dipicolinic acid, its sodium, and its calcium salts," *Applied Spectroscopy*, vol. 54, no. 3, pp. 445–449, 2000.

[9] P. M. Pellegrino, N. F. Fell Jr., and J. B. Gillespie, "Enhanced spore detection using dipicolinate extraction techniques," *Analytica Chimica Acta*, vol. 455, no. 2, pp. 167–177, 2002.

[10] Y. Hathout, B. Setlow, R. M. Cabrera-Martinez, C. Fenselau, and P. Setlow, "Small, acid-soluble proteins as biomarkers in mass spectrometry analysis of bacillus spores," *Applied and Environmental Microbiology*, vol. 69, no. 2, pp. 1100–1107, 2003.

[11] D. Naumann, "Infrared and NIR Raman spectroscopy in medical microbiology," in *Infrared Spectroscopy: New Tool in Medicine*, vol. 3257 of *Proceedings of SPIE*, pp. 245–257, San Jose, Calif, USA, January 1998.

[12] S. Farquharson, L. Grigely, V. Khitrov, W. W. Smith, J. F. Sperry, and G. Fenerty, "Detecting *Bacillus cereus* spores on a mail sorting system using Raman spectroscopy," *Journal of Raman Spectroscopy*, vol. 35, no. 1, pp. 82–86, 2004.

[13] C. W. Garner, P. J. Treado, T. M. Jochem, and G. R. Gilbert, "Demonstration of a robot-based Raman spectroscopy detector for the identification of CBE threat agents," DTIC (Army Report), 2006.

[14] S. Farquharson and W. W. Smith, "Differentiating bacterial spores from hoax materials by Raman spectroscopy," in *Chemical and Biological Point Sensors for Homeland Defense*, vol. 5269 of *Proceedings of SPIE*, pp. 9–15, Providence, RI, USA, October 2003.

[15] K. Maquelin, L. P. Choo-Smith, T. van Vreeswijk et al., "Raman spectroscopic method for identification of clinically relevant microorganisms growing on solid culture medium," *Analytical Chemistry*, vol. 72, no. 1, pp. 12–19, 2000.

[16] C. Shende, F. Inscore, A. Sengupta, and S. Farquharson, "Surface-enhanced Raman spectroscopy: theory and application to the analysis of Chlorpyrifos in orange juice," in *Applications of Vibrational Spectroscopy in Food Science*, E. C. Y. Li-Chan, P. R. Griffiths, and J. M. Chalmers, Eds., pp. 195–209, John Wiley & Sons, West Sussex, UK, 2010.

[17] S. Farquharson, W. W. Smith, S. Elliott, and J. F. Sperry, "Rapid biological agent identification by surface-enhanced Raman spectroscopy," in *Air Monitoring and Detection of Chemical and Biological Agents II*, vol. 3855 of *Proceedings of SPIE*, pp. 110–116, September 1999.

[18] W. G. Murrell, "Chemical composition of spores and spore structures," in *The Bacterial Spore*, G. W. Gould and A. Hurst, Eds., pp. 215–273, Academic Press, London, UK, 1969.

[19] W. H. Nelson, R. Dasari, M. Feld, and J. F. Sperry, "Intensities of calcium dipicolinate and *Bacillus subtilis* spore Raman spectra excited with 244 nm light," *Applied Spectroscopy*, vol. 58, no. 12, pp. 1408–1412, 2004.

[20] S. Farquharson and F. E. Inscore, "Detection of invisible bacilli spores on surfaces using a portable SERS-based analyzer," *International Journal of High Speed Electronics and Systems*, vol. 18, no. 2, pp. 407–416, 2008.

[21] S. Farquharson, F. E. Inscore, A. D. Gift, and C. Shende, "Method for effecting the rapid release of a signature chemical from bacterial endospores, and for detection thereof," U.S. Patent 7, 713, 914, 2010.

[22] S. Farquharson, C. Shende, A. Gift, and F. Inscore, "Detection of bacillus spores by surface-enhanced Raman spectroscopy," in *Bioterrorism*, S. S. Morse, Ed., 2012.

[23] E. Ghiamati, R. S. Manoharan, W. H. Nelson, and J. F. Sperry, "UV resonance Raman spectra of bacillus spores," *Applied Spectroscopy*, vol. 46, no. 2, pp. 357–364, 1992.

[24] S. Farquharson, Y. H. Lee, and C. Nelson, "Material for SERS and SERS sensors and method for preparing the same," U.S. Patent 6, 623, 977, 2003.

[25] S. Farquharson, A. D. Gift, F. E. Inscore, and C. S. Shende, "SERS method and apparatus for rapid extraction and analysis of drugs in saliva," U.S. Patent 7, 393, 691, 2008.

[26] W. H. Woodruff, T. G. Spiro, and C. Gilvarg, "Raman spectroscopy in vivo: evidence on the structure of dipicolinate in intact spores of *Bacillus megaterium*," *Biochemical and Biophysical Research Communications*, vol. 58, no. 1, pp. 197–203, 1974.

[27] J. G. Grasselli, M. K. Snavely, and B. J. Bulkin, *Chemical Applications of Raman Spectroscopy*, chapter 5, John Wiley & Sons, New York, NY, USA, 1981.

[28] J. Bandekar, "Amide modes and protein conformation," *Biochimica et Biophysica Acta*, vol. 1120, no. 2, pp. 123–143, 1992.

[29] J. C. Austin, T. Jordan, and T. G. Spiro, "UVRR studies of proteins and related compounds," in *Biomolecular Spectroscopy*, R. J. H. Clark and R. E. Hester, Eds., vol. 21, John Wiley & Sons, New York, NY, USA, 1993.

[30] C. W. Garner Jr., J. S. Maier, M. P. Nelson et al., "Method for detection of pathogenic microorganisms," US Patent 6, 917, 423, 2005.

[31] R. M. Jarvis, A. Brooker, and R. Goodacre, "Surface-enhanced Raman spectroscopy for bacterial discrimination utilizing a scanning electron microscope with a Raman spectroscopy

interface," *Analytical Chemistry*, vol. 76, no. 17, pp. 5198–5202, 2004.

[32] R. M. Jarvis and R. Goodacre, "Discrimination of bacteria using surface-enhancedRaman spectroscopy," *Analytical Chemistry*, vol. 76, no. 1, pp. 40–47, 2004.

[33] W. R. Premasiri, D. T. Moir, M. S. Klempner, N. Krieger, G. Jones II, and L. D. Ziegler, "Characterization of the surface enhanced Raman scattering (SERS) of bacteria," *Journal of Physical Chemistry B*, vol. 109, no. 1, pp. 312–320, 2005.

[34] J. A. Guicheteau, K. Gonser, and S. D. Christesen, "Raman and surface enhanced Raman of biological material," DTIC, Report No. ADA433013, 2004.

5

A Novel Method for Bacterial UTI Diagnosis Using Raman Spectroscopy

Evdokia Kastanos,[1] Alexandros Kyriakides,[2] Katerina Hadjigeorgiou,[2] and Costas Pitris[2]

[1] *Department of Life and Health Sciences, University of Nicosia, 1700 Nicosia, Cyprus*
[2] *Department of Electrical and Computer Engineering, University of Cyprus, 1678 Nicosia, Cyprus*

Correspondence should be addressed to Costas Pitris, cpitris@ucy.ac.cy

Academic Editor: Jin Zhang

The current state of the art on bacterial classification using Raman and Surface Enhanced Raman Spectroscopy (SERS) for the purpose of developing a rapid and more accurate method for urinary tract infection (UTI) diagnosis is presented. SERS, an enhanced version of Raman offering much increased sensitivity, provides complex biochemical information which, in conjunction with advanced analysis and classification techniques, can become a valuable diagnostic tool. The variety of metal substrates used for SERS, including silver and gold colloids, as well as nanostructured metal surfaces, is reviewed. The challenges in preprocessing noisy and complicated spectra and the various methods used for feature creation as well as a novel method using spectral band ratios are described. The various unsupervised and supervised classification methods commonly used for SERS spectra of bacteria are evaluated. Current research on transforming SERS into a valuable clinical tool for the diagnosis of UTIs is presented. Specifically, the classification of bacterial spectra (a) as positive or negative for an infection, (b) as belonging to a particular species of bacteria, and (c) as sensitive or resistant to an antibiotic are described. This work can lead to the development of novel technology with extremely important benefits for public health.

1. Introduction

Urinary tract infections (UTIs) are some of the most common types of infections in humans with an estimated 34 percent of adults aged 20 or older reported as having had at least one occurrence of a UTI or cystitis. Specifically, over 50% of women and over 13% of men will have a UTI at least once in their lifetime [1, 2]. This results in millions of doctors' visits and a cost of several billion USD every year just in the USA alone [3, 4]. UTI diagnosis is a multistep process which includes the determination of the concentration of pathogens, and the identification of the responsible bacteria, as well as their susceptibility to various antibiotics, the so-called antibiogram. Such assays require repeated culturing of a sample and take over 48 hours in order for bacterial colonies to be grown, counted, and exposed to antibiotics using conventional clinical methods. Since the patient cannot remain untreated during this rather prolonged period before definitive diagnosis is obtained, physicians prescribe broad spectrum antibiotics prior to antibiogram results. This practice has many undesirable consequences, both short term and long term: (i) unsuccessful treatments leading to chronic infections, (ii) increased health care costs, and (iii) increased antibiotic resistance by a growing number of bacterial strains [5–8]. Given these concerns, it is obvious that a rapid and accurate method of UTI diagnosis and bacterial susceptibility to antibiotics would offer significant health benefits. Such methodologies are currently being developed and include the promising application of Raman spectroscopy for the diagnosis of UTIs.

Recently, rapid diagnosis methods based on PCR have been developed in order to bypass the need for culturing [9] as well to identify genes that confer antibiotic resistance [10]. Although such PCR assays are fast and very sensitive, they typically require species and strain-specific probes that may or may not be available for a particular organism. Additionally, amplification methods like PCR suffer from contamination problems, complex interpretation of results, as well as high costs. Mass spectrometry is another method that has been proposed as an alternative approach for bacterial diagnostics without culturing [11]. However, like the PCR approach, mass spectrometry also depends on prior knowledge of the pathogen under study and suffers from increased complexity and cost.

Vibrational spectroscopies, like Raman spectroscopy, have been used, for the last few years, to detect bacteria with minimal sample manipulation [12–14]. Classification of bacterial species, as well as subspecies, has been achieved with great accuracy and speed, especially with surface-enhanced Raman spectroscopy (SERS) [15] which allows enhancement of the inherently weak Raman signal. More recently, it was shown that bacterial susceptibility to antibiotics can also be determined using Raman spectroscopy and SERS [16–18].

1.1. Raman Spectroscopy. Whole organism fingerprinting is a relatively recent approach for the identification of pathogens based on their unique chemical characteristics with minimal sample preparation [19]. Some of the techniques that are being evaluated for such fingerprinting include vibrational spectroscopic methods such as Fourier-Transform Infrared spectroscopy (FTIR), Raman Spectroscopy, and UV resonance Raman spectroscopy (UVRR). Vibrational spectroscopic features arise from the loss or gain of energy by photons which are inelastically scattered from a vibrating sample. This energy change results in an alteration of the photons' wavelength which in the case of organic molecules is directly related to the vibrational states of its chemical bonds. Therefore, quantitative biochemical information can be extracted from the spectra.

The Raman effect is observed when a very small number of photons incident on a molecule (about 1 in 10^7) are inelastically scattered, that is, scattered at different frequencies than the frequency of the incident photons [20]. A Raman spectrum is a plot of the intensity of scattered light versus the energy difference between the incident photons and the Raman-scattered photons, and contains information about the chemical composition and the molecular structure of a substance. A major limitation of Raman is the weakness of the Raman effect which results in very low signals, often below the limit of detection for dilute biological samples. In addition, Raman spectroscopy is not very selective with respect to the various molecules present in complex biological samples, resulting in highly congested spectra. Finally, Raman spectroscopy suffers from a strong fluorescence background which significantly decreases the dynamic range of the measurement. FTIR, on the other hand, is much more sensitive than Raman, but its use in biological systems is very limited due to the strong absorbance of water at those wavelengths. However, several publications have come out in the last few years that show discrimination of bacteria using FTIR and also quite complex classification analysis methods [21, 22]. UVRR is also more sensitive than normal Raman and does not suffer from high fluorescence background. It has already been exploited for the discrimination of UTI bacteria [23]. The highly energetic nature of UV photons, however, can cause photochemical or burning effects on a sample and limits its use as a routine method for bacterial analysis. Additionally, excitation in the UV is not very useful for the discrimination of different species of bacteria as it mostly identifies nucleic acid bases, which are not expected to vary in abundance between different bacterial species. The widespread adoption of UVRR is also limited by the high cost and complexity of the equipment required.

Raman spectroscopy has also been successfully applied to bacterial identification in the configuration of a Raman microscope, which allows collection of spectra from single cells and enhances sensitivity. The group of Hutsebaut et al. have used Raman microscopy to identify samples belonging to 3 different *Bacillus* species [24]. Xie et al. successfully classified 6 different bacterial species [25]. Studies by Harz et al. [26] have shown that discrimination of different Staphylococcus species can be achieved with 95–97% accuracy not only at species but also at strain level. The same group has successfully discriminated 29 strains of bacteria derived from a variety of species [27]. Despite increased sensitivity, studies using micro-Raman spectroscopy have to face a few unique challenges. First, the fact that imaging individual bacteria is a time consuming process that leads to inadequate evaluation of large samples. Second, the increased variability between spectra of individual cells, mostly due to variations in the growth stage of each cell, usually dictates a need for more complex classification methods.

1.2. Surface-Enhanced Raman Spectroscopy. Surface Enhanced Raman Spectroscopy (SERS) is a variation of Raman spectroscopy which offers significant enhancement of the signal (up to 10^{14} times), thus making detection faster, simpler, and more accurate. The enhancement is a result of the effect of plasmon resonance, that is, the unison oscillation of electrons on the surface of a metallic nanostructure as a result of incident light of the right, resonant, frequency. These oscillations produce an enhanced electromagnetic field in the proximity of the surface. If a sample is within a few nanometers (nm) from the nanostructure, it will experience this enhanced field and exhibit a stronger Raman signal [28–31]. The enhancement is such that even single molecules can be detected. However, this enhancement rapidly decreases as the distance between the target sample and the nanoparticle increases and completely disappears within 10–20 nm of separation. For a more detailed discussion of SERS mechanisms, the reader can refer to an extensive volume of literature [28–31]. SERS spectra also exhibit reduced data congestion compared to normal Raman spectra, since the enhancement is significant only for molecules found on, or very close to, the SERS substrate. SERS also significantly improves the dynamic range of the measurement due to the ability of SERS-active substrates to quench fluorescence. Finally, SERS can be performed using simple, mobile, and relatively inexpensive equipment providing a significantly improved signal at high speed and low cost. SERS offers significant advantages for the identification of bacteria species and bacterial function. (a) The enhancement of the Raman signal allows accurate detection of minute quantities of bacteria, therefore, eliminating the need for time and resource consuming culturing processes. (b) In the case of micro-Raman spectroscopy, the enhancement offered by SERS allows identification of even individual bacteria which enhances the selectivity of the identification in the case of samples with multiple constituents. (c) Metal nanoparticles, under certain conditions, are internalized by the bacteria thus providing signal enhancement from both external (cell wall) molecules as well as the internal milieu of the bacterium [32]. This is an important advantage since the sensitivity and selectivity of bacterial classification is significantly improved.

1.2.1. SERS Substrates for Classification of Bacteria. One of the first groups to work on the identification of bacteria by SERS is the group of Efrima et al. [33–36]. The SERS-active substrate they employed was not only silver but also gold colloid. What is unique about their methodology is that they usually produced the colloid in the presence of the bacteria and not mixing pre-formed colloid with the sample. In addition, they demonstrated growth of colloids in the interior or the exterior of bacterial cells obtaining different spectra in each case. As far as the collection of SERS spectra is concerned, they have done it under a large variety of conditions, including various preparations of their substrates, using different fractions of bacterial cells, and different excitations, ranging from the UV to the NIR (250–800 nm). Jarvis et al. is another group who have done a lot of work on SERS and identification of bacteria, and specifically UTI involved bacteria, using silver colloids in combination with different types of Raman microscopes [32, 37–40]. Their group also reported SERS from both intracellular and extracellular bacterial locations [32]. In addition, they were one of the first groups to successfully apply various advanced classification techniques to the SERS spectra. The group of Premasiri et al. have used gold-cluster-covered silicon substrates and a Raman microscope to discriminate SERS spectra of bacteria [41] as well as to classify the bacteria down to the subspecies level [42]. Other groups have shown very high specificity and sensitivity in classification of bacteria, even at the strain level, using silver nanorod arrays as a substrate [43]. Metal nanosurfaces containing various structures have also been used, and some are commercially available like Klarite (Mesophotonics). Klarite gold nanosurfaces do not offer great enhancement for bacteria, however, as the pits of the structure, where most of the enhancement would be expected, are only about 100 nm in diameter and do not fit most bacteria. A very recent study by Walter et al. uses silver colloids and a microfluidic device and greatly minimizes sample volume and exposure time for obtaining SERS spectra of *E. coli*, which are classified with high accuracy [44]. A major limitation of SERS in the identification of bacteria are the spectral variations present, even between spectra from the same sample, since the technique is very sensitive to substrate, sample preparation, as well as local field conditions. Despite these limitations SERS, is definitely much more sensitive for identifying and classifying bacteria than normal Raman and is now the preferred route for developing a rapid method of diagnosis of a UTI.

2. Data Processing, Classification, and Evaluation

Raman spectra from simple and concentrated solutions, of single or few chemical constituents are easy to measure and interpret. However, as the concentration decreases or the samples become more complicated, for example, tissue or bacteria, the extraction of useful information from the weak and noisy Raman spectra becomes more challenging. The two major factors, which make Raman signals difficult to classify, are the fact that they are usually obscured by a large amount of noise and that the signals have a great deal of variance even between spectra which originate from the same sample. Both of these problems (noise and high variance) must be rigorously addressed when attempting to classify the Raman spectra of UTI involved pathogens.

Classification is the exercise of assigning a class to an object. In the case of analyzing Raman spectra for UTI, the objects are the Raman spectra themselves, and the class of each spectrum is the species or subspecies of the bacteria that yielded the specific measurement. The procedure for performing classification can be broken down into four main steps. The first step is preprocessing which attempts to remove noise and to normalize the Raman spectra. The second step is feature creation which uses the denoised and normalized spectra in order to create features which will be fed to the classification algorithm. In statistical terms, the features can be considered to be the "independent variables." The "dependent variable" is the class. The third step in the classification process is the use of the features by the classification algorithm to yield the estimated class of the sample. The final step is the evaluation of the classification to determine its accuracy.

Although these four steps describe the overall picture of the classification process well, it is not always the case that all steps are necessary or even separable. For example, in some procedures, the denoising, normalization, and feature creation can be conceptualized as one step. Furthermore, in one of the most successful techniques presented for the classification of Raman spectra [45], no explicit normalization is performed in the preprocessing step, but instead, the normalization is accomplished by the distance metric used to compare the feature vectors.

2.1. Preprocessing

2.1.1. Noise Removal. The weak nature of the Raman effect results in spectra with very low signal-to-noise ratio (SNR). Noise in Raman spectra takes one of three forms: (i) high-frequency noise, (ii) low-frequency noise, and (iii) cosmic spikes (Figure 1).

High frequency noise comes from the acquisition electronics and other sources of system variation. Median filtering can be used to remove this type of noise [26]. Another technique commonly used is wavelet-based denoising [46]. Both of these techniques work well, and using at least one of them is recommended. Low-frequency noise arises from ambient light entering the spectrograph and fluorescence emission from the sample which, in the case of biological

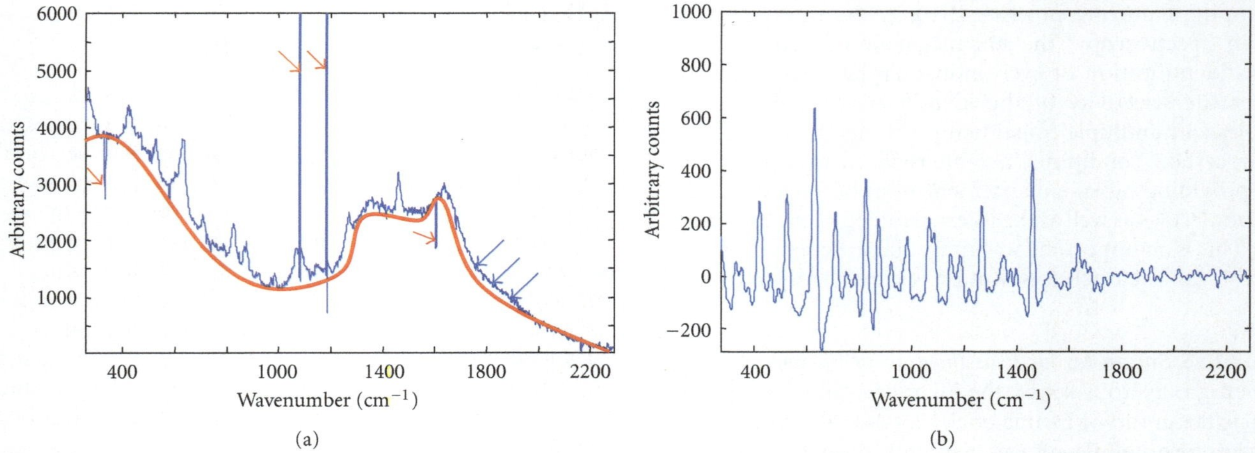

Figure 1: (a) Raman spectrum of fructose exhibiting cosmic spikes (red arrows), background fluorescence (red curve), and high-frequency noise (blue arrows). (b) The same spectrum after preprocessing.

samples, can significantly reduce the dynamic range of the measurement. This is seen as a background baseline which is present in the spectra. Iterative curve fitting or low pass filtering can be used to estimate and remove the background baseline [20]. Removing the low-frequency noise can help improve the accuracy of the classification, but it is not always necessary. Using the first and second derivatives of the spectrum as features for the classification by itself removes the baseline since the derivatives indicate changes and curvature and not background (see Section 2.2 "Feature Creation" below). Cosmic spikes are spurious, very narrow, spikes appearing in the Raman spectra. They are an artefact of the detection electronics. The techniques used to filter out high-frequency noise can also be used to remove cosmic spikes. For example, it has been shown that median filtering is an effective way to remove this type of noise too [17]. Cosmic spikes can produce outliers which can be detrimental to the accuracy of several classification methods. It is therefore highly recommended that cosmic spikes are always removed before initiating any classification process.

2.1.2. Normalization. Raman spectra acquired sequentially or intermittently from exactly the same sample can exhibit a large amount of variance. This is usually in the form of an additive or multiplicative shift. These variations, in certain cases, can adversely affect the classification and should be eliminated by normalizing the spectra. Two of the most common techniques used are normalization using the highest peak [16] and vector normalization [47]. In the first case, all the spectra are modified so that they have the same minimum and maximum values. In the case of vector normalization the samples are considered to be multidimensional vectors and are modified so that their magnitudes are equal.

2.1.3. Considerations and Caveats. It is important to note that the preprocessing step greatly affects the accuracy of the classification results. A sufficient amount of preprocessing must be performed in order to increase the classification accuracy. At the same time, however, it is imperative to understand that too much preprocessing can introduce user bias and overfitting effects, which will decrease the accuracy of the results when truly unknown samples are subsequently classified. A classification algorithm, which is self-normalizing [45] avoids the pitfalls of too much preprocessing and therefore improves the classification accuracy of the technique.

2.2. Feature Creation. Once the noise removal and normalization have been completed, features must be created to be subsequently used by the classification algorithm. The most common technique of feature creation is Principal Component Analysis [48], which uses a linear transformation to represent the original data into a new coordinate space. The dimensions of this new space are in the directions of maximum variance of the features. The first principal component captures the dimension of maximum variance; the second principal component captures the dimension of the second greatest variance, and, similarly, the subsequent principal components capture the dimensions of successively decreasing variances. It is common to choose only a subset of these principal components as features. Usually, the first few principal components are used, although this is not always the best approach [49]. An important benefit of selecting only a subset of the principal components is that the dimensionality of the data is reduced. This can be advantageous for the classification process for several reasons. The reduced number of features might capture the differences between classes more effectively, thus increasing the classification accuracy. In addition, the reduced number of variables might decrease the complexity of the classification, lower the computational cost, and improve the speed of the classification procedure. In methods such as Discriminant Analysis (DA, discussed below), dimensionality reduction is sometimes required in order to remove features which are highly correlated. In the DA algorithm, such features produce a singular matrix with no inverse, and thus DA cannot be performed.

Using the first and second derivatives is also a commonly used technique [50–52] for feature creation. This method has the advantage of capturing the successive intensity differences of the Raman spectrum (first derivative), or the successive rates of change of the differences in the spectrum (second derivative). These measures have proven to be very successful in the analysis of Raman data and have been widely adopted. An additional advantage of taking the derivatives of the spectrum is that the baseline is removed without introducing bias or human error which would occur if the baseline was removed manually.

Another technique, which was shown to be successful in feature creation, is the use of spectral band ratios [45]. Each spectrum is broken down into segments, each consisting of a small band of wavenumbers. The mean intensity of each segment is found, and the ratio of the mean intensity of each segment to every other segment in the spectrum is then calculated. These ratios are then used as features for the classification of the data. This technique has two significant advantages. First, normalization is inherently achieved since the intensity of one part of the spectrum is compared to the intensity of other parts of the same spectrum. In this way, even if the intensity of the whole spectrum uniformly varies, the ratios of intensities remain the same. Second, the number of features is much less than the original number of data points thus achieving a reduction in the dimensionality of the data. The success of this method implies that classification can still achieve good accuracy even when using a low-resolution spectrum, an important deduction supporting the use of clinically viable, commercial, low cost, and low-resolution systems.

2.3. Classification. Accurate classification of Raman spectra is crucial if the technique is ever to have a clinical application and, more importantly, impact disease diagnosis and prognosis. The range of classification techniques in the literature is extensive. However, certain methods appear to perform well when applied to Raman spectra, and those will be covered in this section. Classification can be either supervised or unsupervised. Unsupervised classification methods group the data points into clusters without using any information about the class (label) of each data point. Supervised classification methods on the other hand require a set of data points which are labeled, that is, their class is known *a priory*. These labeled data points are used to train the algorithm, using the known information about the class (label) of each point, to create a model of the data. Once the supervised model has been trained, it can be used to classify data points which were not in the training set, by assigning them to one of the classes described in the training set.

2.3.1. Unsupervised Classification (Clustering). In Raman spectroscopy, hierarchical cluster analysis is the most commonly used unsupervised classification method [53]. Being an unsupervised method, it does not require *a priori* labeling of the data points. During the training phase, clusters are formed and, subsequently, HCA can be used to classify a new data point by determining into which cluster this new data point falls. The results of HCA are easy to interpret and can be easily visualized as a dendrogram (Figure 3). This ease of interpretation has made HCA very popular. However, this method does not provide good classification performance on data points, which are "out-of-sample," that is, not in the training set. HCA is relatively unstable because clusters formed in the lower levels of the hierarchy can constrain the clusters formed at higher levels of the hierarchy. This constrain can make the analysis unreliable.

The most important consideration when performing cluster analysis is the similarity metric (distance metric), which is used to compare data points. The most commonly used distance metric is the Euclidean distance, but this might not be the most appropriate in the case of Raman spectra. A general disadvantage of cluster analysis is the assumption that clusters do actually exist. This assumption is not always true. Furthermore, it is not always clear how many clusters should be formed. The number of clusters is a prominent concern when performing this type of analysis.

2.3.2. Supervised Classification. In supervised classification, the goal is to assign data points, which were not seen before, into classes learned from a training set assuming that each data point belongs to one of the known classes. For example, Raman spectra obtained from bacteria can be labeled and classified based on the species of the bacterium from which they originated. Each spectrum is a data point, and it belongs to one class. Supervised learning requires that the class of some data points is known (labeled data points). Those points are used as a training set to create a model which can then be used to predict the label of unlabeled data points ("out-of-sample" data points) whose class is not known. Semisupervised methods, which do not require all data points to be labeled, do exist. However, in this paper, we will only deal with supervised methods in which all the data points are labeled. Some of the most commonly used supervised learning methods, applicable to the classification of Raman spectra, are listed below including a brief discussion of their advantages and disadvantages.

Nearest Neighbor. The nearest neighbor approach is very simple and does not explicitly require any training. The data points in the training set are used to predict future unlabeled data points by finding the distance between unlabeled data points and the labeled data points in the training set. The unlabeled data point is predicted to belong to the class of the data point in the training set to which it is closest. In the simplest case, the class is determined by the one data point which is closest, the so-called 1-nearest neighbor (1NN). It is possible to perform the classification using more nearest neighbors. For example, if the three data points in the training set closest to the unlabeled data point are used, then the approach is called 3-nearest neighbor (3NN). In the general case, when K nearest neighbors are used, the approach is called K-nearest neighbor (KNN). A disadvantage of this method is that the prediction could be slow because unlabeled data points must be compared to all data points in the training set. Therefore, the prediction speed depends on

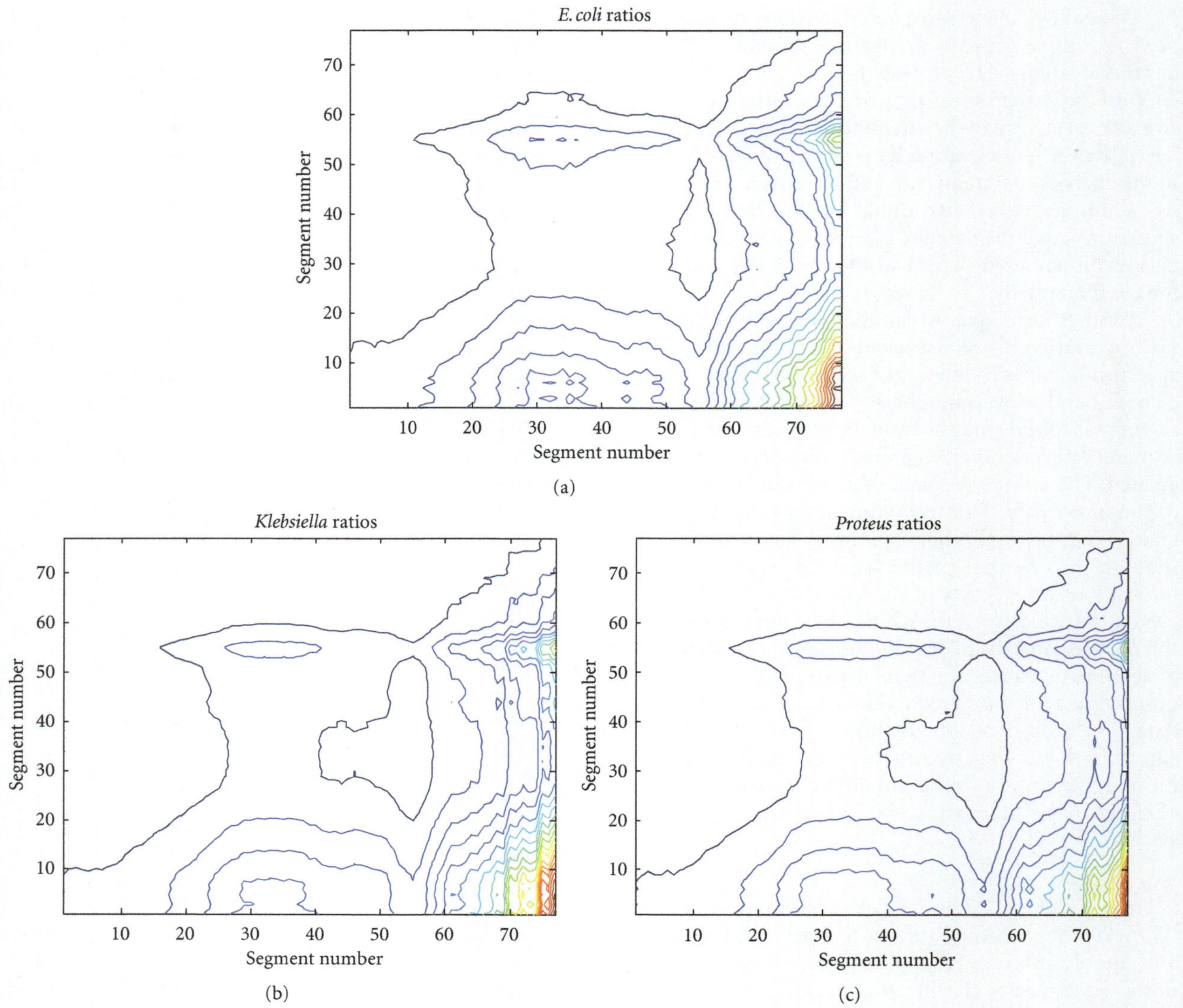

Figure 2: Contour plots of the ratios between different segments of the average Raman spectra from several samples of *E. coli*, *Klebsiella*, and *Proteus*. Differences between the species are evident.

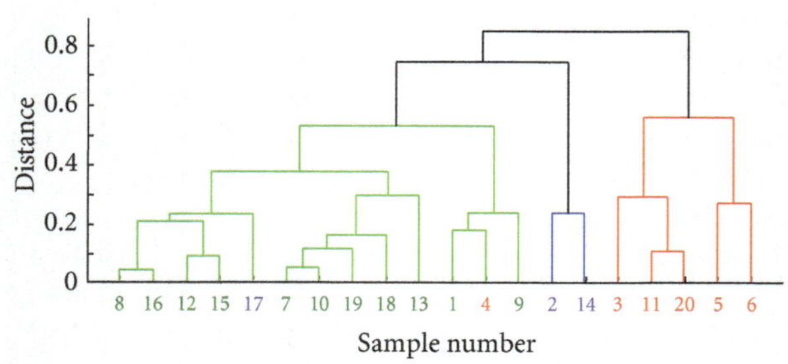

Figure 3: Example of a dendrogram clustering 20 samples according to their "color" property. Below is the sample number indicating the true "color" of each sample. Notice that samples 17 and 4 have been misclassified.

the size of the training set. KNN models have been shown to work well in certain applications although the 3NN approach is usually preferred. It is a very simple technique to implement. The most serious disadvantage of this method is that it is very sensitive to outliers. It is also very sensitive to irrelevant features which might be present in the data points.

Discriminant Analysis. Linear discriminant analysis (LDA) is one of the most commonly-used classification methods, mainly due to its simplicity and low computational cost [54]. It is a parametric technique which assumes that the independent variables (features) follow a multivariate normal distribution. The training data is used to find the parameters of the normal distribution which describes each class. These distributions are then used to create a discriminant model. In LDA, the assumption is that the different classes are described by identical covariance matrices. If this assumption is removed, then the different classes are allowed to have different covariances. In this way, the separating functions become quadratic, and this is called Quadratic Discriminant Analysis (QDA). It is important to choose either LDA or QDA depending on the characteristics of the data. DA is highly sensitive to outliers since they can greatly affect the shape of the calculated distributions. If it is obvious that a data point is an outlier it is best to remove it before carrying out DA. DA is a simple technique that only performs well on data points, which are well separated, and on features which follow a normal distribution. If one of the independent variables (features) is highly correlated with another, or is a function of another set of features, then the calculations for finding the discriminant function will fail. This case is very common when the number of features (independent variables) is much greater than the number of observations (data points). For this reason, it is better to first transform the features before applying DA. The two most commonly used methods for transforming the Raman data are Principal Component Analysis (PCA) and Partial Least squares (PLSs). This leads to the two classification methods known as PC-DFA (Principal Component Discriminant Function Analysis) and PLS-DA (Partial Least Squares Discriminant Analysis).

Principle Components Discriminant Function Analysis (PC-DFA) uses Principal Component Analysis to first transform the data into a new space in order to maximize the variance in each dimension before performing DA. Each dimension is called a principal component. This method can often lead to good results. However, in cases where the variance of the features is not a good criterion for class separability, it is best to use another method, such as PLS-DA.

Partial Least Squares Discriminant Analysis (PLS-DA) is a method which attempts to increase the separation of groups (classes) by representing the data points in a new space [55]. The key difference between PLS and PCA is that PCA aims to find directions of maximum variance, whereas PLS aims to find directions which maximize group separability. In some cases, the two coincide, that is, when group separability is well described by the variance of the features. In other cases, it is best to use PLS instead of PCA. An advantage of PLS-DA is that it can describe complex relationships between features [56]. However, it has been shown that PLS-DA has problems with overfitting [57]. For this reason, it might produce good results when performing cross validation, but it is accuracy on "out-of-sample" data can be low when overfitting on the training data occurs. This is a good example of how cross-validation accuracy fails to capture the expected classification accuracy on future "out-of-sample" data.

Support Vector Machines (SVMs). A Support Vector Machine is a binary classifier which aims to distinguish between two classes of instances by finding the maximum separating hyperplane between them [58]. For this reason it tends to generalize better than the much simpler DA-based approaches. By design, SVMs can only discriminate between two classes. In order to allow for the classification of more than two classes, one can employ more than one SVMs. SVMs, in their simple form, are linear classifiers. It is possible however to create nonlinear SVMs by increasing the dimensionality of the feature space by using the so-called "kernel trick" [59, 60], which employs a kernel function to transform the data. It is thus possible to find a separating hyperplane in higher dimensions where such a hyperplane would not exist in lower dimensions. There are many choices for the type of kernel function to use [60]. The standard choices are the linear kernel (dot-product kernel), the polynomial kernel and the Gaussian Kernel. The best results can be obtained by using an appropriate kernel. It has been shown that for the classification of Raman Spectra, the correlation kernel [45] can give high accuracy on "out-of-sample data."

2.4. Evaluation of the Classification Results. In order to evaluate the performance of a classifier, the most commonly used techniques are cross-validation and performance on a "hold-out set." Cross-validation uses only the training set which is broken down into smaller parts, or "folds." It is often preferred to use 10 folds to perform so-called 10-fold cross-validation. All the folds, except one, are used for training. Testing consists of predicting the class of each data point in the fold which was left out. The cross-validation procedure performs training and testing as many times as there are folds. Each time, a different fold is left out for testing. In this way, a prediction is made on the class of all the data points in the training set. In the extreme case, each fold can consist of only one data point. This is called Leave-One-Out Cross Validation and is most often used when the training set is small. Cross-validation generally results in a good estimate, but this is not always the case. Classification algorithms which do not generalize well can sometimes perform well in cross-validation and not so well on "out-of-sample data" [45]. For this reason it is best to test the classification on a "hold-out set" consisting of data points which were not used for training but whose class label is known. The performance of the classifier on such a test set is a good indication of the performance on future unseen ("out-of-sample") data.

3. Raman Spectroscopy for UTI Diagnosis

For Raman spectroscopy to become a viable clinical tool, the following key diagnostic objectives must be addressed and successfully accomplish:

(1) classification of urine samples as positive or negative for UTI based on the bacterial load;

(2) identification of the pathogen involved in the positive samples;

(3) determination of the antibiogram, that is, antibiotic sensitivity, of the bacteria involved.

Recent research results indicate that it may indeed be possible to achieve all three of the above objectives with Raman or SERS offering rapid and accurate UTI diagnosis using low-cost, turn-key equipment.

3.1. Classification of Bacterial Samples as Positive or Negative. Most of the work published on Raman spectroscopy of bacteria has focused on identification of bacterial species or subspecies, but a few studies are available on quantification of samples using these techniques. Escoriza et al. used Raman spectroscopy to quantify filtered waterborne bacteria, but ran into problems because of the high fluorescence background from the alumina or silver membrane filters they used [61]. There are also some studies on quantification of viruses using SERS [62, 63].

Raman spectroscopy and, in particular, SERS are sensitive enough to detect bacteria at concentrations considered normal and pathologic for UTIs. This can be verified by performing quantification studies of known bacterial concentrations. For example, *E. coli* bacteria were cultured and their concentration in solution was determined using their optical density. Samples were diluted serially with sterile deionized water to obtain the following concentrations: 10^3, 10^4, 10^5, 10^6, 10^7, and 10^8 bacteria/mL. 10 μL of each sample were mixed with an equal volume of concentrated gold nanoparticles, spotted on glass slides, and allowed to dry. SERS spectra were collected from 22 samples using a portable commercial Raman spectrometer at 785 nm excitation wavelength and 4.5 cm^{-1} resolution. Spectra were preprocessed by filtering to remove the background and high-frequency noise. Figure 4 shows the spectra after preprocessing. The feature vector for each sample contained the preprocessed Raman spectrum as well as its first and second derivatives. Each sample was assigned a class of 0 for no infection (concentrations of $\leq 10^4$ bacteria/mL) or 1 for infection (concentration $\geq 10^5$ bacteria/mL.) The outcome showed that ~82% of the samples were correctly classified as negative (class 0) or positive (class 1) for UTI [18].

The same serial dilutions of *E. coli* bacteria (10^3, 10^4, 10^5, 10^6, 10^7, and 10^8 bacteria/mL) were also studies using a 532 nm excitation, 3 cm^{-1} resolution, commercial Raman system. In this case, the bacteria were mixed with silver nanoparticles and spotted on glass slides as before. Spectra were preprocessed by filtering to remove the background and high frequency noise. Figure 5 shows a correlation between the total intensity in the high-wave region of the SERS spectra, from 2900 to 3000 cm^{-1}, with various log concentrations of bacteria indicating an almost linear relationship.

These results are, of course, very preliminary, and much more work is required to develop the technique. They do indicate, however, that it will be possible to examine a urine sample and determine whether it is negative or positive for UTI as well as determine the concentration of bacteria in the urine sample based on its Raman spectral properties. Some preprocessing of urine samples will be done before SERS spectra are collected in order to (a) concentrate samples and make the method more sensitive and (b) remove other types of cells like epithelial or white blood cells that could be present in the sample. Typically, UTIs are caused by a single type of bacteria and it is uncommon to have additional bacterial species in an infected urine culture. A simple 2-step filtration process that would first remove larger, epithelial cells, and WBCs and would secondly concentrate any bacteria on a smaller pore size filter, would allow the isolation of bacteria from urine and the collection of their SERS spectra. In case two or more bacterial species are present in a urine sample the classification algorithm could be trained to detect two or more species at the same time.

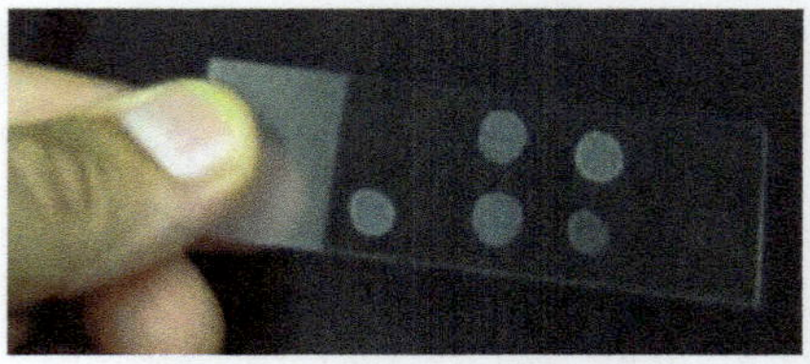

(a)

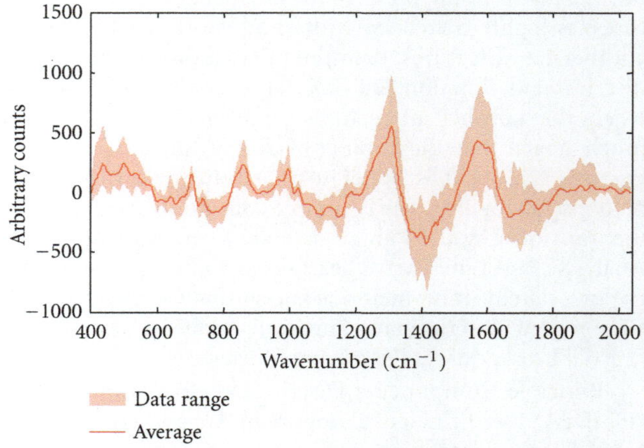

(b)

Figure 4: (a) Bacteria (*E. coli*) and gold nanoparticles spotted on a glass slide for SERS measurement. (b) SERS spectra for different *E. coli* samples after preprocessing, indicating the average spectrum and the range of values, that is, intersample variability.

3.2. Identification of the Pathogen in a Positive Sample. As mentioned in the introduction, several investigators are evaluating the use of Raman spectroscopy as a tool for bacterial classification. Recent studies have shown that employing novel feature creation techniques, such as the ratios of different Raman bands (Figure 2) yields good results even when using a low-cost, low-resolution Raman system [17]. The power of Raman spectroscopy for UTI diagnosis can be verified experimentally. For example, clinical bacterial isolates from patients with UTI ($n = 75$), previously identified by conventional tests as positive, were collected from clinical laboratories. Specifically, spectra from 25 samples of *E. coli*, 25 of *K. pneumonia,* and 25 of *Proteus* spp. were acquired using a portable commercial Raman spectrometer at 785 nm excitation wavelength and 4.5 cm^{-1} resolution. No filtering of the spectra was performed. Each spectrum was broken up into 25 cm^{-1} segments and the mean intensity for each segment was determined. Ratios of each segment's mean intensity to every other segment's mean intensity were calculated. The classification method used was Linear Discriminant Analysis with a Principal Components transformation to reduce the dimensionality of the data. A leave-one-out cross-validation procedure was performed to verify the performance of the classifier. The overall accuracy of the technique was 95%. Very high sensitivity (88–100%) and specificity (94–100%) values for classification of the three species of bacteria were obtained.

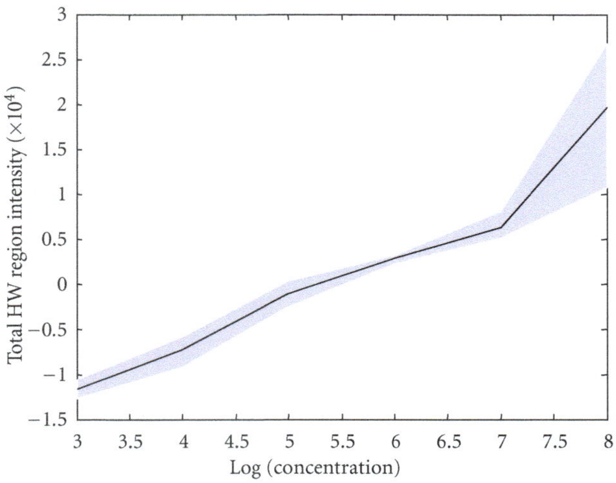

FIGURE 5: Correlation between the total intensity of the 2900–3000 cm^{-1} range of the SERS spectra with various concentrations of *E. coli* samples. The shaded region illustrates the range between minimum and maximum numbers, whereas the black line is the mean value.

The use of SERS can further enhance the sensitivity and specificity of bacterial classification providing accurate results for more than three species at a time. To verify this, clinical bacterial isolates from patients with UTI ($n = 46$) were obtained from clinical laboratories after being identified as positive using conventional tests. Specifically, 10 samples of *E. coli*, 9 of *K. pneumoniae*, 9 of *Proteus* spp., 9 of *Enterococcus* spp., and 9 of *Citrobacter* spp. were obtained for SERS using the same spectrometer as before. Ratio features, PCA, DA, and a leave-one-out cross-validation procedure were performed as before. Classification was originally done for 3 species (*E. coli*, *K. pneumonia*, *Proteus* spp.) and then for all 5 species of bacteria. Table 1 illustrates the results of the classification analysis of the SERS, spectra and Figure 6 shows two examples of the classification. These results suggest that classification of bacteria belonging to 3 classes using SERS is as accurate (93%) as classification using normal Raman spectra (95%). However, much more accurate classification of bacteria belonging to 5 classes is achieved using SERS (91%) compared to using normal Raman spectra (73%). A wavenumber band of 25 cm^{-1} was proved to be a better choice for providing more accurate classification of the 5 classes of bacteria using the ratios method.

TABLE 1: Results of classification analysis of SERS spectra from 3 or 5 classes of bacteria.

Classes	Band size (cm^{-1})	Correct classification
3	50	28/30 (93%)
3	25	28/30 (93%)
5	50	40/46 (87%)
5	25	42/46 (91%)

3.3. Antibiotic Sensitivity of Bacterial Samples. In order for the diagnosis of a UTI to be complete, the antibiotic sensitivity of the involved pathogen must be addressed. Most of the studies, in the past, involved the use of Raman for distinguishing live bacteria from bacteria that were killed by heating [64, 65], desiccation [66, 67], or UV irradiation [68, 69]. More recent work by Liu et al. [16] shows very clear differences between the spectra of bacteria that are sensitive and the spectra of bacteria that are resistant to an antibiotic. However, no classification of samples as sensitive or resistant to an antibiotic was done by this group. The current hypothesis is that exposure of bacteria to an antibiotic to which they are sensitive will cause either external or internal molecular changes, depending on the mechanism of action of the antibiotic. It is expected that the short life cycle of bacteria will allow the visualization of these changes in their Raman spectra within a few hours of exposure to an antibiotic.

In a recent study [17], Raman spectra were collected from 27 strains of bacteria, belonging to the species *E. coli*, *K. pneumonia*, and *Proteus* spp., shown to be sensitive to Ciprofloxacin and resistant to Amoxicillin by conventional antibiograms. The spectra were collected after treatment in the presence or absence of each of each of the two antibiotics for 0, 2, and 4 hours. Spectra were collected using a portable, commercial, Raman spectrometer at 785 nm excitation wavelength and 4.5 cm^{-1} resolution. The spectra were filtered and the fluorescence background was subtracted. The Raman spectrum as well as the first and second derivatives were included in the feature vector. A principal components (PCs) transformation was used, and only PCs describing the highest variance were retained. The data was, then, analyzed using the MANOVA algorithm which calculated a score for each sample which would provide maximum separation between a group of bacteria which were sensitive to an antibiotic with the groups of bacteria which were resistant or were not treated with antibiotic. The study was repeated using SERS with similar results shown in Figure 7; the MANOVA score of bacteria incubated with Ciprofloxacin is significantly lower than that of untreated bacteria or bacteria treated with Amoxicillin even as early as 2 hours after incubation with the antibiotics. These results suggest that SERS, with increased sensitivity and improved speed compared to Raman, could be used to determine the antibiotic susceptibility of bacteria even after a very short treatment with the antibiotic. These studies are currently being expanded to include more bacteria and more antibiotics.

4. Conclusion

In this paper, the current state of the application of Raman spectroscopy in bacterial identification and, more specifically, UTI diagnosis is presented. Raman spectroscopy and SERS, in particular, offer the possibility to develop highly versatile and powerful diagnostic tools. They can provide complex biochemical information which, in conjunction with advanced analysis and classification techniques, can lead to rapid and accurate diagnosis of UTIs. Recent advances in the instrumentation, experimental techniques, analysis,

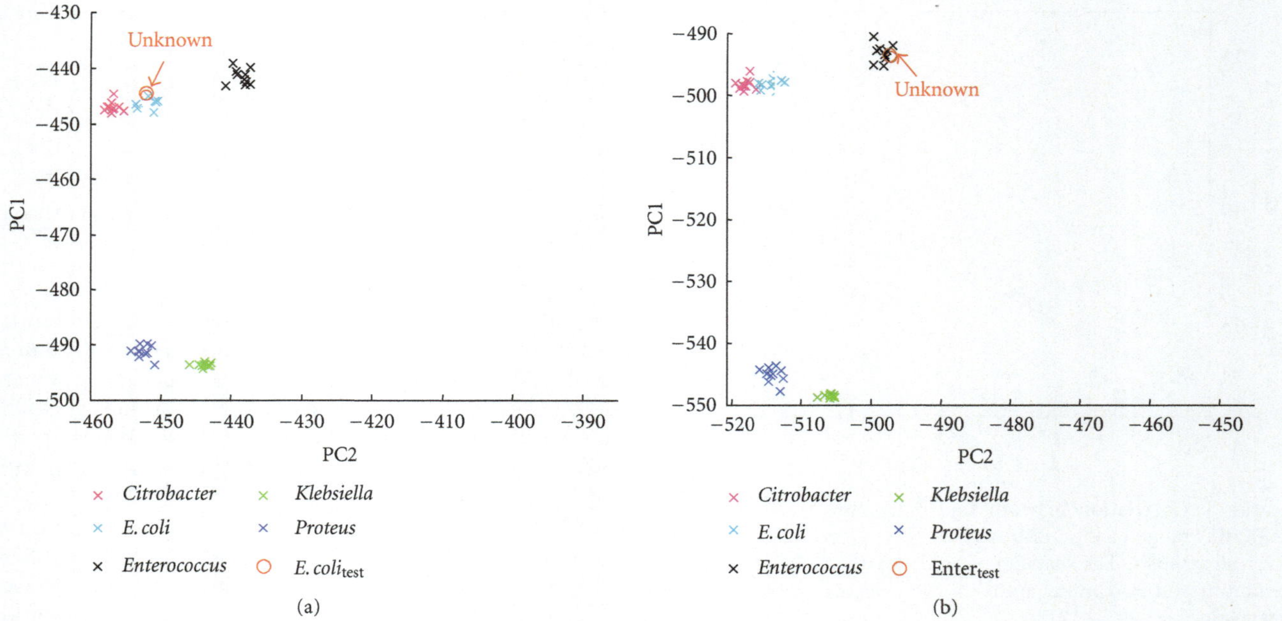

Figure 6: SERS classification of two unknowns, *E. coli*, (a) and *Enterococcus* spp., (b). The first two principal components (PC1 and PC2), that contained the most information, were plotted against each other for visualization purposes. Each of the unknowns clearly falls in the correct species cluster. Note that the plots differ since the training sets differ by one member.

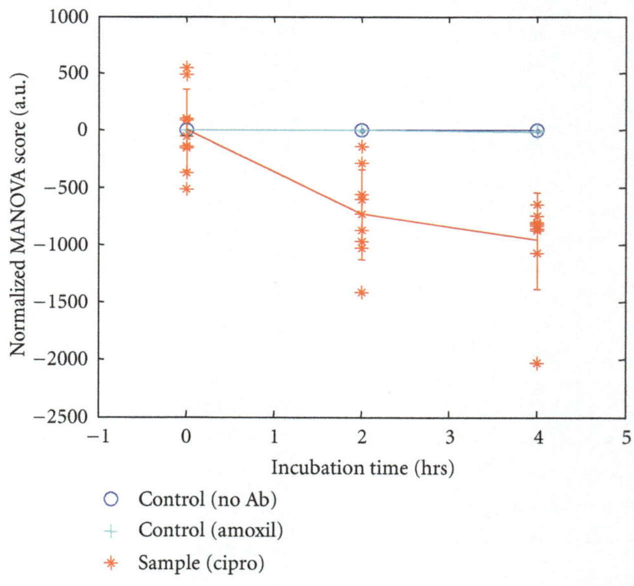

Figure 7: Normalized MANOVA score of SERS spectra of *E. coli*, *K. pneumonia*, and *Proteus* spp. strains after their incubation in the presence of 0.128 μg/mL Ciprofloxacin ($*$), in the presence of 64 μg/mL Amoxicillin (+), or in the absence of any antibiotic (○).

and classification have enabled the fast and accurate quantification and identification of bacterial populations as well as the determination of their sensitivity to antibiotics. These results are still preliminary and must be significantly expanded. However, there is great interest in the area and the research community actively strives to improve the yield and accuracy of Raman-based diagnostics. In the future, a simple, point-of-care, device is envisioned which will be able to give a complete UTI diagnosis, from a single urine sample, within a few hours. Such a technology could have significant short- and long-term benefits for public health by reducing the cost of diagnosis, improving prognosis, and reducing the unnecessary use of antibiotics and, therefore, bacterial resistance.

Acknowledgments

Some of the research presented in this paper was partially supported by the KIOS Research Center for Intelligent Systems and Networks of the University of Cyprus and the Research Promotion Foundation of Cyprus. The authors also wish to thank all their collaborators at P. Hadjikoumis Microbiology Laboratory and the Limassol General Hospital for their support.

References

[1] T. L. Griebling, "Urinary tract infection in women," in *Urologic Diseases in America*, pp. 587–619, DHHS, PHS, NIH, NIDDK, NIH, Washington, DC, USA, 2007.

[2] T. L. Griebling, "Urinary tract infection in women," in *Urologic Diseases in America*, pp. 621–645, DHHS, PHS, NIH, NIDDK, NIH, Washington, DC, USA, 2007.

[3] National Center for Health Statistics, *National Hospital Discharge Survey: 2004 Annual Summary With Detailed Diagnosis and Procedure Data*, DHHS, Centers for Disease Control and Prevention, DHHS, Hyattsville, Md, USA, 2006, 1733.

[4] M. S. Litwin and C. S. Saigal, "Introduction," in *Urologic Diseases in America*, pp. 3–7, DHHS, PHS, NIH, NIDDK, NIH, Washington, DC, USA, 2007.

[5] R. Gruneberg, "Changes in urinary pathogens and their antibiotic sensitivities, 1971–1992," *Journal of Antimicrobial Chemotherapy*, vol. 33, supplement A, pp. 1–8, 1994.

[6] A. Casadevall, "Crisis in infectious diseases: time for a new paradigm?" *Clinical Infectious Diseases*, vol. 23, no. 4, pp. 790–794, 1996.

[7] S. Cosgrove and Y. Carmeli, "The impact of antimicrobial resistance on health and economic outcomes," *Clinical Infectious Diseases*, vol. 36, no. 11, pp. 1433–1437, 2003.

[8] A. Alanis, "Resistance to antibiotics: are we in the post-antibiotic era?" *Archives of Medical Research*, vol. 36, no. 6, pp. 697–705, 2005.

[9] E. Mothershed and A. Whitney, "Nucleic acid-based methods for the detection of bacterial pathogens: present and future considerations for the clinical laboratory," *Clinica Chimica Acta*, vol. 363, no. 1-2, pp. 206–220, 2006.

[10] J. Rolain, M. Mallet, P. Fournier, and D. Raoult, "Real-time PCR for universal antibiotic susceptibility testing," *Journal of Antimicrobial Chemotherapy*, vol. 54, no. 2, pp. 538–541, 2004.

[11] W. Chen, P. Tsai, and Y. Chen, "Functional nanoparticle-based proteomic strategies for characterization of pathogenic bacteria," *Analytical Chemistry*, vol. 80, no. 24, pp. 9612–9621, 2008.

[12] K. Maquelin, L. Choo-Smith, T. Vreeswijk et al., "Raman spectroscopic method for identification of clinically relevant microorganisms growing on solid culture medium," *Analytical Chemistry*, vol. 72, no. 1, pp. 12–19, 2000.

[13] K. C. Schuster, I. Reese, E. Urlaub, J. R. Gapes, and B. Lendl, "Multidimensional information on the chemical composition of single bacterial cells by confocal Raman microspectroscopy," *Analytical Chemistry*, vol. 72, no. 22, pp. 5529–5534, 2000.

[14] K. C. Schuster, E. Urlaub, and J. R. Gapes, "Single-cell analysis of bacteria by Raman microscopy: spectral information on the chemical composition of cells and on the heterogeneity in a culture," *Journal of Microbiological Methods*, vol. 42, no. 1, pp. 29–38, 2000.

[15] K. Kneipp, M. Moskovits, and H. Kneip, "Surface-enhanced Raman scattering," *Physics and Applications Topics in Applied Physics*, vol. 103, pp. 1–464, 2006.

[16] T. T. Liu, Y. H. Lin, C. S. Hung et al., "A high speed detection platform based on surface-enhanced Raman scattering for monitoring antibiotic-induced chemical changes in bacteria cell wall," *Plos One*, vol. 4, no. 5, Article ID e5470, 2009.

[17] E. Kastanos, A. Kyriakides, K. Hadjigeorgiou, and C. Pitris, "A novel method for urinary tract infection diagnosis and antibiogram using Raman spectroscopy," *Journal of Raman Spectroscopy*, vol. 41, no. 9, pp. 958–963, 2010.

[18] E. Kastanos, K. Hadjigeorgiou, A. Kyriakides, and C. Pitris, "Classification of bacterial samples as negative or positive for a UTI and antibiogram using surface enhanced Raman spectroscopy," in *Proceedings of the SPIE*, vol. 7911, 2011.

[19] J. Magee, "Whole-organism fingerprinting," in *Handbook of New Bacterial Systematics*, M. Goodfellow and A. O'Donnell, Eds., pp. 383–427, Academic Press, London, UK, 1993.

[20] A. Mahadevan-Jansen, "Raman spectroscopy from benchtop to bedside," in *Biomedical Photonics Handbook*, T. Vo-Dinh, Ed., CRC Press, London, USA, 2002.

[21] O. Preisner, J. Lopes, and J. Menezes, "Uncertainty assessment in FT-IR spectroscopy based bacteria classification models," *Chemometrics and Intelligent Laboratory Systems*, vol. 94, no. 1, pp. 33–42, 2008.

[22] J. Forrester, N. Valentine, Y. Su, and T. Johnson, "Chemometric analysis of multiple species of Bacillus bacterial endospores using infrared spectroscopy: discrimination to the strain level," *Analytica Chimica Acta*, vol. 651, no. 1, pp. 24–30, 2009.

[23] R. Jarvis and R. Goodacre, "Ultra-violet resonance Raman spectroscopy for the rapid discrimination of urinary tract infection bacteria," *FEMS Microbiology Letters*, vol. 232, no. 2, pp. 127–132, 2004.

[24] D. Hutsebaut, K. Maquelin, P. De Vos, P. Vandenabeele, L. Moens, and G. Puppels, "Effect of culture conditions on the achievable taxonomic resolution of Raman spectroscopy disclosed by three *Bacillus* species," *Analytical Chemistry*, vol. 76, no. 21, pp. 6274–6281, 2004.

[25] C. Xie, J. Mace, M. Dinno et al., "Identification of single bacterial cells in aqueous solution using confocal laser tweezers Raman spectroscopy," *Analytical Chemistry*, vol. 77, no. 14, pp. 4390–4397, 2005.

[26] M. Harz, P. Rösch, K. Peschke, O. Ronneberger, H. Burkhardt, and J. Popp, "Micro-Raman spectroscopic identification of bacterial cells of the genus Staphylococcus and dependence on their cultivation conditions," *Analyst*, vol. 130, no. 11, pp. 1543–1550, 2005.

[27] U. Schmid, P. Rosch, M. Krause, M. Harz, J. Popp, and K. Baumann, "Gaussian mixture discriminant analysis for the single-cell differentiation of bacteria using micro-Raman spectroscopy," *Chemometrics and Intelligent Laboratory Systems*, vol. 96, no. 2, pp. 159–171, 2009.

[28] K. Kneipp, Y. Wang, H. Kneipp et al., "Single molecule detection using surface-enhanced Raman scattering (SERS)," *Physical Review Letters*, vol. 78, no. 9, pp. 1667–1670, 1997.

[29] S. Nie and S. Emory, "Probing single molecules and single nanoparticles by surface-enhanced Raman scattering," *Science*, vol. 275, no. 5303, pp. 1102–1106, 1997.

[30] M. Moskovits, "Surface-enhanced spectroscopy," *Reviews of*

Modern Physics, vol. 57, no. 3, pp. 783–826, 1985.

[31] T. Vo-Dinh, F. Yan, and D. Stokes, "Plasmonics-based nanostructures for surface-enhanced Raman scattering bioanalysis," *Methods in Molecular Biology*, vol. 300, pp. 255–283, 2005.

[32] R. Jarvis, N. Law, I. Shadi, P. O'Brien, J. Lloyd, and R. Goodacre, "Surface-enhanced Raman scattering from intracellular and extracellular bacterial locations," *Analytical Chemistry*, vol. 80, no. 17, pp. 6741–6746, 2008.

[33] S. Efrima and B. Bronk, "Silver colloids impregnating or coating bacteria," *Journal of Physical Chemistry B*, vol. 102, no. 31, pp. 5947–5950, 1998.

[34] L. Zeiri, B. V. Bronk, Y. Shabtai, J. Czege, and S. Efrima, "Silver metal induced surface enhanced Raman of bacteria," *Colloids and Surfaces A*, vol. 208, no. 1–3, pp. 357–362, 2002.

[35] L. Zeiri, B. Bronk, Y. Shabtai, J. Eichler, and S. Efrima, "Surface-enhanced raman spectroscopy as a tool for probing specific biochemical components in bacteria," *Applied Spectroscopy*, vol. 58, no. 1, pp. 33–40, 2004.

[36] S. Efrima and L. Zeiri, "Understanding SERS of bacteria," *Journal of Raman Spectroscopy*, vol. 40, no. 3, pp. 277–288, 2009.

[37] R. Jarvis and R. Goodacre, "Discrimination of bacteria using surface-enhanced Raman spectroscopy," *Analytical Chemistry*, vol. 76, no. 1, pp. 40–47, 2004.

[38] R. Jarvis, A. Brooker, and R. Goodacre, "Surface-enhanced Raman spectroscopy for bacterial discrimination utilizing a scanning electron microscope with a Raman spectroscopy interface," *Analytical Chemistry*, vol. 76, no. 17, pp. 5198–5202, 2004.

[39] R. Jarvis, A. Brooker, and R. Goodacre, "Surface-enhanced Raman scattering for the rapid discrimination of bacteria," *Faraday Discussions*, vol. 132, pp. 281–292, 2006.

[40] R. Jarvis and R. Goodacre, "Characterisation and identification of bacteria using SERS," *Chemical Society Reviews*, vol. 37, no. 5, pp. 931–936, 2008.

[41] W. Premasiri, D. Moir, M. Klempner, N. Krieger, G. Jones, and L. Ziegler, "Characterization of the surface enhanced Raman scattering (SERS) of bacteria," *Journal of Physical Chemistry B*, vol. 109, no. 1, pp. 312–320, 2005.

[42] I. Patel, W. Premasiri, D. Moir, and L. Ziegler, "Barcoding bacterial cells: a SERS-based methodology for pathogen identification," *Journal of Raman Spectroscopy*, vol. 39, no. 11, pp. 1660–1672, 2008.

[43] S. Hennigan, J. Driskell, R. Dluhy et al., "Detection of Mycoplasma pneumoniaein simulated and true clinical throat swab specimens by nanorod array-surface-enhanced raman spectroscopy," *Plos One*, vol. 5, no. 10, Article ID e13633, 2010.

[44] A. Walter, A. Marz, W. Schumacher, P. Rosch, and J. Popp, "Towards a fast, high specific and reliable discrimination of bacteria on strain level by means of SERS in a microfluidic device," *Lab on a Chip*, vol. 11, no. 6, pp. 1013–1021, 2011.

[45] A. Kyriakides, E. Kastanos, K. Hadjigeorgiou, and C. Pitris, "Classification of Raman spectra using the correlation kernel," *Journal of Raman Spectroscopy*, vol. 42, no. 5, pp. 904–909, 2011.

[46] F. Ehrentreich and L. Summchen, "Spike removal and denoising of Raman spectra by wavelet transform methods," *Analytical Chemistry*, vol. 73, no. 17, pp. 4364–4373, 2001.

[47] T. Dörfer, W. Schumacher, N. Tarcea, M. Schmitt, and J. Popp, "Quantitative mineral analysis using Raman spectroscopy and chemometric techniques," *Journal of Raman Spectroscopy*, vol. 41, no. 6, pp. 684–689, 2010.

[48] I. T. Jollife, *Principal Component Analysis*, Springer, New York, NY, USA, 2nd edition, 2002.

[49] I. T. Jolliffe, "A note on the use of principal components in regression," *Journal of the Royal Statistical Society C*, vol. 31, no. 3, pp. 300–303, 1982.

[50] J. R. Ferraro, K. Nakamoto, and C. W. Brown, *Introductory Raman Spectroscopy*, Elsevier Science, New York, NY, USA, 2nd edition, 2003.

[51] Y. L. Loethen, D. Zhang, R. N. Favors, S. B. G. Basiaga, and D. Ben-Amotz, "Second-derivative variance minimization method for automated spectral subtraction," *Applied Spectroscopy*, vol. 58, no. 3, pp. 272–278, 2004.

[52] N. Navas, J. Romero-Pastor, E. Manzano, and C. Cardell, "Raman spectroscopic discrimination of pigments and tempera paint model samples by principal component analysis on first-derivative spectra," *Journal of Raman Spectroscopy*, vol. 41, no. 11, pp. 1196–1203, 2010.

[53] F. B. Baker and L. J. Hubert, "Measuring the power of hierarchical cluster analysis," *Journal of the American Statistical Association*, vol. 70, no. 349, pp. 31–38, 1975.

[54] W. J. Krzanowski, *Principles of Multivariate Analysis: a User's Perspective*, Oxford University Press, New York, NY, USA, 2000.

[55] M. Barker and W. Rayens, "Partial least squares for discrimination," *Journal of Chemometrics*, vol. 17, no. 3, pp. 166–173, 2003.

[56] J. Weston and Watkins, "Support vector machines for multiclass pattern recognition," in *Proceedings of the Seventh European Symposium on Artificial Neural Networks*, Brussels, Belgium, April 1999.

[57] J. A. Westerhuis, H. C. J. Hoefsloot, S. Smit et al., "Assessment of PLSDA cross validation," *Metabolomics*, vol. 4, no. 1, pp. 81–89, 2008.

[58] V. N. Vapnik, *Statistical Learning Theory*, Wiley-Interscience, New York, NY, USA, 1998.

[59] B. Schölkopf, C. J. C Burges, and A. J. Smola, Eds., *Advances in Kernel Methods—Support Vector Learning*, MIT Press, Cambridge, Mass, USA, 1998.

[60] N. Cristianini and J. Shawe-Taylor, *An Introduction to Support Vector Machines: and other Kernel-Based Learning Methods*, Cambridge University Press, Cambridge, UK, 2004.

[61] M. Escoriza, J. VanBriesen, S. Stewart, J. Maier, and P. Treado, "Raman spectroscopy and chemical imaging for quantification of filtered waterborne bacteria," *Journal of Microbiological Methods*, vol. 66, no. 1, pp. 63–72, 2006.

[62] C. Fan, Z. Hu, L. Riley, G. Purdy, A. Mustapha, and M. Lin, "Detecting food—and waterborne viruses by surface-enhanced raman spectroscopy," *Journal of Food Science*, vol. 75, no. 5, pp. M302–M307, 2010.

[63] X. Wang, Y. Li, H. Wang et al., "Gold nanorod-based localized surface plasmon resonance biosensor for sensitive detection of hepatitis B virus in buffer, blood serum and plasma," *Biosensors and Bioelectronics*, vol. 26, no. 2, pp. 404–410, 2010.

[64] M. Baek, W. Nelson, D. Britt, and J. Sperry, "UV-excited resonance raman spectra of heat denatured lysozyme and staphylococcus epidermidis," *Applied Spectroscopy*, vol. 42, no. 7, pp. 1312–1314, 1988.

[65] A. Grow, L. Wood, J. Claycomb, and P. Thompson, "New biochip technology for label-free detection of pathogens and their toxins," *Journal of Microbiological Methods*, vol. 53, no. 2, pp. 221–233, 2003.

[66] L. Robertson, A. Campbell, and H. Smith, "Survival of Cryptosporidium parvum oocysts under various environmental pressures," *Applied and Environmental Microbiology*, vol. 58, no. 11, pp. 3494–3500, 1992.
[67] M. Deng and D. Cliver, "Cryptosporidium parvum studies with dairy products," *International Journal of Food Microbiology*, vol. 46, no. 2, pp. 113–121, 1999.
[68] M. Lorenzo-Lorenzo, M. Ares-Mazas, I. Villacorta-Martinez de Maturana, and D. Duran-Oreiro, "Effect of ultraviolet disinfection of drinking water on the viability of Cryptosporidium parvum oocysts," *Journal of Parasitology*, vol. 79, no. 1, pp. 67–70, 1993.
[69] M. Belosevic, S. A. Craik, J. L. Stafford, N. F. Neumann, J. Kruithof, and D. W. Smith, "Studies on the resistance/reactivation of Giardia muris cysts and Cryptosporidium parvum oocysts exposed to medium-pressure ultraviolet radiation," *FEMS Microbiology Letters*, vol. 204, no. 1, pp. 197–203, 2001.

6

Raman Spectroscopy for Quantitative Analysis of Point Defects and Defect Clusters in Irradiated Graphite

Keisuke Niwase

Hyogo University of Teacher Education, Hyogo 673-1494, Kato, Japan

Correspondence should be addressed to Keisuke Niwase, niwase@hyogo-u.ac.jp

Academic Editor: Maher S. Amer

We report the development of Raman spectroscopy as a powerful tool for quantitative analysis of point defect and defect clusters in irradiated graphite. Highly oriented pyrolytic graphite (HOPG) was irradiated by 25 keV He^+ and 20 keV D^+ ions. Raman spectroscopy and transmission electron microscopy revealed a transformation of irradiated graphite into amorphous state. Annealing experiment indicated a close relation between Raman intensity ratio and vacancy concentration. The change of Raman spectra under irradiation was empirically analyzed by "disordered-region model," which assumes the transformation from vacancy-contained region to disordered region. The model well explains the change of Raman spectra and predicts the critical dose of amorphization, but the nature of the disordered region is unclear. Then, we advanced the model into "dislocation accumulation model," assigning the disordered region to dislocation dipole. Dislocation accumulation model can simulate the irradiation time dependencies of Raman intensity ratio and the c-axis expansion under irradiation, giving a relation between the absolute concentration of vacancy and Raman intensity ratio, suggesting an existence of the barrier on the mutual annihilation of vacancy and interstitial.

1. Introduction

Graphite is one of useful materials for thermal nuclear and fusion reactors due to their outstanding nature with respect to heat load, neutron activation, and so forth. Since the fire of Windscale nuclear reactor occurred in 1957 [1] due to a spontaneous release of stored energy from irradiated graphite, radiation damage has been recognized to be important problems to be solved for irradiated graphite. Also, dimensional changes of growth parallel to c-axis and contraction within the basal planes decrease in c-axis electrical resistivity and so forth occur as serious problems. In spite of a lot of investigations on irradiated graphite [2–33], the natures of point defects and defect clusters, and their reaction kinetics are obscure. This can be attributed to the anisotropy of the graphite lattice which requires a description of the motion of interstitials and vacancies in terms of activation energies parallel and perpendicular to the basal planes, and the instability of vacancy and interstitial clusters [3, 7].

Raman spectroscopy has been used for the characterization of graphite and investigations on the graphitization process and on the change of graphite structure under irradiation. The advantages of this method for the examination of graphite are a selective sensitiveness to the structural changes in the basal plane, such as crystalline size or the formation of a single vacancy or its clusters, and extensive and continuous sensitivity over the range from ordered to amorphized graphite.

In this paper, we review our experimental [8–10, 12, 15] and theoretical studies [18, 19, 26, 29] on irradiated graphite. Our study started from the discovery of irradiation-induced amorphization of graphite, which was judged from the change of TEM diffraction pattern observed along the c-axis [8]. The TEM specimens were investigated by micro-Raman spectroscopy, leading us to directly compare the change of the TEM and the related Raman results [9]. Important Raman results were obtained by annealing of irradiated HOPG at relatively low temperatures from 373 K to 873 K. Relaxation of Raman intensity ratio started to appear at comparatively low temperature of 373 K, which implies a close relationship between Raman intensity ratio and vacancy concentration [10].

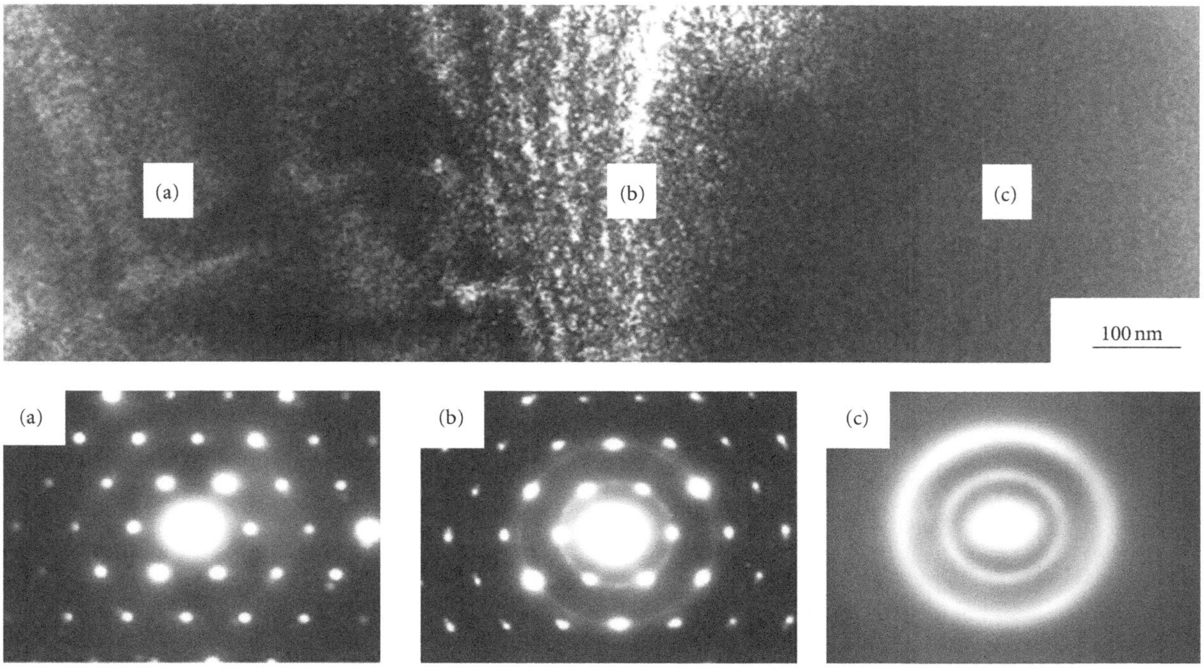

FIGURE 1: TEM image of highly oriented pyrolytic graphite (HOPG) irradiated with 20 keV D^+ at 473 K. The left hand side area (a) of the image is a part of the specimen covered by a mesh wire and the right hand side area (c) is a part irradiated to a dose of 2.5×10^{21} ions/m^2. Selected area diffraction patterns corresponding to the areas (a), (b) and (c) are given below the photograph.

We firstly proposed a semiempirical model of "disordered-region model" to explain the dose and the temperature dependencies of the change of Raman spectra under irradiation [18, 19]. The model assumes the transformation from vacancy-contained region to disordered region and well explains the change of Raman spectra, predicting the critical dose of amorphization. However, the disordered-region model cannot predict the change of the concentration of point defects and defect clusters under irradiation nor give information on the structure of the disordered region. Then, we proposed an advanced model of "dislocation accumulation model," which can simulate the square-root time dependence of Raman intensity ratio under irradiation and gives a relation between the absolute concentration of vacancy and the Raman intensity ratio, suggesting an existence of the barrier on the mutual annihilation of vacancy and interstitial [26]. Thus, we show Raman spectroscopy as a powerful tool for the study of radiation damage of graphite.

2. Experimental Results for Irradiated Graphite

2.1. Ion Irradiation to HOPG Foils

2.1.1. TEM Results and Raman Spectra. Thin foils of highly oriented pyrolytic graphite for transmission electron microscopy (TEM) were irradiated with 25 keV He^+ and 20 keV D^+ at temperatures ranging from room temperature (RT) to 973 K [8]. Figure 1 shows the TEM images and associated diffraction patterns of 20 keV D^+-irradiated HOPG foil. The left-hand side area, marked with (a), was covered by a mesh wire, and, then, there remains the original graphite structure, as confirmed by the [0001] diffraction pattern of a graphite crystalline lattice shown underneath. The right hand side area, (c), irradiated with a dose of 2.5×10^{21} D/m^2, on the other hand, completely loses the extinction contours and coarse dotty contrast, only halo ring being seen in the diffraction pattern. The midst side area, (b), irradiated with an intermediate dose shows a mixed diffraction pattern, spots, and halo rings. Thus, initially perfect crystalline graphite gradually loses atomic ordering with increasing dose and becomes amorphous when the implanted dose is above a critical value. The critical dose for the amorphous becomes larger with the increase of the irradiation temperature.

The amorphization, characterized by the disappearance of the white dotty contrast in the dark field image and the transition from the spots to the halos in the diffraction pattern, is also observed in the case of He^+ irradiation. The appearance of the [0001] diffraction pattern of the irradiated specimens in Figure 1 was classified into three patterns of spot, halo and spot, and halo, which are referred as *s*, *hs*, and

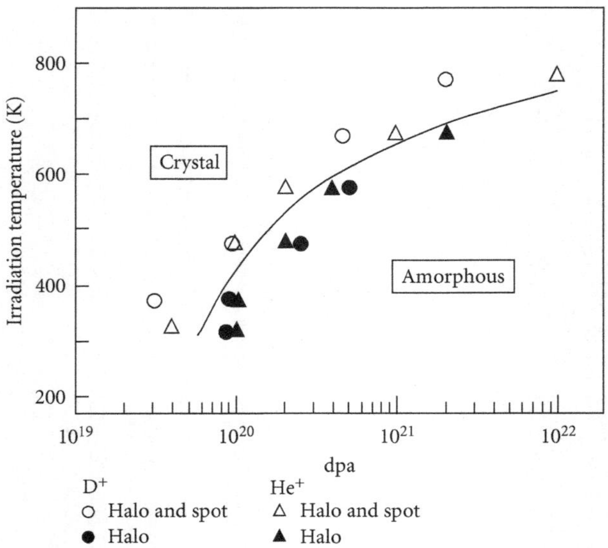

FIGURE 2: Conditions of appearance of TEM diffraction patterns of *hs* (halo and spot) and *h* (halo) by 20 keV D^+ and 25 keV He^+ irradiation. The critical conditions of amorphization are indicated by the solid curve. Note that the critical dpa for the D^+ irradiation almost coincides with that for the He^+ irradiation.

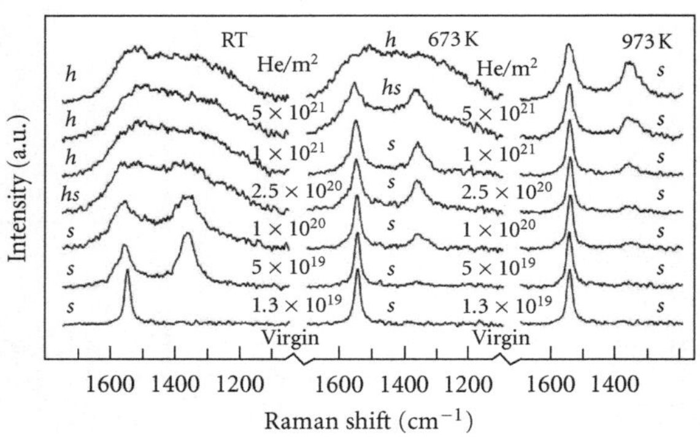

FIGURE 3: Changes in Raman spectra for HOPG foils irradiated with 25 keV He^+ at three different temperatures of RT, 673, and 973 K. The appearance of the TEM diffraction pattern of the foils is denoted by *s* (spot), *hs* (halo and spot), and *h* (halo).

h, respectively and hereafter. The critical dose of He^+ irradiation is much less than that of the D^+ irradiation. Nevertheless, the calculated value of the critical dpa (displacement per atom) for the amorphization by the He^+ irradiation was the same as that for the D^+ irradiation, as shown in Figure 2. This indicates that the amorphization is basically controlled by an atomic displacement and the difference in the nature of the incident particles or the chemical effect plays a minor role.

The TEM specimens were investigated by micro-Raman spectroscopy with a resolution of 2 μm. Figure 3 compares the changes of Raman spectra of HOPG foils irradiated with 25 keV He^+ at RT and 673 and 973 K. The spectrum before irradiation is given at the bottom of the figures, where the Raman-active E_{2g} mode can only be found at 1580 cm^{-1}. After irradiation, a new peak appears around 1355 cm^{-1}, which can be assigned as the maximum phonon density of states, caused by the relaxation of the wave-vector selection rule. The appearances of the [0001] diffraction pattern investigated with TEM for each specimen are referred as *s*, *hs*, and *h*, respectively, in Figure 3. One can see that the Raman spectral change closely correlates with the TEM diffraction change. The onset of amorphization denoted by *hs* corresponds to the onset of remarkable broadening of the two main Raman peaks, while the completion of amorphization denoted by *h* corresponds to a complete overlapping of the two peaks. The effect of irradiation on Raman spectra becomes less appreciable with increasing the irradiation temperature, because of dynamic annealing during irradiation. Increasing the irradiation temperature is seen to delay the onset of amorphization. Contrary to the close relation between the broadening of the Raman peak width and amorphization checked by TEM diffraction pattern, the reduction of the crystalline size, which is calculated from the intensity ratio of I_{1355}/I_{1580} [34], does not coincide with the amorphization process investigated by TEM.

2.1.2. Diagram for the Change of Amorphization. Raman spectra are deconvoluted into several characteristic peaks with a least-square algorithm. The spectra of the irradiated graphite denoted by *s* and *hs* are decomposed into four Lorentzian peaks two dominant Lorentzian peaks at about 1355 and 1580 cm^{-1} and additional two minor Lorentzian peaks at about 1200 and 1500 cm^{-1}, while the spectra denoted by *h* cannot be fitted well by the Lorentizans, but can so by Gaussians.

Figures 4(a) and 4(b) show the changes of the intensity ratio of the two peaks (I_{1355}/I_{1580}) and a full width at half maximum of the 1580 cm^{-1} peak ($FWHM_{1580}$), determined by the fitting analysis, against the dose. A difference can be clearly observed between the two. Below 473 K, the I_{1355}/I_{1580} is already larger than 1 at the initial stage of irradiation. After reaching a maximum, it gradually decreases and levels off. On the other hand, the $FWHM_{1580}$ gradually increases at the initial stage, but it increases dramatically when amorphization starts. The final stage of amorphization, denoted by *h*, corresponds to a value of 180 cm^{-1}. Thus the change of the $FWHM_{1580}$ more closely correlates with the amorphization process than that of I_{1355}/I_{1580}.

Figure 5 shows how the $FWHM_{1580}$ changes in relation to the dose. Linear relationships can be seen at each irradiation temperature, but the increase rate decreases with increasing irradiation temperature. A significant reduction of the increase rate can be found between 573 and 673 K.

Figure 6 shows the relation between the $FWHM_{1580}$ and the I_{1355}/I_{1580} under 25 keV He^+ irradiation. The three linear relations can be seen to correspond to the appearance of the TEM diffraction patterns of *s*, *hs*, and *h*. One should note, however, that several data below 473 K deviated from this relation, indicating different paths from the solid line to the dashed one.

2.1.3. Annealing of Irradiated Graphite. An annealing study was done using Raman spectroscopy in order to examine thermal stability of the defect structures in the HOPG foils

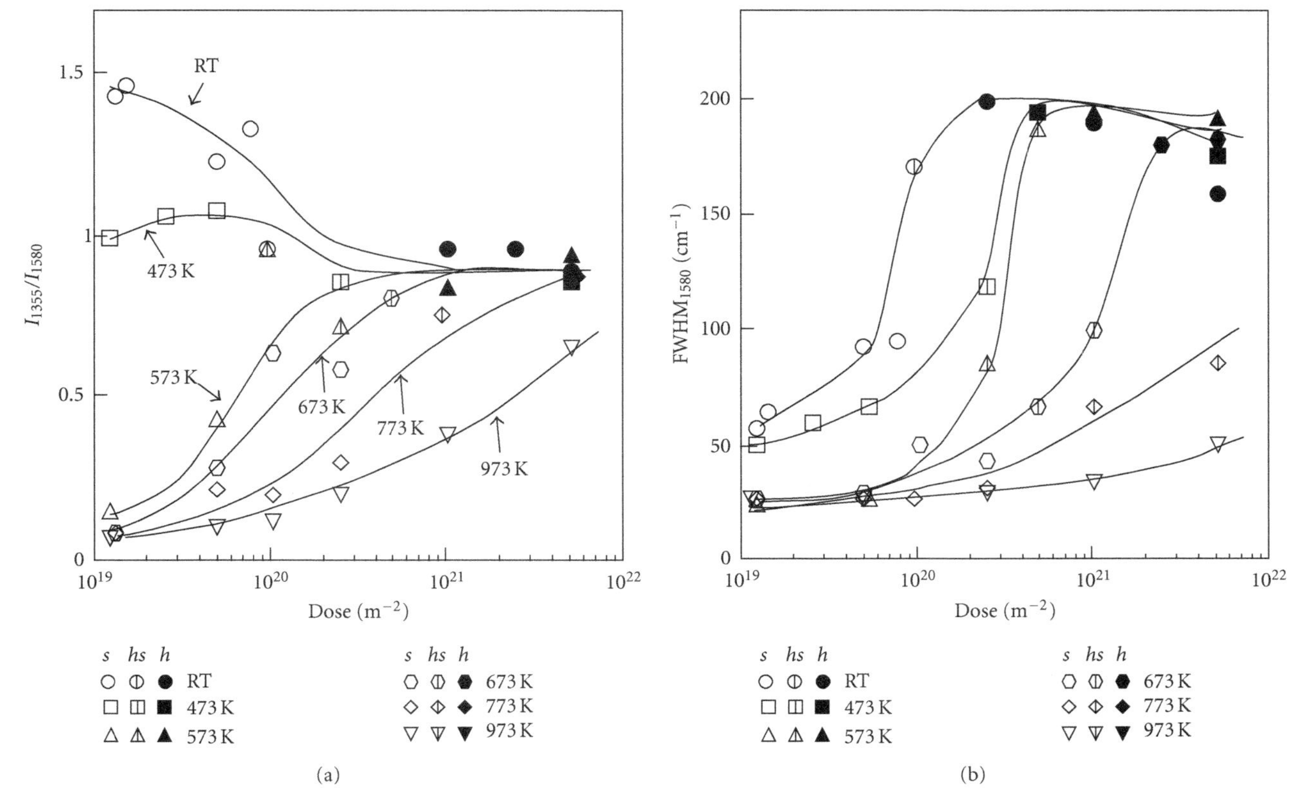

FIGURE 4: (a) The intensity ratio of I_{1355}/I_{1580} and (b) the peak width of 1580 cm^{-1} versus the dose in a logarithmic scale for HOPG foils irradiated with 25 keV He^+ at temperatures ranging from RT to 973 K. The appearance of the TEM diffraction pattern of the foils is denoted by *s* (spot), *hs* (halo and spot), and *h* (halo). The solid curves are guides for eyes.

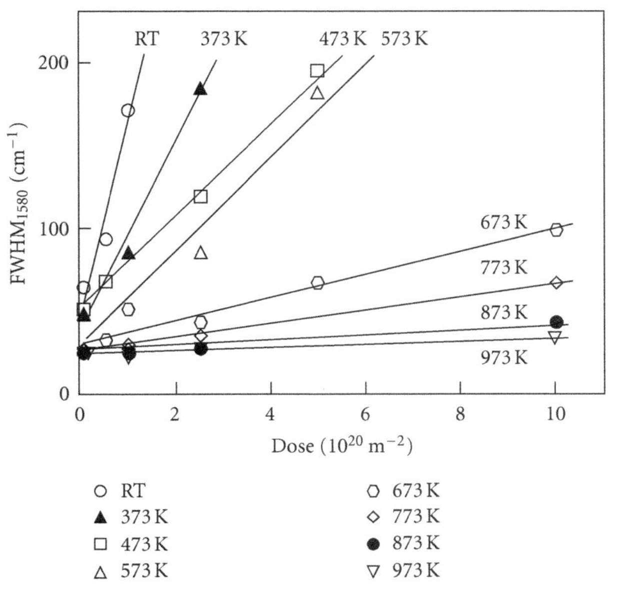

FIGURE 5: Peak width of 1580 cm^{-1} versus the dose in a linear scale for HOPG foils irradiated with 25 keV He^+ at temperatures ranging from RT to 973 K. The dose range corresponds to the amorphizing stage. The peak widths are seen to linearly increase against the dose. The solid lines are guides for eyes.

irradiated with 20 keV D^+ at room temperature and the result is shown in Figure 7. The appearances of the TEM diffraction pattern investigated before annealing are denoted by *s*, *hs*, and *h*. One can clearly distinguish the annealing effect between the unamorphized HOPG (*s*) and amorphized one (*h*). For the unamorphized HOPG (*s*) a remarkable recovery of the Raman spectra appears at annealing temperatures between 373 and 573 K, whereas only a slight change can be seen in the amorphized HOPG (*h*).

2.2. HRTEM Image and Raman Spectra of Neutron-Irradiated Graphite Fiber. A combined study of high-resolution TEM and Raman spectroscopy for neutron-irradiated graphite fibers has given important information on the nature of irradiation-produced defects [12]. Raman spectrum for a dose of 1.4×10^{23} n/m^2 shows a sharp and intense damage peak at 1355 cm^{-1}, whereas that for 1.9×10^{24} n/m^2 is very broad, as respectively shown in Figures 8(c) and 8(d). The two spectra correspond well to those of the ion irradiated HOPG before amorphization (*s*) and in amorphizing process (*hs*), respectively. In spite of the appearance of sharp damage peak in the Raman spectra, the corresponding lattice image in Figure 9 still keeps the long-range order of the basal planes and does not show an apparent reduction of the crystalline

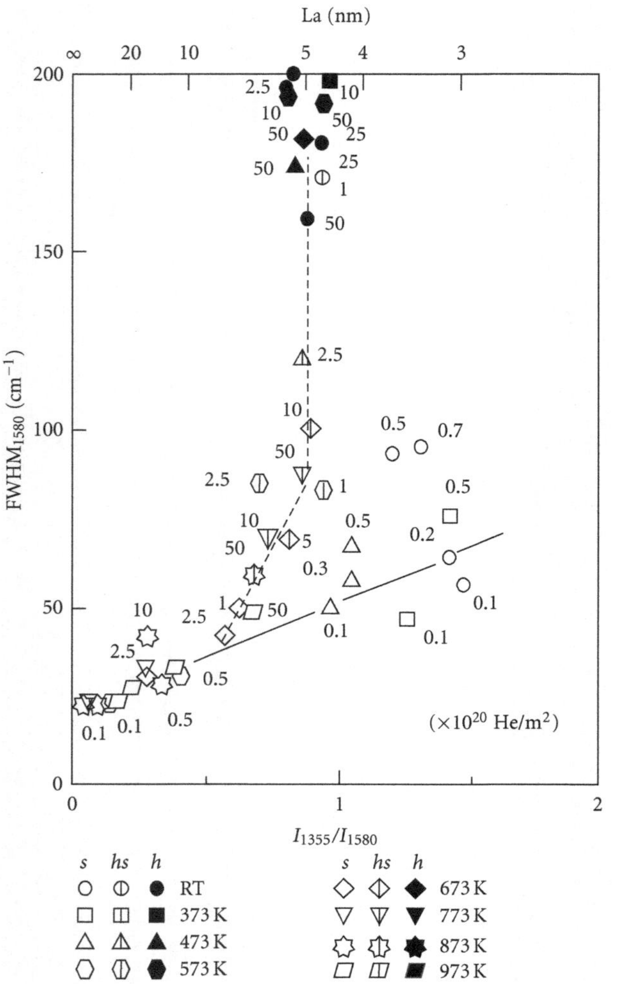

FIGURE 6: Peak width of 1580 cm^{-1} versus intensity ration of I_{1355}/I_{1580} for HOPG foils irradiated with 25 keV He^+ at temperatures ranging from RT to 973 K. Open, half-closed, and closed symbols, respectively, correspond to the appearance of the TEM diffraction patterns of *s* (spot), *hs* (halo and spot), and *h* (halo). The in-plane crystalline size La calculated by Tuinstra and Koenig's relation [34] is also indicated in the figure. The calculated La, especially for the data along the solid line, should not be attributed to the in-plane crystalline size but to the phonon correlation length broken by in-plane defects of vacancies, as described in the text.

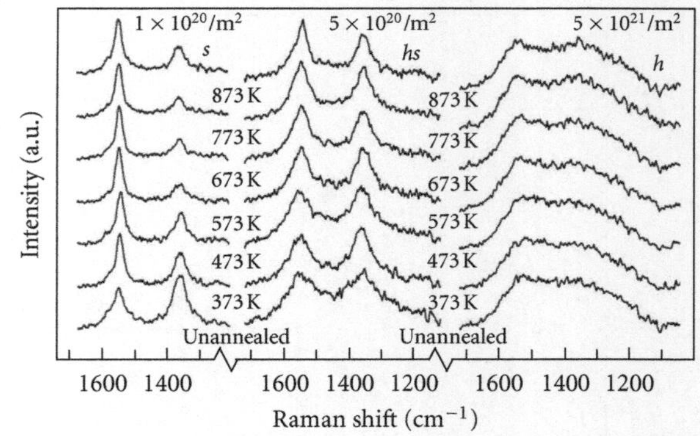

FIGURE 7: Changes in Raman spectra on isochronally annealing in steps of 100 K/1800 s for HOPG foils irradiated with 25 keV D^+ at RT to three different doses. The appearance of the TEM diffraction patterns of the foils before the annealing is denoted by *s* (spot), *hs* (halo and spot), and *h* (halo).

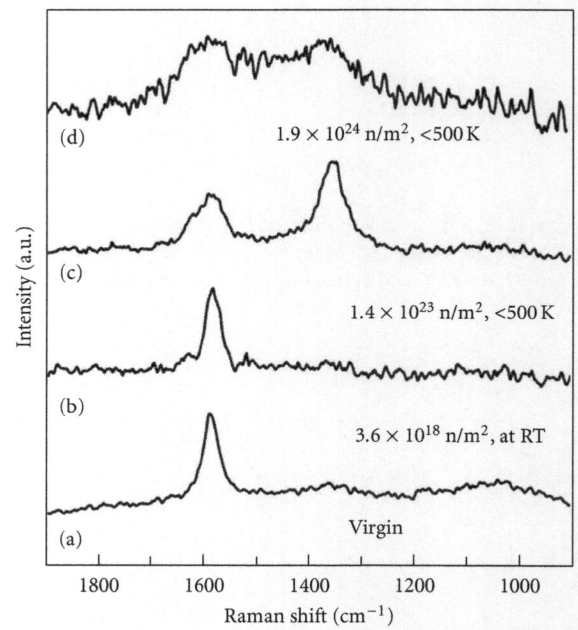

FIGURE 8: Raman spectra of carbon fibers of (a) unirradiated, and irradiated with (b) 14 MeV, 3.6 × 10^{18} n/m^2, (c) 1 MeV, 1.4 × 10^{23} n/m^2, and (d) 1 MeV, 1.9 × 10^{24} n/m^2.

size. Instead, it shows many fine contrasts in the lattice fringe which probably come from the production of in-plane defects. This means that the increase of I_{1355}/I_{1580} is not necessarily owing to the reduction of so-called "crystalline size La" but owing to the production of in-plane defects such as vacancies. Correspondingly, the significant reduction of the intensity of the damage peak at 1355 cm^{-1} for the unamorphized HOPG (s) on the annealing below 573 K in Figure 7 can be well explained by the mutual annihilation of vacancies with mobile interstitials.

The above considerations lead us to conclude that the increase of I_{1355}/I_{1580} in the early stage of the irradiation is originated from the irradiation produced in-plane defect such as vacancy itself. This approves the result by Nakamura and Kitajima, who have explained the change of I_{1355}/I_{1580} of phonon correlation length due to the increase of the concentration of irradiation-produced vacancies [13]. Turbulence and disordering of the basal plane, on the other hand, are clearly seen in the lattice image of the fiber, which shows the broad Raman spectrum in Figure 8. The turbulence and disordering should originate from some irregularity of bonding angle and length and cause the broadening of the Raman spectrum which hardly returns to the sharp peak on annealing below 873 K, as shown in Figure 7.

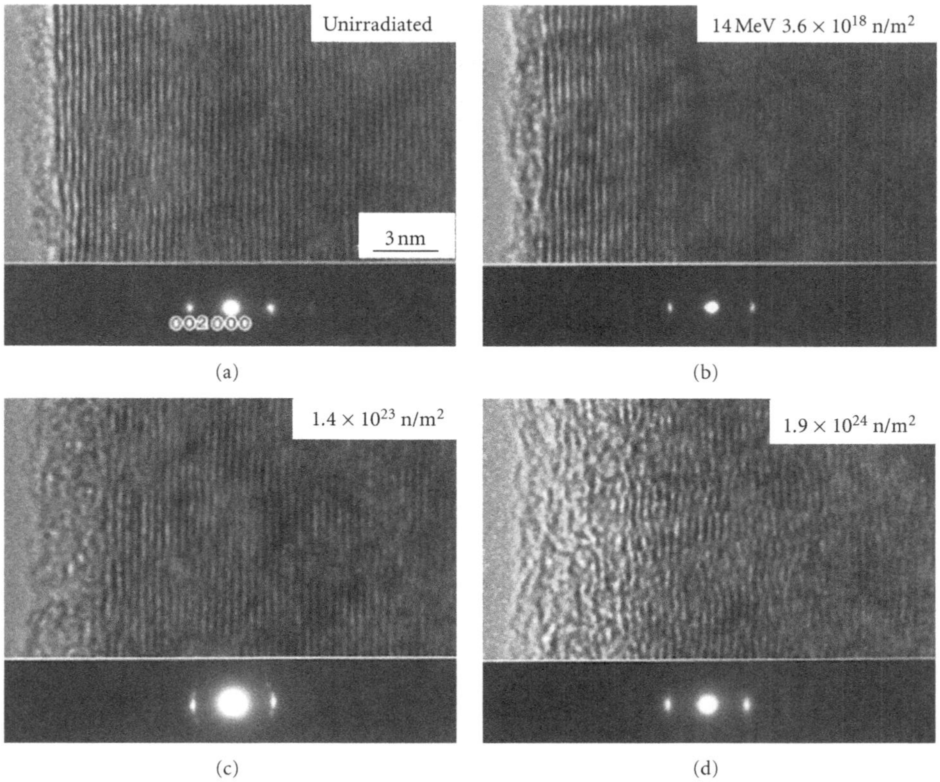

FIGURE 9: Lattice images and diffraction patterns in high-resolution TEM for carbon fibers of (a) unirradiated, and irradiated with (b) 3.6 $\times 10^{18}$ n/m^2, (c) 1.4×10^{23} n/m^2, and (d) 1.9×10^{24} n/m^2.

3. Model for the Change of Raman's Spectra and Irradiation-Induced Amorphization

3.1. Raman's Spectra and Irradiation-Induced Defects. The intensity ratio of I_{1355}/I_{1580} for various grades of graphite has been widely accepted to have an inverse proportionality against a "crystalline size La" [34]. However, we have found a significant reduction of the I_{1355}/I_{1580} for RT-irradiated HOPG on annealing below 573 K, which cannot be explained by the crystalline size change (Figure 7). Reduction of the intensity ratio on annealing only occurred for unamorphized specimens, which correspond to the ones indicated by the solid line in Figure 6. We have concluded that the change of the I_{1355}/I_{1580} under irradiation should not be attributed simply to the crystalline size change but to defects produced in the basal planes such as vacancies. Nakamura and Kitajima, on the other hand, performed a real-time measurement of the change of Raman spectra under He^+ irradiation and have found that the I_{1355}/I_{1580} is proportional to the square root of irradiation dose [13]. They successfully explained the result by the reduction of phonon correlation length due to vacancy formation, giving a relation between the Raman intensity ratio and the concentration of vacancy C_V, as follows:

$$C_V = k\left(\frac{I_{1355}}{I_{1580}}\right)^2, \quad (1)$$

where k is a constant. By replacing the C_V with $C_V{}' = k^{-1}C_V$, we get

$$C_V{}' = \left(\frac{I_{1355}}{I_{1580}}\right)^2. \quad (2)$$

Here, we define the in-plane defect concentration $C_V{}'$, which is expressed by the value of I_{1355}/I_{1580}, as the one in Raman intensity units (R.I. units), and use it hereafter.

Figure 10 shows a schematic drawing of the change in the peak width of 1580 cm^{-1} and the intensity ratio of I_{1355}/I_{1580} on irradiation, in which schematic pictures of graphite layers considered from Raman and HRTEM results are shown. The solid line can be assigned to the formation of in-plane defect such as vacancies, of which existence does not induce the turbulence of the basal planes. The deviation upward from the solid line in the relation between $FWHM_{1580}$ and I_{1355}/I_{1580} (Figure 6) and the appearance of turbulence of the basal plane (Figure 9), on the other hand, corresponds to the start of the amorphization. The values of the I_{1355}/I_{1580} at the deviation points decrease with increasing the irradiation temperature, which are about 1.4 below 373 K, about 1.0 at 473 K, and about 0.5 above 573 K. The decrease of the value should correspond to the decrease of the saturation concentration of vacancy. At temperatures lower than 473 K, the upward deviation from the solid line is seen to start at higher values of I_{1355}/I_{1580} compared to those above 573 K, as shown by the dotted lines. This is probably due to the less

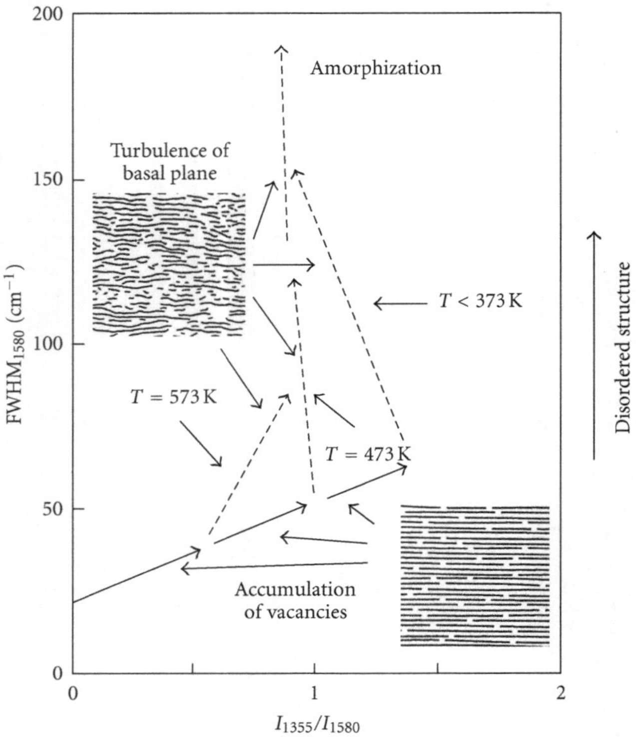

FIGURE 10: Schematic drawing of the change in peak width of 1580 cm^{-1} and intensity ratio of I_{1355}/I_{1580} on irradiation. Schematic pictures of graphite layers considered from Raman and HRTEM results are shown in the figure. The solid lines can be assigned to the formation of in-plane defects such as vacancies. The upward deviation from the solid line corresponds to the turbulence and disordering of the basal plane. At temperatures lower than 473 K, the upward deviation from the solid line is seen to start at higher values of I_{1355}/I_{1580} compared to those above 573 K, as shown by the dotted lines. This is probably due to the less annihilation of in-plane defects with mobile interstitials.

annihilation of vacancies with mobile interstitials at lower irradiation temperatures.

3.2. Semiempirical Model (Disordered-Region Model). The Raman intensity ratio of I_{1355}/I_{1580} and the peak width of FWHM_{1580} give different information on the formation of vacancies and disordered regions. The change of Raman spectra is now considered with the following assumptions.

(a) Single vacancies are saturated to the concentration of $C_{VS}(T)$ at an irradiation temperature T to a dose t_1.

(b) Graphite structural regions with single vacancies (G regions) gradually transform into disordered regions (D regions). The transformation rate is proportional to the saturated concentration of single vacancies $C_{VS}(T)$.

(c) The Raman intensity ratio (I_{1355}/I_{1580}) and the peak width of the G-peak (FWHM_{1580}) are expressed by the sum of Raman components from the G and D regions.

(d) The I_{1355}/I_{1580} has a square-root dependence on the concentration of vacancies and D regions. The FWHM_{1580} has a square-root dependence on the concentration of vacancies, but has a proportional dependence on the concentration of the D regions.

According to (a) and (b), the concentration of the D regions, C_D, at a dose of t_2 is given by

$$C_D(t_2) = k\, C_{VS}(T)(t_2 - t_1), \tag{3}$$

where k is a reaction constant. The value of C_D before the saturation of the in-plane defects is negligibly small after several repetition of t_1. From (c) and (d) the intensity ratio and the peak width of the 1580 cm^{-1} peak at t_2 are, respectively, expressed as

$$I_{1355}/I_{1580}(t_2) = C_V(t_2)^{1/2} + A\, C_D(t_2)^{1/2}, \tag{4}$$

$$\text{FWHM}_{1580}(t_2) = E + BC_V(t_2)^{1/2} + D\, C_D(t_2), \tag{5}$$

where A, B, and C are constant parameters and E is the peak width of virgin HOPG at room temperature (Figure 12). One should note that C_V and C_D are not expressed in atomic concentration, but in R.I. units, which are defined in the last section. From (3)–(5), the FWHM_{1580} is expressed as

$$\begin{aligned}\text{FWHM}_{1580}(t_2) &= E + B\left[\frac{I_{1355}}{I_{1580}}(t_2) - Ak^{1/2}C_{VS}(T)^{1/2}(t_2 - t_1)^{1/2}\right] \\ &\quad + DkC_{VS}(T)(t_2 - t_1).\end{aligned} \tag{6}$$

Now, in comparing the model with experimental results, it is clear that (6) cannot predict the changes of Raman spectra only by itself, as it contains three variable parameters of $\text{FWHM}_{1580}(t_2)$, $I_{1355}/I_{1580}(t_2)$, and $C_{VS}(T)$. Then we use the experimental results of I_{1355}/I_{1580} (Figure 4(a)) to give information on the changes of concentration of C_V and C_D under irradiation. All the constant parameters k, A, B, D, and E can be definitely obtained from the experimental results to be 0.2, 2/3, 30, 200, and 22, respectively. With these parameters and the experimental results of I_{1355}/I_{1580}, the disordered-region model can predict the dose and temperature dependencies of FWHM_{1580}, using the saturated concentration of single vacancies, $C_{VS}(T)$, as a fitting parameter at each irradiation temperature.

Figure 11 compares the experimental and predicted results of the dose dependence of the FWHM_{1580} at each irradiation temperature. Symbols represent the experimental results, while solid curves are the calculated results. The theoretical curves closely match the experimental results. The gradual increase of the FWHM_{1580} before amorphization and the remarkable increase after the onset of amorphization are well predicted. The relation between the FWHM_{1580} and

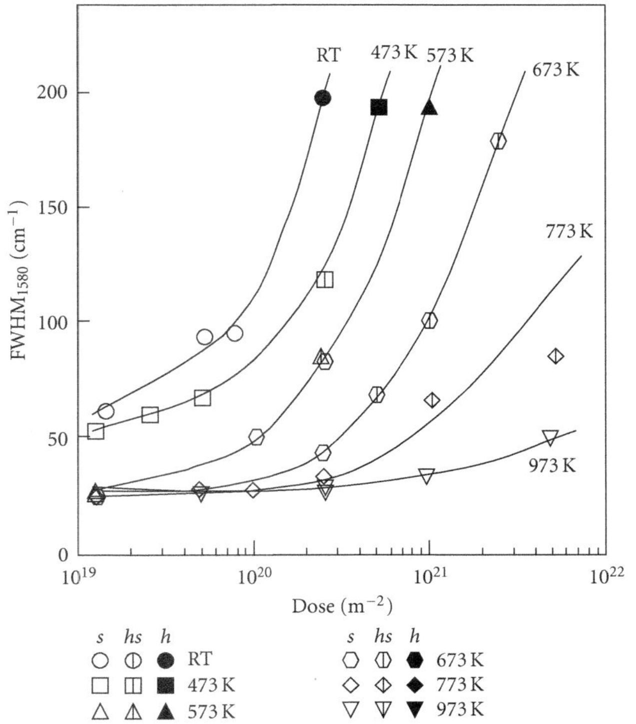

Figure 11: Comparison of the peak width of 1580 cm^{-1} calculated by the present model with the experimental result. The calculated values are shown as solid curves.

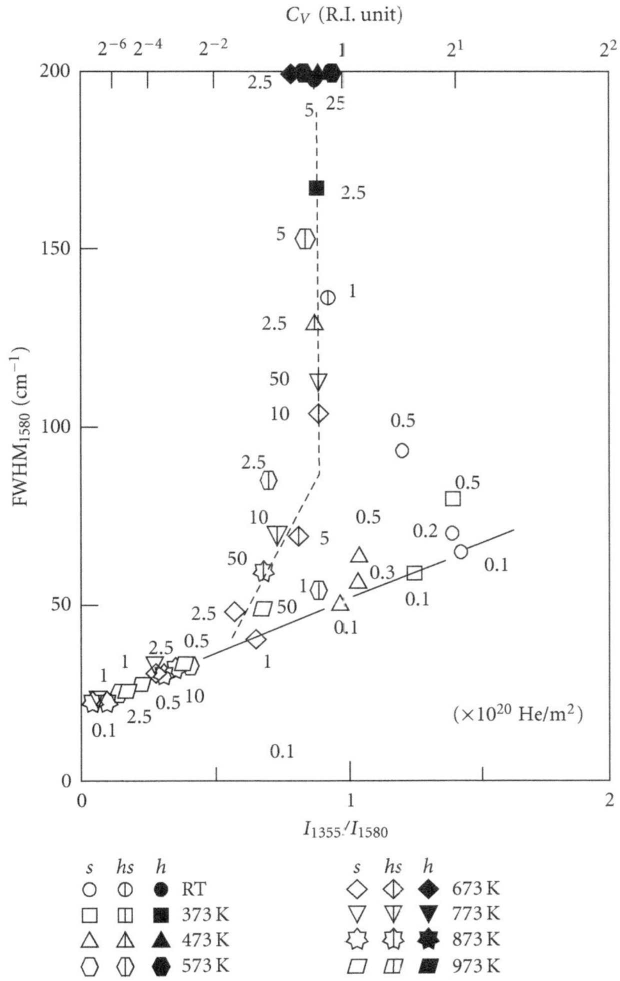

Figure 12: Relation between the peak width of 1580 cm^{-1} and the intensity ratio of I_{1355}/I_{1580} predicted by the present model. The value of C_V are expressed in R.I. units.

the I_{1355}/I_{1580} shown in Figure 6 is also well predicted by the model [18, 19].

In this disordered-region model, we can clearly define the critical dose of amorphization (t_c), that is, the dose at which the D regions reach a critical amount of A, thereby causing the disappearance of diffraction spots and the appearance of the halo in the TEM diffraction pattern of the irradiated graphite. Using (3) and neglecting the saturation time t_1 since $t_1 \ll t_c$, the amorphization condition can be expressed as

$$C_{VS}\, t_C = A', \tag{7}$$

where A' is $k^{-1}A$. This indicates that the critical dose of amorphization, t_C, is inversely proportional to the saturated concentration of single vacancies, C_{VS}.

Figure 13 shows the Arrhenius plot of the ${C_{VS}}^{-1}$, of which value is determined as the fitting parameter in the model calculation. We can find two linear relations, showing two activation energy of 0.083 eV below 573 K and 0.58 eV [19] above 573 K. It is interesting to compare the change of ${C_{VS}}^{-1}$ with the TEM diffraction pattern on the amorphization. Notably, two lines parallel to the Raman results can be drawn passing between the two points denoted by *hs* and *h* at each temperature, as shown in Figure 13. The consistency of the temperature dependencies of the critical dose of amorphization given by two different methods of Raman spectroscopy and TEM strongly supports the disordered-region model.

3.3. Kinetic Model (Dislocation Accumulation Model). The disordered-region model can predict the temperature dependence of the critical dose for amorphization and give an insight into the mobilities of point defects [18, 19], but it cannot predict the change in the defect concentration under irradiation nor give information on the structure of the disordered region. Then, we have advanced the "disordered region model" into a "dislocation accumulation model" by assigning the disordered region to a dislocation dipole [26].

The dislocation accumulation model [26] is a kinetic model, taking into account the point and defect clusters shown in Figure 14. The reaction kinetics can be summarized as follows. An incident atom knocks a host atom, resulting in the formation of an interstitial and a vacancy (Frenkel pair). The fraction of atomic sites at which the Frenkel pair production can take place gradually decreases with increasing the concentration of vacancy and dislocation dipoles. Single interstitials move two-dimensionally parallel to the basal

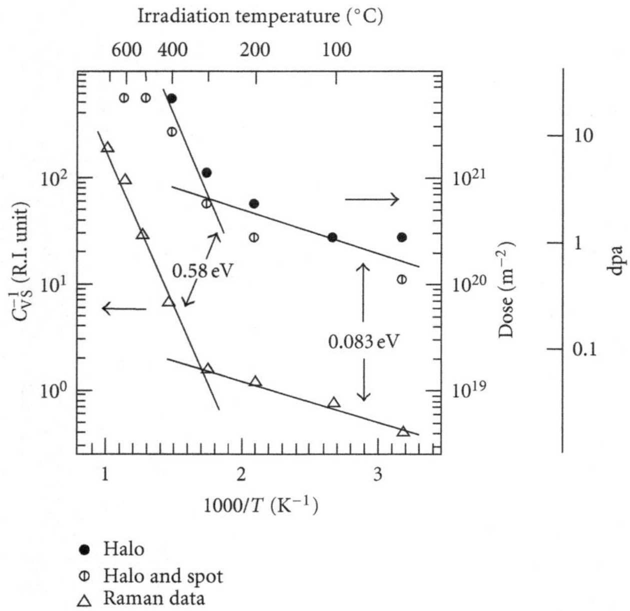

FIGURE 13: Arrhenius plot of ${C_{VS}}^{-1}$ determined by the fitting analysis. The values of ${C_{VS}}^{-1}$ are expressed in R.I. Units. This plot corresponds to the critical dose of amorphization as expressed by (7). Two linear relations are clearly seen, giving activation energy of 0.083 eV below 573 K and 0.58 eV above 573 K. The TEM results on the change of diffraction pattern from halo and spot to halo can be seen to give two similar linear relations.

plane. Other movements of interstitials and vacancies parallel to the basal plane and along the c-axis are neglected owing to the high activation energies. An interstitial in a recombination volume around a vacancy annihilates with the vacancy. A barrier exists for the mutual annihilation. A pair of interstitials is thermally stable once formed. We neglect the formation of interstitial clusters except for di-interstitials as we suppose that the interstitial clusters are not dominant for the total reaction kinetics at comparatively low temperatures. Divacancies formed by knock-ons, occurring at sites close to single vacancies, reconstruct to eliminate dangling bonds as shown in the left-hand side area of Figure 15. The reconstructed divacancies do not annihilate with interstitials and become the nucleus of the dislocation dipoles. Dislocation dipoles can grow by successive knock-ons close to the ends of dislocation dipoles as shown in the right-hand side area of Figure 15. The dislocation dipoles cannot annihilate with interstitials.

In Figure 16, the real-time change in the Raman intensity ratio of the original graphite peak and the defect peak, I_D/I_G, under ion irradiation at room temperature for two different production rates of the Frenkel pair 7.0×10^{-7} and 6.7×10^{-6} dpa/s [13] is compared with the dislocation accumulation model [26]. The experimental results of I_D/I_G exhibit a square-root dependence against irradiation dose. To show the square-root dependence, the vacancy concentration C_V should increase proportionally with the irradiation dose; that is, the rate of mutual annihilation of a vacancy and an interstitial should be negligibly small. By utilizing the proper value of $q = 5.0 \times 10^{-4}$, which is a recombination probability relating to the barrier for mutual annihilation [26], the calculated results can be fitted to the experimental results as shown in Figure 16. The downward deviation from the square-root dose dependence for the upper result, which comes from the effect of mutual annihilation of interstitial and vacancy, can be simulated in the calculation. The dose range of Raman measurements done by Nakamura and Kitajima [13] corresponds to the initial stage where C_V increases proportionally to the irradiation time; that is, $C_V = Pt$. This is the reason why Raman intensity ratio exhibited the square-root dependence. To obtain the linear relationship, a very small value of the recombination probability between a vacancy and an interstitial, that is, a strong barrier for the mutual annihilation, is needed.

The existence of the strong barrier for the mutual annihilation was supported by another simulation [26] for the dose dependence of the c-axis expansion given by Koike and Pedraza [17] with the dislocation accumulation model. Figure 17 compares the dimensional change in highly oriented pyrolytic graphite (HOPG) along the c-axis during 300 keV electron irradiation at room temperature with the results calculated by the present theory using several values of q. The dimensional change along the c-axis, $\Delta L_c/L_{co}$, for the calculations is assumed to increase proportionally with the accumulation of di-interstitials. J is a parameter representing the expansion rate per atom by the formation of di-interstitials. The agreement is quite acceptable with the values of $q = 5.0 \times 10^{-4}$ and $J = 4.2$ for the two stages of the initial fast elongation up to a dose of 0.2 dpa and the subsequent linear elongation up to 4.0 dpa. The value of the expansion rate $J = 4.2$ is acceptable in its order of magnitude compared to the case of no barrier ($q = 1$) where J has to take an unacceptable high value of 10^3 to accord with the experimental result of the dimensional change. Nevertheless, $J = 4.2$ is larger than the value $J = 1$ expected for the case when di-interstitials do not induce excess expansion. This may occur in cases where the di-interstitials effectively expand the lattice or dislocation dipoles induce buckling of the basal planes. One should note that the saturation of vacancy concentration assumed in the disordered-region model [18, 19] is modified to the peak formation of vacancy concentration in the dislocation accumulation model [26].

Utilizing (1) with a proper parameter of k, we can quantitatively deduce the concentration of vacancy C_V from Raman intensity ratio in the case where most of defects are single vacancies. Previously, we estimated the value of k as 0.0016 based on the in situ Raman measurement for 3 keV He^+ irradiation [13], where Raman intensity was corrected with the observed Raman intensity, as the damaged region is shallower than the region where Raman scattering is detectable. To obtain the precise value of k, it should be determined in the case where single vacancies are homogeneously distributed in the Raman detectable region. In this sense, the 25 keV He^+ irradiation should be better than the 3 keV He^+ irradiation for the determination of k, because all Raman scattering comes from the damage region.

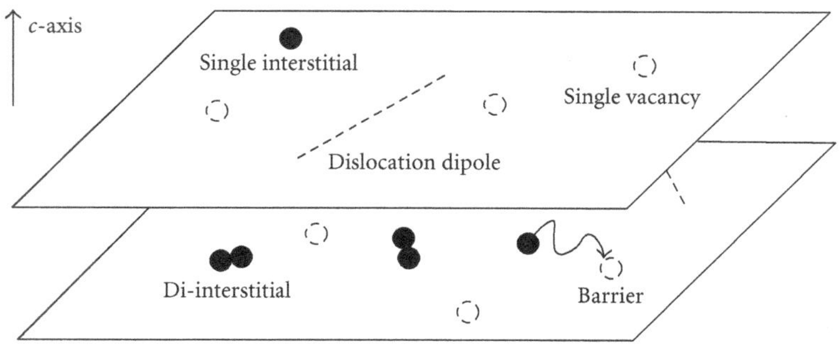

FIGURE 14: Point defects, defect clusters, and their reactions assumed in the dislocation dipole accumulation model. A recombination barrier exists between an interstitial and vacancy.

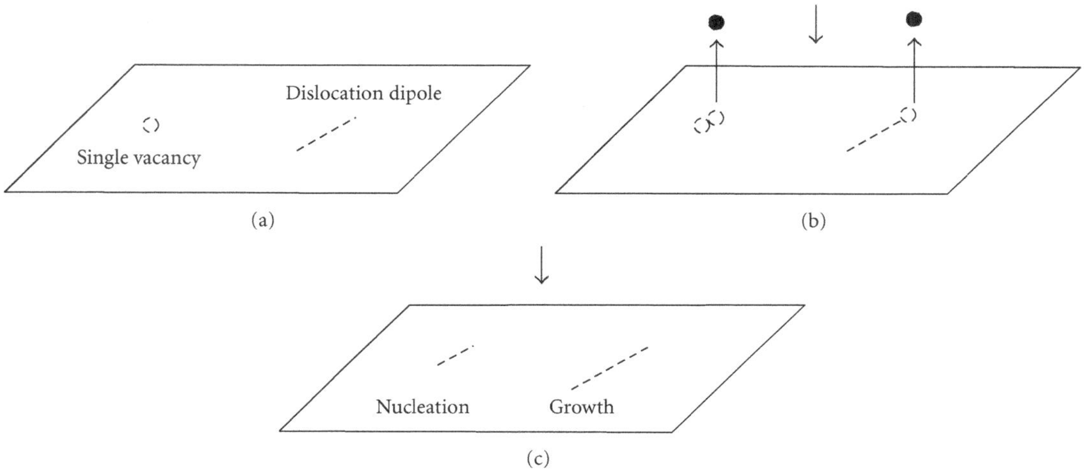

FIGURE 15: Nucleation and growth of dislocation dipole. Divacancies formed by knock-ons, occurring at sites close to single vacancies, reconstruct to eliminate dangling bonds as shown in the left-hand side area. The reconstructed divacancies are the nucleus of the dislocation dipoles. Dislocation dipoles can grow by successive knock-ons close to the ends of dislocation dipoles as shown in the right-hand side area.

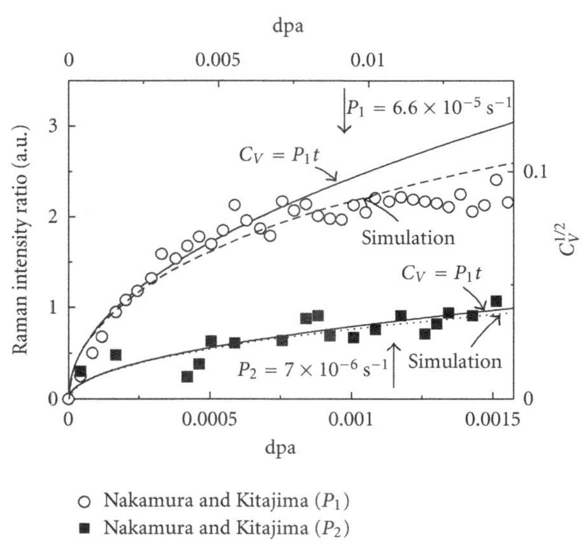

FIGURE 16: The real-time changes of Raman intensity ratio for 3 keV Ar^+ ion irradiation at room temperature at fluxes of 7.0×10^{-7} and 6.0×10^{-6} dpa s^{-1} [13].

Taking the Raman intensity ratio of 1.5 and dpa of 0.05 at the dose of 0.13×10^{20} He/m^2 for the room temperature irradiation and assuming that all the vacancies formed by irradiation remained in the specimen, we obtained the value of k as 0.022. Then (1) can be expressed as

$$C_V = 0.022 \left(\frac{I_{1355}}{I_{1580}}\right)^2. \tag{8}$$

Validity of this equation would be discussed elsewhere.

3.4. Nucleation and Growth of Dislocation Dipole. Dislocation accumulation model assumes the nuclei of dislocation dipoles are divacancies and they grow by successive knock-ons close to the ends as shown in Figure 15. However, the model does not give any information on the structure of dislocation dipoles as simply shown by broken lines in Figures 14 and 15. So, here, we discuss the formation process of dislocation dipoles.

Figures 18(a) and 18(b) show two possibilities of the nucleation process of dislocation dipole, that is, the reconstruction of divacancies. The left one transforms to two

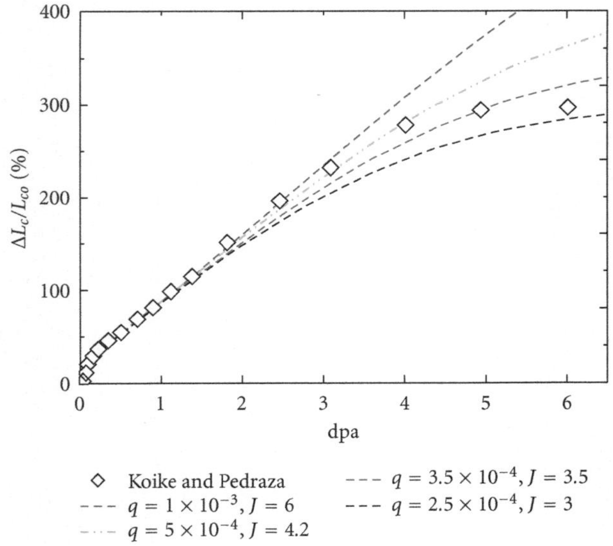

Figure 17: The dimensional change along the c-axis under 300 keV electron irradiation at room temperature [17] and the simulation with the dislocation accumulation model with several values of q. The dimensional change is assumed to increase proportionally with the amount of di-interstitials accumulated between the basal planes. The values of J correspond to the c-axis expansion ratio per atom of di-interstitials.

eightfold rings, sitting opposite to each other, and the right one to eightfold ring and two fivefold rings. The right one should be energetically favorable [27]. If knock-on occurs at sites denoted by dark spots in Figure 18(b), both of them would grow to the same structure consisting of eight-, seven-, and fivefold rings as shown in Figure 18(c). Successive knock-ons at the ends of the dislocation dipoles induce their growth along the zigzag line.

A long dislocation dipole generated from nine vacancies (Figure 19(a)) is shown in Figure 19(b), where a vacancy line reconstructs to a dislocation dipole to make an energetically favorable structure by eliminating dangling bonds. In the reconstructed structure of Figure 19(b), one dangling bond remains at a site denoted by a circle. The dislocation dipole can grow by knock-on at the ends of dislocation dipole denoted by symbols in Figure 19(b). Neighboring eight- and sixfold rings at the left-hand side end of the dislocation dipole in Figure 19(b) transforms to neighboring seven- and fivefold rings in Figure 19(c) after a knock-on of an atom denoted by a circle. On the other hand, the neighboring seven- and fivefold rings at the right-hand side end of the dislocation dipole in Figure 19(b) can grow into the neighboring eight- and sixfold rings in Figure 19(c) after a knock-on of an atom denoted by a triangle. The two processes can repeat alternately, leading the growth of the dislocation dipole. The contraction in the basal plane along the direction denoted by the broken line in Figure 19(a) occurs. We should note that the dislocation dipole grows just at the ends where reconstruction can occur.

3.5. Accumulation of Dislocation Dipole and Broadening of Raman Spectra. The dislocation accumulation model depicts the following picture of the disordering process of graphite under irradiation, taking into account the accumulation of dislocation dipoles. In the dose range before approaching the peak of vacancy concentration, the long-range order of the graphite sheet along the c-axis and the sequence of the 6-member rings in the basal planes remain as the dislocation dipole formation is negligibly small. With increasing dose, the formation of dislocation dipoles by knock-on process becomes significant due to an increase of vacancy concentration. The accumulation of dislocation dipoles destroys the topological long-range order in the basal plane by introducing, rotation, and buckling of the graphite sheets, leading to the fragmentation of the basal planes. In the early stage of this process, the irradiation-induced disordered structure would still reflect the original layered structure of HOPG. In fact, in spite of the TEM diffraction pattern of graphite observed along the c-axis, which shows a halo when irradiated to a dose of about 1 dpa, the sequence of the basal planes still remains in the HREM images although the lattice image is broken up into small segments [12, 17, 30, 33]. The disordered graphite in this stage can be used as a raw material to generate amorphous diamond by shock compression [35]. In the highly accumulated state of the dislocation dipoles, nanocrystallization should become considerable and then the long-range order of the layered structure would disappear as shown by the HREM images of HOPG [17]. The formation of dislocation dipoles may play an important role in the generation of the concentric graphitic structure of "bucky onions" under electron irradiation [36], because the formation of the dislocation dipoles induces curvature of the graphitic layers.

One should remember that the remarkable broadening of Raman spectrum appears when disordering occurs under irradiation. This should be due to the life time broadening of phonons originated in the lattice distortion around the dislocation dipole. In contrary to this, the remarkable broadening does not occur in the initial stage of irradiation where single vacancies are dominantly formed. This suggests that the lattice distortion around single vacancy is not so strong to induce the remarkable life time broadening.

4. Conclusion

Raman spectral changes and associated TEM diffraction changes of highly oriented pyrolytic graphite (HOPG) under 25 keV He^+ and 20 keV D^+ irradiation have been systematically investigated in terms of irradiation dose and temperature. Remarkable broadening of Raman peaks and appearance of halo rings in the TEM diffraction pattern indicate amorphization of the irradiated specimens after a critical dose. To explain the change of Raman spectra in terms of defect formation, we proposed "disordered-region model" at first and then advanced the model into "dislocation accumulation model." These model calculations gave us important information on the critical dose of amorphization, a close relation between Raman intensity ratio and the vacancy concentration, an existence of barrier

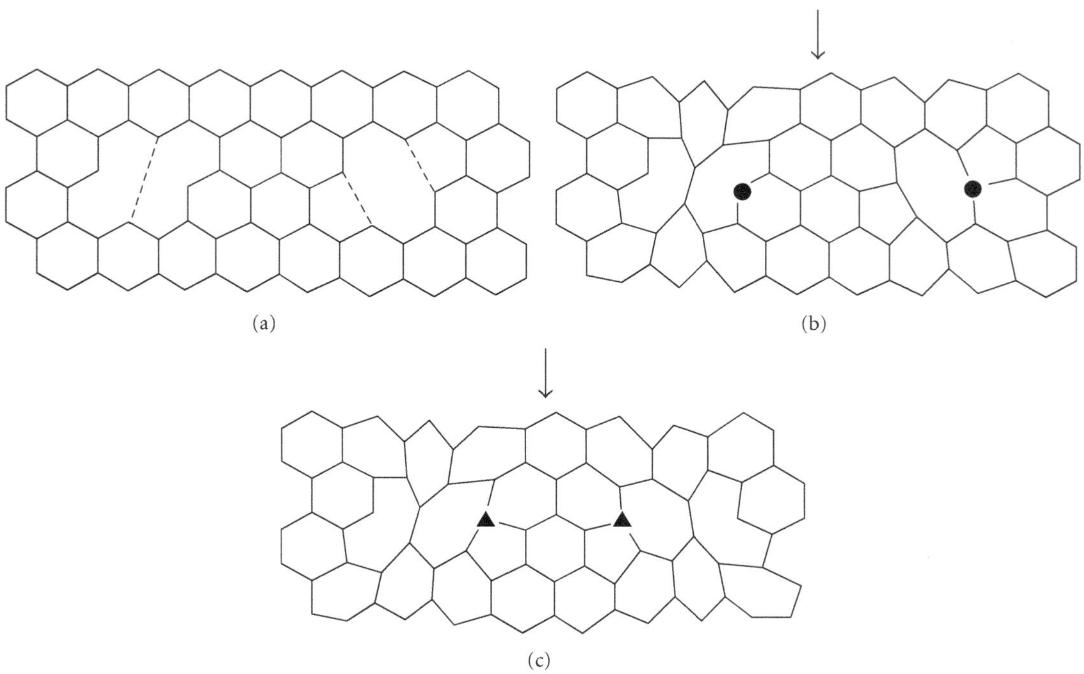

FIGURE 18: A possible nucleation process of dislocation dipole by the reconstruction of di-vacancy. The left one transforms to two eightfold rings, sitting opposite to each other, and the right one to eightfold ring and two fivefold rings. If knock-on occurs at sites denoted by circles in (b), both of them can transform to the same structure consisting of eightfold, sevenfold, and fivefold rings as shown in (c).

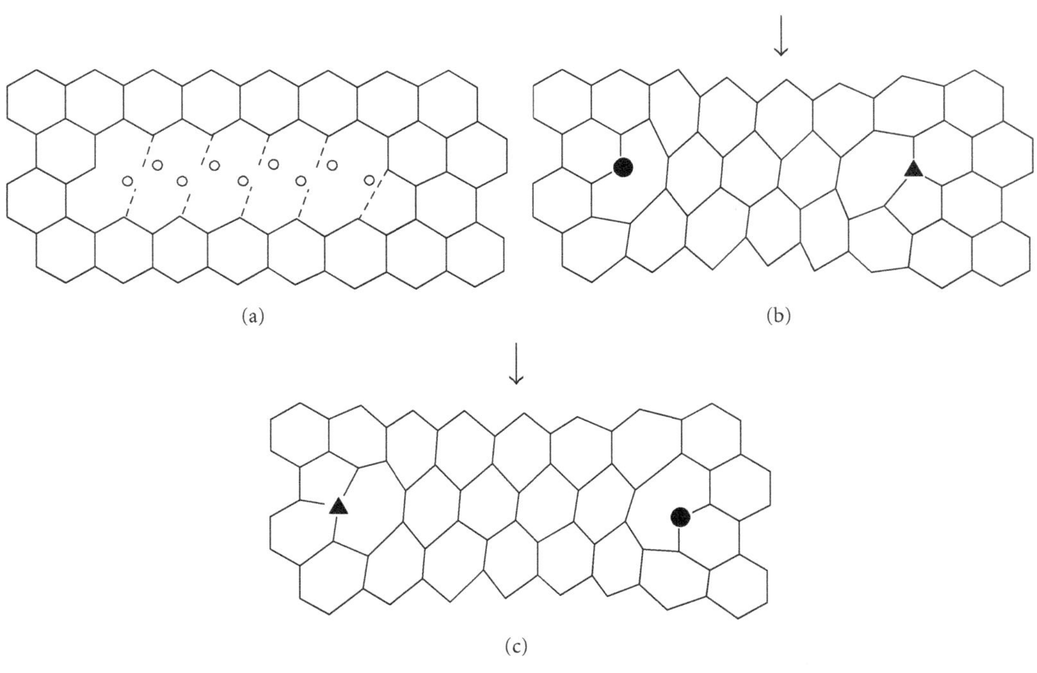

FIGURE 19: A long dislocation dipole generated from 9 vacancies (light spots). The dislocation dipole can grow by knock-on at the ends of dislocation dipole denoted by symbols in (b). Neighboring eightfold and sixfold rings at the left-hand side end of the dislocation dipole in (b) transforms to neighboring sevenfold and fivefold rings in (c) after a knock-on of an atom denoted by a circle. The neighboring sevenfold and fivefold rings can grow into the neighboring eightfold and sixfold rings as shown at the right-hand side area by knocking-on the atom denoted by a triangle. The contraction in the basal plane along the directions denoted by the broken lines in (a) occurs.

on the mutual annihilation, and so forth. Thus, Raman spectroscopy can be a powerful tool for the analysis of point defect and defect clusters in irradiated graphite.

Acknowledgments

The author would like to thank many collaborators shown in the following references. This work was supported by the JSPS Grant-in-Aid (no. 23560787).

References

[1] L. Arnold, *Windscale, 1957: Anatomy of a Nuclear Accident*, Palgrave Macmillan, London, UK, 1995.

[2] W. N. Reynolds, *Radiation Damage in Graphite*, Pergamon, New York, NY, USA, 1965.

[3] P. A. Thrower, *Chemistry and Physics of Carbon*, vol. 5 of *edited by P. L. Walker*, Dekker, New York, NY, USA, 1969.

[4] P. A. Thrower and R. M. Mayer, "Point defects and self-diffusion in graphite," *Physica Status Solidi A*, vol. 47, no. 1, pp. 11–37, 1978.

[5] B. T. Kelly, *Physics of Graphite*, Applied Science, London, UK, 1981.

[6] B. S. Elman, M. S. Dresselhaus, G. Dresselhaus, E. W. Maby, and H. Mazurek, "Raman scattering from ion-implanted graphite," *Physical Review B*, vol. 24, no. 2, pp. 1027–1034, 1981.

[7] T. Iwata, "Fine structure of Wigner energy release spectrum in neutron irradiated graphite," *Journal of Nuclear Materials*, vol. 133-134, pp. 361–364, 1985.

[8] K. Niwase, M. Sugimoto, T. Tanabe, and F. E. Fujita, "Electron microscope study of radiation damage in graphite produced by D^+ and He^+ bombardment," *Journal of Nuclear Materials*, vol. 155–157, no. 1, pp. 303–306, 1988.

[9] K. Niwase, K. Nakamura, I. Tanaka, Y. Miyamoto, and T. Tanabe, "Laser Raman microprobe analysis of the modified surface of He^+-irradiated graphite," *Journal of Nuclear Materials*, vol. 179–181, no. 1, pp. 214–217, 1991.

[10] K. Niwase, T. Tanabe, and I. Tanaka, "Annealing experiment of ion-irradiated graphite by laser Raman spectroscopy," *Journal of Nuclear Materials*, vol. 191–194, pp. 335–339, 1992.

[11] K. Nakai, C. Kinoshita, and A. Matsunaga, "A study of amorphization and microstructural evolution of graphite under electron or ion irradiation," *Ultramicroscopy*, vol. 39, no. 1–4, pp. 361–368, 1991.

[12] T. Tanabe, S. Muto, and K. Niwase, "On the mechanism of dimensional change of neutron irradiated graphite," *Applied Physics Letters*, vol. 61, no. 14, pp. 1638–1640, 1992.

[13] K. Nakamura and M. Kitajima, "Ion-irradiation effects on the phonon correlation length of graphite studied by Raman spectroscopy," *Physical Review B*, vol. 45, no. 1, pp. 78–82, 1992.

[14] M. S. Dresselhaus and R. Kalish, *Ion Implantation in Diamond, Graphite and Related Materials*, Springer, Berlin, Germany, 1992.

[15] K. Niwase and T. Tanabe, "Defect structure and amorphization of graphite irradiated by D^+ and He^+," *Materials Transactions*, vol. 34, no. 11, pp. 1111–1121, 1993.

[16] B. T. Kelly, *Materials Science and Technology*, vol. 10, VCH, New York, NY, USA, 1994.

[17] J. Koike and D. F. Pedraza, "Dimensional changes in highly oriented pyrolytic graphite due to electron-irradiation," *Journal of Materials Research*, vol. 9, no. 7, pp. 1899–1907, 1994.

[18] K. Niwase, "Irradiation-induced amorphization of graphite," *Physical Review B*, vol. 52, no. 22, pp. 15785–15798, 1995.

[19] K. Niwase, "Erratum: irradiation-induced amorphization of graphite," *Physical Review B*, vol. 56, no. 9, pp. 5685–5685, 1997.

[20] T. Tanabe, T. Maruyama, M. Iseki, K. Niwase, and H. Atsumi, "Radiation damage of graphite: degradation of material parameters and defect structures," *Fusion Engineering and Design*, vol. 29, pp. 428–434, 1995.

[21] T. Tanabe, "Radiation damage of graphite—degradation of material parameters and defect structures," *Physica Scripta T*, vol. 64, pp. 7–16, 1996.

[22] H. Abe, H. Naramoto, A. Iwase, and C. Kinoshita, "Effect of damage cascades on the irradiation-induced amorphization in graphite," *Nuclear Instruments and Methods in Physics Research, Section B*, vol. 127-128, pp. 681–684, 1997.

[23] S. Muto and T. Tanabe, "Damage process in electron-irradiated graphite studied by transmission electron microscopy. I. High-resolution observation of highly graphitized carbon fibre," *Philosophical Magazine A*, vol. 76, no. 3, pp. 679–690, 1997.

[24] S. Muto, S. Horiuchi, and T. Tanabe, "Local structural order in electron-irradiated graphite studied by high-resolution high-voltage electron microscopy," *Journal of Electron Microscopy*, vol. 48, no. 6, pp. 767–776, 1999.

[25] T. Meguro, A. Hida, M. Suzuki et al., "Nanoscale modification of electronic states of graphite by highly charged Ar-ion irradiation," *Journal of Vacuum Science and Technology B*, vol. 19, no. 6, pp. 2745–2748, 2001.

[26] K. Niwase, "Irradiation-induced amorphization of graphite: a dislocation accumulation model," *Philosophical Magazine Letters*, vol. 82, no. 7, pp. 401–408, 2002.

[27] R. H. Telling, C. P. Ewels, A. A. El-Barbary, and M. I. Heggie, "Wigner defects bridge the graphite gap," *Nature Materials*, vol. 2, no. 5, pp. 333–337, 2003.

[28] A. Hashimoto, K. Suenaga, A. Gloter, K. Urita, and S. Iijima, "Direct evidence for atomic defects in graphene layers," *Nature*, vol. 430, no. 7002, pp. 870–873, 2004.

[29] K. Niwase, "Formation of dislocation dipoles in irradiated graphite," *Materials Science and Engineering A*, vol. 400-401, no. 1-2, pp. 101–104, 2005.

[30] A. Asthana, Y. Matsui, M. Yasuda, K. Kimoto, T. Iwata, and K. I. Ohshima, "Investigations on the structural disordering of neutron-irradiated highly oriented pyrolytic graphite by X-ray diffraction and electron microscopy," *Journal of Applied Crystallography*, vol. 38, no. 2, pp. 361–367, 2005.

[31] T. Hirai, J. Compan, K. Niwase, and J. Linke, "Laser Raman microprobe analysis of graphite exposed to edge plasma in the TEXTOR tokamak," *Journal of Nuclear Materials*, vol. 373, no. 1–3, pp. 119–122, 2008.

[32] T. Hirai, J. Compan, and K. Niwase, "Micro-Raman study on structure evolution of graphite exposed to intense thermal shock load," *Advanced Materials Research*, vol. 59, pp. 66–70, 2009.

[33] C. Karthik, J. Kane, D. P. Butt, W. E. Windes, and R. Ubic, "In situ transmission electron microscopy of electron-beam induced damage process in nuclear grade graphite," *Journal of Nuclear Materials*, vol. 412, no. 3, pp. 321–326, 2011.

[34] F. Tuinstra and J. L. Koenig, "Raman spectroscopy of graphite," *Journal of Chemical Physics*, vol. 53, no. 3, pp. 1126–1130, 1970.

[35] K. Niwase, K. G. Nakamura, M. Yokoo, K. I. Kondo, and T. Iwata, "Pathway for the transformation from highly oriented pyrolytic graphite into amorphous diamond," *Physical Review Letters*, vol. 102, no. 11, Article ID 116803, 2009.

[36] D. Ugarte, "Formation mechanism of quasi-spherical carbon particles induced by electron bombardment," *Chemical Physics Letters*, vol. 207, no. 4–6, pp. 473–479, 1993.

7

Study of the Halogen Bonding between Pyridine and Perfluoroalkyl Iodide in Solution Phase Using the Combination of FTIR and ^{19}F NMR

Briauna Hawthorne,[1] Haiyan Fan-Hagenstein,[1] Elizabeth Wood,[2] Jessica Smith,[2] and Timothy Hanks[2]

[1] *Chemistry Department, Claflin University, Orangeburg, SC 29115, USA*
[2] *Chemistry Department, Furman University, Greenville, SC 29613, USA*

Correspondence should be addressed to Haiyan Fan-Hagenstein; hfan-hagenstein@claflin.edu

Academic Editor: Rolf W. Berg

Halogen bonding between pyridine and heptafluoro-2-iodopropane (iso-C_3F_7I)/heptafluoro-1-iodopropane (1-C_3F_7I) was studied using a combination of FTIR and ^{19}F NMR. The ring breathing vibration of pyridine underwent a blue shift upon the formation of halogen bonds with both iso-C_3F_7I and 1-C_3F_7I. The magnitudes of the shifts and the equilibrium constants for the halogen-bonded complex formation were found to depend not only on the structure of the halocarbon, but also on the solvent. The halogen bond also affected the C_α-F (C-F bond on the center carbon) bending and stretching vibrations in iso-C_3F_7I. These spectroscopic effects show some solvent dependence, but more importantly, they suggest the possibility of intermolecular halogen bonding among iso-C_3F_7I molecules. The systems were also examined by ^{19}F NMR in various solvents (cyclohexane, hexane, chloroform, acetone, and acetonitrile). NMR dilution experiments support the existence of the intermolecular self-halogen bonding in both iso-C_3F_7I and 1-C_3F_7I. The binding constants for the pyridine/perfluoroalkyl iodide halogen bonding complexes formed in various solvents were obtained through NMR titration experiments. Quantum chemical calculations were used to support the FTIR and ^{19}F NMR observations.

1. Introduction

Halogen bonding, a noncovalent interaction between a halogen atom acting as an electron acceptor and an electron-rich Lewis base, has been known for nearly 150 years [1–6]. However, only recently has halogen bond started to attract wide-spread attention among scientists in a diversity of fields. Metrangolo and coworkers have reviewed the basic concepts as well as the major applications of the halogen bond [7, 8]. Examples include a wide range of applications in separation science, synthesis of liquid crystals and electronic materials, and the assembly of functional super molecules [9–18]. Recently, Meyer and Dubois [19] highlighted the application of halogen bonding to the synthesis of functional materials such as liquid crystal, nonlinear optical, magnetic conducting material, and halogen bonding based surface modification. A review by Beale et al. [20] provides an informative summary of solution thermodynamics and applications of halogen bonding. Of particular importance are their presentations of detailed thermodynamic parameters measured for the halogen bonding systems made of organic donors and halogen anions in the solution phase and which emphasized their importance in the anion recognition. Erdélyi [21] reviewed the nature of halogen bonding in solution phase and summarized the techniques that have been used to study the halogen bond in solution.

Using a jet cooled molecular beam, combined with the Fourier-transform microwave spectroscopy, Legon has investigated the geometry and the stability of some halogen bonding complexes in the gas phase through the analysis of their rotational spectra [22, 23]. This research has allowed the detailed comparison of halogen and hydrogen bonding. It was pointed out that halogen bonds can be as strong as or slightly weaker than hydrogen bonds, depending on the structures of the participating component molecules. While hydrogen bonds can adopt a variety of angles, halogen bonds are mostly linear relative to the halogen, the donor atom, and their covalently bonded substrates. This is due to the large electrostatic anisotropy of the halogen, which is highly localized along the extension of the carbon-halogen bond.

Halogen bond was first recognized in dihalide-containing complexes, but it is now understood that other halogen containing species can also act as electron acceptors [24–33]. When the halogens are bound to an electron-withdrawing moiety, the positive potential on the halogen atom is enhanced and the potential for halogen bonding is increased. In addition, halogen bonding ability increases with increasing the size of the halide because of the increase in the polarizability. Thus, iodine forms the strongest halogen bonds. For these reasons, perfluoroalkyl iodides have been often chosen to be the halogen donors in most halogen bonding studies. Fluorinated organoiodides have several scientific and industrial uses, including pharmaceuticals and nanoscience applications [34–39]. For example, halogen bond between perfluorobenezene iodide and pyridine containing molecules has recently been used to assemble gold nanoparticles [40].

As it is experimentally challenging to directly measure the strength of the halogen bond in a particular system, most research in the field has been focused on the geometry and structural impact. Like hydrogen bond [41, 42], halogen bond often leads to observable changes in the vibrational spectra of the participating molecules. Focusing on the halogen bond impact on the perfluoroalkyl iodide, Messina et al. [43] led the study of the systems between α, ω-diiodoperfluoroalkanes, and other electron donors using FTIR and Raman spectroscopy. According to the noisy light Coherent Anti-Stokes Raman Scattering Spectroscopy ($I^{(2)}$CARS) [44], halogen bond between pyridine and perfluoroalkyl iodide causes the ring breathing vibration of pyridine to be blue shifted. The magnitude of the shift was comparable with that observed in the hydrogen bonding system between pyridine and water. Using FTIR and Raman spectroscopy, Van der Veken et al. [45–48] have studied the effect of halogen bond between CF_3X (X = Cl, Br, I) and various electron lone pair donors on the vibrations of the participants. The complexation enthalpies were determined based on the temperature dependence of the bond association constants. An *ab initio* calculation at the MP2/aug-cc-pVDZ(-PP) level was also performed to be compared with the spectroscopic observation. In addition to Raman and FTIR spectroscopy, ^{19}F NMR is another sensitive method to detect the halogen bond between perfluoroalkyl halides (PFC-iodide or bromide) and Lewis bases [49–52]. The change in the chemical shift upon halogen bonding was found to be the greatest on the alpha fluorine signals. Thus, both vibrational spectroscopy and ^{19}F NMR are useful tools for the investigation of halogen bond in solution.

Considering that halogen bond is one of the strongest noncovalent intermolecular forces, the thermodynamics in the halogen bonding formation have also been received attention. Fan and coworkers [44] studied the pyridine and perfluoroalkyl iodides. Their observations led to a hypothesis that there existed some self-intermolecular halogen bond among iso-C_3F_7I molecules in the liquid phase, while this self-intermolecular halogen bond did not exist or at least was very weak for 1-C_4F_9I and 1-$C_6F_{13}I$.

It has been long believed that the magnitude of the vibrational frequency shift in the hydrogen bonded complex is proportional to the strength of the hydrogen-bond [53]. In other words, the magnitude of the vibrational frequency shift is correlated with the complexation enthalpies. It is reasonable to apply the same principle to the halogen bond. Therefore, FTIR results provide information about enthalpy change in the halogen bond formation. On the other hand, ^{19}F NMR titration experiments provide information about binding constants and thereby changes in the Gibb's free energy during halogen bond formation. As a result, the comparison between the FTIR and ^{19}F NMR results makes it possible to derive the contribution of entropy to the halogen bond formation.

Using the combination of ^{19}F NMR and theoretical calculations, Sarwar and coworkers [52] examined the thermodynamics of halogen bonding in iodoperfluoroarenes involved solution system with respect to substituent, structural, and solvent effects. The experimental results were compared with the density functional theory (DFT) calculations [52] in order to be able to model the supermolecular interactions for drug design.

Whether in a liquid phase binary system or in solution, halogen bonding may play very important roles in defining the structure of the liquid and solution. Inspired by previous work, the present paper seeks to determine if halogen bond exists among molecules of perfluoroalkyl iodides. Based on the hypothesis by Fan et al. [44], the intermolecular halogen bond among iso-C_3F_7I molecules, a secondary iodide, should be stronger than those formed among molecules of primary iodides (e.g., 1-C_4F_9I, 1-$C_6F_{13}I$). In order to test this argument, the present work uses a combination of FTIR and ^{19}F NMR to quantify the halogen bonding between pyridine and iso-C_3F_7I/1-C_3F_7I in various solvents. First, the strength of halogen bond is compared between pyridine/iso-C_3F_7I and pyridine/1-C_3F_7I. Second, for the same halogen bonding system, the strength is compared in different solvents. Third, a series of dilution experiments were performed using both FTIR and ^{19}F-NMR to gather evidence of intermolecular self-halogen bond. Finally, the thermodynamics of the halogen bond for the system between pyridine and perfluoroalkyl iodides in solution phase was studied using the combination of the results from FTIR and ^{19}F NMR measurements.

2. Experimental Section

2.1. Chemicals and Regents. All reagents and solvents were purchased from commercial houses. All solvents were purchased as anhydrous or used immediately after dispensing from a solvent purification system.

2.2. FTIR and ^{19}F NMR Measurement. FTIR spectra were recorded on a PerkinElmer Paragon 500 FTIR spectrometer. All the measurements were carried out at the room temperature, and the corresponding solvents were used to record the background spectra. A liquid sample holder with a NaCl windows was used, and the resolution of the FTIR spectrometer was set at $1\,cm^{-1}$ during experiments. ^{19}F NMR spectra were recorded on a Varian 400MR 400 MHz High-Field Superconducting NMR spectrometer. 1.00 M KF solution in D_2O was used in each measurement as an external reference for ^{19}F NMR.

2.3. Theoretical Calculations. All computations were performed with the Gaussian 03 or Gaussian 09 [54, 55] suite of programs. The interaction energies and geometrical parameters in solution phase were computed using the DFT method with the functional M052X and the aug-cc-pVDXZ basis set for all atoms except iodine, where a pseudopotential was required (aug-cc-pVDXZ-PP) [56]. Geometry optimizations for monomer and dimer of iso-C_3F_7I, 1-C_3F_7I, as well as the hydrogen bonding complex between iso-C_3F_7I/1-C_3F_7I and $CHCl_3$ in vacuum phase were performed with density function theory (DFT) using 3-21G-B3LYP, CalcAll, and the tight convergence criteria. The calculations of ^{19}F NMR were performed using the optimized geometry.

3. Result and Discussion

3.1. Comparison of Halogen Bonding Strength and Solvent Effect between Pyridine/Iso-C_3F_7I and Pyridine/1-C_3F_7I. According to previous work using $I^{(2)}$CARS, halogen bond between pyridine and perfluoroalkyl iodides in the liquid phase binary system led the ring breathing vibration frequency of pyridine blue shift by 7.8, 7.6, and $9.7\,cm^{-1}$ for 1-C_4F_9I, 1-$C_6F_{13}I$, and iso-C_3F_7I, respectively [44]. Using FTIR, the present work showed similar impact of halogen bond between pyridine and iso-C_3F_7I/1-C_3F_7I in the solution phase. The ring breathing vibration of pyridine was found to be blue shifted on the formation of halogen bonding complex with iso-C_3F_7I and 1-C_3F_7I. According to Table 1, the blue shift of $9 \pm 1\,cm^{-1}$ observed for pyridine/iso-C_3F_7I in acetonitrile (0.4 M) is comparable with the shift observed for the same donor/receiver in the binary system. The difference in the magnitude of blue shift between liquid binary system and the solution phase is caused by the solvent rather than the concentration of the solute. As shown in Table 1, for the same solvent, the ring breathing vibration of pyridine shows greater blue shift for the pyridine/iso-C_3F_7I system than that for the pyridine/1-C_3F_7I system. This indicates that pyridine forms stronger halogen bond with iso-C_3F_7I than with 1-C_3F_7I regardless of the solvent used. This is also consistent with the ^{19}F NMR results by Metrangolo et al. [49, 50] using perfluoroalkyl iodide involved in a halogen bonding system where secondary iodide invariably was found to give larger change in the F chemical shift than a primary iodide. It was pointed out [57] that perfluorinated chains were stronger electron-withdrawing residues and more sterically demanding than fluorine atoms. Consequently, the carbon connected to iodine in iso-C_3F_7I is more electropositive than the carbon connected to iodine in 1-C_3F_7I. The same principle can be applied to the liquid phase binary system described by Fan et al. [44].

Table 1: The magnitudes of the frequency blue shift in the ring breathing vibration of pyridine on the formation of halogen bonding with iso-C_3F_7I/1-C_3F_7I in various solvents (the concentrations of pyridine and perfluoroalkyl iodide are 0.2 M in all the solutions).

Solvent	Iso-C_3F_7I	1-C_3F_7I
Cyclohexane	$8 \pm 1\,cm^{-1}$	$6 \pm 1\,cm^{-1}$
Hexane	$8 \pm 1\,cm^{-1}$	$6 \pm 1\,cm^{-1}$
Chloroform	$5 \pm 1\,cm^{-1}$	$4 \pm 1\,cm^{-1}$
Acetone	$9 \pm 1\,cm^{-1}$	$7 \pm 1\,cm^{-1}$
Acetonitrile	$9 \pm 1\,cm^{-1}$	$6 \pm 1\,cm^{-1}$

For the same system, the magnitude of the frequency blue shift depended on the solvents used. The magnitude of shifts seems to be associated with the polarity of the solvent as well as the potential competitive intermolecular interactions between solvent molecules and the halogen bonding precursors (Table 1). Chloroform, acetone, and acetonitrile can all provide an electron lone pairs for the potential halogen bond with perfluoroalkyl iodides. On one hand, chloroform has been observed to form hydrogen bonds with pyridine [42]. Moreover, chloroform can potentially form hydrogen bonds with the fluorine in perfluoroalkyl iodide [58]. The combination of these effects may explain the smallest blue shift of ring breathing vibration of pyridine in chloroform among all the solvents used. Except for chloroform, the ring breathing vibration of pyridine in its halogen bonding complexes with perfluoroalkyl iodides shows larger blue shifts in polar solvents, indicating that the halogen bonding complexes between pyridine and perfluoroalkyl iodides can be stabilized by the polar solvents.

3.2. Impact of Halogen Bonding on Iso-C_3F_7I. The halogen bond between pyridine and a perfluoroalkyl iodide not only leads to the vibration frequency shift in pyridine molecule but also affects the vibrations of the perfluorocompounds. Previous research [43] indicated that the C-I stretching and C_α-F stretching vibration frequency experienced a red shift while a perfluoroalkyl iodide is engaged in halogen bonding. When the iso-C_3F_7I was mixed with pyridine solution in cyclohexane, the infrared spectra were recorded. Figure 1 shows that in cyclohexane solution, the vibration of iso-C_3F_7I at $1113\,cm^{-1}$ and $1168\,cm^{-1}$ were both red shifted on the formation of halogen bond with pyridine. Based upon quantum mechanical frequency calculations, we assigned the vibration of iso-C_3F_7I observed at $1113\,cm^{-1}$ to the C_α-F stretching involved vibration (predicted to come $\sim 1106\,cm^{-1}$). The vibration at $1168\,cm^{-1}$ was assigned to the C_β-F stretching involved vibration (predicted to come at $1165\,cm^{-1}$). Calculations (DFT and MP_2), performed by Valerio et al., indicated that the halogen bond between NH_3 and CF_3I led to the elongation of C-F bond. This was attributed to charge-transfer involving antibonding LUMO orbitals in the CF_3I molecule [57]. This explains why the frequency of C_α-F stretching exhibited some red shift when the iodine in iso-C_3F_7I is engaged in the halogen bond with pyridine. The frequency red shift observed the for C_β-F involving stretching indicated that the fluorine connected to C_β might play certain roles in the

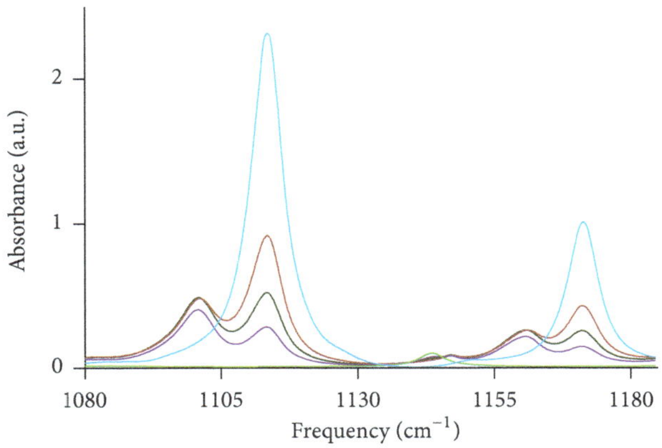

FIGURE 1: FTIR absorbance of C_α-F stretching involved vibration in iso-C_3F_7I upon the formation of halogen bonding with pyridine in cyclohexane (X_{py} and X_{pfc} denote the molar fractions of pyridine and that of iso-C_3F_7I).

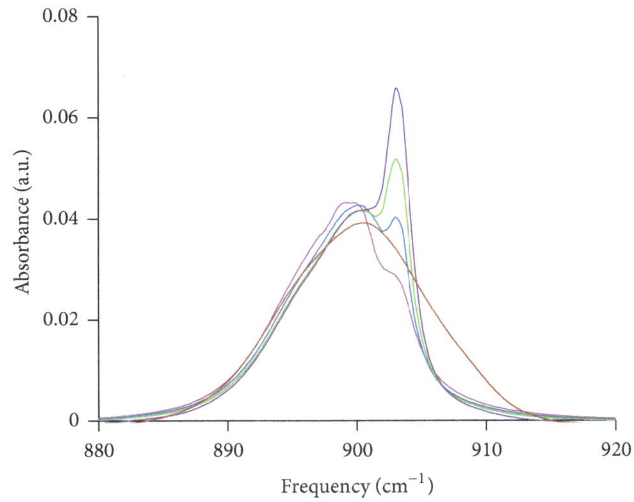

FIGURE 2: FTIR absorbance of C_α-F bending involved vibration in iso-C_3F_7I upon dilution with cyclohexane.

halogen bonding complexes. Similar shift patterns about C_α-F were observed for the systems acetone/iso-C_3F_7I, triethylamine/iso-C_3F_7I, and acetonitrile/iso-C_3F_7I in cyclohexane, indicating that acetone, acetonitrile, and triethylamine can form halogen bond with iso-C_3F_7I. However, the mixture of chloroform and iso-C_3F_7I did not show the red shift in the C_α-F stretching vibration. This suggests that the halogen bond between iso-C_3F_7I and $CHCl_3$ is weaker than both halogen bond between iso-C_3F_7I and iso-C_3F_7I and the hydrogen bond between iso-C_3F_7I and $CHCl_3$.

While the vibration of iso-C_3F_7I was significantly affected by halogen bond, the FTIR spectra of various 1-C_3F_7I halogen bonding systems did not show similar shifts. We suggest that the vibrations of 1-C_3F_7I are not as sensitive as those of iso-C_3F_7I to halogen bond formation, as discussed below.

3.3. Dilution of Iso-C_3F_7I with Various Solvents. The effect of halogen bonding on the C_α-F stretching involved vibration of iso-C_3F_7I does not provide direct evidence for the intermolecular self-halogen bonding among the molecules of iso-C_3F_7I. Therefore, the dilution experiments with some halogen bonding inactive solvents were performed.

The intermolecular self-halogen bond between two iso-C_3F_7I molecules can be expressed with the following equation (1), where one molecule uses its iodine atom and the other uses its alpha fluorine atom to form the halogen bond:

$$\text{iso-}C_3F_7I + \text{iso-}C_3F_7I \rightleftharpoons \text{iso-}C_3F_7I \cdots \text{iso-}C_3F_7I. \quad (1)$$

The dilution test will cause the equilibrium above to shift to left; that is, the amount of iso-C_3F_7I engaged in the self-halogen bond will decrease and the amount disengaged from the self-halogen bond will increase upon dilution. The dilution test was performed first with cyclohexane, and the results were summarized in Figure 2. According to the geometry optimization calculations for iso-C_3F_7I in vacuum (DFT/3-21GB3LYP/CalcAll), the vibration absorption at ~900 cm^{-1} shown in Figure 2 is a C_α-F bending involved vibration. This peak is very broad in pure iso-C_3F_7I, resolving into two components as the dilution was carried out. Moreover, the component at higher frequency becomes more intense as the dilution proceeds. A theoretical calculation was performed for a dimer of iso-C_3F_7I in vacuum. The results showed that the C_α-F bending involved vibration at ~900 cm^{-1} split into two (899.95 cm^{-1} and 905.08 cm^{-1}). The one at the lower frequency corresponds to the C_α-F engaged in the halogen bonding, and the one at the higher frequency is correlated with the C_α-F free of the halogen bonding. This is perfectly consistent with the observation in Figure 2. As the dilution proceeds, the amount of nonhalogen bonded C_α-F increases and leads to a relative intensity increase of the component at the higher frequency.

Figure 3 shows dilution of iso-C_3F_7I in various solvents. Both acetone and acetonitrile form halogen bonds with iso-C_3F_7I; consequently the spectra in these solvents show primarily isolated fluorocarbon with little or no indication of self-halogen bond. Hexane, on the other hand, does not have a significant influence on the equilibrium between the self-halogen bonded complex and the isolated iso-C_3F_7I. As a result, the dilution with hexane exhibited mainly the C_α-F bending vibration in an aggregation of iso-C_3F_7I. In chloroform, the C_α-F bending involved vibration peaked at the same position with that in hexane. This suggests that chloroform formed a hydrogen bond with the alpha fluorine in iso-C_3F_7I. The hydrogen bond between iso-C_3F_7I and chloroform is supported by the geometry optimization calculation for the iso-C_3F_7I-chloroform system in vacuum.

The vibrational spectra of 1-C_3F_7I were not only insensitive to the halogen bond between 1-C_3F_7I and various halogen bonding acceptors, they were also insensitive to the dilution tests. Theoretical calculation suggests that the self-halogen bond in vacuum between two 1-C_3F_7I molecules

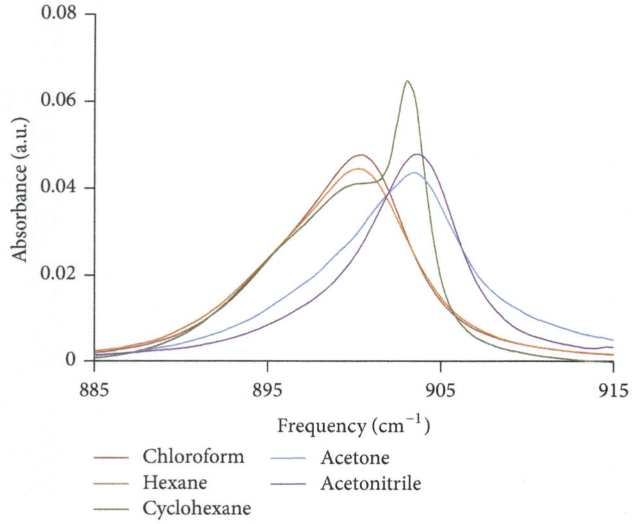

FIGURE 3: FTIR absorbance of C_α-F bending involved vibration in iso-C_3F_7I upon dilution with various solvents (the concentrations of iso-C_3F_7I in various solvents are 0.06 M).

(3.68 kcal/mol) is weaker than that between two iso-C_3F_7I molecules (4.86 kcal/mol). In addition, while there is only one alpha fluorine in iso-C_3F_7I, there are two in 1-C_3F_7I. We suggest that this further decreases any impact that the weaker halogen bond might have on the C-F_α bending frequency, rendering it undetectable in the FTIR spectrum at the present resolution.

3.4. ^{19}F NMR Dilution Experiments. The ^{19}F NMR spectra of both iso-C_3F_7I and 1-C_3F_7I were collected in various solvents. A sealed capillary tube filled with KF/D_2O solution and placed in the NMR sample tube was used as an internal reference for each measurement. The ^{19}F NMR spectrum of iso-C_3F_7I is straightforward. While a triplet at −79 ppm represents the resonance of fluorine in the CF_3group, the pentet at −153 ppm is the resonance of the alpha fluorine. Two triplets at −83 ppm and −122 ppm and nonets at −67 ppm were identified in the ^{19}F NMR spectrum of 1-C_3F_7I. The assignment of this spectrum is rather tricky. Generally, the coupling constant J_{ij} decreases dramatically as the number of chemical bonds separating nuclei i and j increases. In the case of 1-C_3F_7I, however, the alpha fluorine shows a strongest coupling with the gamma fluorine in CF_3 group [59]. When combined with the coupling of the beta fluorine in CF_2 group, the resulting nonet was assigned as the alpha fluorine (the fluorine germinal to iodine) resonance. On the other hand, the coupling constant $J_{\beta\alpha}$ is much larger than the coupling constant $J_{\beta\gamma}$, Thus, the beta fluorine appears as triplet at −122 ppm [59].

The dilution of iso-C_3F_7I and 1-C_3F_7I was performed in hexane, cyclohexane, chloroform, acetone, and acetonitrile. The ^{19}F NMR spectra were collected over the concentration range of 0.04 M to 4 M. The plot of $\Delta\delta = \delta_{\text{solution}} - \delta_{\text{pure iso-}C_3F_7I/1\text{-}C_3F_7I}$ versus the concentration in hexane, cyclohexane, chloroform, acetone, and acetonitrile is shown in Figures 4, 5, 6, 7, and 8, respectively. In allvspace0.4ptthe solvents, the chemical shift of beta and gamma fluorine showed a consistent upfield shift of NMR frequency upon dilution. In the dilution with hexane, the chemical shift of all alpha, beta, and gamma fluorine displayed a similar trend. In the dilutions with other solvents, the change in the chemical shift of alpha fluorine(s) exhibited a different trend from that of beta and gamma fluorine. While the upfield shift of beta and gamma fluorine upon dilution can be attributed to the impact of screening constant σ_0, determined mainly by the regular solvent effect, the trend in the chemical shift of alpha fluorine indicates that the local diamagnetic screen constant σ_d is responsible for the shielding parameter change of the alpha fluorine upon dilution.

The local diamagnetic screen constant σ_d depends on the electronic structure in the immediate vicinity of the fluorine atom, so the chemical shift in the ^{19}F NMR signal should experience the greatest change if the fluorine atom directly participates in the halogen bond or is next to the halogen bonding donor iodine. Previous work on ^{14}N NMR indicated that the donation of the electron lone pair by nitrogen in the halogen bond relationship led to more shielding of nitrogen [49]. Similarly, the alpha F, when involved in self-halogen bond as an electron lone pair donor, will be subject to more shielding. This is consistent with the results from theoretical calculations (NMR calculations for a dimer in vacuum, based on the optimized geometry), which showed that the fluorine engaged in halogen bonding is more shielded than the fluorine free of halogen bonding. The difference in the shielding parameters between the F involved in halogen bond and the F free of halogen bond is similar for iso-C_3F_7I and 1-C_3F_7I. This also explains the similar results observed in ^{19}F NMR dilution tests between iso-C_3F_7I and 1-C_3F_7I. The unified dilution trends among fluorine chemical shifts in the dilution with hexane indicated that dispersion forces were insufficient to break down the self-halogen bonded complex. On the other hand, cyclohexane provides stronger dispersion forces, because the ring shape allows for a larger area of contact. Thus, the dilution of both 1-C_3F_7I and iso-C_3F_7I using cyclohexane leads to the dissociation of self-halogen bonded complex. As the dilution proceeded, more alpha fluorine disengaged the halogen bonding and became less shielding, leading to a downfield shift of the NMR frequency. This downfield shift because of the dissociation of self-halogen bonding complex counteracted the up-field shift from the general dilution effect. As a result, the $\Delta\delta$ ($\Delta\delta = \delta_{\text{solution}} - \delta_{\text{pure 1-}C_3F_7I}$ or $\Delta\delta = \delta_{\text{solution}} - \delta_{\text{pure iso-}C_3F_7I}$) values for the alpha fluorine in both 1-C3F7I and iso-C3F7I are smaller than those of beta and gamma fluorine when they are diluted with cyclohexane.

As for the dilution with chloroform, it was proposed earlier and by theoretical calculation that chloroform formed hydrogen bond with alpha fluorine in iso-C_3F_7I. Additionally, our calculations showed that hydrogen bonds can also be formed between the alpha fluorine in 1-C_3F_7I and chloroform. Moreover, formation of hydrogen bond made the alpha fluorine less shielding, causing a downfield shift of the NMR

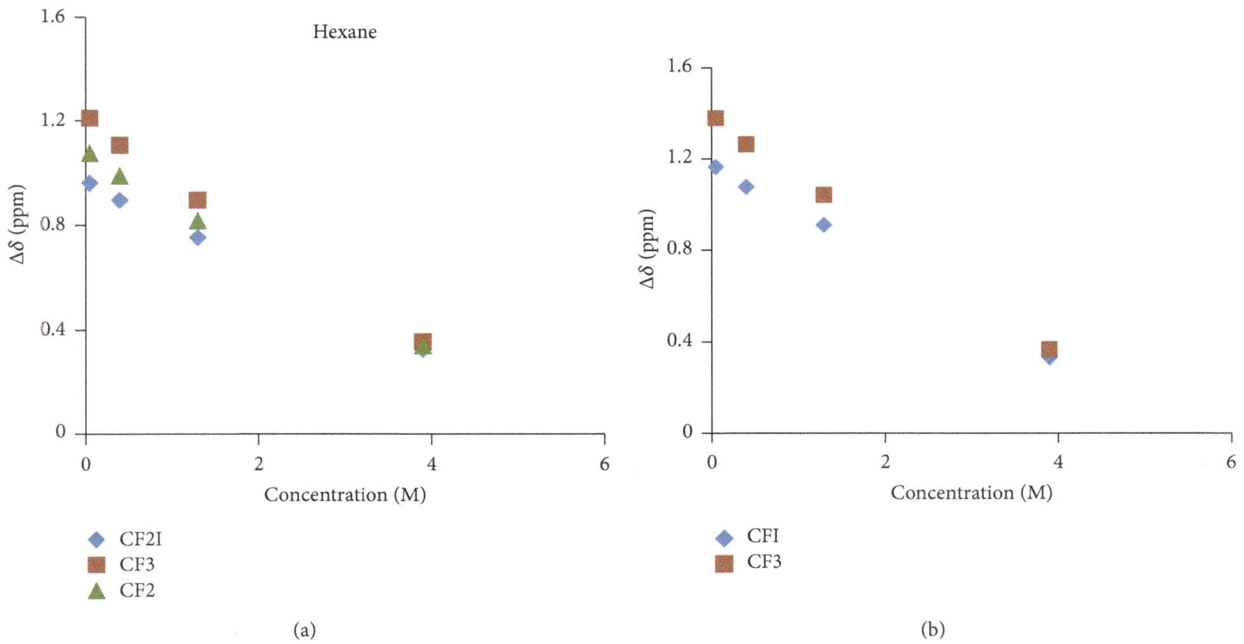

FIGURE 4: Change in the chemical shift of ^{19}F in the perfluoroalkyl iodides upon dilution with hexane (from Figure 4 to Figure 8, the upper panel is for 1-C_3F_7I, the lower panel is for iso-C_3F_7I, $\Delta\delta = \delta_{\text{solution}} - \delta_{\text{pure1-C3F7I}}$ for 1-C_3F_7I, and for iso-C_3F_7I, $\Delta\delta = \delta_{\text{solution}} - \delta_{\text{pure iso-C3F7I}}$).

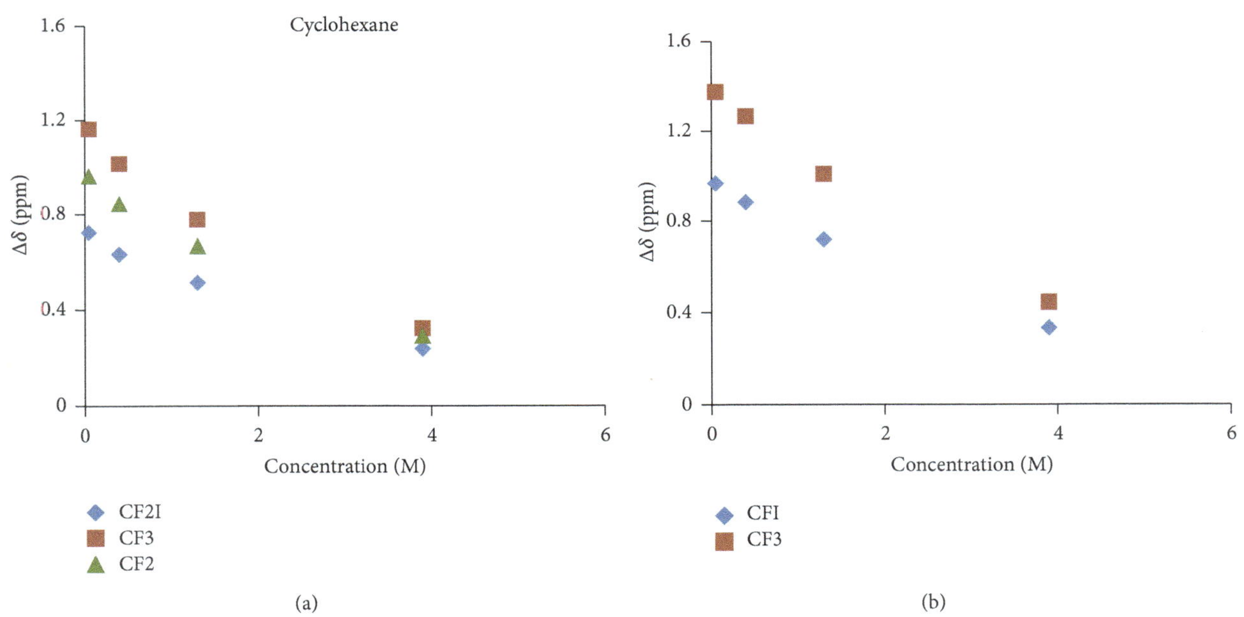

FIGURE 5: Change in the chemical shift of ^{19}F in the perfluoroalkyl iodides upon dilution with cyclohexane.

frequency. The same effect is brought by the dissociation of self-halogen bonded complex. As a result, the ^{19}F NMR dilution tests in chloroform showed some similar alpha fluorine trend to the ones in cyclohexane. In the ^{19}F NMR study of 2′-F-2′-deoxyarabinoflavoproteins [60], the chemical shift of fluorine moves to a higher frequency when the fluorine atom donated its electron lone pair in hydrogen bond.

While the dilution was performed using acetone and acetonitrile, the solvent molecules are not only able to dissociate the self-halogen boned complex, but also to form halogen bond with the iodine in both iso-C_3F_7I and 1-C_3F_7I. Both of these effects make the alpha fluorine less shielding and a larger downfield shift of the NMR frequency. As a result, the $\Delta\delta$ value became negative upon dilution.

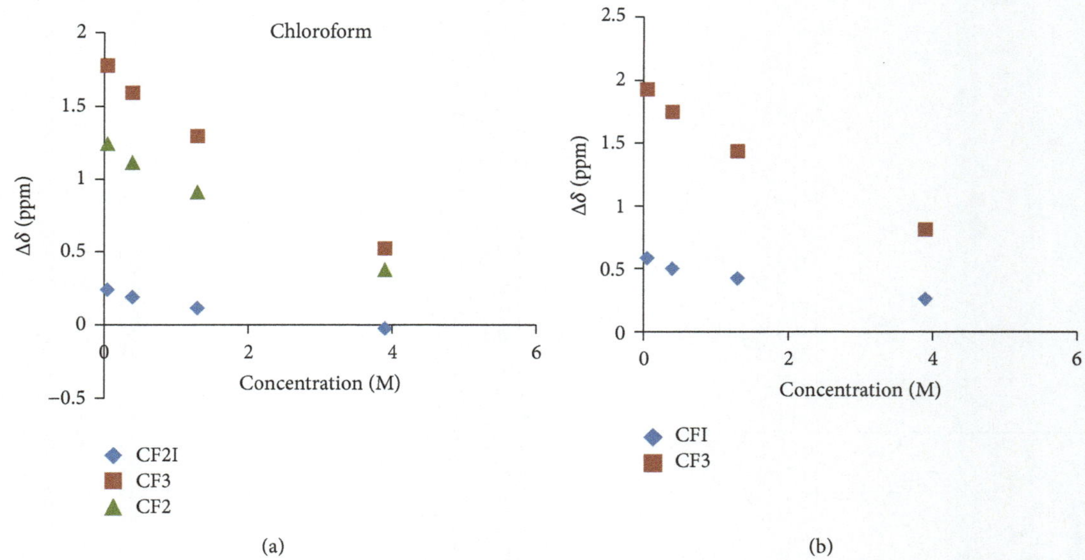

FIGURE 6: Change in the chemical shift of ^{19}F in the perfluoroalkyl iodides upon dilution with chloroform.

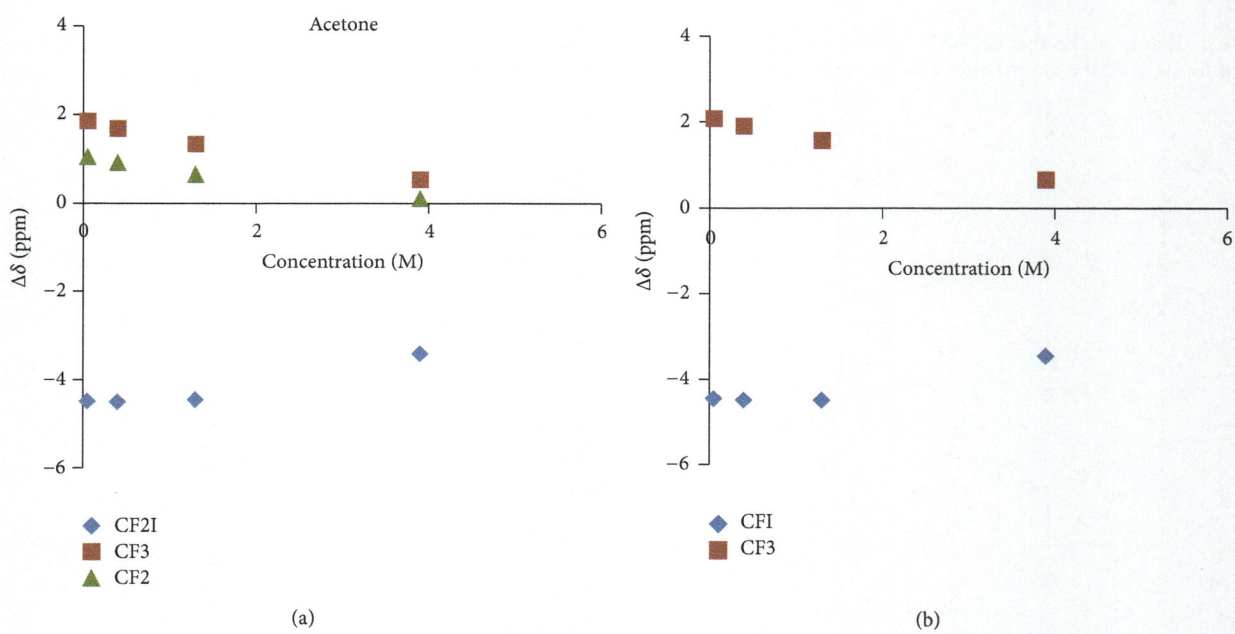

FIGURE 7: Change in the chemical shift of ^{19}F in the perfluoroalkyl iodides upon dilution with acetone.

3.5. ^{19}F NMR Titration and the Halogen Bonding Association Constants for Pyridine/Iso-C_3F_7I(1-C_3F_7I) in Various Solvents. ^{19}F NMR titration experiments were performed to determine the equilibrium constants for the halogen bonding complex formed between pyridine and perfluoroalkyl iodides in various solvents (Table 2). These data indicate that the association between pyridine and perfluoroalkyl iodides is more favorable in nonpolar solvents such as hexane and cyclohexane. According to Table 1, the halogen bond formed between pyridine and perfluoroalkyl iodides caused the largest blue shift of pyridine ring breathing vibration in acetone. However, the halogen bonding association constants obtained in acetone and acetonitrile are among the lowest values. In Table 1, the magnitudes of the frequency shift in the ring breathing vibration of pyridine are greater in polar solvents than those in nonpolar solvents. This indicates that the enthalpy change for the halogen bonding formation is more negative in polar solvents than in non-polar solvents. On the other hand, in Table 2, the halogen bonding association constants are greater in nonpolar solvents than those in polar solvents. In other words, the changes in Gibb's free energy are more negative in nonpolar solvents than those in polar solvents. The only way to rationalize this observation is to consider the contribution of entropy change in the formation of halogen bonds.

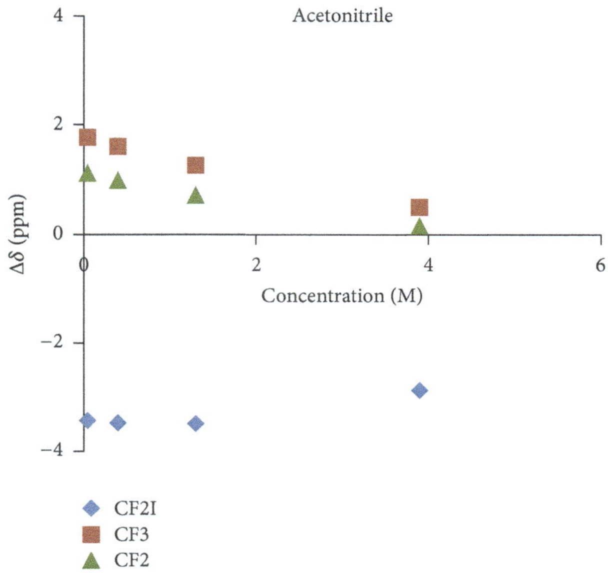

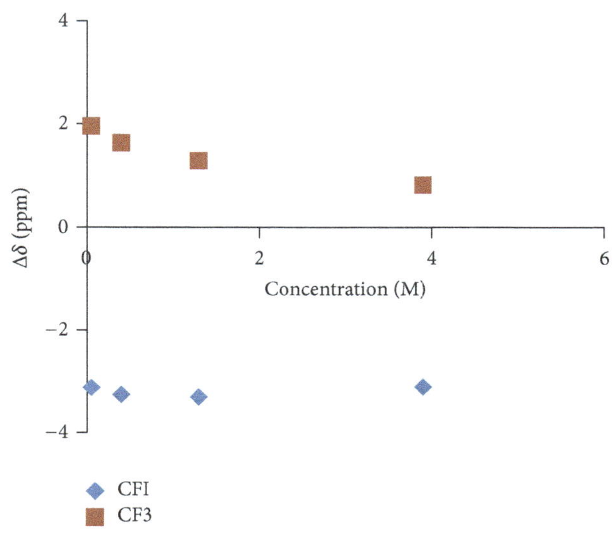

FIGURE 8: Change in the chemical shift of ^{19}F in the perfluoroalkyl iodides upon dilution with acetonitrile.

The overall entropy change in the formation of halogen bond between pyridine and perfluoroalkyl iodides in different solvents depends on whether the molecules of perfluoroalkyl iodides stay as an aggregation through self-halogen bond or isolated through hydrogen/halogen bonding with solvent molecules. In the non-polar solvents like cyclohexane and hexane, where perfluoroalkyl iodides mostly stay as an aggregation, the formation of halogen bond between pyridine and perfluoroalkyl iodides results in a significant increase in entropy. In polar solvents like chloroform, acetone, and acetonitrile, where both perfluoroalkyl iodides and pyridine have tendency to form halogen or hydrogen bond with solvent molecules, the formation of halogen bond between pyridine and perfluoroalkyl iodides does not necessarily lead to an increase in entropy and sometimes can even lead to an entropy decrease. After all, in the nonpolar solvents, the contribution from the positive entropy change compensates or even overrides the less negative enthalpy change in the formation of halogen bond between pyridine and perfluorocompound. Thus, the overall larger halogen bonding association constants were observed in the nonpolar solvents than in the polar solvents.

4. Conclusion

The combination of FTIR and ^{19}F NMR indicated that intermolecular self-halogen bond among the molecules of iso-C_3F_7I and 1-C_3F_7I is present in several solvents. The halogen bond between pyridine and perfluoroalkyl iodide not only affects some of the vibrational frequencies of pyridine but also impacts the vibrational frequencies of iso-C_3F_7I. As an electron lone pair donor, the alpha fluorine in perfluoroalkyl iodides is more shielded when it is engaged in the halogen bond, but less shielded when it is engaged in hydrogen bonding. The halogen bonded complexes of pyridine and perfluoroalkyl iodides, though stronger in polar solvents, showed significantly lower association constants in acetone and acetonitrile than in non-polar solvents such as hexane and cyclohexane. The entropy change makes a favorite contribution to the halogen bond between pyridine and perfluoroalkyl iodides in the non-polar solvent such as hexane and cyclohexane.

TABLE 2: Halogen-bonding association constants for pyridine/iso-C_3F_7I and pyridine/1-C_3F_7I systems in various solvents.

Solvent	Iso-C_3F_7I	1-C_3F_7I
Cyclohexane	21.5 ± 5.4	10.2 ± 2.0
Hexane	13.0 ± 2.7	8.4 ± 1.7
Chloroform	3.8 ± 0.9	2.6 ± 0.7
Acetone	3.1 ± 0.8	1.9 ± 0.6
Acetonitrile	3.6 ± 0.8	2.4 ± 0.6

Acknowledgments

This material is based in part upon work supported by the National Science Foundation under Grant no. 0648964. Any opinions, findings, and conclusions or recommendations expressed in this material are those of the author(s) and do not necessarily reflect the views of the National Science Foundation.

References

[1] F. Guthrie, "On the iodide of iodammonium," *Journal of the Chemical Society*, vol. 16, pp. 239–244, 1863.

[2] R. S. Mulliken, "Structures of complexes formed by halogen molecules with aromatic and with oxygenated solvents," *Journal of the American Chemical Society*, vol. 72, no. 1, pp. 600–608, 1950.

[3] R. S. Mulliken, "Molecular compounds and their spectra. II," *Journal of the American Chemical Society*, vol. 74, no. 3, pp. 811–824, 1952.

[4] R. S. Mulliken, "Molecular compounds and their spectra. III. The interaction of electron donors and acceptors," *The Journal of Physical Chemistry*, vol. 56, no. 7, pp. 801–822, 1952.

[5] O. Hassel, J. Hvoslef, E. Vihovde, V. Hadler, and N. A. Sörensen, "The structure of bromine 1,4-dioxanate," *Acta Chemica Scandinavica*, vol. 8, p. 873, 1954.

[6] O. Hassel, "Structural aspects of interatomic charge-transfer bonding," *Science*, vol. 170, no. 3957, pp. 497–502, 1970.

[7] P. Metrangolo and G. Resnati, *Halogen Bonding: Fundamentals and Applications*, Springer, Berlin, Germany, 2008.

[8] P. Metrangolo, H. Neukirch, T. Pilati, and G. Resnati, "Halogen bonding based recognition processes: a world parallel to hydrogen bonding," *Accounts of Chemical Research*, vol. 38, no. 5, pp. 386–395, 2005.

[9] P. Metrangolo, Y. Carcenac, M. Lahtinen et al., "Nonporous organic solids capable of dynamically resolving mixtures of diiodoperfluoroalkanes," *Science*, vol. 323, no. 5920, pp. 1461–1464, 2009.

[10] A. G. Orpen, D. Braga, J. S. Miller, G. R. Desiraju, and S. L. Price, "Innovation in crystal engineering," *CrystEngComm*, vol. 4, no. 83, pp. 500–509, 2002.

[11] G. M. J. Schmidt, "Photodimerization in the solid state," *Pure and Applied Chemistry*, vol. 27, no. 4, pp. 647–678, 1971.

[12] P. Metrangolo, G. Resnati, T. Pilati, R. Liantonio, and F. Meyer, "Engineering functional materials by halogen bonding," *Journal of Polymer Science A*, vol. 45, no. 1, pp. 1–15, 2007.

[13] H. L. Nguyen, M. B. Hursthouse, A. C. Legon, D. W. Bruce, and P. N. Horton, "Halogen bonding: a new interaction for liquid crystal formation," *Journal of the American Chemical Society*, vol. 126, no. 1, pp. 16–17, 2004.

[14] P. Metrangolo and G. Resnati, "Halogen bonding: a paradigm in supramolecular chemistry," *Chemistry*, vol. 7, no. 12, pp. 2511–2519, 2001.

[15] D. W. Bruce, "The materials chemistry of alkoxystilbazoles and their metal complexes," *Advances in Inorganic Chemistry*, vol. 52, pp. 151–159, 2001.

[16] A. Sun, J. W. Lauher, and N. S. Goroff, "Preparation of poly(diiododiacetylene), an ordered conjugated polymer of carbon and iodine," *Science*, vol. 312, no. 5776, pp. 1030–1034, 2006.

[17] C. Wilhelm, S. A. Boyd, S. Chawda et al., "Pressure-induced polymerization of diiodobutadiyne in assembled cocrystals," *Journal of the American Chemical Society*, vol. 130, no. 13, pp. 4415–4420, 2008.

[18] F. C. Pigge, V. R. Vangala, P. P. Kapadia, D. C. Swenson, and N. P. Rath, "Hexagonal crystalline inclusion complexes of 4-iodophenoxy trimesoate," *Chemical Communications*, no. 39, pp. 4726–4728, 2008.

[19] F. Meyer and P. Dubois, "Halogen bonding at work: recent applications in synthetic chemistry and materials science," *CrystEngComm*, vol. 15, no. 16, pp. 3058–3071, 2013.

[20] T. M. Beale, M. G. Chudzinski, M. G. Sarwar, and M. S. Taylor, "Halogen bonding in solution: thermodynamics and applications," *Chemical Society Reviews*, vol. 42, no. 4, pp. 1667–1680, 2013.

[21] M. Erdélyi, "Halogen bonding in solution," *Chemical Society Reviews*, vol. 41, no. 9, pp. 3547–3557, 2012.

[22] A. C. Legon, "The halogen bond: an interim perspective," *Physical Chemistry Chemical Physics*, vol. 12, no. 28, pp. 7736–7747, 2010.

[23] A. C. Legon, "Prereactive complexes of dihalogens XY with Lewis bases B in the gas phase: a systematic case for the halogen snalogue B···XY of the hydrogen bond B···HX," *Angewandte Chemie*, vol. 38, no. 18, pp. 2686–2714, 1999.

[24] R. N. Hazeldine, "Reactions of fluorocarbon radicals. Part XII. The synthesis of fluorocarbons and of fully fluorinated iodo-, bromo-, and chloroalkaness," *Journal of the Chemical Society*, vol. 75, pp. 3761–3768, 1953.

[25] N. F. Cheetham and A. D. E. Pullin, "Interaction between tertiary amines and perfluoro-organo-halides," *Chemical Communications*, no. 18, pp. 418–421, 1965.

[26] N. F. Cheetham and A. D. E. Pullin, "A gas-phase donor-acceptor complex," *Chemical Communications*, no. 5, pp. 233–234, 1967.

[27] N. F. Cheetham and A. D. E. Pullin, "Donor-acceptor complexes formed by perfluoro-organo bromides and iodides with nitrogenous and other bases. I. Vapour pressure measurements," *Australian Journal of Chemistry*, vol. 24, pp. 479–487, 1971.

[28] A. Misha and A. D. E. Pullin, "Donor-acceptor complexes formed by perfluoro-organo bromides and iodides with nitrogenous and other bases. II. Band shapes and widths in the absorption spectrum of gaseous $CF_3I\text{-}N(CH_3)_3$," *Australian Journal of Chemistry*, vol. 24, no. 12, pp. 2493–2507, 1971.

[29] N. F. Cheetham, I. J. McNaught, and A. D. E. Pillin, "Donor-acceptor complexes formed by perfluoro-organo bromides and iodides with nitrogenous and other bases. III. Qualitative examination of condensed phase spectra of CF_3I and CF_3Br and of their complexes with trimethylamine and other bases," *Australian Journal of Chemistry*, vol. 27, no. 5, pp. 973–985, 1974.

[30] N. F. Cheetham I, J. McNaught, and A. D. E. Pullin, "Donor-acceptor complexes formed by perfluoro-organo bromides and iodides with nitrogenous and other bases. IV. Analysis of the infrared spectra of $CF_3I.N(CH_3)_3$ and $CF_3Br.N(CH_3)_3$ and related complexes," *Australian Journal of Chemistry*, vol. 27, no. 5, pp. 987–1007, 1974.

[31] J. McNaught and A. D. E. Pullin, "Donor-acceptor complexes formed by perfluoro-organo bromides and iodides wlth nitrogenous and other bases. V. Comparison of liquid phase complexes of CFI_3, C_2F_5I or C_3F_7I with Nme_3, NEt_3 or NPr_3; infrared, far-infrared and N.M.R. spectra," *Australian Journal of Chemistry*, vol. 27, no. 5, pp. 1009–1015, 1974.

[32] A. C. Legon, D. J. Millen, and S. C. Rogers, "Microwave spectrum of a gas-phase charge-transfer complex," *Chemical Communications*, no. 14, pp. 580–581, 1975.

[33] A. L. Allred and D. W. Larsen, "Halogen complexes. III. The association of 2,4,6-trimethylpyridine and trifluoroiodomethane," *The Journal of Physical Chemistry*, vol. 69, no. 7, pp. 2400–2401, 1965.

[34] P. Auffinger, F. A. Hays, E. Westhof, and P. S. Ho, "Halogen bonds in biological molecules," *Proceedings of the National Academy of Sciences of the United States of America*, vol. 101, no. 48, pp. 16789–16794, 2004.

[35] L. K. Steinrauf, J. A. Hamilton, B. C. Braden, J. R. Murrell, and M. D. Benson, "X-ray crystal structure of the Ala-109 → Thr variant of human transthyretin which produces euthyroid hyperthyroxinemia," *The Journal of Biological Chemistry*, vol. 268, no. 4, pp. 2425–2430, 1993.

[36] E. I. Howard, R. Sanishvili, R. E. Cachau et al., "Ultrahigh resolution drug design I: details of interactions in human aldose reductase-inhibitor complex at 0.66 Å," *Proteins*, vol. 55, no. 4, pp. 792–804, 2004.

[37] C. Yabe-Nishimura, "Aldose reductase in glucose toxicity: a potential target for the prevention of diabetic complications," *Pharmacological Reviews*, vol. 50, no. 1, pp. 21–34, 1998.

[38] F. A. Hays, J. M. Vargason, and P. S. Ho, "Effect of sequence on the conformation of DNA holliday junctions," *Biochemistry*, vol. 42, no. 32, pp. 9586–9597, 2003.

[39] D. A. Kraut, M. J. Churchill, P. E. Dawson, and D. Herschlag, "Evaluating the potential for halogen bonding in the oxyanion hole of ketosteroid isomerase using unnatural amino acid mutagenesis," *ACS Chemical Biology*, vol. 4, no. 4, pp. 269–273, 2009.

[40] T. Shirman, T. Arad, and M. E. van der Boom, "Halogen bonding: a supramolecular entry for assembling nanoparticles," *Angewandte Chemie*, vol. 49, no. 5, pp. 926–929, 2010.

[41] E. R. Berg, D. D. Green, D. C. Moliva A, B. T. Bjerke, M. W. Gealy, and D. J. Ulness, "Ion-pair interaction in pyridinium carboxylate solutions," *The Journal of Physical Chemistry A*, vol. 112, no. 5, pp. 833–838, 2008.

[42] H. Fan, D. A. Moliva, J. K. Elison et al., "Effects of hydrogen bonding on the ring breathing mode of pyridine in pyridine/chloroform and pyridine/bromoform systems," *Chemical Physics Letters*, vol. 479, no. 1–3, pp. 43–46, 2009.

[43] M. T. Messina, P. Metrangolo, W. Navarrini, S. Radice, G. Resnati, and G. Zerbi, "Infrared and Raman analyses of the halogen-bonded non-covalent adducts formed by α,ω-diiodoperfluoroalkanes with DABCO and other electron donors," *Journal of Molecular Structure*, vol. 524, no. 1–3, pp. 87–94, 2000.

[44] H. Fan, J. K. Eliason, C. D. Moliva et al., "Halogen bonding in iodo-perfluoroalkane/pyridine mixtures," *The Journal of Physical Chemistry A*, vol. 113, no. 51, pp. 14052–14059, 2009.

[45] D. Hauchecorne, N. Nagels, B. J. van der Veken, and W. A. Herrebout, "C-X$\cdots\pi$ halogen and C-H$\cdots\pi$ hydrogen bonding: interactions of CF_3X (X = Cl, Br, I or H) with ethene and propene," *Physical Chemistry Chemical Physics*, vol. 14, no. 2, pp. 681–690, 2012.

[46] D. Hauchecorne, B. J. van der Veken, A. Moiana, and W. A. Herrebout, "The C-Cl$\cdots$N halogen bond, the weaker relative of the C-I and C-Br$\cdots$N halogen bonds, finally characterized in solution," *Chemical Physics*, vol. 374, no. 1–3, pp. 30–36, 2010.

[47] D. Hauchecorne, A. Moiana, B. J. van der Veken, and W. A. Herrebout, "Halogen bonding to a divalent sulfur atom: an experimental study of the interactions of CF_3X (X = Cl, Br, I) with dimethyl sulfide," *Physical Chemistry Chemical Physics*, vol. 13, no. 21, pp. 10204–10213, 2011.

[48] D. Hauchecorne, R. Szostak, W. A. Herrebout, and B. J. van der Veken, "C-X$\cdots$O halogen bonding: interactions of trifluoromethyl halides with dimethyl ether," *ChemPhysChem*, vol. 10, no. 12, pp. 2105–2115, 2009.

[49] M. T. Messina, P. Metrangolo, W. Panzeri, E. Ragg, and G. Resnati, "Perfluorocarbon-hydrocarbon self-assembly. Part 3. Liquid phase interactions between perfluoroalkylhalides and heteroactom containing hydrocarbons," *Tetrahedron Letters*, vol. 39, no. 49, pp. 9069–9072, 1998.

[50] C. Cavallotti, P. Metrangolo, F. Meyer, F. Recupero, and G. Resnati, "Binding energies and ^{19}F nuclear magnetic deshielding in paramagnetic halogen-bonded complexes of TEMPO with haloperfluorocarbons," *The Journal of Physical Chemistry A*, vol. 112, no. 40, pp. 9911–9918, 2008.

[51] D. Hauchecorne, B. J. van der Veken, W. A. Herrebout, and P. E. Hansen, "A ^{19}F NMR study of C-I$\cdots\pi$ halogen bonding," *Chemical Physics*, vol. 381, no. 1–3, pp. 5–10, 2011.

[52] M. G. Sarwar, B. Dragisic, L. J. Salsberg, C. Gouliaras, and M. S. Taylor, "Thermodynamics of halogen bonding in solution: substituent, structural, and solvent effects," *Journal of the American Chemical Society*, vol. 132, no. 5, pp. 1646–1653, 2010.

[53] C. Laurence, M. Queignec-Cabanetos, T. Dziembowska, R. Queignec, and B. Wojtkowiak, "1-iodoacetylenes. 1. Spectroscopic evidence of their complexes with Lewis bases. A spectroscopic scale of soft basicity," *Journal of the American Chemical Society*, vol. 103, no. 10, pp. 2567–2573, 1981.

[54] M. J. Frisch, G. W. Trucks, H. B. Schlegel et al., Gaussian 03, Revision E.01, Gaussian, Wallingford, Conn, USA, 2004.

[55] M. J. Frisch, G. W. Trucks, H. B. Schlegel et al., Gaussian 09, Revision A.1, Gaussian, Wallingford, Conn, USA, 2009.

[56] K. A. Peterson, D. Figgen, E. Goll, H. Stoll, and M. Dolg, "Systematically convergent basis sets with relativistic pseudopotentials. II. Small-core pseudopotentials and correlation consistent basis sets for the post-d group 16–18 elements," *Journal of Chemical Physics*, vol. 119, no. 21, pp. 11113–11123, 2003.

[57] G. Valerio, G. Raos, S. V. Meille, P. Metrangolo, and G. Resnati, "Halogen bonding in fluoroalkylhalides: a quantum chemical study of increasing fluorine substitution," *The Journal of Physical Chemistry A*, vol. 104, no. 8, pp. 1617–1620, 2000.

[58] J. A. K. Howard, V. J. Hoy, D. O'Hagan, and G. T. Smith, "How good is fluorine as a hydrogen bond acceptor?" *Tetrahedron*, vol. 52, no. 38, pp. 12613–12622, 1996.

[59] E. Pitcher, A. D. Buckingham, and F. G. A. Stone, "Spectroscopic studies on organometallic compounds. V. Fluorine nuclear magnetic resonance spectra of some perfluoroalkyl and perfluoroacyl metal compounds," *The Journal of Chemical Physics*, vol. 36, no. 1, pp. 124–129, 1962.

[60] Y. V. S. Murthy and V. Massey, "^{19}F NMR studies with 2′-F-2′-deoxyarabinoflavoproteins," *The Journal of Biological Chemistry*, vol. 271, no. 33, pp. 19915–19921, 1996.

8

On Applicability of a Miniaturised Laser Ablation Time of Flight Mass Spectrometer for Trace Elements Measurements

Marek Tulej, Andreas Riedo, Maria Iakovleva, and Peter Wurz

Space Research & Planetary Sciences, Institute of Physics, University Bern, Sidlerstrasse 5, 3012 Bern, Switzerland

Correspondence should be addressed to Marek Tulej, marek.tulej@space.unibe.ch

Academic Editor: Michael Balogh

We present results from mass spectrometric analysis of NIST standard materials and meteoritic samples conducted by a miniaturised laser ablation mass spectrometer designed for space research. The mass analyser supports investigation with a mass resolution ($m/\Delta m$) $\approx$ 500–600 and dynamic range within seven decades. Nevertheless, to maintain an optimal spectral quality laser irradiances lower than $\sim$1 GW/cm^2 are applied so far which results in a spread of RSC values. To achieve the quantitative performance of mass analyser, various effects influencing RSC factors have to be investigated. In this paper we investigate influence of laser irradiance, sampling procedure and plasma chemistry on the quantitative elemental and isotopic analysis. The studies indicate necessity for accurate control of laser characteristics and acquisition procedure. A relatively low irradiance applied causes a negligible sample damage and allows for accumulation of large number of waveforms from one sample location. The procedure yields statistically well averaged data and allows a sensitive in-depth analysis. The quantitative analyses of isotopic composition can be performed with accuracy and precision better as 1% and 2%, for isotopic patterns of elements and clusters, respectively. The numerical integration methods would be preferred to achieve more accurate results. The measurements of Allende sample yield detection of Pb isotopic pattern, nevertheless cluster species are readily observed in spectrum and make the elemental analysis of other trace elements difficult due to isobaric interferences. These detections are of a considerable interest because of possible application of the instrument for *in situ* elemental and isotopic analysis and radiometric dating of solids.

1. Introduction

The elemental and isotopic composition of solid materials is a subject of considerable interest to various research fields and a wide range of quite different applications [1]. From the elemental composition, the bulk chemical composition of material can be inferred. Major and minor elements can be used for determination of the material structure and chemistry. Studies of elemental abundances of rocks or soils can provide, for instance, an insight to modal mineralogy [2]. Solid materials contain frequently a low abundant fraction of elements, trace, or ultratrace elements. Investigations of these elements, their distribution in the material, and their temporal abundance changes are subject of gradually increasing interest. Analysis of trace elements can be used for diagnostic of material purity, its heterogeneity, and for process control during manufacturing the high purity materials, for example, semiconductors or special ceramics. By investigation of such processes as corrosion, catalysis, or adhesion effects on the surfaces, the elemental impurities at $\sim$ppm level can play important role in chemical and physical processes [3]. The detection and monitoring of trace elements are used in control of pollution of environmental samples, air, water, plants, animals, and soils that poison agricultural or manufactured products [4]. Studies of isotopes of trace and ultratrace elements (e.g., P, S, Mg, Ca, K, Fe, Cu, Mn) having a vital role in living organisms are of considerable interest and can be used to investigate many biological processes and also the influence of the environment on biological functions [5]. Investigation of trace elements in geological and planetary samples yields unique information on transformation processes on the Earth, planets, and other solar bodies at the scale of the age of our solar system. The element fractionations resulting from melting and crystallisation of rocks can yield information on planetary differentiation, and isotopic fractionations can be used for a diagnostic of the formation conditions such as volatility-controlled or rare earth element fractionation. Dating of samples and identification of possible biological activities can be derived from the analysis of the details of the isotope variation. Exploration of the surface composition of planets, their moons, and asteroids is planned in most of future planetary missions. Development of instrumentation capable of sensitive and quantitative measurements of elemental and isotopic composition of solid materials is of considerable interest and drives also our present investigations [6].

So far, mostly spectroscopic methods have been used for conducting the chemical analysis of the planetary materials. X-ray fluorescence (XRF) spectroscopy was applied for the detection of elements on the Viking landers on Mars. All Mars rovers to date have also used this technique for conducting the chemical analysis of rocks and soils. This spectroscopic approach yields sufficiently accurate data for determination of the abundances of major and some minor elements. Frequently, however, elements heavier than sodium can be only well determined and the abundance of lighter elements is inferred after extensive background and interelement absorption corrections [7]. The γ-rays and neutron spectroscopic techniques were used for investigation of the elemental abundances of the surface of the Moon, Mars, and asteroids from orbiting spacecraft. These studies brought, for instance, a sensitive detection of potassium and thorium, elements occurring typically in minor and trace abundances. Nevertheless, sensitivity of the element detection by γ-ray and neutron spectrometry depends on counting statistics, and measurement duration in space exceeds months. The spectroscopic instruments have significantly contributed to the analysis of planetary surfaces providing, for example, characterisation of volcanic rocks and sedimentary deposits on Mars or a global database of element abundances on the Moon [8]. Nevertheless, for the measurements of trace elements, analyses of radiogenic isotopes or study of complex organic matter on planetary surfaces and other more sensitive techniques have to be developed. The mass spectrometric approaches are clearly the methods of choice and are preferred because of their higher sensitivity, simplicity of operation, and straightforward data analysis. The time-of-flight mass analysers (TOF MS) are the most convenient in this respect. They can be used to investigate solid materials *in situ* once combined with an appropriate laser ablation ion source.

The mass spectrometric approaches are well established analytical methods in investigation of solid materials. They are most frequently used in laboratory practice for sensitive, accurate, and precise analysis of the elemental composition and determination of trace and ultratrace elements in variety of solid or liquid samples [1]. Developments since 2000 show, however, that the laser-based techniques such as laser ionisation (ablation) mass spectrometry (LIMS) and laser ablation inductively coupled plasma mass spectrometry (LA-ICP-MS) replace systematically other, more expensive approaches and this is in great part due to the progress in development of lasers and fast electronics. LA-ICP-MS is well recognised and frequently used for analysis of ultratrace elements with the detection sensitivities corresponding to sub-ppb g/g. At present stage of the development of this instrument, its application for *in situ* investigation of planetary surfaces is very challenging owing to high complexity of design and operation. Conversely, LIMS or laser ablation mass spectrometry is clearly the method of choice due to its simplicity and robustness.

LIMS uses a laser ablation ion source. Laser radiation is used to ablate untreated samples of any solid material including metals, powders, or glass, and there is no need for special sample preparation procedures (e.g., ovens, chemical agents). The risk of sample loss or its contamination is thus minimised. The ablation process is induced by focusing the laser radiation to a spot of several micrometers allowing investigations of chemical composition with a high spatial (vertical and lateral) resolution and low sample consumption [9–12]. Coupling of laser ablation source with the time of flight mass spectrometer is usually preferred and offers several advantages. The TOF mass spectrometer is highly amenable to pulsed ion sources and its dominant property is fast measurement of transient signals, high sensitivity, and good mass resolution. TOF mass analysers yield a complete mass spectrum for each laser pulse. An ablation/ionisation event generated at each laser pulse is well spatially and temporary localised and initiate clock for counting the time of ion flight from the source to the detector. The mass spectrum is achieved by recording the temporary-resolved signal generated by ion bunches striking an ion detector. The last decade has shown that TOF mass analysers can be miniaturised to 120 mm long cylinder of 50 mm diameter. The weights of the instrument including electronics would be about ~1.5 kg yet maintaining the performance of larger laboratory systems [13, 14]. They can be also operated in extreme environments of temperature, pressure, and radiation. The ROSINA-Rosetta Orbiter Spectrometer is a good example of potential capabilities of a TOF analyser for space research. This reflectron-type time-of-flight mass spectrometer (RTOF) is designed for investigation of charged and neutral gases. It enables studies with mass resolution up to few thousandths, and detection sensitivities better than ~ppm g/g [15]. Further reduction of the size and weight by factor of ~4 could be recently achieved [16]. Several miniaturised TOF mass analysers have been also developed for carrying out the chemical analysis of solids [13, 14, 17] with the LAZMA instrument [18] being part of a payload accepted for missions to Phobos (PHOBOS GROUND) and our Moon (LUNA RESOURCE, LUNA GLOB) [17–20]. These mass analysers are equipped with a laser ablation ion source. This novel approach in space research promises improvements of the elemental analysis *in situ* on solar system body surfaces providing increased elemental sensitivity and complementing results obtained by spectroscopic methods.

LIMS is a sensitive technique, nevertheless, it is still considered to be semiquantitative [11, 21, 22], in spite of its long development history [23]. Complexities introduced by the ablation process make the quantitative analysis sometimes difficult because of ion yield dependence on laser irradiance, wavelength, pulse duration, or physical properties of the material (morphology, homogeneity). Molecular clusters, frequently accompanying elements in the mass spectra, are formed by plasma chemistry. In addition, a broad and element-specific kinetic energy distribution of ions generated in the ablation process can limit the instrumental mass resolution and cause severe problems for optimal ion detection [24]. In recent developments, control of the plasma process has been partially achieved and advancements in theoretical studies are continued to progress [21]. Achieving the control over the kinetic energy spread of ions generated by ablation is linked with historical development of this technique. Initial success has been achieved by the implementation of a reflectron and other energy focusing devices in the ion-optical system, which resulted in the development of laser-microprobe mass analyser (LAMMA) and later, another versatile system LIMA 2A [25]. The latter system enabled routine qualitative mass spectrometric analysis. Improvements of the design and functionality were also reported after a miniaturisation of laser ablation TOF mass analyser initiated in 1990s by filtering ions of specific energies (energy windowing) [17], or by just applying appropriate ion optics and reflectron combined with ring-shaped detector [13]. Significant improvements of the instrumental performance have been recently achieved by application of novel laboratory LIMS instrument utilising collisional cooling of ions in He buffer gas combined with an orthogonal extraction [26]. The quantitative analysis of the sample composition is, however, not straightforward and requires a careful control of experimental conditions for laser ablation and additional studies of appropriate standard reference materials. The scaling factors have to be achieved for each element and they form the basis for conducting the quantitative elemental analysis of solid samples. Progress in developments and application of laser ablation mass spectrometry in the last decade has been well described in recent comprehensive review articles [21, 22].

Our laboratory develops a miniaturised laser ablation reflectron-type TOF mass spectrometer (LMS) for investigation of chemical composition of surfaces of solar system bodies. Present laboratory efforts are focused on investigation of performance of the instrument. This report extends and follows our recent publication summarising the studies on NIST standard materials, minerals, and meteoritic samples ablated by a nanosecond IR laser radiation [27]. The choice of the IR laser is (although not generally preferred) motivated by simplicity and robustness of this laser. The studies of NIST reference materials, minerals and meteoritic samples have proved high sensitivity of this approach [27]. Nevertheless, low sensitivity in the detection of some nonmetallic elements such as P or S remains and RSCs factors for large number of elements differ clearly from one. To resolve some of possible sources of elemental (isotopic) fractionation, the effects of sampling from one laser spot location, dependence of ion yield on laser irradiance, and formation of clusters are investigated in the present study.

2. Experimental

2.1. Experimental Setup and Principles of Operation. Figure 1 displays a schematic construction of our laser ablation mass analyser. The design and construction details of the mass spectrometer used in the present study can be found in the previous publications [13, 27]. The instrument is located within a UHV chamber that is routinely pumped down to the 10^{-7} mbar range. The sample is placed on X, Y, Z translational stage operated with a position accuracy and resolution in the micrometer range just below the entrance port of the mass analyser. Radiation of a Q-switched Nd:YAG laser system (1064 nm, $\tau \sim 4$ ns, repetition rate 20 Hz) is used to ablate the surface material. The laser radiation is focused to a spot size of about 10 μm at the sample surface by an optical system consisting of beam expander and a doublet lens ($f/200$, NA = 4). The latter is located just above the reflectron. The laser beam is guided via dielectric mirror system and is aligned colinearly with the ion-optical axis of the TOF mass spectrometer. After entering the mass analyser through the window mounted on the top of the reflectron, it travels along the reflectron and central hole (Ø 6.4 mm) of the detector assembly, ion optical components, all way down until it reaches the sample. The material ablated from the surface forms a hot plasma plume containing atomised and ionised species. Only the fraction of positive ions enters the mass analyser through the conical extraction electrode. After entering to the interior of the analyser, ions are further accelerated, focused, and collimated by an electrostatic immersion lens. Subsequently, ions fly through the field-free region and reflectron where they are reflected towards the multichannel plate (MCP) detector. A pair of multichannel plates arranged in a chevron configuration is used to generate the signal after the ions strike the MCP plate. The ions arrive at the MCP detector in a sequence of times proportional to the square root of their mass-to-charge ratio (m/q). The electron current generated by MCP plates is collected by four concentric anode rings and sent to ADC-card.

The laser system initiates the experimental cycle and triggers the data acquisition. The mass spectra acquired from two different anode rings are fed to two channels of a PCI standard high speed ADC-card (Agilent, U1082A, AP240 with averager firmware). This card supports two channels, each with a sampling rate of 1 GS/s with an analog bandwidth of 1 GHz and vertical resolution of 8 bits for a single shot in a single channel mode. The mass spectra were recorded within a mass range of 0–250 amu/q, which corresponds to the flight times of up to ~20 μs.

Untreated sample pieces of solid material are placed on the sample holder before closing and evacuating the vacuum chamber. The measurements in space could be performed using robotic arm and a rotating wheel sample provider [20]. The sample is positioned at a fixed distance (~3 mm) from the entrance to avoid sample contamination by contact with the outer electrode. The laser fluence was controlled by a polarisation-sensitive attenuator, and the laser irradiances within the range of 0.1–1 GW/cm^2 were applied to ablate the surface. The performance of the mass analyser is limited

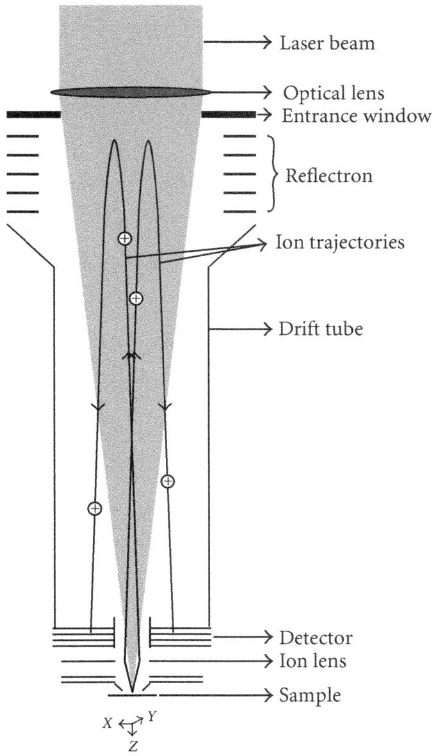

FIGURE 1: Schematic overview of the laser ablation mass spectrometer used in this study. The instrument is of cylindrical symmetry, 120 mm long and its diameter is 50 mm. The laser beam enters from the top and is focussed on the sample placed on XYZ translational stages at the bottom. Positive ions leaving an ablation plume enter the interior of the analyser through the nose-piece electrode for mass analysis (see text for details).

by charged effects when an excessive amount of ions is generated during ablation process. The space charge effects, such as Coulombic repulsion or sample charging, can cause a deterioration of the instrumental mass resolution.

The mass spectra are measured using signals generated at two of four available anode rings. These are fed into an 8 bit ADC card and accumulated by the averager firmware with up to 64′000 triggers per segment. The averager allows the total vertical resolution of 24 bits corresponding to a maximum of 7 decades of dynamic range per channel. Nevertheless, owing to noise present in the card circuit effective signal to noise of ~4-5 decades per channel can be only achieved and to increase the detection sensitivity two or more channels of different signal gains have to be used. In standard acquisition mode, individual spectra from 64′000 laser shots are added *on line,* for each channel, which are kept at different gain levels. The individual spectra from 1′000 laser shots were acquired in order to control homogeneity of the sample. To maintain linearity, the amplitude range of the high-gain (HG) channel overlaps with the range of the low gain (LG) channel. The effective dynamic range of the detection is increased by factor of about ~140. The LG channel is maintained at the gain at which major elements can be measured at full dynamic range supported by this channel and care is taken to avoid saturation effects. In the HG channel, the minor and trace species can be recorded in with better vertical resolution at an expense of saturation of mass peaks corresponding to major elements.

2.2. Mass Spectrometric Analysis. The TOF mass spectra are measured on a linear time scale t. The transformation into mass scale is made with fixed scaling factors by the relation: $m = c(t - t_0)^2$, where m (amu) is mass t (ns) denotes time of flight and c, t_0 are calibration constants, which are determined experimentally [28]. Because the values of c and t_0 constants depend on the voltage settings of the reflectron and the other ion optical components, they can be subject of minor changes. These constants are obtained as parameters in fitting of the flight times for elements identified in the spectrum. The instrumental mass resolution is typically about $m/\Delta m = 500$–600. The peak intensities are determined by a direct integration of the peak area with background corrections, which considers the signal baseline next to the individual mass peak. The direct integration is preferred over a curve fitting because asymmetric peak shapes can be observed for some of the intense isotopic components of the major and minor elements. The curve fitting procedure (e.g., Gauss function) would underestimate the abundances of these species. Nevertheless, the exponentially modified Gauss function (EMG) can also be used to model the experimental mass peaks. Signal ringing due to the small impedance mismatch or MCP feedback effects can sometimes be observed for the intense mass peaks along with a periodic high-frequency noise which is collected by ADC card when a large number of waveforms is accumulated. The numerical filtering of these effects can be, if necessary, performed in a straightforward manner.

The mass spectrum resulting from the accumulation of 60′000 waveforms is measured within seven decades of dynamic range for each LG and HG channel. However, three decades are frequently covered by background noise and effective dynamic range of the signal is about four decades for each channel. By combining the LG and HG spectra, an effective dynamic range of 6-7 decades is obtained. The calculations of atomic abundances for major elements are made by integration of the mass peaks from LG spectrum. An independent check from unsaturated mass peaks in the HG spectrum is done by applying the isotopic normalisation procedure. The isotopic normalisation procedure is also used to determine the abundances of elements in case of isobaric interferences and abundances of the trace elements for which only the most intense isotopic components are measured. The performance of the mass analyser is studied using certified NIST standard reference materials (SRM661, SRM664). Because the abundances of elements in the NIST samples are given as weight percentages, $w\%$, they have to be calculated as atomic percentages, $a\%$, for the comparison with our mass spectrometric results. The atomic abundances are subsequently referenced to the most abundant element in the sample. The comparison of the values determined from measurements and abundances quoted by NIST yields the relative sensitivity coefficients (RSCs), which are a measure of elemental fractionation effects.

3. Results and Discussion

3.1. NIST Standard Reference Materials. The initial measurements of NIST samples (SRM661 and SRM664) have shown high performance of the instrument in terms of mass calibration, mass resolution, and sensitivity [27]. The studies were conducted with laser irradiances (<1 GW/cm^2) somewhat lower than usually applied in laser ablation mass spectrometry (3–8 GW/cm^2) [1]. At these irradiances, relatively slow deepening of the ablation crater and long temporal stability of the ion signal are observed allowing measurement of highly averaged mass spectra from one location. A statistically averaged spectrum, high mass resolution, and measurements of intense and stable ion beam enables preparation of sensitive measurements.

In spite of high sensitivity and reproducibility of the measurements, the quantitative analysis shows that the detection efficiency of different elements can vary significantly.

The correlation between a low detection efficiency and high ionisation potential (IP) could be made for a large fraction of species including nonmetallic elements. Moreover, large RSCs have also been observed, for instance, for Pb and Bi. These could be explained by an enhancement of their concentration in the plasma plume owing to their low boiling temperatures. Furthermore, the detection saturation effects can cause a less sensitive detection of subsequently incoming ions. For example, reduced detection efficiency of Ni, Co elements are observed after detection of Fe ions when studying the NIST SRM661. Figure 2 shows the typical correlation plot between the abundances of the elements quoted by NIST and measured by our instrument. The abundances of metallic elements such as Ti, V, Fe, Mn, and Cr, correlate closely to the quoted values whereas the abundances of nonmetallic elements including C, Si, S, and P can be 100–1000 times lower than the values quoted by NIST.

A number of reasons can contribute to an inaccurate determination of elemental abundances including details of experimental alignment, sampling method, or a poor control over the laser fluence (irradiance). The time-resolved analysis of the ion yield of individual element (isotopic component) and elemental abundances has been performed in parallel with analysis of the evolution of the overall peak envelope. The investigation was conducted by collecting 60 individual spectra at one sample location where each individual spectrum is obtained by adding 1000 waveforms. A sequential analysis of these 60 individual spectra has been performed and the results are displayed in Figure 3. Figure 3(a) shows the temporal variation of the intensity of typical mass peak in the spectrum. An increase of intensity is observed during the first few thousandths of laser shots followed by its abrupt decrease and relatively slow decay until the end of the measurements. Only, small changes of the peak widths are observed and the peak shape evolves from nearly symmetric Gauss-like shape (initially optimised by tuning the voltage settings on the ion optics and reflectron) to the enveloped represented best by the exponentially modified

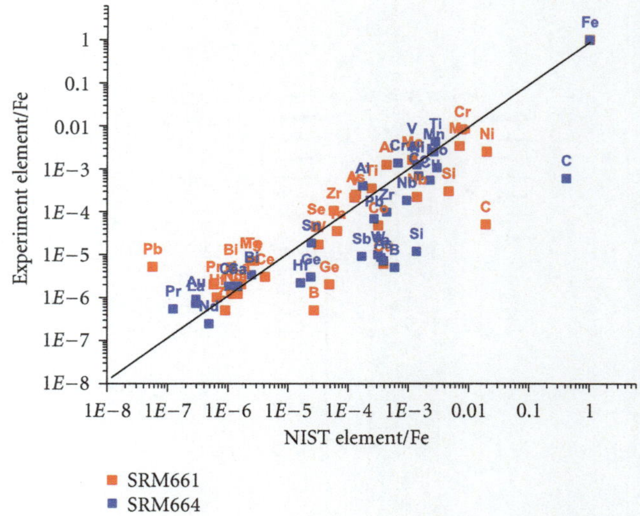

Figure 2: The correlation plot between the elemental abundances of elements quoted by NIST and values determined from the measurements. The measurements underestimate the abundances of nonmetallic elements and elements with high IP's. The abundances of elements possessing low boiling temperatures are, however, clearly overestimated. The measurement of Ni, Co, or Cu abundances suffers from the detector saturation effects caused by saturation detection of ^{54}Fe.

Gauss function with the longer tail directed towards longer flight times. The variations of the peak width and the parameters defining the exponentially-modified Gauss are displayed in the next three panels of Figure 3, respectively. Figure 4 summarises the results from the analysis of the ion yields of individual elements (isotopic components) measured during the initial 15′000 laser shots. First panel to the left displays variation of the elemental ion yield determined after each 1000 laser shots. During the first 5′000 laser pulses, considerable changes of the ion yields for the individual isotopes are observed. The yields of all isotopes are observed to be higher initially and after approximately 5′000 they all drop down but not at the same rates. For some of the elements, the relative yields are observed to reverse with duration of the measurements. This is observed clearly for ^{12}C and ^{98}Mo or ^{48}Ti and ^{51}V. The plot of cumulative sums indicates that the ion yield increases at a different rate for individual elements. The rate of ion yield change tends to stabilise after 10′000–12′000 laser shots. This can be more clearly seen in Figure 4(c) where individual rates of the ion yield change are normalised by the rate of the ion yield of ^{54}Fe.

These results indicate importance of controlling of the measuring procedure for reproducibility of the measurements. Too short measurements would result in the determination of very different elemental abundances than the long ones. A comparable rate of ion yield increase achieved by long sampling procedure guarantees also the reproducibility of the results. Although the elemental abundances measured for NIST standards still differ from the quoted values the determination of RSC factors for the individual elements

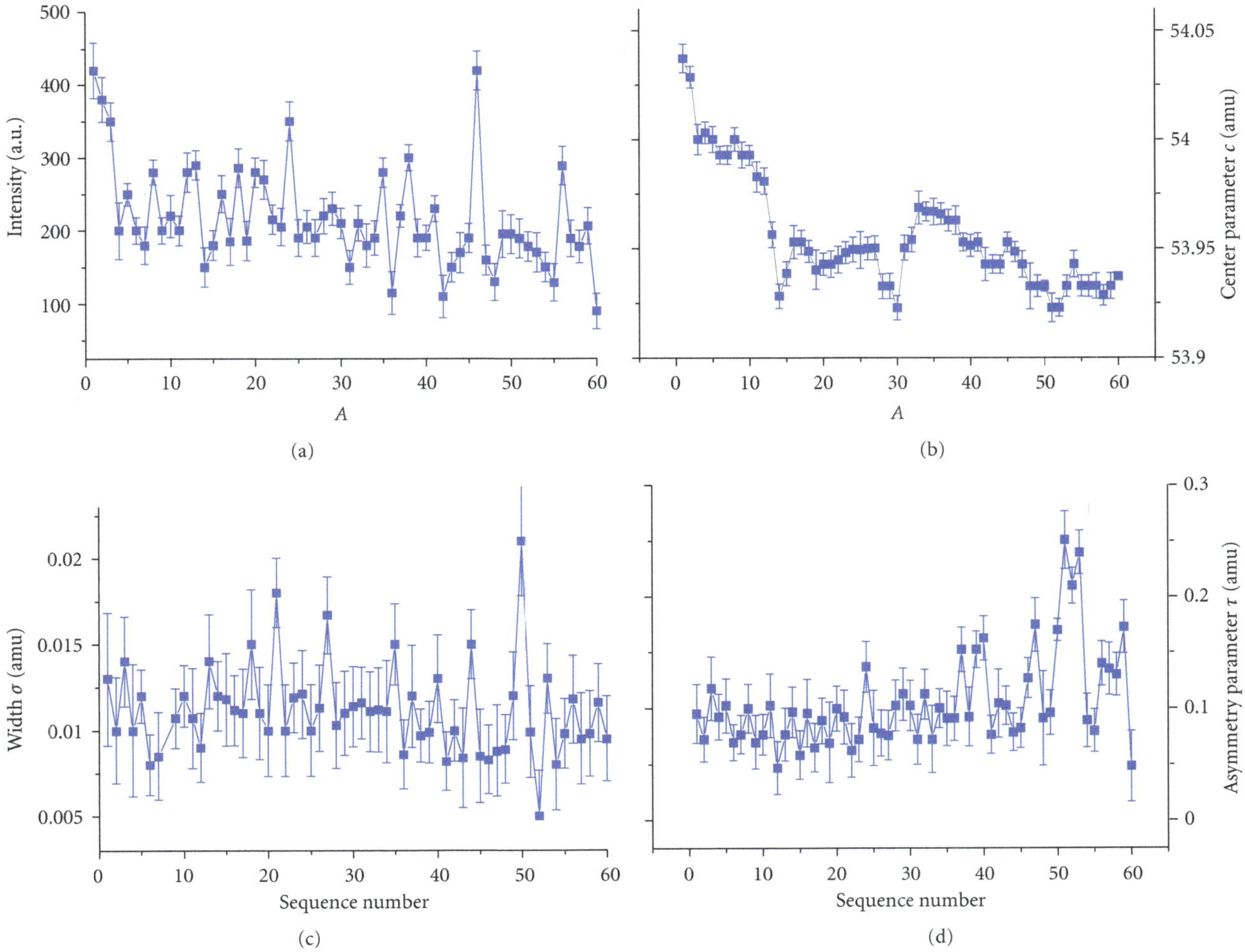

FIGURE 3: The temporal evolution of the intensity and shape of a typical mass peak during the measurements performed from one sample location. ((a) and (c)) display changes of the peak intensity and its width during the measurements. ((b) and (d)) display temporal changes of the mass peak envelope; the peak centre and exponential decay constant define the exponentially modified Gauss function used to model the experimental mass peak.

can be made on reproducible manner and applied to the quantitative measurements of unknown samples.

Observation of intense spectra during first ~5000 laser pulses can be understood when one assumes that initially the measurements were conducted at higher laser irradiances. Taking into account sharp focus conditions applied in our studies (diameter of laser spot is ~8 μm), a decrease of the peak intensities after 5000 shots may indicate that the ablation front is moving out from the laser focus because of the removal of material from the sample. Due to a deepening of the ablation crater, the effective laser spot diameter interacting with surface becomes larger and, in consequence, laser irradiance decreases. A sharp change of the laser irradiance with number of laser pulses can be expected at ablation conditions where the laser focus is located at the level of the irradiated sample surface. With deepening of the crater, the laser irradiance decrease may cause variation of the ion yield of particular elements (ion yields of particular elements are generally irradiance dependant) and yield time-dependent fractionation of their abundances. Clearly, the measurements performed during this phase can be largely uncertain. The changes of the laser spot size seem to be less dramatic for the rest period of the measurement duration. The observation of a change of the mass resolution and peak shape after this initial phase complements this conclusion. It is well known that at the irradiances used in this study, the ion yield of most elements will depend on laser irradiance [1, 21]. The saturation of the ion yield for most of the elements is observed for irradiances larger than few dozens of GW/cm^2 which corresponds to temperature of the plasma about ~50′000 K [26]. Although sharp changes of the laser irradiance at the sample surface may explain our observation, other effects can also contribute, for example, the appearance of the melting rim. Segregation of the elements during the melting and cooling phases clearly can occur and distort the measurement. In any case, these observations indicate

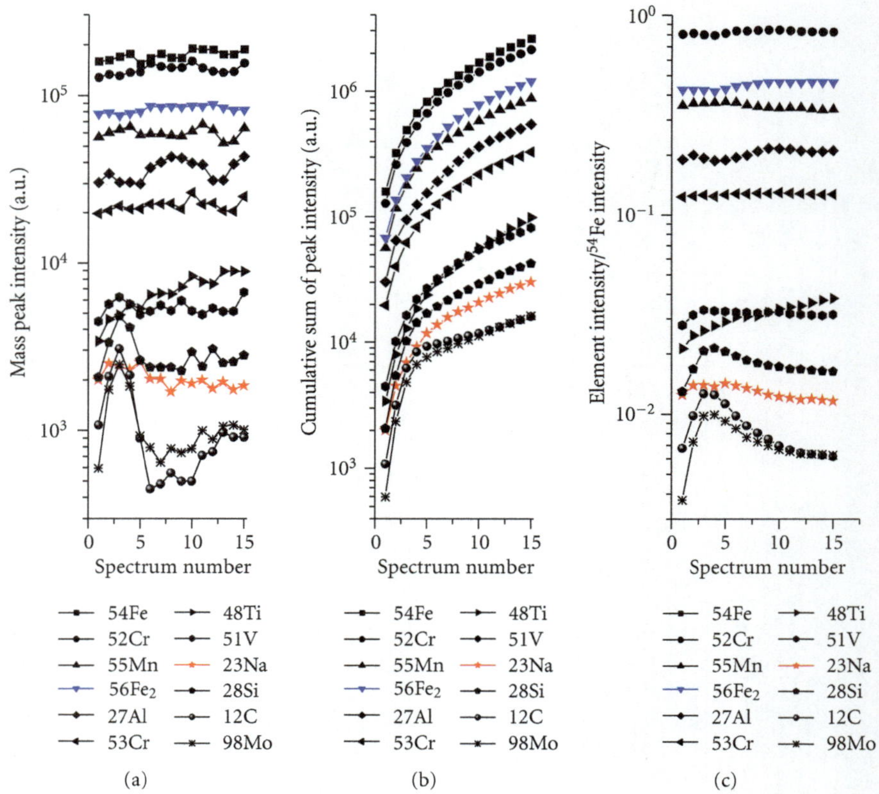

FIGURE 4: The temporal variation of ion yields of individual isotopic mass peaks recorded after each 1000 laser pulses (a). (b) displays a cumulative sum obtained by accumulation of individual ion yields and reflect the rate of ion yield increase during the measurements. The cumulative sums of the ion yields of the individual isotopic components of elements and clusters normalised to the cumulative sum of the ion yield of ^{54}Fe are displayed in (c).

that due to large changes in ion yield observed during the initial few thousandths of laser shots may introduce errors in the abundance determination. It seems, however, that these effects are become well averaged after 15′000 laser pulses. Nevertheless, too long laser sampling from one location may exert material changes due to long-term melting and recrystallisation (melting rims). These can result in changes of the material properties or possible sample damage. Apart to the control of sampling duration, the control over the sample position against the laser focus and laser fluency is essential for achieving quantitative and reproducible results.

Figure 5 displays part of two mass spectra of the NIST SRM661 reference sample obtained in the mass region where B, C, P, and S should be detected. The mass spectra are recorded at irradiances of ~0.3 and ~1 GW/cm^2, respectively. An enhancement of the intensity of the mass peaks of B and C can be observed comparing the low and high irradiance spectra. Isotopic components of ^{10}B, ^{11}B, ^{12}C, and ^{13}C can be seen as well demonstrating a large dynamic range in these spectra. The abundance of the element B in SRM661 quoted by NIST is 5 ppm, corresponding to ^{10}B and ^{11}B isotope abundances of 1 and 4 ppm, respectively. The abundance of C is 0.39% by weight, corresponding to ^{12}C and ^{13}C isotopic abundances of 3900 and 42 ppm, respectively. The isotopic ratios for ^{10}B and ^{11}B isotope determined from the spectra are 21.0% and 79.0%, respectively, compared to the tabulated values of 19.9% and 80.1%. For ^{12}C, and ^{13}C these values are 99.27% and 0.73%, respectively, compared to the tabulated values of 98.9% and 1.07%.

The RSC factors for B and C elements determined from the spectrum in lower panel are of the order of ~0.01 and ~0.001, respectively. These factors are about one order of magnitude larger when the higher irradiance is applied. Other trends of ion yield dependence is observed in a previous report [29]. With increase of the laser fluence, lower abundances of atomic ions were observed but an increase in abundance molecular species composed from these elements.

The RSC factors for S and P are similar to that of C (a rough estimation of RSC factor for S from the studies of Galena sample yields RSC~0.001 at ~0.3 GW/cm^2 [27]). The S and P elements have not been detected in our previous measurements of the NIST samples due to their low abundances (~150 ppm) and low RSC. At higher laser irradiances applied in the present study, a small mass peak of ^{32}S can be measured on the extensively perturbed background curve caused by the saturation detection of Al peak. These studies indicate that the RSCs for the nonmetallic elements can be still improved by applying higher laser irradiances. Unfortunately, at present stage of

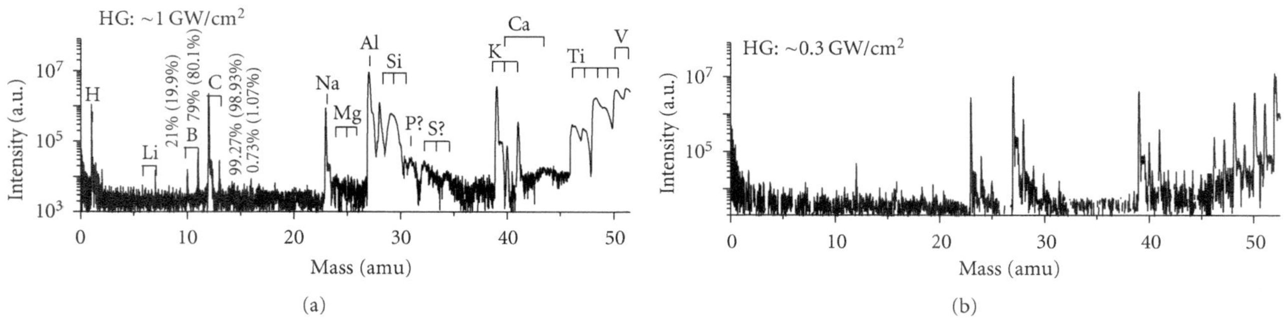

FIGURE 5: Part of two mass spectra obtained at different laser irradiances. The spectrum in (a) shows an increase of the detection efficiency for C and B elements. The mass peak broadening observed in (a) is attributed to space charge effects resulting from the increased ion abundances.

the instrument development, an increase of laser fluence can cause a large decrease of spectral quality. Broadening of the most intense mass lines of the major and minor elements (see Ti, V mass peaks in the upper panel) is clearly observed. Notice that the mass peaks corresponding to low abundant species are still considerable narrow. The broadening is likely caused by space-charge effects (Coulombic repulsion) and could not be compensated by focussing capabilities of the reflectron. The threshold of the ion yield saturation for most of elements is expected for irradiances an order of magnitude larger than applied in an present investigations (~10 GW/cm^2) [30]. The ion yield of most of elements is observed to saturate for most of materials by applying even higher irradiances [21]. Existence of threshold energy is in accordance to several experimental and theoretical studies. In case of metallic samples, the energy threshold depends on the cohesive energy of the metall [31].

Thermal processes induced during ablation of the material by nanosecond lasers cause vaporisation, atomisation, and ionisation of the elements owing to high plasma temperature. However, during the expansion of the plasma plume cooling, recombination processes begin to play a role and molecular fraction can be built up. At relatively low laser irradiances $\ll$0.1 GW/cm^2 molecules or clusters residing on the surface of a solid material can be also removed via desorption into gas phase. Recently, our instrument has been, for instance, used to measure a fraction of carbonaceous contamination on charge state conversion surfaces [32]. In the present investigations, the mass spectra of SRM664 sample have been recorded at quasidesorption (about 10 MW/cm^2) and ablation (~0.4 GW/cm^2) conditions (Figure 6). The spectra were taken in LG channel, each time from a fresh surface location. Intense mass peaks of metallic oxides (MeO_x), clusters, and cluster oxides were measured. The application of higher laser irradiance removes rapidly the oxidation layer, and the characteristic isotopic pattern of low abundant Fe_2 cluster can be only seen. Other metallic clusters and sometimes residual abundances of iron oxides can also be observed in the spectra recorded in HG channel (~100 times amplified signal). These clusters can be built up via plasma chemistry. The cluster abundances do not change significantly with an increase of the laser fluence. Only for irradiances close to ~10 GW/cm^2, a considerable reduction of cluster contribution was observed [33]. Nevertheless, even for much higher irradiances, clusters are not disappearing and can cause isobaric interferences [29]. Due to deficiency of oxygen in the SRM standards, the molecular fraction containing various oxides is reduced providing basis for the analysis of elements in the mass range of 63–160 amu. Noticeable is also a decrease of abundances of Na and K. These elements are not present in the NIST sample but frequently are observed as a contamination of sample surface.

The relative abundances of oxides and clusters observed in our studies are qualitatively consistent with previous reports. By using IR laser, a thermal ablation mechanism is appropriate for the description of ablation process, which explains formation of clusters and reduced abundances of oxides. The laser fluence change may exert competitive effects on cluster formation. Increase of the laser fluence can cause an increase of the plasma densities and the rate of cluster formation. On the other hand, the plasma temperature increase can result in a destabilisation of clusters or even disable larger cluster formation [34]. For low laser irradiances (<10^7 W/cm^2), the clusters can be formed via desorption/ionisation mechanism, but at larger irradiances, plasma chemistry is essential [35, 36]. When UV radiation will be used for ablation, further reduction of cluster abundances is to be expected owing to an enhancement of the rate of photodissociation processes induced by more energetic UV photons [37]. A complete absence of clusters is observed when a short pulse duration laser beams are applied [38].

The assignment of the mass peaks in the spectra of SRM661 and SRM664 is straightforward for masses smaller than 63 amu. For mass peaks of heavier elements, one has to consider possible interferences caused by presence of clusters and oxides. To assess the contribution of clusters in the spectrum, a good isotopic signature of the cluster mass peaks is necessary. Previous studies show a good performance of the instrument for investigation of the isotope ratios of elements [27]. The isotopic ratios could be determined with

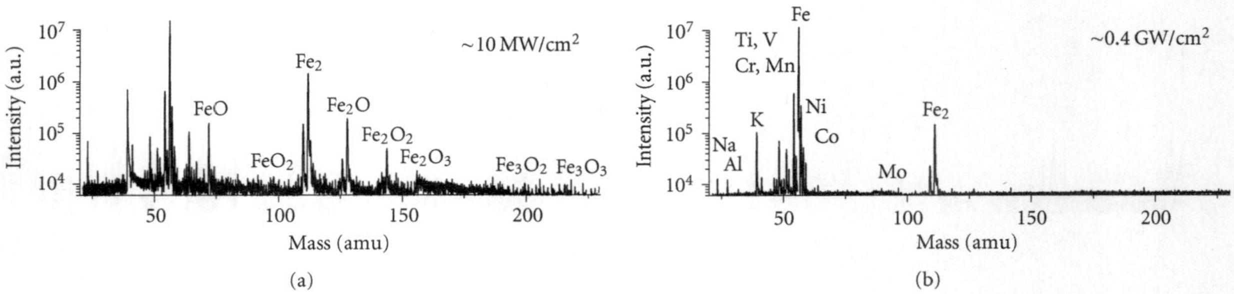

FIGURE 6: The mass spectrum of SRM664 sample recorded in quasidesorption (a) and ablation mode (b). Both, elements and clusters can be measured and assigned considering their isotopic pattern.

the uncertainty better than 1%. This knowledge can be used for determination of elemental abundances in case of isobaric interferences caused by an overlap with other elements or clusters by taking into account the standard isotopic normalization procedure. However, the detection of trace elements, for which sometimes only the most intense isotopic components can be observed, has to be considered with a care owing to multiplicity of possible isotopic components of clusters (see, e.g., next sub-section). Figure 7 shows the mass spectra of Fe, Mo, and Pb, their oxides/sulphides, and clusters. The measurements of Fe species were conducted on NIST SRM664 sample (in quasidesorption mode), studies of Mo species-on high purity Mo material, and Pb clusters were measured using Galena sample. In case of Fe species, a small contamination due to abundances of other minor metallic elements may interfere with the Fe isotopic lines and isotopic Pb ratios are affected by a fine natural fractionation, respectively.

Table 1 shows the calculated and measured isotopic abundances of element, oxide/sulphide, and clusters. The abundance ratios of isotopic components for cluster, oxides, or sulphides are obtained using their natural abundance ratios. The relative abundances of diatomic clusters are calculated using multi-nominal expansions, which are well-known expressions [39]. The isotopic abundances are normalised to 100%, their values are given in Table 1. Usually they do not differ by more than 2% of the values calculated from the appropriate expressions. The calculated isotopic percentages are compared with measured ones and represented as bar spectra in Figure 7. The errors of isotopic fractions of oxides/sulphides and clusters are somewhat larger than those for elements, which may be partially attributed to accuracy of the integration method and possible fractionation caused by plasma chemistry. In principle, the mass peaks corresponding to elements could be extracted from the spectra contaminated by diatomic clusters by subtracting cluster isotopic pattern modelled from the terrestrial isotopic composition of corresponding element. One has to be aware, however, that the fractionation effects due to chemical recombination reactions in plasma may be large and are species- and cluster-specific. Fractionation effects caused by chemistry can be clearly observed in the relative abundance distribution of various oxides and clusters. They do not follow strictly the elemental abundance distribution. The knowledge of the isotopic distribution of clusters can be helpful for conducting the quantitative control and is of considerable interest in speciation analysis [36].

3.2. Allende Sample. In our previous study, the applicability of the instrument to study of planetary samples has been demonstrated by measurements of the elemental composition of the sample of Allende meteorite [27]. Light elements (Li, Na, Mg, Al, K, Ca) and heavier elements (Ti, V, Cr, Mn, Fe, Ni, Co) could be measured. With a careful control of the experimental conditions and calibration of elemental abundances by appropriate standard, the diagnostic of meteoritic material and its assignment to appropriate meteoritic class can be made [27]. Allende belongs to the group of carbonaceous chondrites (CV3) having primitive chemical composition, meaning that its composition changed only slightly from the time of their formation [40]. The studies of trace elements distributions in highly heterogeneous structure of the Allende meteorite (chondrules, calcium aluminium-rich inclusions (CAIs), fine-grained regions, and several mineral phases) are of considerable interest. They can allow inferring the roles of various types of elemental fractionations in the solar system, including sorting and segregation of chondrite components. Our investigations of standard materials show clearly possibilities for detection of elements abundant at sub-ppm and have stimulated present investigations.

Figure 8 displays a typical mass spectrum of Allende sample. The spectrum can be measured within 6-7 decades of the effective dynamic range, and the analysis of element abundances can be made with the support of the experimental RSC factors for almost all elements up to 63 amu [27]. Nevertheless, the assignment of heavier elements can be difficult due to presence of abundant metallic oxides and clusters. These can be identified in the spectrum taking into account their characteristic isotopic patterns and the mass peaks of iron species clearly dominate in the spectrum (see Figure 9). Although assignment of most of mass peaks can be made relatively easy studying isotopic patterns of metallic clusters and their oxides the contribution of the elements could not be safely elucidated due to spectral complexity. Thus, with the exception of Pb, the identification of other elements could not be made safely this time.

TABLE 1: Measured and calculated isotope abundance ratio of Fe, Mo, and Pb oxides/sulphides and their diatomic clusters determined from the experiment and calculated from the multi-nominal expansions. The values marked by (∗) are adopted from the calculation.

Element /cluster	Exp. abundance %	Nat. abundance %	Element /cluster	Exp. abundance %	Nat. abundance %
Fe			Pb		
54	5.834	5.845	204	1.5	1.4
56	91.773	91.754	206	24.6	24.1
57	2.111	2.119	207	21.6	22.1
58	0.282*	0.282	208	52.4	52.4
FeO			PbS		
70	5.45	6.12	236	1.419	1.331
71	0.32	0.48	237	0.010*	0.010
72	88.22	88.93	238	24.415	22.972
73	5.70	2.56	239	21.768	21.192
74	0.93	0.94	240	52.273	50.923
75	0.49*	0.49	241	1.334*	1.334
76	0.48*	0.48	242	2.9	2.231
			244	0.002*	0.002
			245	0.005*	0.005
Fe_2			Pb_2		
108	0.66	0.40	408	0.018*	0.018
110	10.30	10.73	410	0.63*	0.63
111	0.41	0.31	411	0.57*	0.57
112	82.95	83.86	412	5.607	6.75
113	4.71	3.93	413	9.479	9.89
114	0.97	0.62	414	29.00	35.16
115	0.07*	0.07	415	19.456	21.50
116	0.06*	0.06	418	36.457	25.48
Mo			Mo_2		
92	14.78	14.77	184	1.945	2.079
94	8.66	9.23	186	2.234	2.598
95	16.32	15.90	187	4.293	4.476
96	18.02	16.68	188	5.415	5.508
97	9.61	9.56	189	5.396	5.489
98	23.41	24.19	190	11.72	12.15
100	9.20	9.67	191	7.27	6.74
			192	12.23	17.22
			193	11.33	10.37
			194	10.55	10.26
			195	8.18	7.34
			196	9.89	8.65
			197	1.64	1.76
			198	6.56	4.46
			200	1.34	0.89
MoO					
108	14.40	14.734	114	25.47	24.169
109	0.006*	0.006	115	0.029*	0.029
110	8.90	9.238	116	9.810	9.696
111	15.90	15.865	117	0.004*	0.004
112	16.60	16.664	118	0.020*	0.020
113	9.30	9.576			

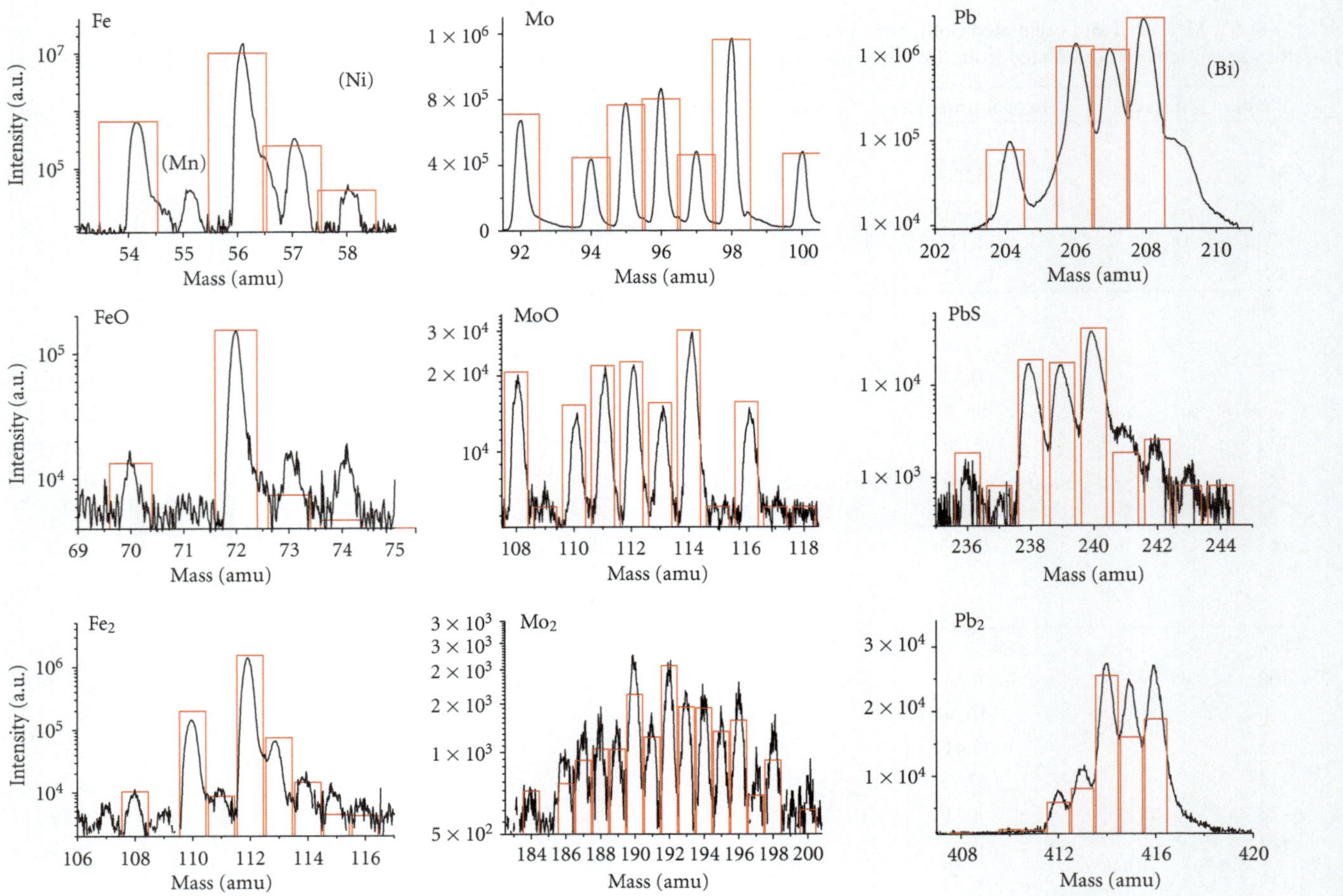

FIGURE 7: Isotopic pattern of Fe, Mo, and Pb elements and their oxides and clusters. Bar spectra of oxides and clusters represent isotopic pattern calculated from natural isotopic distribution of accompanying elements. The isotopic abundances of elements are usually reproduced with 1% accuracy whereas the isotopomers of oxides and clusters can be measured generally within 2% of the calculated value.

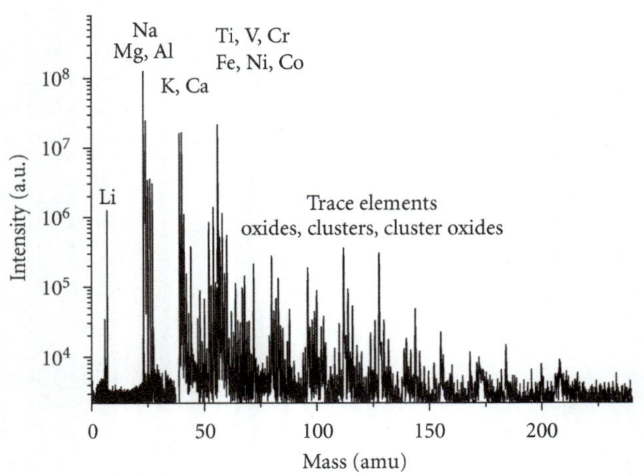

FIGURE 8: A typical mass spectrum of Allende sample, which can be measured at various sample locations. Similar mass peak distributions, relatively insensitive to laser irradiance increase are observed in the mass range corresponding to trace elements and are attributed to metallic oxides, clusters, and cluster oxides.

The observation of abundant metallic clusters and oxides in the studied meteoritic sample is not surprising. Owing to high affinity of metallic elements to oxygen, their oxides can be formed relatively easy due to plasma chemistry. In contrast to NIST standard materials, where oxygen content in the sample is very low, the detection of trace elements can be made with a high confidence whereas the detection of trace elements in mineral samples can be more difficult due to isobaric interferences caused by presence of clusters. Abundances of oxides detected in present study could not be reduced by applying higher laser irradiances. Other laser ablation studies show as well that the formation of clusters and oxides is not avoidable and results from variety of chemical processes which can take place in plasma [41–45].

The trace element analysis by laser ablation mass spectrometry can be sometimes difficult for complex mineral samples and isobaric interferences due to presence of oxides and clusters can limit detection efficiency. An ablation process induced by ns-lasers is complex owing to several competing processes and rich plasma chemistry. Nevertheless, ablation studies by lasers with pulse durations in

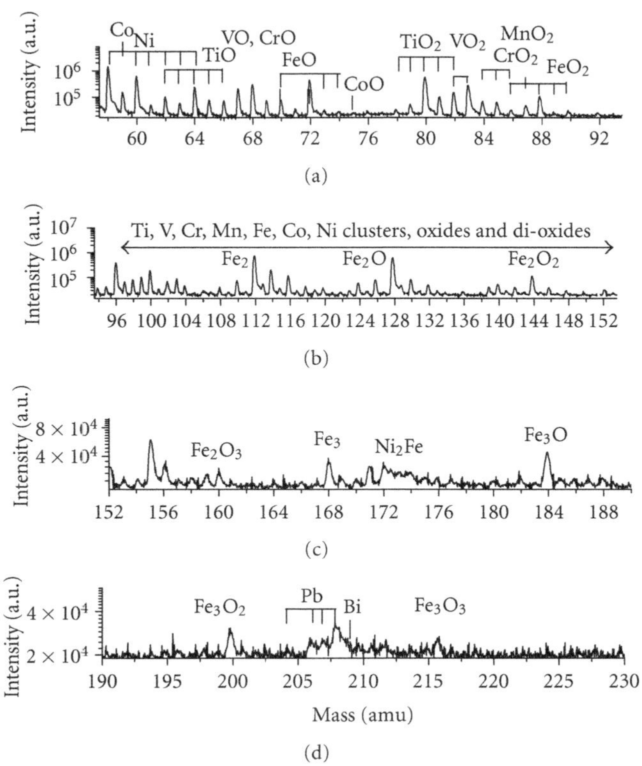

FIGURE 9: The assignment of the mass spectrum recorded in the mass range corresponding to heavier trace elements. Appearance of the most mass peaks is consistent with the observation of clusters and their oxides, which can be confirmed considering their characteristic isotopic patterns.

picosecond or femtosecond range are expected to improve the performance of our analyser. A femtosecond-laser-induced ablation process can result in a significant reduction or even absence of interfering clusters and oxide.

4. Summary

A miniaturised laser ablation time of flight mass spectrometer designed for planetary research has been used to study the elemental and isotopic composition of untreated solid samples. The instrument is capable of recording well-calibrated mass spectra and its mass resolution can be higher than ~600. A high sensitivity of the measurements could be achieved in studies of NIST standard samples and the detection of trace elements at a sub-ppm level could be performed. For reproducibility of the measurements, however, well-defined measuring procedure has to be maintained. Present studies investigate influence of the sampling duration and the laser irradiance on reproducibility of the measurements. Appropriate control over the number of laser pulses used for the ablation is essential for reproducibility of the results and determination of relative sensitivity factors in studies of standard reference materials. These can be subsequently used for the quantitative measurement of elemental composition of unknown samples. However, for complex samples containing several metallic elements and abundant in oxygen/sulphur, severe contamination in mass range 63–200 is to be expected due to formation of oxides and clusters. These can make sometimes studies of heavier trace elements difficult due to isobaric interferences.

The isotopic abundance ratio can be measured typically within 1% of the value given from the standard isotopic distribution. The abundances of isotopomers of clusters can be measured within 2% of the value calculated from multinominal expansions indicating a good quantitative performance of the instrument for conducting the isotopic measurements.

Present study provides hints for the improvements of the quantitative capabilities of the instrument. The control over the experimental conditions such sample alignment and standard instrumental performance in terms of mass resolution is essential. Nevertheless, one of the main reason for a semiquantitative performance of the instrument seems to be the application of ns-resolved infrared laser radiation for the ablation of sample. The elemental fractionation effects in the detection of nonmetallic elements are relatively large and readable abundances of clusters and oxides formed via plasma chemistry are observed. The activities aimed at a more precise control over the laser fluence/irradiance, better accommodation of the plasma plume at higher irradiances are now undertaken along with application of other laser systems including also short-pulse duration lasers.

Acknowledgments

Technical support from Harald Mischler is gratefully acknowledged. This work is supported by the Swiss National Science Foundation.

References

[1] J. S. Becker, *Inorganic Mass Spectrometry*, John Wiley & Sons, 2007.

[2] K. E. Jarvis and J. G. Williams, "Laser ablation inductively coupled plasma mass spectrometry (LA-ICP-MS): a rapid technique for the direct, quantitative determination of major, trace and rare-earth elements in geological samples," *Chemical Geology*, vol. 106, no. 3-4, pp. 251–262, 1993.

[3] V. Hoffmann, M. Kasik, P. K. Robinson, and C. Venzago, "Glow discharge mass spectrometry," *Analytical and Bioanalytical Chemistry*, vol. 381, no. 1, pp. 173–188, 2005.

[4] C. J. Koester and A. Moulik, "Trends in environmental analysis," *Analytical Chemistry*, vol. 77, no. 12, pp. 3737–3754, 2005.

[5] R. Lobinski, C. Moulin, and R. Ortega, "Imaging and speciation of trace elements in biological environment," *Biochimie*, vol. 88, no. 11, pp. 1591–1604, 2006.

[6] P. Wurz, D. Abplanalp, M. Tulej et al., "In situ mass spectrometric analysis in planetary science," *Solar System Research*. In press.

[7] R. Rieder, T. Economou, H. Wänke et al., "The chemical composition of martian soil and rocks returned by the mobile alpha proton x-ray spectrometer: preliminary results from the x-ray mode," *Science*, vol. 278, no. 5344, pp. 1771–1774, 1997.

[8] H. Y. McSween Jr., R. L. McNutt Jr., and T. H. Prettyman, "Spacecraft instrument technology and cosmochemistry," *The Proceedings of the National Academy of Sciences of the United States of America*, pp. 1–6, 2011.
[9] L. A. McDonnell and R. M.A. Heeren, "Imaging mass spectrometry," *Mass Spectrometry Reviews*, vol. 26, no. 4, pp. 606–643, 2007.
[10] R. Huang, B. Zhang, D. Zou, W. Hang, J. He, and B. Huang, "Elemental imaging via laser ionization orthogonal time-of-flight mass spectrometry," *Analytical Chemistry*, vol. 83, no. 3, pp. 1102–1107, 2011.
[11] J. S. Becker and H. J. Dietze, "State-of-the-art in inorganic mass spectrometry for analysis of high-purity materials," *International Journal of Mass Spectrometry*, vol. 228, no. 2-3, pp. 127–150, 2003.
[12] L. Matus, H. M. Seufert, and K. P. Jochum, "Microanalysis of geological samples by laser plasma ionization mass spectrometry (LIMS)," *Fresenius' Journal of Analytical Chemistry*, vol. 350, no. 4-5, pp. 330–337, 1994.
[13] U. Rohner, J. A. Whitby, and P. Wurz, "A miniature laser ablation time-of-flight mass spectrometer for in situ planetary exploration," *Measurement Science and Technology*, vol. 14, no. 12, pp. 2159–2164, 2003.
[14] U. Rohner, J. A. Whitby, P. Wurz, and S. Barabash, "Highly miniaturized laser ablation time-of-flight mass spectrometer for a planetary rover," *Review of Scientific Instruments*, vol. 75, no. 5, pp. 1314–1322, 2004.
[15] H. Balsiger, K. Altwegg, P. Bochsler et al., "Rosina—Rosetta orbiter spectrometer for ion and neutral analysis," *Space Science Reviews*, vol. 128, no. 1–4, pp. 745–801, 2007.
[16] D. Abplanalp, P. Wurz, L. Huber et al., "A neutral gas mass spectrometer to measure the chemical composition of the stratosphere," *Advances in Space Research*, vol. 44, no. 7, pp. 870–878, 2009.
[17] W. B. Brinckerhoff, G. G. Managadze, R. W. McEntire, A. F. Cheng, and W. J. Green, "Laser time-of-flight mass spectrometry for space," *Review of Scientific Instruments*, vol. 71, no. 2 I, pp. 536–545, 2000.
[18] G. G. Managadze and I. Y. Shutyaev, *Chemical Analysis*, vol. 124, John Wiley & Sons, New York, NY, USA, 1993.
[19] L. M. Zelenyi and A. V. Zakharov, "Phobos-grunt project: devices for scientific studies," *Solar System Research*, vol. 44, no. 5, pp. 359–361, 2010.
[20] G. G. Managadze, P. Wurz, R. Z. Sagdeev et al., "Study of the main geochemical characteristics of phobos' regolith using laser time-of-flight mass spectrometry," *Solar System Research*, vol. 44, no. 5, pp. 376–384, 2010.
[21] Y. Lin, Q. Yu, W. Hang, and B. Huang, "Progress of laser ionization mass spectrometry for elemental analysis—a review of the past decade," *Spectrochimica Acta B*, vol. 65, no. 11, pp. 871–883, 2010.
[22] Q. Yu, L. Chen, R. Huang et al., "Laser ionization time-of-flight mass spectrometry for direct elemental analysis," *Trends in Analytical Chemistry*, vol. 28, no. 10, pp. 1174–1185, 2009.
[23] R. E. Honig and J. R. Woolston, "Laser-induced emission of electrons, ions, and neutral atoms from solid surfaces," *Applied Physics Letters*, vol. 2, no. 7, pp. 138–139, 1963.
[24] N. C. Fenner, "Ion energies in the plasma produced by a high power laser," *Physics Letters*, vol. 22, no. 4, pp. 421–422, 1966.
[25] M. Southon, M. Witt, A. Harris, E. Wallach, and J. Myatt, "Laser-microprobe mass-analysis of surface layers and bulk solids," *Vacuum*, vol. 34, no. 10-11, pp. 903–909, 1984.
[26] J. He, R. Huang, Q. Yu, Y. Lin, W. Hang, and B. Huang, "A small high-irradiance laser ionization time-of-flight mass spectrometer," *Journal of Mass Spectrometry*, vol. 44, no. 5, pp. 780–785, 2009.
[27] M. Tulej, M. Iakovleva, I. Leya, and P. Wurz, "A miniature mass analyser for in-situ elemental analysis of planetary material-performance studies," *Analytical and Bioanalytical Chemistry*, vol. 399, no. 6, pp. 2185–2200, 2011.
[28] S. Scherer, K. Altwegg, H. Balsiger et al., "A novel principle for an ion mirror design in time-of-flight mass spectrometry," *International Journal of Mass Spectrometry*, vol. 251, no. 1, pp. 73–81, 2006.
[29] X. Wang, S. Amoruso, A. Tortora et al., "Analysis of charged fragments emitted during excimer laser ablation of YNi2B2C borocarbide targets by time-of-flight mass spectrometry," *Applied Surface Science*, vol. 186, no. 1–4, pp. 303–308, 2002.
[30] L. Torrisi, "Fractional ionization in plasmas produced by pulsed laser ablation," *Radiation Effects and Defects in Solids*, vol. 157, no. 3, pp. 347–356, 2002.
[31] L. Torrisi, G. Ciavola, S. Gammino et al., "Metallic etching by high power Nd:yttrium-aluminum-garnet pulsed laser irradiation," *Review of Scientific Instruments*, vol. 71, no. 11, pp. 4330–4334, 2000.
[32] A. Riedo, P. Wahlström, J. A. Scheer, P. Wurz, and M. Tulej, "Effect of long duration UV irradiation on diamondlike carbon surfaces in the presence of a hydrocarbon gaseous atmosphere," *Journal of Applied Physics*, vol. 108, no. 114915, 2010.
[33] S. Amoruso, A. Amodeo, V. Berardi, R. Bruzzese, N. Spinelli, and R. Velotta, "Laser produced plasmas in high fluence ablation of metallic surfaces probed by time-of-flight mass spectrometry," *Applied Surface Science*, vol. 96–98, pp. 175–180, 1996.
[34] J. Álvarez-Ruiz, M. López-Arias, R. De Nalda, M. Martín, A. Arregui, and L. Bañares, "Generation of CdS clusters using laser ablation: the role of wavelength and fluence," *Applied Physics A*, vol. 95, no. 3, pp. 681–687, 2009.
[35] V. Ignatova, L. Van Vaeck, R. Gijbels, and F. Adams, "Molecular speciation of inorganic mixtures by Fourier transform laser microprobe mass spectrometry," *International Journal of Mass Spectrometry*, vol. 225, no. 3, pp. 213–224, 2003.
[36] B. Yan, L. Li, Q. Yu, W. Hang, J. He, and B. Huang, "High irradiance laser ionization mass spectrometry for direct speciation of iron oxides," *Journal of the American Society for Mass Spectrometry*, vol. 21, no. 7, pp. 1227–1234, 2010.
[37] R. Torres and M. Martin, "Laser ablation and time-of-flight mass-spectrometric study of SiO," *Applied Surface Science*, vol. 193, no. 1–4, pp. 149–155, 2002.
[38] R. Hergenröder, O. Samek, and V. Hommes, "Femtosecond laser ablation elemental mass spectrometry," *Mass Spectrometry Reviews*, vol. 25, no. 4, pp. 551–572, 2006.
[39] R. K. Boyd, C. Basic, and R. A. Bethem, *Trace Quantitative Analysis by Mass Spectrometry*, John Wiley & Sons, 2008.
[40] H. Y. McSween Jr. and G. R. Huss, *Cosmochemistry*, Cambridge University Press, Cambridge, UK, 2010.
[41] F. Aubriet, C. Poleunis, J. F. Muller, and P. Bertrand, "Laser ablation and secondary ion mass spectrometry of inorganic transition-metal compounds. Part I: comparison between static ToF-SIMS and LA-FTICRMS," *Journal of Mass Spectrometry*, vol. 41, no. 4, pp. 527–542, 2006.
[42] L. Van Vaeck, A. Adriaens, and F. Adams, "Microscopical speciation analysis with laser microprobe mass spectrometry and static secondary ion mass spectrometry," *Spectrochimica Acta B*, vol. 53, no. 2, pp. 367–378, 1998.
[43] B. Maunit, A. Hachimi, P. Manuelli, P. J. Calba, and J. F. Muller, "Formation of iron oxides clusters induced by

resonant laser ablation/ionization," *International Journal of Mass Spectrometry and Ion Processes*, vol. 156, no. 3, pp. 173–187, 1996.

[44] A. T. De Ville D'Avray, E. E. Carpenter, C. J. O'Connor, and R. B. Cole, "Characterization of ferrite nanoparticles by laser desorption/ionization mass spectrometry," *European Journal of Mass Spectrometry*, vol. 4, no. 6, pp. 441–449, 1998.

[45] F. Aubriet and J. F. Muller, "About the atypical behavior of CrO_3, MoO_3, and WO_3 during their UV laser ablation/ionization," *Journal of Physical Chemistry A*, vol. 106, no. 25, pp. 6053–6059, 2002.

Spectrophotometric Method for Determination of Five 1,4-Dihydropyridine Drugs Using *N*-Bromosuccinimide and Indigo Carmine Dye

Mohamed A. El Hamd,[1] Sayed M. Derayea,[2] Osama H. Abdelmageed,[3] and Hassan F. Askal[4]

[1] *Department of Pharmaceutical Analytical Chemistry, Faculty of Pharmacy, Al-Azhar University, Assiut 71524, Egypt*
[2] *Department of Pharmaceutical Analytical Chemistry, Faculty of Pharmacy, Minia University, Minia 61519, Egypt*
[3] *Department of Pharmaceutical Chemistry, Faculty of Pharmacy, King Abdulaziz University, Jeddah 21589, Saudi Arabia*
[4] *Department of Pharmaceutical Analytical Chemistry, Faculty of Pharmacy, Assiut University, Assiut 71526, Egypt*

Correspondence should be addressed to Mohamed A. El Hamd; abo_elhamd2007@yahoo.com

Academic Editor: Hakan Arslan

Indirect spectrophotometric method is described for quantification of five of 1,4-dihydropyridine (1,4-DHP) drugs using *N*-bromosuccinimide (NBS) with the aid of indigo carmine (INC) dye. The method is based on addition of known excess of NBS to an acidified solution of 1,4-DHP drugs and determining the residual of NBS through its ability to bleach the colour of the used dye; the amount of NBS that reacted corresponded to the amount of drugs. Beer's law is obeyed in the concentration range 1.25–13.00 μg/mL. Good correlation coefficients (0.998-0.999) were found between the absorbance values and the corresponding concentrations. Limits of detections ranged from 0.141 to 0.500 μg/mL. The proposed method was successfully applied to the analysis of dosage forms; percent of recoveries ranged from 97.31 to 99.46% without interference from any common excipients. The statistical comparison by Student's t-test and variance ratio F-test showed no significant difference between the proposed and official or reported methods.

1. Introduction

1,4-DHP derivatives are vasodilators with a great selectivity for vascular smooth muscles; they are primarily used for treatment of some of cardiovascular diseases such as hypertension, angina, and cardiac arrhythmias. Recently, they were used for the treatment of other pathological states, such as seizures and cerebral ischemic disorders [1, 2].

Many analytical methods were developed for determination of 1,4-DHP drugs, namely, nifedipine (NIF), nicardipine (NIC), nimodipine (NIM), felodipine (FEL), and amlodipine (AML) (Figure 1), in their pharmaceutical formulations and in biological fluids, such as titrimetric methods [3, 4], spectrometric methods (spectrophotometry [5–13] or spectrofluorimetry [12, 14–20]), electrochemical methods [21–23], liquid chromatographic methods [24–28], and gas chromatographic methods [29–32].

Classical spectrophotometric measurements and their applications are considered as a routine analytical method in most quality control laboratories. Subsequently, the aim of this study to develop a simple, sensitive, and cost-effective method for determination of these drugs in pure and dosage forms using spectrophotometric technique. The method utilized NBS-INC reagents (Figure 2) and their developed offer the advantage of simplicity, sensitivity, speed, accuracy, and precision without the need for costly equipment/chemicals.

2. Experimental

2.1. Instrumentation. Absorbance measurements were made on Shimadzu 1601PC, UV-visible spectrophotometer (Shimadzu, Tokyo, Japan), Jenway 6305, UV-visible spectrophotometer (Jenway LTD, U.K).

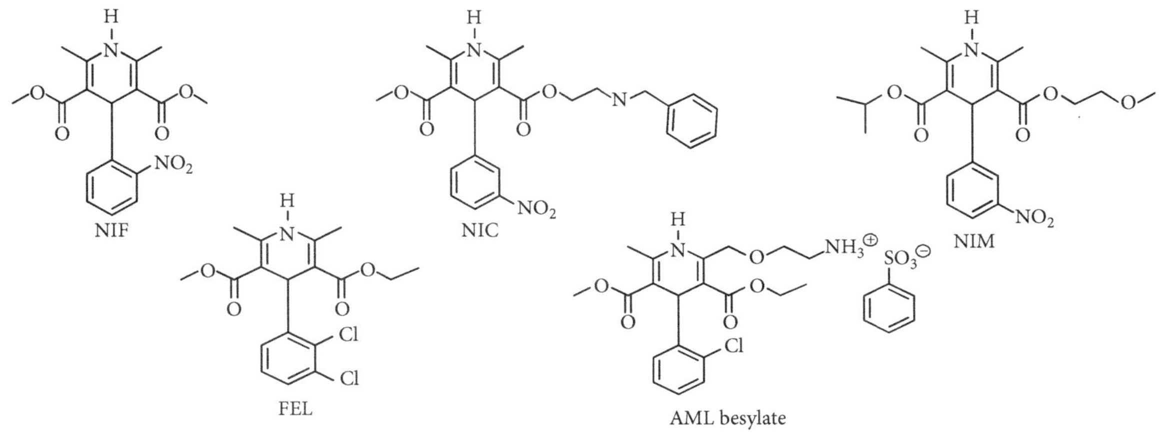

FIGURE 1: Chemical structure of the investigated 1,4-DHP drugs.

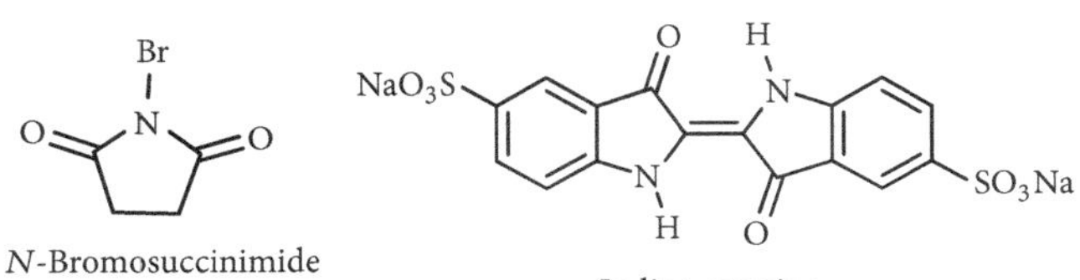

FIGURE 2: Chemical structure of the chemical reagents.

2.2. Reagents and Solutions. All chemicals, solvents, and reagents were of analytical grade, and their solutions were prepared and diluted in distilled water.

2.2.1. NBS Reagent. NBS reagent (Aldrich Co Ltd., Gillingham-Dorst, Germany) was prepared fresh daily in distilled water as a concentration 0.02% w/v.

2.2.2. INC Dye. INC dye (Winlab Co., UK) was prepared fresh daily in distilled water as a concentration 0.07% w/v.

2.2.3. Reference Standard of Pure Cited Drugs. Reference standard of pure cited drugs (NIF, atenolol, NIC, NIM, FEL, AML besylate, and metoprolol) was generously supplied by their respective manufacturers in Egypt, without any conflicts of interests in our submitted paper.

Stock of 1,4-DHP standard solutions was prepared fresh daily by dissolving 50 mg of each of them in 2.0 mL of perchloric acid (70%, 11.6 M) using 100 mL calibrated flask; the contents of the flask were swirled then completed to the volume with diluted aqueous 1.0 M perchloric acid. The working standard solutions were prepared by further dilution with distilled water to obtain the final aliquots concentration.

2.3. Procedure for Calibration Curves. An aliquot of 1.0 mL of the standard or sample solution was transferred into a 10 mL calibrated flask. 1.0 mL of NBS (0.02% w/v) was added and mixed well. The reaction was allowed to proceed for 25 min at room temperature. 1.0 mL of INC (0.07% w/v) was added. The resulting solution was allowed to stand for further 10 min. The solution was diluted to the mark with distilled water, and the absorbance was measured at 607 nm against a blank experiment treated similarly.

2.4. General Pretreatment Procedures for Tablets and Capsules

2.4.1. Pretreatment Procedure for the Assay of Tablet and Capsule Samples. An amount equivalent to 50 mg of active ingredient from 20 finally powdered tablets or 20 mixed capsules was dissolved in 10.0 mL chloroform in a beaker. The contents of the beaker were swirled, sonicated for about 5 min, then filtered to 100 mL calibrated flask through a Whatman no. 42 filter paper, and washed with small amount of chloroform. The filtrate was evaporated to dryness, and the residue was dissolved in 2.0 mL of perchloric acid and then completed to the marke by diluted 1.0 M perchloric acid to give 0.5 mg/mL of sample solution then the working solutions were treated as mentioned above.

2.4.2. Pretreatment Procedure for the Assay of Tablet and Capsule Samples Containing Two Drugs. 20 tablets (Logimax tablets, containing FEL and metoprolol) or capsules (Tenolat SR capsules, containing NIF and atenolol) were finely powdered. An accurately quantity equivalent to 50 mg of the active ingredient was dissolved in about 10.0 mL of ethylether, or in 10.0 mL of chloroform for tablets and capsules in a beaker, respectively. The contents of the beaker were swirled, sonicated for 5 min, and then filtered. The filtrate was evaporated to dryness, and then the procedure was completed as mentioned before.

3. Results and Discussion

NBS is an oxidizing agent and perhaps the most important positive bromine containing organic compound, used for spectrophotometric determination of many pharmaceutical compounds [33–35]. A close examination of the literature search presented in the introduction reveals that NBS has not yet been used for the spectrophotometric determination

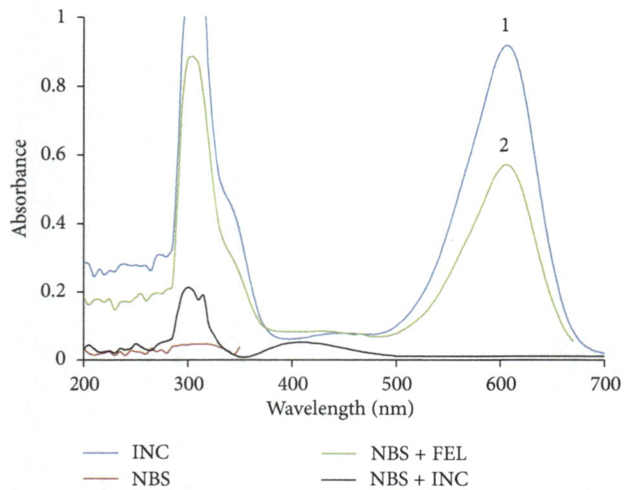

FIGURE 3: Absorption spectra of 0.07% w/v INC (1) in the absence and (2) in the presence of 0.2% w/v NBS with FEL, 7.0 μg/mL.

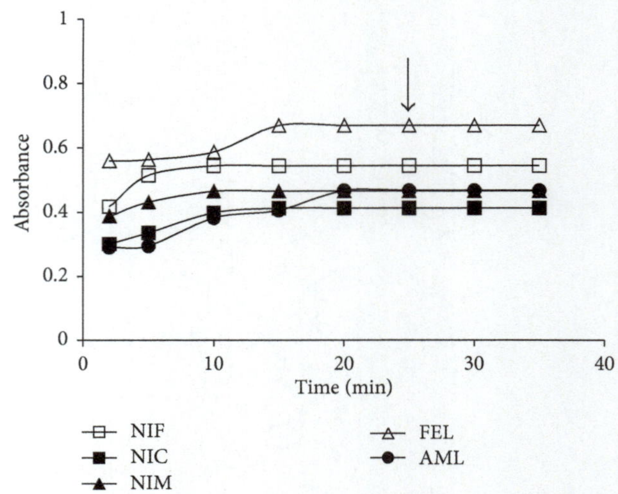

FIGURE 4: Effect of time on the reaction between drugs, 7 μg/mL and NBS/INC.

of these drugs. The present work involves the bromination of the investigated drugs by NBS followed by determination of surplus NBS after allowing the bromination reaction to complete. The unreacted NBS is determined by reacting with a fixed amount of INC dye and measured at 607 nm, (Figure 3) using FEL as a representative example.

3.1. Optimization of the Reaction Variables

3.1.1. Reagent Concentration. A preliminary experiments were performed to determine the optimal concentration of NBS and INC dye. According to the above-mentioned reactions, NBS solution should be added in excess to react with the investigated drugs. By measuring the excess reagent, the consumed reagent would correspond to the amount of the drug. The highest concentration of each reagent, which gives the highest absorption value within the practical sensitivity range of absorption values (≈0.9), was found to be 0.02% and 0.07% w/v for NBS and INC, respectively.

3.1.2. Type and Concentration of the Acid. Perchloric acid was found to be the desirable medium for bromination of the investigated drugs by NBS; additionally it contributed in drugs solubility and also increased NBS potential for INC oxidation.

3.1.3. Reaction Time and the Reaction Product Stability. For a quantitative reaction between the investigated drugs and NBS, contact time of 20 min was found necessary in the proposed method, and constant absorbance readings were obtained when contact times were extended up to 45 min (Figure 4). Another standing time for 10 min was necessary for the bleaching of the dye colour by the residual NBS. The measured colour was found to be stable for several hours in the presence of the reaction products.

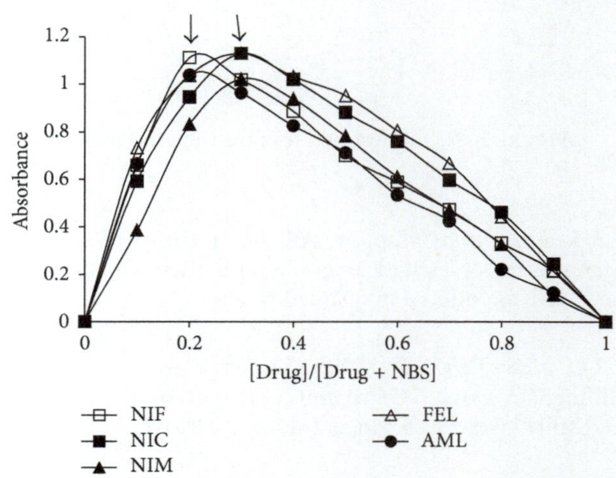

FIGURE 5: Molar ratios of NBS and investigated drugs.

3.2. Reaction Stoichiometry and Suggested Mechanisms

3.2.1. Reaction Stoichiometry. Job's method of continuous variation [36] was employed to establish the stoichiometry of the reaction; master equimolar solutions 1.5×10^{-3} M for both NBS (≈0.02%, w/v, fixed system between NBS and INC) and the drugs were prepared. Series of 10.0 mL portions of the master solutions were made up involving different complimentary proportions (0.00 : 0.10, 0.10 : 0.90,. . ., 0.90 : 0.10, 0.10 : 0.00) using 10 mL volumetric flasks, mixed well, and then subjected to the recommended procedure.

The results which we obtained revealed that the ratio between NBS and NIF or AML was 1 : 4. The ratio between NBS and NIC or NIM or FEL was 1 : 2 (Figure 5).

3.2.2. Suggested Reaction Mechanism between the Studied Drugs and NBS. A suggested bromination reaction between

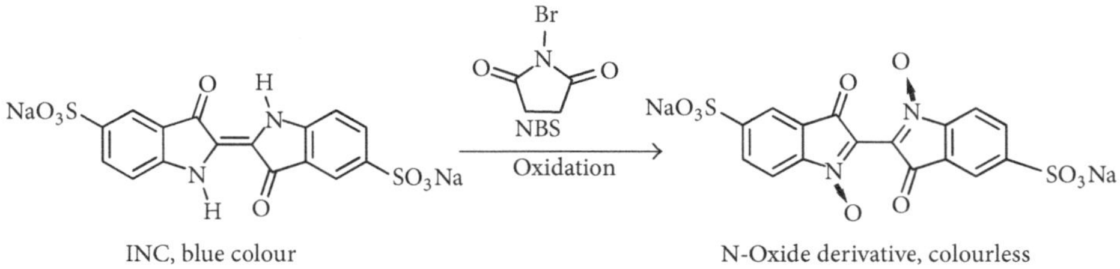

SCHEME 1: Suggested reaction mechanism between INC and NBS.

TABLE 1: Quantitative parameters and statistical data for determination of the studied drugs using NBS/INC method.

Drug	Linear range (μg/mL)	Intercept ± SD[1]	Slope ± SD[1]	r	$\varepsilon \times 10^4$ (L mol^{-1} cm^{-1})	LOD (μg/mL)	LOQ (μg/mL)
NIF	1.25–12.00	0.041 ± 0.008	0.067 ± 0.001	0.998	2.511	0.376	1.253
NIC	1.50–13.00	0.003 ± 0.009	0.056 ± 0.001	0.997	2.945	0.498	1.660
NIM	1.50–13.00	0.070 ± 0.010	0.060 ± 0.001	0.997	2.746	0.500	1.666
FEL	1.25–10.00	0.076 ± 0.004	0.085 ± 0.001	0.999	3.914	0.141	0.470
AML	1.50–13.00	0.046 ± 0.004	0.059 ± 0.001	0.999	3.728	0.203	0.677

[1]Mean of three determinations.

TABLE 2: Assay of five replicate samples of the studied drugs by NBS/INC method at one concentration level.

Drug	Conc. (μg/mL)	Absorbance 1	2	3	4	5	Mean ± SD	RSD
NIF	7.50	0.587	0.582	0.589	0.583	0.586	0.585 ± 0.003	0.440
NIC	10.0	0.611	0.613	0.612	0.615	0.617	0.614 ± 0.002	0.351
NIM	8.00	0.548	0.546	0.547	0.551	0.549	0.548 ± 0.001	0.314
FEL	5.00	0.521	0.525	0.523	0.528	0.526	0.525 ± 0.002	0.461
AML	8.00	0.527	0.533	0.531	0.529	0.532	0.530 ± 0.003	0.406

some of the studied drugs and bromate/bromide mixture was reported previously in [3, 4]. In the proposed method NBS was used as a safe brominating reagent; on the other hand the proposed method was carried out using *N*-chlorosuccinamide (NCS) reagent (Aldrich Co., Ltd., Gillingham-Dorst, Germany), which has nearly the same oxidation potential as NBS. The results revealed that NCS incapable to react with the studied drugs through an oxidation reaction that indicate the reaction pathway is a bromination rather than oxidation.

3.2.3. Suggested Reaction Mechanism between INC and NBS. INC has a blue colour in aqueous solution showed λ_{max} at 607 nm due to its highly conjugated structure (Figure 2) and NBS in acid medium has an oxidation power sufficient to oxidize INC to give colourless N-oxide derivative, the suggested mechanism is given in Scheme 1.

3.3. Validation of Proposed Methods. The developed method was fully validated according to USP XXVI [39] validation guidelines and International Conference on Harmonization (ICH) [40] guidelines.

3.3.1. Linearity Range, Detection, and Quantification Limits. Under the specified optimum reaction conditions, the calibration curves for the investigated drugs were constructed by analyzing a series of different concentrations of the standard solutions.

Linear relationships were found between the absorbance intensities and the concentrations of drugs, as indicated by the high correlation coefficients (r) which are obtained (0.998-0.999), in the concentration range of 1.25–13.00 μg/mL. The limits of detection (LOD) and limits of quantitation (LOQ) were determined according to the IUPAC definitions [41] using the formula LOD or LOQ = $\kappa SD_{a/b}$, where $\kappa = 3$ for LOD and 10 for LOQ, SD_a is the standard deviation of the intercept, and b is the slope. The regression equations for the results were derived using the least square method; the obtained results were summarized in Table 1.

3.3.2. Precision. The precision of the proposed method was determined by carrying out replicate analysis of five separate sample solutions at one concentration level. The relative standard deviations (RSD) did not exceed 2%, indicating the good repeatability of the proposed method, Table 2.

The interday precision was assessed by analyzing the same sample, as triplicate, in five separate days.

3.3.3. Robustness and Ruggedness. Robustness was examined by evaluating the influence of small variation in some experimental parameters on the method's suitability and specificity.

Table 3: Robustness studies for NBS/INC method.

Variation	Recovery% ± SD[a]				
	NIF[b]	NIC[b]	NIM[b]	FEL[b]	AML[b]
No variation	99.41 ± 0.65	97.22 ± 1.07	99.80 ± 0.33	99.71 ± 1.16	99.80 ± 0.23
NBS conc.					
0.019% w/v	99.32 ± 0.36	97.82 ± 0.38	97.82 ± 0.79	99.62 ± 0.54	97.44 ± 0.34
0.021% w/v	97.25 ± 0.63	99.38 ± 0.53	99.55 ± 0.74	98.55 ± 0.63	99.26 ± 0.22
INC conc.					
0.069% w/v	97.90 ± 1.75	99.30 ± 0.67	99.67 ± 0.78	97.71 ± 0.89	98.71 ± 1.33
0.071% w/v	98.15 ± 0.95	97.18 ± 0.15	99.82 ± 0.39	98.27 ± 0.29	99.10 ± 0.28
Reaction time of NBS					
With INC					
9 min	98.16 ± 0.69	99.15 ± 0.16	97.41 ± 0.38	98.77 ± 0.19	99.92 ± 0.19
11 min	99.20 ± 1.05	98.44 ± 0.67	98.56 ± 0.86	97.21 ± 0.98	100.22 ± 1.05
With drug					
24 min	99.73 ± 0.72	99.70 ± 1.38	98.28 ± 1.02	100.26 ± 1.15	97.78 ± 0.15
26 min	98.12 ± 0.98	97.51 ± 1.15	99.21 ± 0.73	99.71 ± 0.65	98.74 ± 1.65

[a]Mean of three determinations.
[b]Drugs concentration 10.0 μg/mL.

Table 4: Ruggedness studies for NBS/INC method.

Drugs[b]	Recovery% ± SD[a]				
	Instrument		Interday variation		
	Shimadzu	Jenway	Day 1	Day 2	Day 3
NIF	99.72 ± 0.85	98.90 ± 0.64	100.10 ± 0.85	100.31 ± 0.68	99.51 ± 1.30
NIC	100.31 ± 0.78	99.37 ± 0.69	98.51 ± 0.82	99.20 ± 1.02	99.74 ± 0.65
NIM	101.81 ± 0.82	100.62 ± 1.15	100.20 ± 1.06	99.80 ± 0.75	100.26 ± 0.60
FEL	99.40 ± 0.96	100.21 ± 1.18	101.50 ± 0.51	100.42 ± 1.40	99.80 ± 1.04
AML	100.30 ± 1.32	99.80 ± 0.76	99.52 ± 0.97	100.68 ± 0.59	99.60 ± 1.15

[a]Mean of three determinations.
[b]Concentration of the drugs is 10.0 μg/mL.

Table 5: Application NBS/INC method for 1,4-DHP drugs' tablets and capsules.

Product	Recovery% ± SD[a]		F-value[b]	t-value[b]
	Proposed method	Official or reported method[c]		
Epilate capsules	98.62 ± 0.05	99.52 ± 0.11	0.20	1.54
Epilate retard tablets	99.21 ± 0.04	98.51 ± 0.13	0.10	1.63
Tenolat SR capsules*	100.33 ± 0.85	100.52 ± 0.66	2.00	1.11
(Pelcard SR capsules)[c]	97.31 ± 0.47	99.74 ± 0.11	0.05	2.00
Nimotop tablets	99.46 ± 0.20	98.61 ± 0.19	0.25	1.82
Plendil tablets	99.24 ± 0.07	99.24 ± 0.13	0.82	0.41
Plentopine tablets	98.39 ± 0.13	99.22 ± 0.11	0.72	1.86
Logimax tablets*	99.22 ± 0.15	99.34 ± 0.17	0.10	0.24
(Alkapress tablets)[c]	99.34 ± 0.10	99.21 ± 0.12	0.17	0.17
(Myodura tablets)[c]	99.26 ± 0.05	98.31 ± 0.11	0.21	1.45
(Amlodipine tablets)[c]	98.38 ± 0.04	97.20 ± 0.12	0.11	1.85
(Regcor tablets)[c]	98.43 ± 0.11	99.24 ± 0.16	0.47	0.47
(Vasonorm tablets)[c]	97.72 ± 0.11	98.41 ± 0.19	2.44	0.54

[a]Mean of five determinations.
[b]Theoretical values for F and t at 95% confidence limit ($n = 5$) were 6.39 and 2.78, respectively.
[c]Reported methods [37, 38].
*Tablets and capsules that contain two drugs.

The changed parameters were the concentration of analytical reagents and the reaction time. In these experiments, one parameter was changed whereas the other was kept unchanged, and the recovery percentages were calculated each time. The obtained results revealed that none of these variables significantly affect the original data of the proposed method (Table 3).

Ruggedness for the proposed method was tested by applying the assay of the investigated drugs using the same experimental conditions with other different instrument at two different laboratories and different elapsed time. The results obtained were found to be reproducible, as RSD did not exceed 2% (Table 4).

3.3.4. Accuracy and Analysis of Tablets and Capsules. Different pharmaceutical formulations including tablets and capsules were analysed for their contents of 1,4-DHP drugs by the proposed, official [42, 43] and reported methods [37, 38] (Table 5). The results which are obtained from the proposed method were compared with those obtained from either official or reported methods with respect to the accuracy (t-test) and precision (F-test). No significant differences were found between the calculated and theoretical values of t- and F-tests at 95% confidence level proving similar accuracy and precision in the analysis of the investigated drugs in their dosage forms as in pure form.

4. Conclusion

In this work a validated method was described; all the analytical reagents used are safe, inexpensive, have good shelf life and available in the most analytical laboratories. Interferences from other co-formulated drugs and common excipients can be eliminated by a simple one-step extraction procedure.

References

[1] Y. Ikegaya, N. Nishiyama, and N. Matsuki, "L-type Ca^{2+} channel blocker inhibits mossy fiber sprouting and cognitive deficits following pilocarpine seizures in immature mice," *Neuroscience*, vol. 98, no. 4, pp. 647–659, 2000.

[2] Y. Murai, H. Uneyama, H. Ishibashi, K. Takahama, and N. Akaike, "Preferential inhibition of L- and N-type calcium channels in the rat hippocampal neurons by cilnidipine," *Brain Research*, vol. 854, no. 1-2, pp. 6–10, 2000.

[3] B. Kanakapura, C. Umakanthappa, and N. Paregowda, "Titrimetric and spectrophotometric assay of felodipine in tablets using bromate-bromide, Methyl Orange and Indigo Carmine reagents," *Journal of the Serbian Chemical Society*, vol. 70, no. 7, pp. 969–978, 2005.

[4] B. Kanakapura, C. Umakanthappa, and N. Paregowda, "Titrimetric and modified spectrophotometric methods for the determination of amlodipine besylate using bromate-bromide mixture and two dyes," *ScienceAsia*, vol. 32, no. 3, pp. 271–278, 2006.

[5] M. Nasr-Esfahani, M. Moghadam, and G. Valipour, "Rapid and efficient aromatization of Hantzsch 1,4-dihydropyridines with potassium peroxomonosulfate catalyzed by manganese(III) Schiff base complexes," *Journal of the Iranian Chemical Society*, vol. 5, no. 2, pp. 244–251, 2008.

[6] S. B. Wankhede, K. C. Raka, S. B. Wadkar, and S. S. Chitlange, "Spectrophotometric and HPLC methods for simultaneous estimation of amlodipine besilate, losartan potassium and hydrochlorothiazide in tablets," *Indian Journal of Pharmaceutical Sciences*, vol. 72, no. 1, pp. 136–140, 2010.

[7] M. A. Rontogianni, C. K. Markopoulou, and J. E. Koundourellis, "HPLC and chemometrically-assisted spectrophotometric estimation of two binary mixtures for combined hypertension therapy," *Journal of Liquid Chromatography and Related Technologies*, vol. 29, no. 18, pp. 2701–2719, 2006.

[8] N. Rahman and S. N. Hejaz Azmi, "New spectrophotometric methods for the determination of nifedipine in pharmaceutical formulations," *Acta Biochimica Polonica*, vol. 52, no. 4, pp. 915–922, 2005.

[9] N. Rahman and S. N. H. Azmi, "Validated spectrophotometric method for the assay of nifediplne in bulk and commercial dosage forms," *ScienceAsia*, vol. 32, no. 4, pp. 429–435, 2006.

[10] B. Kanakapura, C. Umakanthappa, and N. Paregowda, "Spectrophotometric and high performance liquid chromatographic determination of amlodipine besylate in pharmaceuticals," *ScienceAsia*, vol. 31, pp. 13–21, 2005.

[11] K. R. Mahadik, G. B. Byale, H. N. More, and S. S. Kadam, "Spectrophotometric estimation of nifedipine and its formulation," *East-Pharm*, pp. 34121–34122, 1991.

[12] H. F. Askal, O. H. Abdelmageed, S. M. S. Ali, and M. A. El Hamd, "Spectrophotometric and spectrofluorimetric determination of 1, 4- dihydropyridine drugs using potassium permanganate and cerium (iv) ammonium sulphate," *Bulletin of Pharmaceutical Sciences*, vol. 33, no. 2, pp. 201–215, 2010.

[13] M. A. El Hamd, S. M. Derayea, O. H. Abdelmageed, and H. F. Askal, "A novel spectrophotometric method for determination of five 1, 4-dihydropyridine drugs in their tablets and capsules using vanillin reagent," *The American Journal of Analytical Chemistry*, vol. 4, pp. 148–156, 2013.

[14] H. H. Abdine, "Spectrofluorimetric determination of amlodipine," *Mansoura Journal of Pharmaceutical Sciences*, vol. 25, no. 1, pp. 31–38, 2009.

[15] A. B. Tabrizi, "A new spectrofluorimetric method for determination of nifedipine in pharmaceutical formulations," *Chemia Analityczna*, vol. 52, no. 4, pp. 635–643, 2007.

[16] M. I. Walash, F. Belal, N. El-Enany, and A. A. Abdelal, "Kinetic spectrofluorometric determination of certain calcium channel blockers via oxidation with cerium (IV) in pharmaceutical preparations," *International Journal of Biomedical Science*, vol. 5, no. 2, pp. 146–157, 2009.

[17] F. Belal, A. A. Al-Majed, S. Julkhuf, and N. Y. Khalil, "Spectrofluorometric determination of nimodipine in dosage forms and human urine," *Pharmazie*, vol. 58, no. 12, pp. 874–876, 2003.

[18] S. M. Al-Ghannam and A. M. Al-Olyan, "Spectrofluorometric determination of nicardipine, nifedipine and isradipine in pharmaceutical preparations and biological fluids," *Central European Journal of Chemistry*, vol. 6, no. 2, pp. 222–228, 2008.

[19] H. M. Abdel-Wadood, N. A. Mohamed, and A. M. Mahmoud, "Validated spectrofluorometric methods for determination of amlodipine besylate in tablets," *Spectrochimica Acta A*, vol. 70, no. 3, pp. 564–570, 2008.

[20] S. M. Derayea, H. F. Askal, O. H. Abdelmageed, and M. A. El Hamd, "Spectrophotometric determination of amlodipine and nicardipine in pharmaceutical formulations via binary complex

formation with eosin Y," *Journal of Applied Pharmaceutical Science*, vol. 2, no. 6, pp. 84–89, 2012.

[21] R. J. B. Diez-Caballero, L. L. de la Torre, J. F. A. Valentin, and A. A. Garcia, "Adsorptive stripping voltammetry for the determination of nifedipine in human serum," *Talanta*, vol. 36, no. 4, pp. 501–504, 1989.

[22] M. M. Ghoneim, A. Tawfik, and P. Y. Khashaba, "Cathodic adsorptive stripping square-wave voltammetric determination of nifedipine drug in bulk, pharmaceutical formulation and human serum," *Analytical and Bioanalytical Chemistry*, vol. 375, no. 3, pp. 369–375, 2003.

[23] N. Özaltin, C. Yardimci, and I. Süslü, "Determination of nifedipine in human plasma by square wave adsorptive stripping voltammetry," *Journal of Pharmaceutical and Biomedical Analysis*, vol. 30, no. 3, pp. 573–582, 2002.

[24] M. V. Vertzoni, C. Reppas, and H. A. Archontaki, "Sensitive and simple liquid chromatographic method with ultraviolet detection for the determination of nifedipine in canine plasma," *Analytica Chimica Acta*, vol. 573-574, pp. 298–304, 2006.

[25] X.-D. Wang, J.-L. Li, Y. Lu et al., "Rapid and simultaneous determination of nifedipine and dehydronifedipine in human plasma by liquid chromatography-tandem mass spectrometry: application to a clinical herb-drug interaction study," *Journal of Chromatography B*, vol. 852, no. 1-2, pp. 534–544, 2007.

[26] H. S. Abou-Auda, T. A. Najjar, K. I. Al-Khamis, B. M. Al-Hadiya, N. M. Ghilzai, and N. F. Al-Fawzan, "Liquid chromatographic assay of nifedipine in human plasma and its application to pharmacokinetic studies," *Journal of Pharmaceutical and Biomedical Analysis*, vol. 22, no. 2, pp. 241–249, 2000.

[27] A.-E. F. Nassar, "Online hydrogen-deuterium exchange and a tandem-quadrupole time-of-flight mass spectrometer coupled with liquid chromatography for metabolite identification in drug metabolism," *Journal of Chromatographic Science*, vol. 41, no. 8, pp. 398–404, 2003.

[28] L. H. Migliorança, R. E. Barrientos-Astigarraga, B. S. Schug, H. H. Blume, A. S. Pereira, and G. De Nucci, "Felodipine quantification in human plasma by high-performance liquid chromatography coupled to tandem mass spectrometry," *Journal of Chromatography B*, vol. 814, no. 2, pp. 217–223, 2005.

[29] M. T. Rosseel and M. G. Bogaert, "Determination of nifedipine in human plasma by capillary gas chromatography with nitrogen detection," *Journal of Chromatography*, vol. 279, pp. 675–680, 1983.

[30] J. Martens, P. Banditt, and F. P. Meyer, "Determination of nifedipine in human serum by gas chromatography-mass spectrometry: validation of the method and its use in bioavailability studies," *Journal of Chromatography B*, vol. 660, no. 2, pp. 297–302, 1994.

[31] A. T. Wu, I. J. Massey, and S. Kushinsky, "Capillary column gas chromatographic method using electron-capture detection for the simultaneous determination of nicardipine and its pyridine metabolite II in plasma," *Journal of Chromatography B*, vol. 59, no. 1, pp. 65–73, 1987.

[32] R. Nishioka, I. Umeda, N. Oi, S. Tabata, and K. Uno, "Determination of felodipine and its metabolites in plasma using capillary gas chromatography with electron-capture detection and their identification by gas chromatography-mass spectrometry," *Journal of Chromatography*, vol. 565, no. 1-2, pp. 237–246, 1991.

[33] N. K. Mathur and C. K. Narang, *Determination of Organic Compounds with N-Bromosuccinamide and Allied Reagents*, Academic Press, London, UK, 1975.

[34] G. A. Saleh, "Two selective spectrophotometric methods for the determination of amoxicillin and cefadroxil," *Analyst*, vol. 121, no. 5, pp. 641–645, 1996.

[35] I. H. Refaat, *Contribution To the Analysis of Some Phenothiazines*, Assiut University, Assiut, Egypt, 1983.

[36] P. Job, *Advanced Physicochemical Experiments*, Analytical Chemistry, Oliner and Boyd, 1964.

[37] H. Huang and H. Li, "Ultra-violet spectrophotometric determination of the content of nicardipine preparations," *Yaowu Fenxi Zazhi*, vol. 10, no. 6, pp. 359–360, 1990.

[38] K. Basavaiah, U. Chandrashekar, and H. C. Prameela, "Sensitive spectrophotometric determination of amlodipine and felodipine using iron(III) and ferricyanide," *Farmaco*, vol. 58, no. 2, pp. 141–148, 2003.

[39] D. C. Washington, *The American Pharmaceutical Association*, vol. 1, 31th edition, 2008.

[40] M. Haase, "Stability testing on vaccines: results of activities of the international conference on harmonization (ICH) of technical requirements for registration of pharmaceuticals for human use," *Biologicals*, vol. 22, no. 4, pp. 373–375, 1994.

[41] IUPAC, "Nomenclature, symbols, units and their usage in spectrochemical analysis. II. Data interpretation analytical chemistry division," *Spectrochimica Acta B*, vol. 33, p. 242, 1978.

[42] *The British Pharmacopoeia, Majesty's Stationary Office*, vol. 1, London, UK, 2009.

[43] *The British Pharmacopoeia, Majesty's Stationary Office*, vol. 2, London, UK, 2009.

10

Probing Local Environments by Time-Resolved Stimulated Emission Spectroscopy

Ana Rei,[1] Graham Hungerford,[2] Michael Belsley,[1] M. Isabel C. Ferreira,[1] and Peter Schellenberg[1]

[1] *Centro de Física, Universidade do Minho, Campus de Gualtar, 4710-057 Braga, Portugal*
[2] *HORIBA Jobin Yvon IBH Ltd., Skypark 5, 45 Finnieston Street, Glasgow G3 8JU, UK*

Correspondence should be addressed to Peter Schellenberg, peter.schellenberg@fisica.uminho.pt

Academic Editor: A. M. Brouwer

Time-resolved stimulated emission spectroscopy was employed to probe the local environment of DASPMI (*4-(4-(dimethylamino)styryl)-N-methyl-pyridinium iodide*) in binary solvents of different viscosity and in a sol-gel matrix. DASPMI is one of the molecules of choice to probe local environments, and the dependence of its fluorescence emission decay on viscosity has been previously used for this purpose in biological samples, solid matrices as well as in solution. The results presented in this paper show that time-resolved stimulated emission of DASPMI is a suitable means to probe the viscosity of local environments. Having the advantage of a higher time resolution, stimulated emission can provide information that is complementary to that obtained from fluorescence decay measurements, making it feasible to probe systems with lower viscosity.

1. Introduction

Fluorescent stilbenoid dyes are among the classes of chromophores most widely used in staining biological samples to be investigated by fluorescence microscopy methods [1–4]. Stilbenoid dyes show complex excited state dynamics, such as isomerisation processes, which are often accompanied by the emergence of charge transfer states [2, 5–9]. An important consequence of these dynamics is that their fluorescence properties are particularly sensitive to the local environment, which not only increases the contrast between sample features, but also delivers specific information on local properties. In the particular case of DASPMI (*4-(4-(dimethylamino)styryl)-N-methyl-pyridinium iodide*), solvent polarity as well as viscosity can influence the fluorescence properties of this dye. To a good approximation, the peak of the emission spectrum can be used as an indicator for the polarity of its surroundings, while the fluorescence lifetime varies with the viscosity of the local environment [8, 9].

It has been previously demonstrated that the fast molecular dynamics processes in DASPMI, such as the rearrangements of the dye upon transitions to intramolecular charge transfer states, are sensitive to the local viscosity [2, 10, 11]. In particular, the fluorescence lifetime of DASPMI is quite sensitive to the local viscosity, making fluorescence lifetime measurements a method of choice to probe the microheterogeneous environments in confining media such as sol-gel based or biological systems [10, 12, 13]. Previous studies have shown that when DASPMI is incorporated in sol-gel-derived media, the lifetime of the fastest decay component can increase up to a hundred fold while the lifetime of the longer decay components tend to stabilize [13].

Sol-gel-derived media are promising materials for applications as optical biosensors [14–16]. Their porosity, robustness, transparency in the near UV and visible part of the spectrum along with the inherent flexibility of the sol-gel method have allowed the encapsulation of a wide range of biomolecules, including proteins, DNA, and even whole cells [17, 18], and their structure allows for a limited substance mobility within the amorphous porous network. Other directions of research are motivated by the requirement to produce new materials to be used in electronics, communication, energy, and other high-technology fields [19, 20]. The encapsulation of enzymes is particularly attractive because of the efficiency, specificity, and selectivity of these biocatalysts. There are many publications on the successful entrapment of enzymes and for a review see [21]; however, some major issues still remain, explaining why this is still an intense field of investigation. Knowledge about the microenvironment composition inside the pores and about the molecular mobility of the enzyme and substrates/products of reaction is crucial to comprehend the enzyme kinetics in confining media. In this frame, measurements concerning the viscosity evolution in a sol-gel sample would be important for characterizing mass transport inside the matrix [14, 22].

In the present work, we report on complementary measurements employing femtosecond time-resolved stimulated emission to elucidate the DASPMI kinetics in homogeneous solution and a microheterogeneous sol-gel medium. As the experimental time resolution in these techniques is of order 100 fs, the technique could be employed to probe faster decay processes compared to fluorescence lifetime measurements. This would allow one to explore the faster kinetics affiliated with lower viscosity matrices, extending the range of the technique.

In this work, the principal applicability of the method is tested. Apart from measurements on matrices produced by the sol-gel path, we probe glycerol-water mixtures of various viscosities. As was the case for the fluorescence decay signals using fluorescence upconversion experiments [23, 24], we are able to identify wavelength-dependent kinetics. With the increased time resolution of an ultrafast laser system, we can clearly identify dynamic processes in DASPMI on a subpicosecond scale.

2. Experimental Setup

The pump-probe measurements were carried out using the output of a Coherent Legend Ti : sapphire-chirped pulse regenerative amplifier system seeded by a Coherent Mira Ti : sapphire oscillator which in turn was pumped by a 5 W frequency-doubled Nd-laser. The system generated a 1 kHz pulse train at a wavelength of 800 nm with a FWHM of around 120 fs and pulse energies of 2 mJ. A small portion of these amplified pulses were used to generate the excitation pulses centred at 400 nm by frequency doubling in a BBO crystal. The majority of the pulse intensity is used to generate the probe beam by pumping an ultrafast optical parametric amplifier (OPA) from Light Conversion (TOPAS). The time delay Δt between the probe and the excitation beam was scanned with a delay line (Standa, Lithuania 8MT167-100) with minimum step sizes corresponding to a delay of around 4 fs. The signals were collected with a fast PIN-biased photodiode (Thorlabs DET10A/M). A modular boxcar integrator system SR-240/SR-245/SR-250/SR-280 (Stanford Research Systems) was used for sampling and digitizing the signals. For controlling the experiment; a homemade labview program was employed. The data were fitted with the Fluofit program (Picoquant Berlin) employing an artificially created gaussian profile deconvolution function to take into account the finite temporal width of the excitation pulses. Its width was set to 120 fs FWHM, which was derived from the experimental autocorrelation function of the Ti : Sa amplifier output. The pump and probe beams were made to intersect at an angle of approximately 15 degrees, and the liquid solutions were placed in a 1 mm glass cuvette centred on the crossing of the pump and probe beams.

The DASPMI dye was purchased from Molecular Probes (Invitrogen, SA), the Glycerol was from Riedel-de Haën, and both were used without further purification.

The sol-gel media were produced as previously described [25], based on the method presented by Flora and Brennan [26], in the form of a monolith. Briefly, the sol was made by mixing 9 mL of tetraethyl orthosilicate (TEOS) (Aldrich) with 3 mL of acidic water. This mixture was sonicated for 1 hour and stored at −18°C for about a month. The matrices were produced by taking 2 mL of the sol and mixing it with 2 mL of phosphate buffer solution (pH 7) containing the fluorophore DASPMI. The samples, prepared between two glass plates, were about 1-2 mm thick. The outer edges of the glass plates were sealed with epoxy to prevent solvent evaporation from the freshly prepared gel. The doped sol-gel matrices were used within hours of preparation. Because of photobleaching of the dye, a new probing position was used for each time delay point. Since the samples contained small spatial inhomogeneities, this caused a noticeable increase in the noise level of the sol-gel measurements as compared with those carried out in solution.

3. Results and Discussion

The dynamics of DASPMI kinetics were probed by stimulated emission spectroscopy in various glycerol : water mixtures, in pure water, and in matrices produced via the sol-gel process for probe wavelengths of 560 nm to 620 nm. The recovered parameters are shown in Table 1 and a majority of the decays are multiexponential in nature.

To exemplify, an illustration of a selected set of decays is given in Figure 1. It shows the time-resolved stimulated emission kinetics of DASPMI in different media at a probe wavelength of 560 nm. The dramatic shortening of the emission decay time upon lowering the viscosity is immediately obvious.

The strong dependence of the stimulated emission decay on the probe wavelength can be seen in Figure 2 for the lowest viscosity solvent (water). In addition to a shortening of the decay time by about a factor of 100, we also observe a rise time component for probe wavelengths of 600 nm and 620 nm. Tentatively this may also be present at 580 nm, although it was not possible to obtain a satisfactory fit for a temporally increasing component of the data available at this wavelength. This rise time component may be attributed to a fast excited state relaxation of excited Frank-Condon states, for example, due to a twisted geometry [11], and/or due to solvent relaxation in the excited state [27] particularly as a response to its CT-character. This view is supported by the dramatic increase of the rise time when going from low-viscosity water solution to higher-viscosity glycerol mixtures (Table 1). Interestingly, we could not observe such a rise time in the sol-gel matrixes, which may be attributed to the absence of this motional degree of freedom, although it is

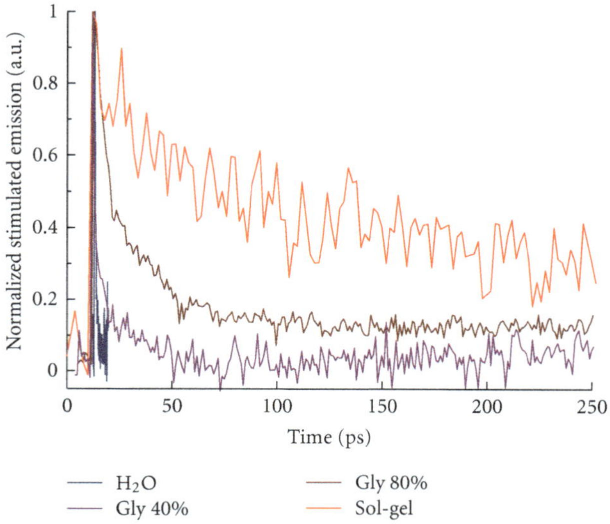

FIGURE 1: Stimulated emission decay curves of DASPMI in different matrices at a probe wavelength of 560 nm.

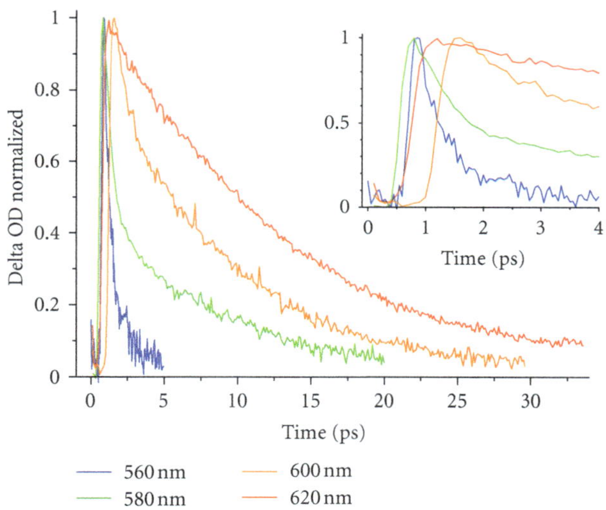

FIGURE 2: Stimulated emission decay curves of DASPMI in water at different probe wavelengths.

also possible that this component evades observation due to the lower signal to noise ratio.

Table 1 summarises all experimentally determined stimulated emission decay fit values for water, two different water : glycerol mixtures (40% and 80% glycerol, expressed in mass percent) and the sol-gel system. Empty fields indicate the absence of a decay constant in that time range. The time ranges are arbitrarily chosen and do not necessarily represent the presence or absence of a particular molecular process. It should be borne in mind that, in highly viscous media, inhomogeneous broadening of the kinetic components can occur, leading to dispersive kinetics. Such processes are not explicitly considered here but may contribute to the complex kinetics observed. The timescale of the parameters presented in this table also shows that the use of an ultrafast system allows an increase in the time resolution for emission studies, as well as the possibility for exploring other excited state kinetic processes.

The values shown in Table 1 emphasise that studies with higher time resolution compared to fluorescence decay are more sensitive to probe local viscosity alterations in low-viscosity environments. It should also be noted that, at a viscosity of 45.9 cp [28] represented by a 80% Glycerol sample, the data approximately match those obtained with previous fluorescence decay measurements [10]. We note, however, that the fastest decay component is shorter in the stimulated emission experiments.

Following the assessment of Strehmel et al. [11], based on fluorescence decay data, the dynamics of DASPMI can be attributed to several processes: the direct transition from the first excited state to the ground state, a relaxation probably due to a planar charge transfer state, and a slow relaxation related to a twisted charge transfer state.

It can be seen from these experiments that in solution the decay time of the shortest lived component, which is attributed to the planar state of DASPMI, increases significantly upon augmenting the viscosity up to 3.1 cp. Further increases in viscosity by more than an order of magnitude only change the decay for a factor of two. In contrast, a similar proportional increase in the decay time in the longer lived (here τ_3) component attributed to a planar CT state is seen at higher viscosities (on passing from 3.18 to 45.9 cP). In the time-resolved stimulated emission of DASPMI doped into sol-gel-derived media, only a single component is observed. However, in these studies, we could not follow the ultrafast decay kinetics with high precision but instead were limited to the intermediate time regime. This was because, in order to avoid photobleaching of the sample, different probe regions were selected. This led to increased point-to-point fluctuations due to macroscopic sample inhomogeneities. Furthermore, the number of laser pulses that could be used to probe the solid medium was limited by photobleaching, which, in turn, reduced the measurement statistics.

In previous time-resolved fluorescence measurements, the decay kinetics has been reported to be multiexponential, with three different lifetime components [10, 29]. In comparison with those studies, it is possible that the lifetime component found here corresponds to the fast kinetic component in those experiments, but the dynamics of DASPMI in the confined environment may also be due to processes not observed in the homogeneous liquid phase. An indication for this is that the time constant for the direct transition from S_1 to S_0 is only weakly dependent on the viscosity, as witnessed by the small increase observed when going from 3.1 cp to 46 cp. In the sol-gel media, however, the time constant is between 84 ps and 186 ps, much greater than the less than 20 ps observed in a solution with a viscosity of 46 cp. In fact, a strong increase of the lifetime of DASPMI upon inclusion into sol-gel matrices has been observed before [13], and our data give strong evidence that DASPMI is not localized in microcavities of low viscosity solvent. This is particularly noticeable, as the hydrogels used in this study were still in their initial stages of aging, at which the pores network is still forming.

Table 1: Stimulated emission decay times and relative amplitudes for various different water : glycerol mixtures and for the sol-gel matrix at the different probe wavelengths. A negative amplitude indicates a rising rather than a decaying component. The sol-gel data could be fitted to a single exponential function with a residual amount around 25–30% that did not decay in the time window of the experiment. The viscosity values are taken from [28].

Sample	Probe λ (nm)	$\tau 1$ (ps)	$\alpha 1$	$\tau 2$ (ps)	$\alpha 2$	$\tau 3$ (ps)	$\alpha 3$
Water (0.89 cp)	560			0.53	1		
	580			0.67	0.59	9.86	0.41
	600	0.25	−1	0.57	0.54	8.46	0.46
	620	0.2	−1			12.1	1
Glycerol 40% (3.18 cp)	560			7.91	1		
	580			1.7	0.59	34.3	0.41
	600	6.0	−1	6.1	0.41	69.9	0.59
	620	5.1	−1	8.3	0.60	38.5	0.40
Glycerol 80% (45.9 cp)	560			14.2	1		
	580			18.9	0.12	255.7	0.88
	600					203.2	1
	620	4.73	−1			254	1
Sol-gel	560					84	0.73
	580					91	0.74
	600					128	0.75
	620					186	0.70

Also, note that the longer lifetimes are not seen in these stimulated emission experiments, as the time range was limited due to the length of the optical delay available for these studies. However, the presence of longer excitation dynamics components can be deduced from a residual amplitude of around 25 to 30% representing excited states which have not decayed within the time window of the experiment.

In the solutions of different viscosity, the deviation in the shorter lifetimes, as compared to those measured by previous fluorescence lifetime measurements [10] could be due to a lower time resolution in time correlated single photon-counting experiments. In the previous measurements, the observed lifetime is derived after deconvolution of a much longer instrument response function. Alternatively the transitions observed in stimulated emission may be different ones compared to those in fluorescence decay. While fluorescence is dominated by the decay of the relaxed excited state, stimulated emission processes can probe the molecule as it evolves from the initial excitation to the relaxed state, thereby providing complementary information on the probe environment.

4. Conclusion

Time-dependent stimulated emission was employed to demonstrate the potential of using DASPMI in matrices to characterise local viscosities, both in solution and sol-gel-derived media. It can be shown that time-resolved stimulated emission experiments are promising for probing local environments in microheterogeneous environments, like polymers and other amorphous systems and biological samples. The sensitivity of the effect can be compared to that of the corresponding method of employing fluorescence lifetime spectroscopy techniques. However, the time scale for stimulated emission experiments is much faster, as it is basically limited by the pulse width of the laser employed, which in our case was around 100 fs. Consequently much faster processes can be accurately probed. In DASPMI, the time evolution of the excited state scales with the viscosity of the solvent and an increased temporal resolution allows one to greatly increase the range over which the local viscosity can be probed. Therefore, time-resolved stimulated emission spectroscopy could be a suitable complementary method to time-resolved fluorescence techniques by which to investigate lower viscosity samples.

It is known from previous work on sol-gel systems that there are kinetic processes on longer time scales as accessible in the present stimulated emission setup. In future experiments, it is crucial to identify the physical processes associated with the time constants observed in sol-gel matrices. It is not unreasonable to associate the subnanosecond decay in sol-gel material with the faster reconfiguration processes in lower viscosity solvents as observed in the stimulated emission experiments presented here and in previous studies using fluorescence upconversion [23, 27]. This would provide additional insight into the interaction of DASPMI in the sol-gel-derived microheterogeneous environment.

Acknowledgments

The authors acknowledge financial support from Fundação para a Ciência e a Tecnologia (FCT) for the reequipment Grant REEQ/25/FIS/2005. A. Rei was supported from FCT through the Ph.D. Grant SFRH/BD/27933/2006.

References

[1] R. Ramadass and J. Bereiter-Hahn, “Picosecond fluorescence decay microscopy of DASPMI stained mitochondria in living cells,” *European Journal Of Cell Biology*, vol. 84, pp. 107–108, 2005.

[2] R. Ramadass and J. Bereiter-Hahn, "Photophysical properties of DASPMI as revealed by spectrally resolved fluorescence decays," *Journal of Physical Chemistry B*, vol. 111, no. 26, pp. 7681–7690, 2007.

[3] R. Ramadass and J. Bereiter-Hahn, "How DASPMI reveals mitochondrial membrane potential: fluorescence decay kinetics and steady-state anisotropy in living cells," *Biophysical Journal*, vol. 95, no. 8, pp. 4068–4076, 2008.

[4] R. Ramadass and J. Bereiter-Hahn, "Excited state photophysical properties of DASPMI in solvents and living cells," *Biophysical Journal*, p. 326A, 2007.

[5] D. Bingemann and N. P. Ernsting, "Femtosecond solvation dynamics determining the band shape of stimulated emission from a polar styryl dye," *The Journal of Chemical Physics*, vol. 102, no. 7, pp. 2691–2700, 1995.

[6] M. Dekhtyar and W. Rettig, "Charge-transfer transitions in twisted stilbenoids: interchangeable features and generic distinctions of single- and double-bond twists," *Journal of Physical Chemistry A*, vol. 111, no. 11, pp. 2035–2039, 2007.

[7] W. Rettig, B. Strehmel, and W. Majenz, "The excited states of stilbene and stilbenoid donor-acceptor dye systems. A theoretical study," *Chemical Physics*, vol. 173, no. 3, pp. 525–537, 1993.

[8] M. Sczepan, W. Rettig, A. I. Tolmachev, and V. V. Kurdyukov, "The role of internal twisting in the photophysics of stilbazolium dyes," *Physical Chemistry Chemical Physics*, vol. 3, no. 17, pp. 3555–3561, 2001.

[9] M. J. van der Meer, H. Zhang, W. Rettig, and M. Glasbeek, "Femto- and picosecond fluorescence studies of solvation and non-radiative deactivation of ionic styryl dyes in liquid solution," *Chemical Physics Letters*, vol. 320, no. 5-6, pp. 673–680, 2000.

[10] A. Rei, G. Hungerford, and M. I. C. Ferreira, "Probing local effects in silica sol-gel media by fluorescence spectroscopy of p-DASPMI," *Journal of Physical Chemistry B*, vol. 112, no. 29, pp. 8832–8839, 2008.

[11] B. Strehmel, H. Seifeit, and W. Rettig, "Photophysical properties of fluorescence probes. 2. A model of multiple fluorescence for stilbazolium dyes studied by global analysis and quantum chemical calculation," *Journal of Physical Chemistry B*, vol. 101, no. 12, pp. 2232–2243, 1997.

[12] D. Jóźwik, E. Miller, B. Wandelt, and S. Wysocki, "The styrylpyridine dye for the silane sol-gel transition studies by time-dependent fluorescence," *Spectrochimica Acta A*, vol. 64, no. 5, pp. 1125–1132, 2006.

[13] G. Hungerford, M. R. Pereira, J. A. Ferreira et al., "Probing Si and Ti based sol-gel matrices by fluorescence techniques," *Journal of Fluorescence*, vol. 12, no. 3-4, pp. 397–417, 2002.

[14] G. Hungerford, A. Rei, M. I. C. Ferreira, K. Suhling, and C. Tregidgo, "Diffusion in a sol-gel-derived medium with a view toward biosensor applications," *Journal of Physical Chemistry B*, vol. 111, no. 13, pp. 3558–3562, 2007.

[15] D. Avnir, O. Lev, and J. Livage, "Recent bio-applications of sol-gel materials," *Journal of Materials Chemistry*, vol. 16, no. 11, pp. 1013–1030, 2006.

[16] S. Hudson, J. Cooney, and E. Magner, "Proteins in mesoporous silicates," *Angewandte Chemie*, vol. 47, no. 45, pp. 8582–8594, 2008.

[17] T. Besanger, Y. Zhang, and J. D. Brennan, "Characterization of fluorescent phospholipid liposomes entrapped in sol-gel derived silica," *Journal of Physical Chemistry B*, vol. 106, no. 41, pp. 10535–10542, 2002.

[18] W. Jin and J. D. Brennan, "Properties and applications of proteins encapsulated within sol-gel derived materials," *Analytica Chimica Acta*, vol. 461, no. 1, pp. 1–36, 2002.

[19] H. Dislich, "Sol-Gel 1984 → 2004 (?)," *Journal of Non-Crystalline Solids*, vol. 73, no. 1–3, pp. 599–612, 1985.

[20] S. Sakka, "Sol-gel synthesis of glasses: present and future," *American Ceramic Society Bulletin*, vol. 64, no. 11, pp. 1463–1466, 1985.

[21] A. C. Pierre, "The sol-gel encapsulation of enzymes," *Biocatalysis and Biotransformation*, vol. 22, no. 3, pp. 145–170, 2004.

[22] G. Hungerford, A. Rei, M. I. C. Ferreira, C. Tregidgo, and K. Suhling, "Molecular diffusion within sol-gel derived matrices viewed via fluorescence recovery after photobleaching," *Photochemical & Photobiological Sciences*, vol. 6, no. 8, pp. 825–828, 2007.

[23] A. M. Jonkman, P. van der Meulen, H. Zhang, and M. Glasbeek, "Subpicosecond solvation relaxation of DASPI in polar liquids," *Chemical Physics Letters*, vol. 256, no. 1-2, pp. 21–26, 1996.

[24] H. Wang, H. Zhang, W. Rettig, A. I. Tolmachev, and M. Glasbeek, "Femtosecond dynamics of the S_2 and S_1 fluorescence of ionic styryl dyes in polar solvents," *Physical Chemistry Chemical Physics*, vol. 6, no. 13, pp. 3437–3446, 2004.

[25] G. Hungerford, A. Rei, and M. I. C. Ferreira, "Use of fluorescence to monitor the incorporation of horseradish peroxidase into a sol-gel derived medium," *Biophysical Chemistry*, vol. 120, no. 2, pp. 81–86, 2006.

[26] K. K. Flora and J. D. Brennan, "Characterization of the microenvironments of PRODAN entrapped in tetraethyl orthosilicate derived glasses," *Journal of Physical Chemistry B*, vol. 105, no. 48, pp. 12003–12010, 2001.

[27] P. van der Meulen, H. Zhang, A. M. Jonkman, and M. Glasbeek, "Subpicosecond solvation relaxation of 4-(dicyanomethylene)-2-methyl-6-(*p*-(dimethylamino)styryl)-4H-pyran in polar liquids," *Journal of Physical Chemistry*, vol. 100, no. 13, pp. 5367–5373, 1996.

[28] R. C. Weast, in *Handbook Of Chemistry And Physics*, CRC, Boca Raton, Fla, USA, 57th edition, 2000.

[29] G. Hungerford, A. Allison, D. McLoskey, M. K. Kuimova, G. Yahioglu, and K. Suhling, "Monitoring Sol-to-Gel transitions via fluorescence lifetime determination using viscosity sensitive fluorescent probes," *Journal of Physical Chemistry B*, vol. 113, no. 35, pp. 12067–12074, 2009.

11

Crystal Defects and Cation Redistribution Study on Nanocrystalline Cobalt-Ferri-Chromites by Positron Annihilation Spectroscopy

Kunal B. Modi,[1] Nimish H. Vasoya,[2] Vinay K. Lakhani,[3] Tushar K. Pathak,[4] and P. M. G. Nambissan[5]

[1] *Department of Physics, Saurashtra University, Rajkot, Gujarat 360005, India*
[2] *Om Shanti Engineering College, Morbi Road, Rajkot, Gujarat 360030, India*
[3] *Department of Physics, Bahauddin Science College, Junagadh, Cujarat 362001, India*
[4] *Government Engineering College, Kalawad Road, Rajkot, Gujarat 360005, India*
[5] *Applied Nuclear Physics Division, Saha Institute of Nuclear Physics, Kolkata, West Bengal 700064, India*

Correspondence should be addressed to Kunal B. Modi; kunalbmodi2003@yahoo.com

Academic Editor: Hicham Fenniri

Positron lifetime and Doppler broadening measurements were carried out on nanocrystalline (grain size ~60–65 nm) samples of the Cr^{3+}-substituted cobalt ferrite system with general chemical formula $CoCr_xFe_{2-x}O_4$ (x = 0.0–2.0) synthesized by the coprecipitation technique. The results indicated selective trapping of positrons in large vacancy clusters initially at the tetrahedral (A-) sites and then with Cr^{3+}-substitution up to concentration (x) = 0.7, at the octahedral (B-) sites. The results are consistent with the cation distribution determined from X-ray diffraction line intensity calculations, which indicated partial inversion of the inverse spinel ferrite, subsequent stabilization over a range of substitution (x = 0.7 to 1.7), and finally the full inversion to the normal spinel chromite ($CoCr_2O_4$, x = 2.0). In the intermediate range of substitution, lattice contraction prevented a fraction of Co^{2+} ions released from the (B-) sites from entering the tetrahedral sites, and these vacancies at the (A-) sites trapped positrons. Although the samples were composed of nanocrystalline grains, only an insignificant fraction of positrons were diffused and annihilated at the grain surfaces, since the grain sizes and the thermal diffusion length of positrons nearly overlapped.

1. Introduction

Crystalline materials with the characteristic spinel structure, comprising of well-designated tetrahedral (A-) and octahedral (B-) sites, constitute a very interesting class of condensed matter systems evoking interest even from the very fundamental science viewpoint [1, 2]. Cobalt ferrite ($CoFe_2O_4$), Cobalt chromite ($CoCr_2O_4$), and their solid solutions with a typical crystalline structure XY_2O_4 are candidate materials in this class, where X is normally a divalent and Y is a trivalent ion. These materials have attracted a large number of chemists, physicists, and metallurgists to study their different aspects, including the structure and properties, using both theoretical modeling and wide variety of experimental tools [3–5]. Ferrites composed of nanometer-sized particles elevate this interest to new dimensions as the very large network of interfaces will then play a decisive role in controlling the atomic transport and spatial rearrangement of atoms within the structure. So far as the tools for investigating such details are concerned, conventional experimental methods such as X-ray diffraction (XRD) and transmission electron microscopy (TEM) have helped to obtain substantial information on a macroscopic scale. But defect-specific probes such as positron annihilation spectroscopy are needed to pinpoint their role in the postsynthesis and characterization treatments such materials have to undergo. We present in this paper the results of our investigation carried out on $CoFe_2O_4$, both on the nascent mother sample and samples in which the Fe^{3+} ions are replaced by Cr^{+3} ions, that is, $CoCr_xFe_{2-x}O_4$, x = 0.0–2.0. The purpose of the work is double sided. Initially we will present the important observations which were obtained from more conventional experimental tools such as XRD, TEM, and energy dispersive analysis of X-rays (EDAX) and highlight the changes occurring at the different stages of Cr^{3+}-substitution. The emphasis will be thereafter diverted to demonstrate the ability of positron annihilation techniques to sense such changes. The latter is of importance since such works are scarcely available in the literature so far, and secondly it offers a viable investigative probe for such studies which are highly essential in the current scenario of novel materials and arising challenges in their understanding.

According to the existing literature, cobalt ferrite ($CoFe_2O_4$) is an inverse spinel and taken to be collinear ferrimagnet [6], while cobalt chromite, $CoCr_2O_4$, is a normal spinel with a canted ferromagnetic structure, and its Curie temperature is 97 K [7]. Previous studies of magnetic properties and Mossbauer spectroscopy on mixed cobalt-ferri-chromites, $CoCr_xFe_{2-x}O_4$, of coarse-grain composition indicated that canting of magnetic structure is observed when Co^{2+} is present at the tetrahedral (A-) site [8, 9]. While magnetization measurements on the same system could be explained by Neel's model as far as the series remained inverse spinel, they deviated significantly when they began to have normal spinel structure [10]. Recently, structural and magnetic properties of nanocrystalline $CoCr_xFe_{2-x}O_4$ ($0 \le x \le 1.0$) system prepared by the sol-gel autocombustion route have been studied by Tolesha et al. [11].

The redistribution of cations when one species is replaced by ions of neighboring elements in the periodic table has been of tremendous significance in modifying the properties as well as in giving rise to new phenomena and processes [6, 8, 9]. The transformation of a nearly complete inverse spinel ferrite, $CoFe_2O_4$, to a normal spinel chromite, $CoCr_2O_4$, through successive replacement of the Fe^{3+} ions by Cr^{3+} ions has been found to generate drastic changes in the cationic distribution in the structure, owing to lattice contraction as well as the likely presence of vacancy-type structural defects. Although substitution effects generally prompt such redistributions, concomitant lattice contraction or expansion can also influence it, and quite often it will result in the generation of structural defects in the form of unoccupied lattice sites. The latter are potential sites for investigation by positron annihilation experiments. Such a defect-sensitive spectroscopic probe will be of immense benefit as it can pinpoint the origin and evolution of such defects and their dominating role over the redistribution of ions in the lattice. As is popularly known, the positron lifetimes and Doppler broadening of the electron-positron annihilation gamma ray spectrum are directly related to the electron density and momentum distribution in a material, and hence, the information carried by the signals of annihilation can unravel the material properties in the atomic scale [12].

2. Experimental Procedure

The spinel ferrite system, $CoCr_xFe_{2-x}O_4$, with variable compositions with the range of x = 0.0, 0.1, 0.3, 0.7, 1.1, 1.3, 1.5, 1.7, 1.8, 1.9, and 2.0 were prepared by air oxidation of an aqueous suspension containing Co^{2+}, Fe^{2+}, and Cr^{3+} cations in stoichiometric proportions. The starting solutions was prepared by mixing 50 mL of aqueous solutions of $CoSO_4{\cdot}7H_2O$, $FeSO_4{\cdot}7H_2O$, and $Cr_2(SO_4)_3{\cdot}6H_2O$ in proper proportions. A 2 M solution of NaOH was prepared as a precipitant. The starting solution (pH ~ 3.5) was added into the precipitant so that the solubility product constants for the hydroxides of the cation are exceeded and their sequential precipitation can be avoided. The suspension (pH = 10.5) thus formed, containing dark green intermediate precipitates, was then heated and kept at 333 K for 1 hour. During the heating, oxygen gas was bubbled uniformly into the suspension to stir it and to promote oxidation reaction until all the intermediate precipitates changed into dark brownish precipitates of the spinel ferrite. The samples were filtered, washed by acetone several times, and dried at 473 K under vacuum.

The formation of the ferrite powders by the oxidation method consists of oxidation by air (O_2) bubbling through an aqueous solution containing ferrous ions (Fe^{2+}) and other divalent ions (M^{2+}) after an alkaline solution (ROH) has been added. This is according to the chemical reaction $Fe^{2+} + M^{2+} + ROH + O_2 \rightarrow M^{2+}(Fe^{3+})_2O_4$ where ROH is NaOH, KOH, NH_4OH, and so forth. Ferrite powder of high homogeneity and purity and with sufficient control over the particle size can be achieved by this method.

The stoichiometry of the powdered samples was confirmed from EDAX measurements. The compositional values were determined within an accuracy of 1%. X-ray diffraction data were collected using a Philips (PW 1710) automated X-ray powder diffractometer with Cu K_α radiation, graphite monochromator, and Xe-filled proportional counter. Data were collected in the angle range 5°–80°. TEM images of the samples were taken using a TECNAIK 20 (Philips) microscope operated at 200 kV. For the TEM observations, the powder was first dispersed in amyl acetate by ultrasonication, and then the suspensions were dropped on a copper grid with a carbon film. The grain sizes and shapes were determined from the TEM pictures.

Positron annihilation studies of the powdered samples were performed in the conventional way. The radioactive isotope ^{22}Na in the form of $^{22}NaCl$ dissolved in dilute HCl was deposited and dried on a thin (~2 μm) Ni foil and folded to form the source of positrons. It was kept immersed in the volume of the sample taken in a glass tube. Care was taken to ensure that the powdered sample surrounded the source from all sides sufficiently enough to ensure annihilation of positrons within it. Further, the sample with the source embedded in it was maintained under dry vacuum conditions during the experiments. This is done to eliminate the possibility of the trapping of positrons and quenching of positronium (the metastable bound state of the positron and an electron) by the air and gases otherwise getting trapped within the powdered sample [13]. The positron lifetime measurements were carried out using a slow-fast coincidence spectrometer, having a time resolution (FWHM) of 200 ps for the gamma rays from ^{60}Co source under experimental conditions. For Doppler broadening measurements, a high pure germanium (HPGe) detector with resolution 1.14 keV at 511 keV was used. This and another detector with identical resolution were used on either side of the source-sample assembly for coincidence Doppler broadening spectroscopy (CDBS) studies. The CDBS experiments helped to identify the elemental environment around the positron trapping sites in materials by virtue of the ability to record the gamma ray energy distribution spectra of positron annihilation with the high momentum core electrons, without the nuclear background. The details about these measurements and the methods of data analysis to extract the relevant information can be seen in the paper by Asoka-Kumar et al. [14].

3. Results and Discussion

3.1. Characterization of the Samples. After obtaining the final products, it was essential to check the chemical composition of each sample. The reason for making EDAX characterization was to ratify the purity and surety of the chemical composition. Two representative EDAX patterns, namely, of $x = 0.3$ and 1.8 compositions are shown in Figure 1. The results of EDAX confirm the expected stoichiometry with small deficiencies of Cr^{3+} ions. Further, the incorporation of Cr^{3+} in the place of Fe^{3+} was indicated by the intensities of the respective peaks in the two patterns. No traces of impurities were found. The EDAX results suggested that the precursors had fully undergone the chemical reaction to form ferrite material of the expected composition.

The particle size and morphology of all the compositions were studied by means of transmission electron microscopy. Typical images for $x = 0.0, 1.0$, and 1.6 compositions are shown in Figure 2. It can be seen that the average particle size is in the range 60–65 nm.

The $CoCr_xFe_{2-x}O_4$ samples were characterized by XRD to ascertain the single-phase structure formation and to deduce cell edgeparameter and cation distribution and grain size verification. Typical XRD patterns of $CoCr_xFe_{2-x}O_4$ samples with $x = 0.0$, 0.7, 1.1, 1.3, 1.7, and 2.0 are shown in Figure 3. The background noise and the broadness of the peaks are characteristics of particles with nanometer dimensions since there is not a sufficient number of crystallographic planes to result in sharp diffraction lines. The XRD patterns also showed that all the samples have the monophase spinel structure. No extra lines corresponding to any other phase or nonreacted ingredients were detected. The diffraction patterns could be indexed for a face centered cubic (fcc) structure [15]. The lattice constant for each composition was determined by using the "Powder-X" software [16]. The concentration dependence of the lattice constant (a) determined from the X-ray data is presented in Figure 4. The lattice constant remains more or less constant initially but rapidly decreases for higher concentrations of Cr^{3+}. The observed change in lattice constant value with Cr^{3+}-content (x) is attributed to the small difference in the ionic radii of the constituent cations, Fe^{3+} (0.640 Å) and Cr^{3+} (0.630 Å), and the change in the distribution of cations among the available (A-) and (B-) sites of the spinel lattice.

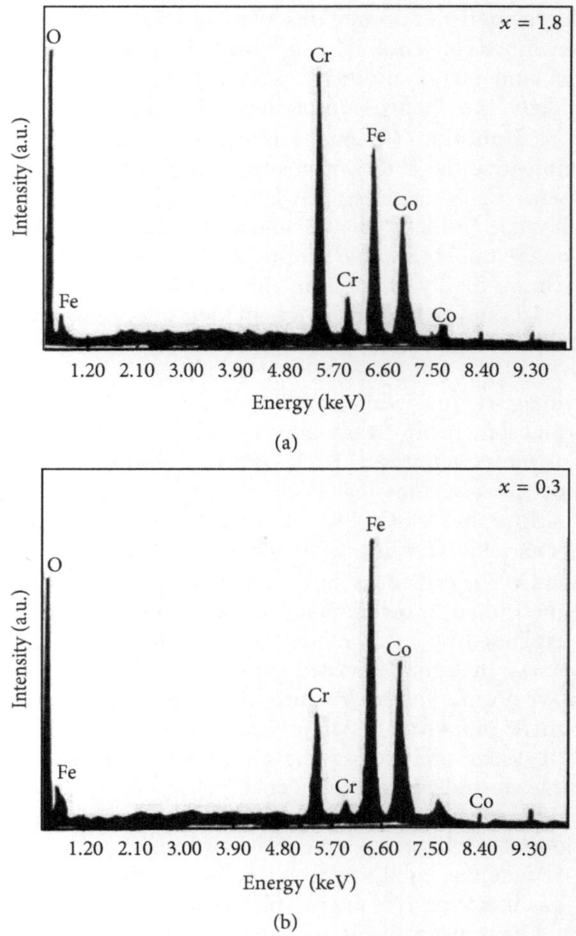

FIGURE 1: The EDAX spectrum of the Cr^{3+}-substituted samples, at concentrations $x = 0.3$ and 1.8.

In order to determine the cation distribution, the XRD line intensity calculations were made using the formula suggested by Buerger [17]:

$$I_{hkl} = |F_{hkl}|^2 PL. \quad (1)$$

Here, I_{hkl} is the relative integrated intensity, F_{hkl} is the structure factor, P is the multiplicity factor, and $L = (1 + \cos^2 2\theta)/(\sin^2\theta \cos\theta)$ is the Lorentz polarization factor.

According to Ohnish and Teranishi [18], the intensity ratios of planes I_{220}/I_{440} and I_{400}/I_{422} are considered to be sensitive to the cation distribution. There exists distinct contrast in the atomic scattering factors of Cr^{3+} or Fe^{3+} and Co^{2+} cations present in the system. This makes the determination of the cation distribution quite reliable. Any alteration in the distribution of cations causes a significant change in the XRD intensity ratios. Therefore, in the process of arriving at the final cation distribution, the site occupancy of all the cations was varied for many combinations, and those that agreed with the experimental intensity ratios are shown in Table 1. The final cation distributions were deduced simultaneously by considering the Bragg plane ratios, the fitting of the magnetization data at 80 K, and the ion distribution

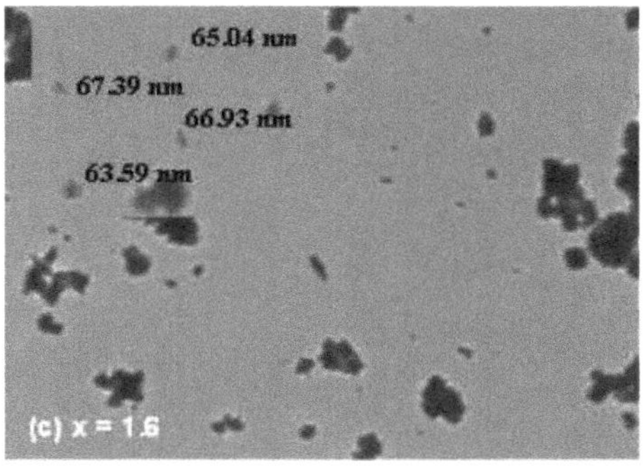

(a)

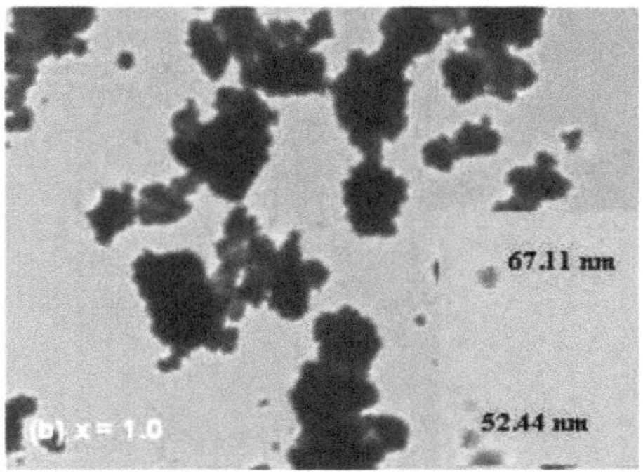

(b)

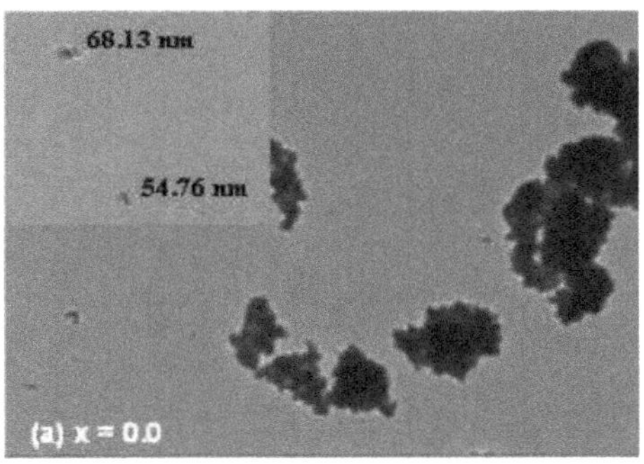

(c)

FIGURE 2: Transmission electron micrographs of $CoCr_xFe_{2-x}O_4$ system for (a) $x = 0.0$, (b) 1.0, and (c) 1.6 compositions. The insets show the enlarged images of grains in each case.

parameters of Fe^{3+} among the (A-) and (B-) sites of spinel lattice derived from Mossbauer spectral analysis [19].

3.2. Positron Lifetimes in the Unsubstituted Sample, $CoFe_2O_4$ ($x = 0.0$). The positron lifetime spectra were analyzed using the PALSfit computer program developed by the Risoe group [20]. The spectra of all the samples were fitted to obtain variances of fit within satisfactory limits (1.07 ± 0.12). The fits yielded three distinct lifetimes τ_1, τ_2, and τ_3 in all the cases, and their magnitudes, as discussed below, were the characteristics of positron trapping in specific sites within the spinel structure or positronium formation within the grain boundaries. In the $CoFe_2O_4$ ($x = 0.0$) sample, the intermediate lifetime τ_2 was found as high as 356 ps with relative intensity $I_2 = 48.2\%$. The normal interpretation

TABLE 1: The cationic distribution in the samples at different Cr^{3+} concentrations (x).

x	Cation distribution
0.0	$(Fe^{3+}{}_{0.9}Co^{2+}{}_{0.1})^A\ [Co^{2+}{}_{0.9}Fe^{3+}{}_{1.1}]^B\ O_4{}^{2-}$
0.1	$(Fe^{3+}{}_{0.8}Co^{2+}{}_{0.2})^A\ [Co^{2+}{}_{0.8}Cr^{3+}{}_{0.1}Fe^{3+}{}_{1.1}]^B\ O_4{}^{2-}$
0.3	$(Fe^{3+}{}_{0.6}Co^{2+}{}_{0.4})^A\ [Co^{2+}{}_{0.6}Cr^{3+}{}_{0.3}Fe^{3+}{}_{1.1}]^B\ O_4{}^{2-}$
0.7	$(Fe^{3+}{}_{0.3}Co^{2+}{}_{0.7})^A\ [Co^{2+}{}_{0.3}Cr^{3+}{}_{0.7}Fe^{3+}{}_{1.0}]^B\ O_4{}^{2-}$
1.1	$(Fe^{3+}{}_{0.3}Co^{2+}{}_{0.7})^A\ [Co^{2+}{}_{0.3}Cr^{3+}{}_{1.1}Fe^{3+}{}_{0.6}]^B\ O_4{}^{2-}$
1.3	$(Fe^{3+}{}_{0.3}Co^{2+}{}_{0.7})^A\ [Co^{2+}{}_{0.3}Cr^{3+}{}_{1.3}Fe^{3+}{}_{0.4}]^B\ O_4{}^{2-}$
1.5	$(Fe^{3+}{}_{0.3}Co^{2+}{}_{0.7})^A\ [Co^{2+}{}_{0.3}Cr^{3+}{}_{1.5}Fe^{3+}{}_{0.2}]^B\ O_4{}^{2-}$
1.7	$(Fe^{3+}{}_{0.3}Co^{2+}{}_{0.7})^A\ [Co^{2+}{}_{0.3}Cr^{3+}{}_{1.7}]^B\ O_4{}^{2-}$
1.8	$(Co^{2+}{}_{1.0})^A\ [Cr^{3+}{}_{1.8}Fe^{3+}{}_{0.2}]^B\ O_4{}^{2-}$
1.9	$(Co^{2+}{}_{1.0})^A\ [Cr^{3+}{}_{1.9}Fe^{3+}{}_{0.1}]^B\ O_4{}^{2-}$
2.0	$(Co^{2+}{}_{1.0})^A\ [Cr^{3+}{}_{2.0}]^B\ O_4{}^{2-}$

for the observation of such a well-resolved longer lifetime with appreciable intensity is the presence of vacancy-type crystalline defects within the material since positrons get trapped in such lower-than-average-electron-density sites. This is a reasonable assumption since it is nearly impossible to synthesize ferrites with fully occupied crystalline structure. Besides, those positrons managing to diffuse out to the vacancies on the nanocrystalline grain interfaces may also contribute to this component. The reason is that the thermal diffusion lengths of positrons in oxide materials are typically about 50–60 nm [21, 22]. Hence, a small fraction of positrons would inevitably diffuse and migrate to the surfaces of the nanocrystals (which are of sizes about 60–65 nm) before their annihilation. Despite prolonged heating, the grain dimensions could not be increased to more than the above limit. On the other hand, the diffusion lengths in the present case could be shortened due to the trapping of positrons by vacancy clusters if present within the nanoparticles. The positron lifetime in the perfect crystalline sample (τ_f, for which no theoretical value is available) can be calculated using the trapping model equation [23]:

$$\frac{1}{\tau_f} = \frac{I_1}{\tau_1} + \frac{I_2}{\tau_2} + \frac{I_3}{\tau_3}. \qquad (2)$$

Substituting the experimental values of the positron lifetimes and their intensities of the $CoFe_2O_4$ sample in the above equation, we obtain $\tau_f = 199$ ps. The shorter lifetime component τ_1 is obviously less than this value in all the cases due to admixing with the Bloch state residence time of trapped positrons [12]. A small contribution coming to τ_1 from parapositronium atoms of lifetime 125 ps is ignored as the intensity of this component, one-third that of the orthopositronium intensity I_3, is negligibly small.

The annihilation characteristics of positrons diffusing out to the grain surfaces are also reflected in the variation of the longest lifetime τ_3 and its intensity I_3. The magnitude of this lifetime (1.8–2.1 ns) is typical of the "pick-off" annihilation of orthopositronium atoms formed at the interfacial regions of the grains [12]. Although positronium formation is not significant enough to alter the interpretations in metallic oxides, it has been found still relevant enough to force a

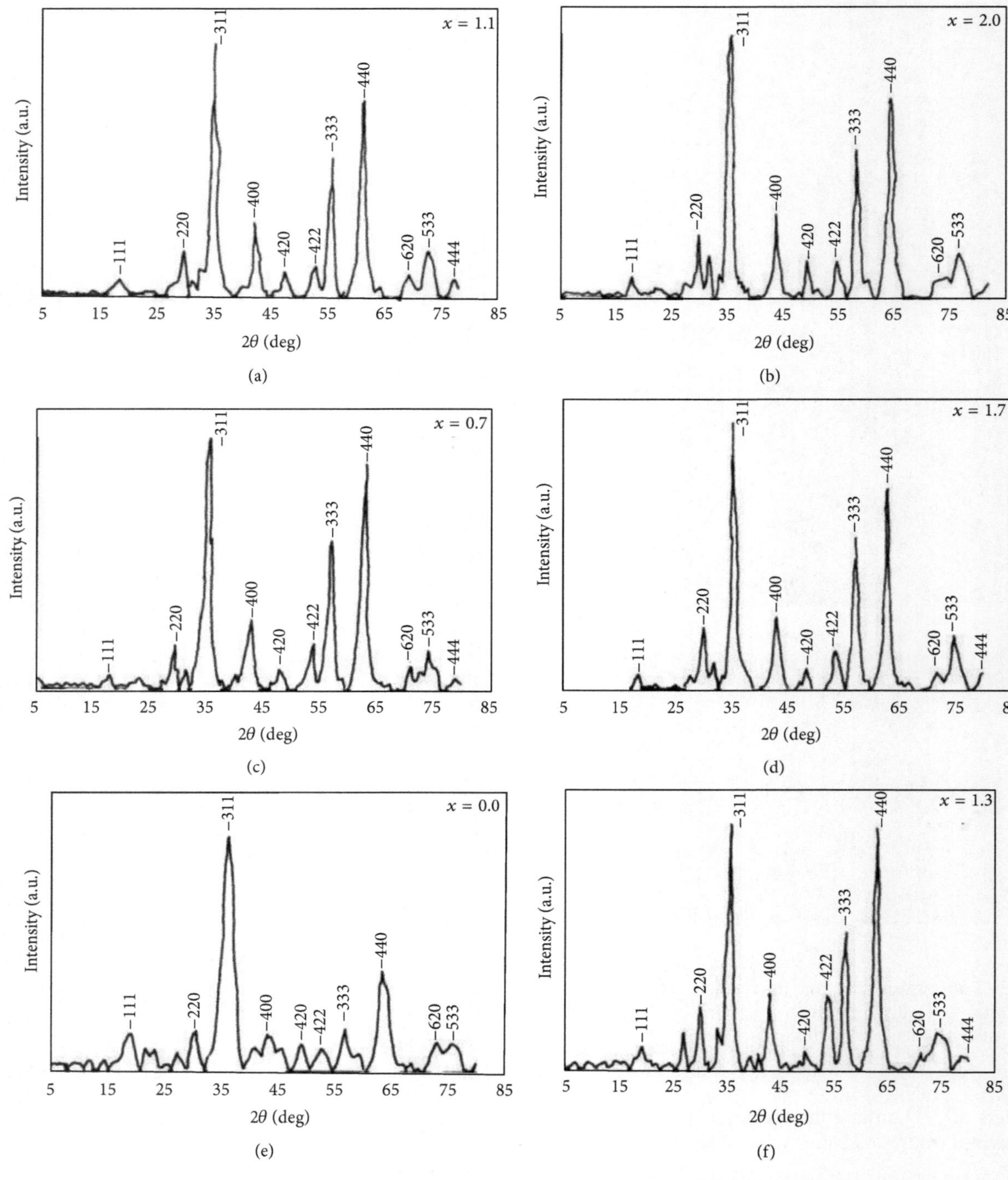

FIGURE 3: Typical X-ray diffraction patterns of the $CoCr_xFe_{2-x}O_4$ system for different Cr^{3+} concentration (x).

three-component analysis of the positron lifetime spectra of nanocrystalline materials [24–26], and the intensity I_3, despite being relatively small (0.8–1.4%), indicates the presence of large free volume regions in the intergranular regions of materials when composed by nanometer-sized particles or grains.

3.3. Results of Cr^{3+} Substitution. Figures 5 and 6 describe the changes occurring in the positron annihilation parameters as a result of Cr^{3+} substitution for Fe^{3+} in $CoFe_2O_4$. A close look into the trends of variation helps to identify three distinct stages of defects evolution and/or structural variations. In the first stage spreading over the concentration $x = 0.1$ to 0.7, the two positron lifetimes τ_1 and τ_2 show remarkable increase in the initial stage and attain saturation. The longer lifetime τ_3 and its intensity I_3 show characteristic decrease that will be discussed later. The second stage of variation is marked for $x > 0.7$ till 1.7 during which the lifetimes decrease and the

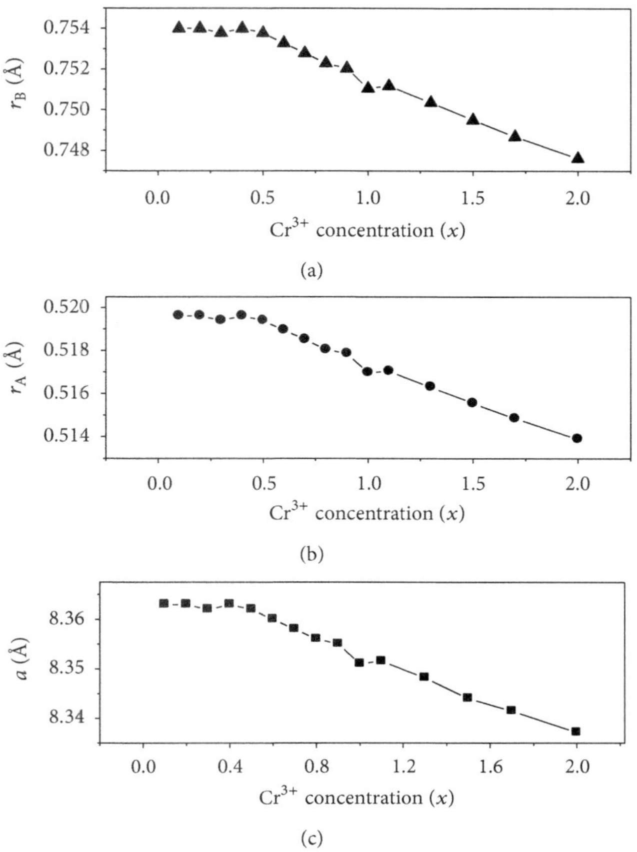

FIGURE 4: The lattice constant (a) and the radii of the tetrahedral (r_A) and octahedral (r_B) sites versus Cr^{3+} concentration (x).

intensity sharply rises. All these trends are just reversed once again in the last stage from $x = 1.8$ to 2.0.

The variation of the different positron annihilation parameters with Cr^{3+}-substitution is thus highly complex in nature, as the potential trapping centers might have changed during the different stages of substitution, due to not only the arrival of a new element but also the relative displacement they may cause in the positioning of the other ions already present in the crystalline structure. Certain information in this direction is available from the results of CDBS measurements shown in Figures 7(a) and 7(b). The data has been analyzed using the usual quotient spectral method in which the projected one-dimensional spectra on the $((E_1 - E_2)/2)$ axis of the counts in the window $((E_1 + E_2)/2) = 511 \pm 1.2$ keV are peak normalized and divided by that of a pure reference sample (Si single crystals) [14, 27]. The choice of Si to serve as a reference is not unjustified since it is not a constituent of the material at any stage in this investigation, and the purpose is to magnify the differences in shapes of the momentum distribution curves for easy understanding and interpretation. In Figure 7(a), the ratio curves of the samples with the two extreme compositions, $CoFe_2O_4$ ($x = 0.0$) and $CoCr_2O_4$ ($x = 2.0$), are shown together with the identical curves obtained for the constituent elemental samples. Figure 7(b) illustrates the curves obtained similarly for the Cr^{3+}-substituted samples of a few representative concentrations. The ratio curves of the samples are found having a characteristic peaks at $p_L = 10.3 \times 10^{-3}\, m_0c$ (where m_0 is the electron mass and c is the velocity of light). The peaks of the ratio curves of the three constituent metals, that is, Co, Fe, and Cr, appear at $15.0 \times 10^{-3}\, m_0c$, $12.2 \times 10^{-3}\, m_0c$, and $11.3 \times 10^{-3}\, m_0c$, respectively, but with decreasing amplitudes. This observation is consistent with the decreasing number of d-electrons and decreasing radius of the $3d$-shell. Thus the peak of the ratio curve of either the pure or any of the Cr^{3+}-substituted samples does not coincide with those of the elemental curves as an ample proof to suggest that positrons are not trapped in oxygen vacancies, a fact even otherwise vindicated by their positive charge that will repel positrons. On the other hand, trapping takes place in the cationic vacancies, and the peak at $p_L = 10.3 \times 10^{-3}\, m_0c$ common to all the samples and irrespective of the Cr^{+} concentration (x) indicates the encirclement of the defects by oxygen ions. In several of our recent studies on nanocrystalline oxide semiconductors, we have similarly obtained the peak due to annihilation with oxygen electrons at $p_L = 10.3 \times 10^{-3}\, m_0c$ [28–30].

The identical elemental environment around the positron trapping sites at all concentrations of Cr^{3+} substitution is further verified from the S versus W plot shown in Figure 8 that is normally used to identify the changes in the predominant type of positron trapping defects at different stages of variation of the experimental parameters. The S and W parameters have been derived from the CDB spectra as the counts falling under segments, respectively, from 0 to $3.75 \times 10^{-3}\, m_0c$ and from $7.5 \times 10^{-3}\, m_0c$ to $12.25 \times 10^{-3}\, m_0c$ normalized by the total counts accumulated under 0 to $37.5 \times 10^{-3}\, m_0c$. The S-W plot is linear, and all the points lie essentially on a straight line. This indicates that positrons essentially encounter similar elemental environments irrespective of the cationic redistribution. This is a further credence to the argument that the defects which trap positrons are surrounded by oxygen ions, and therefore, the traps are none other than the cationic vacancies. But there are variations in the intensity of annihilation with the oxygen electrons, as indicated by the individual variations of the S and W parameters with Cr^{3+} concentration (x) shown in Figures 9(a) and 9(b). (For the sake of clarity, the curves of not all the samples are shown in Figure 7(a) or Figure 7(b), but the peak coordinates of all the curves are shown in Figure 8). From Figures 9(a) and 9(b) also, we can distinguish from one another basically three regions, the demarcation being identical to that mentioned in the case of the positron lifetime results. The first two regions ($x = 0.0$–0.7 and $x = 0.7$–1.7) are characterized by a fall and a rise of the peaks of the curves, and the last stage is marked by again a fall. It can be argued that although the annihilation environment of positrons essentially remains identical, they are trapped at different stages of Cr^{3+} substitution by the defects situated at different sites in the lattice structure. These points are further discussed in detail afterwards.

The insensitivity of the CDB spectra to the oxygen vacancies can be explained on the basis of the results of positron lifetime measurements as well. The difference between the lifetime characteristic of defects (i.e., τ_2) and the bulk lifetime τ_f is normally considered as an indication of the size of the defect. The enhancement in the positron lifetime due to trapping in monovacancies is ~40–80 ps in typical metals and alloys [31]. Assuming that $\tau_2 = 356$ ps is an upper limit of the positron lifetime in vacancy clusters in the unsubstituted sample, $\tau_2 - \tau_f = 356 - 199 = 157$ ps will correspond to defects much larger than monovacancies. Theoretical estimations in Fe, which is normally bcc in structure but a constituent of the present samples, have shown the enhancement of positron lifetime in a neutral vacancy cluster composed of 4-5 neighboring monovacancies as 152 ps [32]. Considering these facts, the positron trapping site in the undoped alloy can be conceived to be a vacancy cluster composed of the monovacancy created by the absence of a doubly ionized cation and four of its coordinated oxygen ions. Based on X-ray diffraction, magnetization and Mossbauer results, Mohan et al. [10] have shown that the $CoFe_2O_4$ ($x = 0.0$) is a nearly complete inverse spinel and the ionic distribution in it is of the form $(Fe^{3+}{}_{0.9}Co^{2+}{}_{0.1})^A[Co^{2+}{}_{0.9}Fe^{3+}{}_{1.1}]^B O_4{}^{2-}$. The absence of a Co^{2+} ion with four oxygen neighbor ions will give rise to a neutral pentavacancy cluster in which positrons can be trapped and annihilated. The other possibility of the absence of a trivalent cation with four neighboring oxygen ions cannot be ruled out as it would have enhanced positron trapping due to surplus negative charge, and hence, it is necessary to point out whether the said vacancy cluster is centered at the (A-) site or at the (B-) site. This can be answered by looking at the effects of Cr^{3+} substitution on the cationic redistribution. A schematic diagram showing the two neighboring octants of a normal spinel structure is shown in Figure 10. At the very onset of substitution, a drastic drop in the intensity of the peak in the CDB spectra (represented by the W parameter) is observed (Figures 7(b) and 9(b)), and it indicates the diminishing positron annihilation probability with oxygen electrons. The ionic distributions obtained from the X-ray diffraction peak intensity analysis for samples with the different concentrations of the Cr^{3+} ions are given in Table 1. Thus, for example, the distribution is $(Fe^{3+}{}_{0.8}Co^{2+}{}_{0.2})^A[Co^{2+}{}_{0.8}Cr^{3+}{}_{0.1}Fe^{3+}{}_{1.1}]^B O_4{}^{2-}$ for $x = 0.1$. This implies that the substituted Cr^{3+} ions initially replace an equal number of Fe^{3+} ions from the (B-) sites, but simultaneously an equal number of Fe^{3+} ions from the (A-) sites move over to the (B-) sites in exchange of Co^{2+} ions from the (B-) sites to the (A-) sites. In effect, an inversion of the spinel structure is prompted as a result of the substitution process. Hence, as shown in Table 1, the number of Fe^{3+} ions at the (A-) sites decreases whereas that at the (B-) sites remains unaltered till $x = 0.7$ compositions. Since CDB spectra indicate diminishing annihilation with oxygen electrons and the positron lifetime τ_2 increases from 356 ps to 374 ps, it is reasonable to argue that the defects in the sample with $x = 0.1$ are larger in size and increasingly deficient in oxygen ions than those in the unsubstituted ($x = 0.0$) sample. In other words, the defects were centered at the (A-) sites in the unsubstituted sample and at the (B-) sites in the substituted samples. We attribute the second positron lifetime component τ_2 to such large vacancy clusters.

As already stated, the substitution or doping resulted in sharp rises in the two lifetimes, τ_1 and τ_2. The intensity I_2, however, did not show any change. It is, therefore, a local effect in which the vacancy cluster has undergone an increase in size. From EDAX studies, we have estimated the actual concentration of the Cr^{3+} ions effectively substituted in the crystallites (Figure 1). It has been found that the (B-) sites suffered from nonstoichiometric deficiencies of Cr^{3+} ions, and hence, the Co^{2+} ions transferred to the (A-) sites are also less in number than that predicted by the formula $(Fe^{3+}{}_{0.9-x}Co^{2+}{}_{0.1+x})^A[Co^{2+}{}_{0.9-x}Cr^{3+}{}_{x}Fe^{3+}{}_{1.1}]^B O_4{}^{2-}$. The result is that the vacancies so created will add to the existing vacancy clusters, resulting in further increase in their size and thereby enhancing the positron lifetimes. However, the deficiency decreases on subsequent doping, and therefore, the lifetime τ_2 and intensity I_2 remain rather unchanged in the range of concentration from 0.1 to 0.7.

On the other hand, the concentration of Fe^{3+} ions at the (B-) sites does not change despite the substitution by Cr^{3+} ions till $x = 0.7$ due to the simultaneous inversion process in which the Co^{2+} ions migrate to the (A-) sites [10]. This means that, the Cr^{3+} ions occupy the (B-) sites. The Cr^{3+} substitution (x) above this value does not indicate further inversion, and hence, the effective number of Fe^{3+} ions at the (B-) sites starts decreasing (Table 1), whereas at $x = 0.7$ the peak in the CDB spectrum appears the sharpest, and the positron lifetimes start decreasing above this concentration, indicating a reduction in the size of the vacancy clusters. The intensity I_2 increases (Figure 5). The fact that positrons are now annihilating at sites which are smaller in size can be understood as follows. The lattice constant of the $CoCr_xFe_{2-x}O_4$ samples steeply reduces in samples with increased Cr^{3+} concentration (Figure 4). The contraction of the lattice can be attributed to the slightly smaller ionic radius of Cr^{3+} (0.630 Å) compared to that of Fe^{3+} (0.640 Å). Using the experimentally found values of the lattice constant (a) and the oxygen positional parameter (u) [1], it is possible to calculate the radii of the tetrahedral and octahedral sites, r_A and r_B, respectively, using the relations [1]

$$r_A = \sqrt{3}\left(u - \frac{1}{4}\right)a - R_0, \quad (3a)$$

$$r_B = \left(\frac{5}{8} - u\right)a - R_0. \quad (3b)$$

Here, R_0 is the radius of the oxygen ion (taken as 1.32 Å), and u is taken as = 0.379 considering that $CoCr_xFe_{2-x}O_4$ is fully inverse at $x = 0.0$ composition [1, 2]. The site radii for the different compositions estimated from the above equations are also shown in Figure 4. Increasing the concentration of Cr^{3+} will result in filling the vacancies at the octahedral sites, and, owing to the positive charge, positrons are trapped in a reducing number in the Cr^{3+}-vacancy complexes so formed. On the other hand, while the octahedral sites were large enough to accommodate the Co^{2+} ions, the radii of the tetrahedral sites are magnitude-wise smaller than its ionic radius (0.740 Å), and hence, it is likely that a fraction of

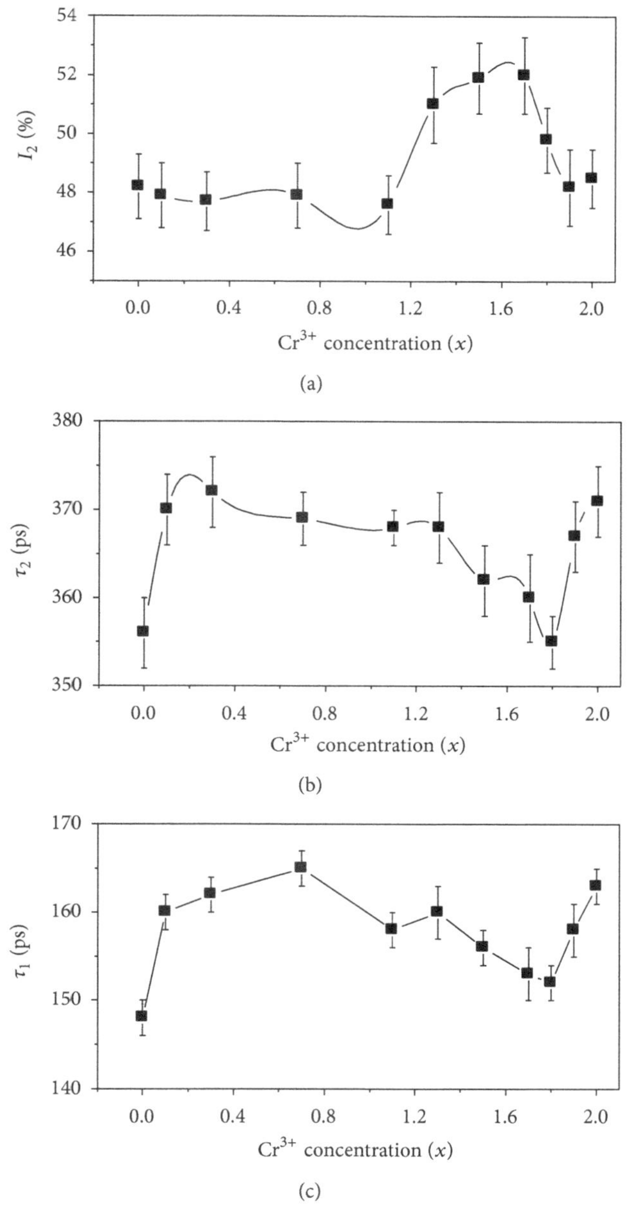

FIGURE 5: The positron lifetimes, τ_1 and τ_2, and intensity I_2 versus Cr^{3+} concentration (x).

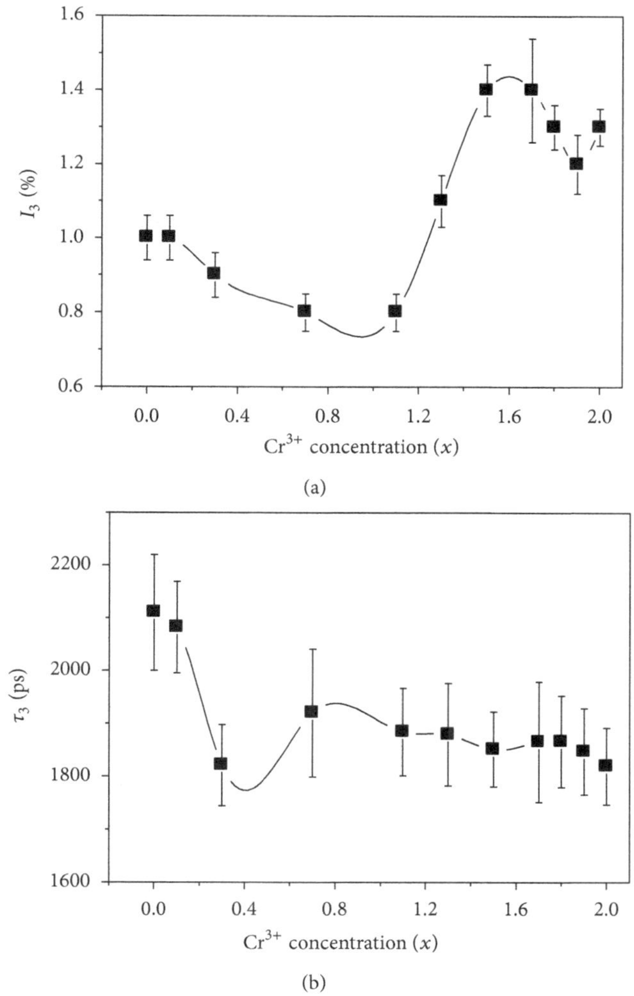

FIGURE 6: The orthopositronium lifetime τ_3 and intensity I_3 versus Cr^{3+} content (x).

the tetrahedral sites are unoccupied during the inversion of the spinel structure at concentrations $x < 0.7$. As a result, positrons now that are no more trapped at the octahedral vacancies will move over to the vacancies at the tetrahedral sites and get trapped there. The fast decreasing positron lifetimes support this argument since the tetrahedral vacancy clusters are smaller than the octahedral ones. The availability of additional trapping sites expectedly increases the intensity I_2 as well. Considering that positrons are now getting trapped in the vacancy clusters present at the tetrahedral sites, the lattice contraction would directly influence their lifetime, and the observations in the range mentioned above are in accordance with the same. A careful study of the ionic distribution shown in Table 1 makes us realize that from $x = 0.7$ onwards till $x = 1.7$, the occupancies of Fe^{3+} and Co^{2+} at the (A-) sites remain unchanged, whereas the Cr^{3+} ions directly replace the Fe^{3+} ions at the (B-) sites. At $x = 1.7$, all the Fe^{3+} ions at the (B-) sites are replaced by Cr^{3+} ions (Table 1).

The last stage in the variation of the positron annihilation parameters versus the Cr^{3+} concentration (x) is observed between $x = 1.8$ and 2.0. During $x = 0.7$ to 1.7, the cation distribution at the (A-) sites remained unaltered as $Fe^{3+}{}_{0.3}Co^{2+}{}_{0.7}$, while Cr^{3+} ions monotonically replaced the Fe^{3+} ions at the (B-) sites with the Co^{2+} concentration at the (B-) sites remaining unaltered as 0.3. As the lattice contraction continues, the transformation of the spinel structure from the inverse to the normal configuration that had started during $x = 0.1$–0.7 and discontinued during $x = 0.7$–1.7 gets completed. From octahedral site stabilization energy considerations, it is known that cobaltchromite ($CoCr_2O_4$) is a normal spinel [10]. In earlier positron annihilation studies

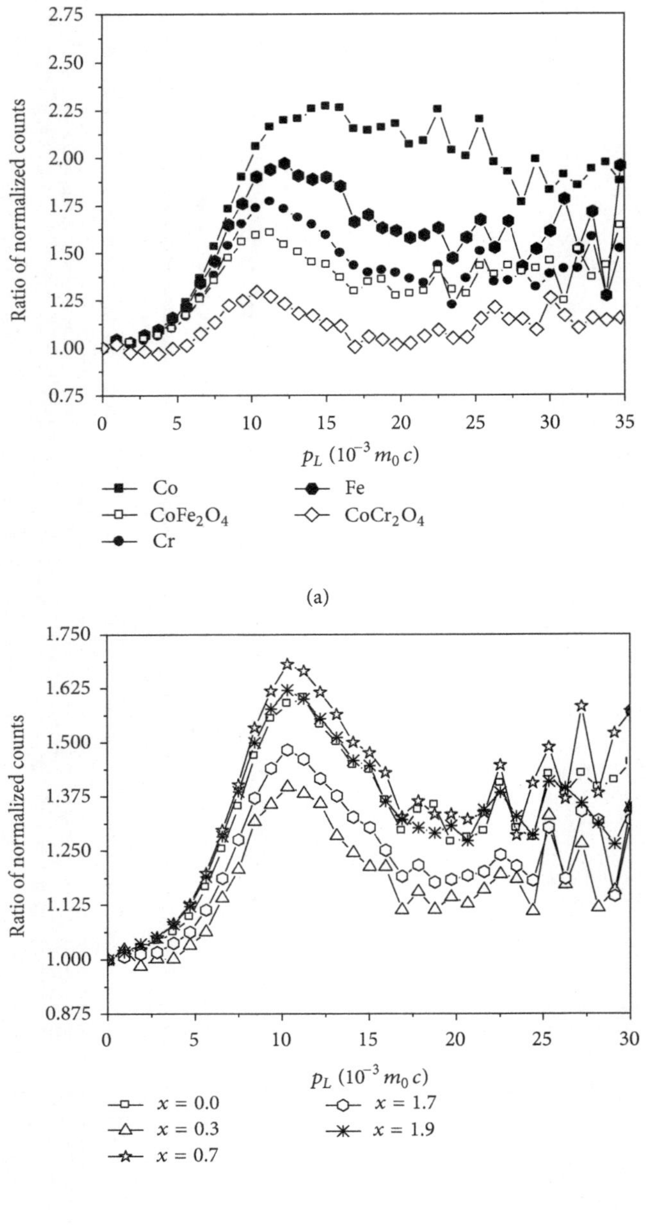

FIGURE 7: The ratio curves generated from the coincidence Doppler broadening spectra of the different samples—(a) elements Co, Fe, and Cr besides $CoCr_xFe_{2-x}O_4$ of $x = 0.0$ and 2.0; (b) $CoCr_xFe_{2-x}O_4$ of $x = 0.0$, 0.3, 0.7, 1.7, and 1.9. All the spectra had been peak normalized and then divided by that of a pure reference Si sample in order to generate the ratio curves.

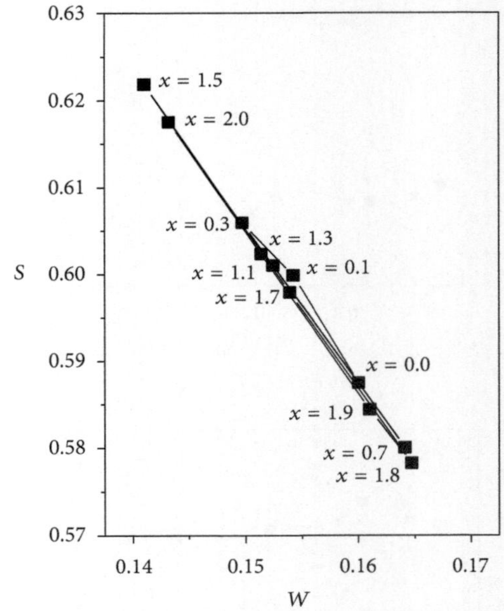

FIGURE 8: The S-W plot of the Cr^{3+}-substituted samples.

of nanocrystalline $ZnFe_2O_4$ [24] and $NiFe_2O_4$ [25, 33], the positron lifetimes had been observed to decrease when a normal spinel ferrite transforms to an inverse spinel, and conversely they increase when the transformation is just the opposite. These observations had been verified through Mossbauer spectroscopic studies too [25, 33]. Since the two positron lifetimes, τ_1 and τ_2, drastically increase during $x = 1.8$ and 2.0 (Figure 5), this stage is attributed to the total transformation of the partly inverse $CoCr_xFe_{2-x}O_4$ to the fully normal $CoCr_2O_4$. Note further that unlike during $x = 0.1$–0.7 when the intensity I_2 did not show any change, it decreases in the final stage of inversion indicating the full occupancy of the (A-) sites by Co^{2+} ions. The fact remains that the spinel structures normally suffer from nonstoichiometric disorders, and therefore, vacancy clusters are inherently inbuilt in the structure. The large value of τ_2 with still an appreciable intensity I_2 supports this argument.

As has been already pointed out, the longest lifetime τ_3 and its intensity I_3 are due to the nanocrystalline dimensions of the samples, and they result from positronium atoms annihilating at the intergranular region. Hence, they need not necessarily reflect the effects of any change in the vacancy cluster dynamics within the grains. Yet τ_3 shows a sudden decrease during the initial stage $x = 0.1$ to 0.7 and then remains constant (Figure 6). The intensity I_3 gradually falls during this stage but shows a characteristic rise during the second stage $x = 0.7$ to 1.7 and then remains constant (Figure 6). The initial fall can be attributed to small traces of Cr^{3+} ions, unsuccessful in being incorporated into the spinel structure and hence left to remain in trace amounts within the intergranular region. EDAX analysis also indicated frustration even within the lattice due to the failure in the complete substitution of Fe^{3+} by the Cr^{3+} ions in the system. In the latter stage (i.e., $x > 0.7$), the lattice contraction has expectedly resulted in a decrease by 0.3% in the grain size, and thereby, the number of positrons reaching out on the grain surfaces has slightly increased.

Although it is known that the magnetic properties of the samples undergo rapid and interesting changes during the Cr^{3+} substitution, correlating such changes to the behavior of positron annihilation parameters is never straightforward

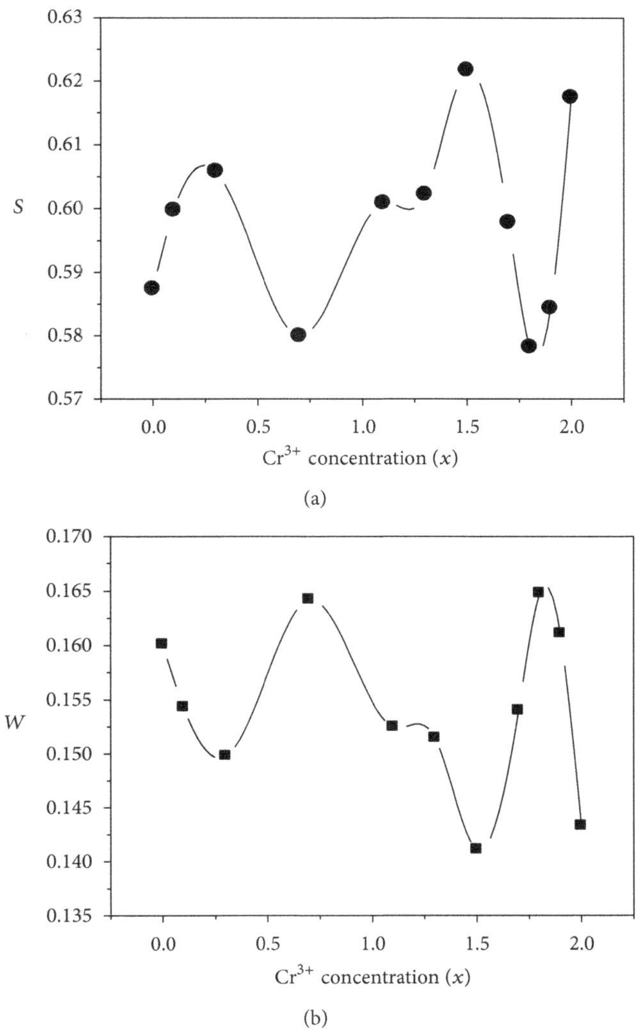

FIGURE 9: The S and W parameters versus Cr^{3+} concentration (x).

and is not attempted here. The structural properties and their changes, as depicted by the positron annihilation parameters and their variations, may influence the magnetic properties, which need to be investigated by appropriate experimental methods.

4. Conclusions

In understanding the effects of Cr^{3+}-substitution in place of Fe^{3+} in $CoFe_2O_4$ studied by positron annihilation spectroscopy, we have offered the physical interpretation of the results in terms of three distinct stages of defect evolution and interaction. First, the unsubstituted ferrite ($CoFe_2O_4$, $x = 0.0$) itself was found to contain large vacancy clusters. These clusters are identified as being present at the (A-) sites with the divalent Co^{2+} ion and four of its coordinated oxygen ions making way for such very strong trapping centers for positrons. At the onset of Cr^{3+}-substitution (x), the positron lifetimes increased due to the transfer of positron trapping into defects to the (B-) sites. In the second stage from $x = 0.7$ to $x = 1.7$, a concomitant lattice contraction influenced the positron annihilation characteristics. This contraction is attributed to the slightly smaller ionic radius of Cr^{3+} than that of Fe^{3+}. There is also a change in the positron trapping sites from the vacancy clusters at the (B-) sites back to those at the (A-) sites. The last stage is marked by a full inversion of the structure to that of a normal spinel chromite, and the positron annihilation parameters depicted this stage with a characteristic reversal of the trend of variation with the Cr^{3+} concentration.

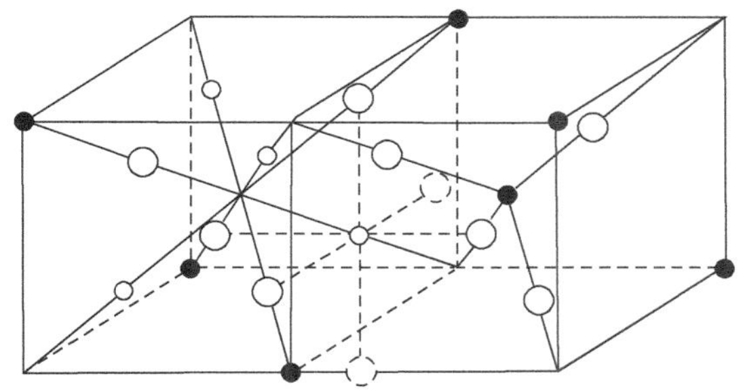

FIGURE 10: Schematic representation of the two neighboring octants of a normal spinel structure. The large circles represent oxygen ions. The small solid and open circles represent the cations at the tetrahedral and octahedral sites, respectively, as illustrated in [33].

Finally, we conclude that the positron lifetime measurements complemented by results from coincidence Doppler broadening spectroscopy can be a viable alternative experimental tool to monitor the generation and evolution of structural disorders in XY_2O_4 (where X and Y are divalent and trivalent metals, resp.) systems during physical treatments like doping and grain size reduction. Positron annihilation parameters are seen to sense, directly or indirectly, physical phenomena of different kinds and implications like the redistribution of cations, lattice contraction or expansion, and structural transformations in certain cases [24–26].

Acknowledgments

The authors are grateful to Professor, Indranil Das of Saha Institute of Nuclear Physics (SINP), Kolkata, for providing the reference Co and Cr samples for CDBS experiments. Some useful discussions with Dr. N. N. Mondal and a graphical illustration (Figure 10) prepared by Mr. Pradipta K. Das are also gratefully acknowledged. One of the authors (Kunal B. Modi) is thankful to Professor H. H. Joshi for providing some of the ferrite samples.

References

[1] J. Smit and H. P. J. Wijn, *Ferrites—Physical Properties of Ferromagnetic Oxides in Relation to Their Technical Applications*, N. V. Philips Gloeilampenfabrieken, Eindhoven, The Netherlands, 1959.

[2] F. Scordari, *Fundamentals of Crystallography*, edited by C. Giacovazzo, Oxford University Press, New York, NY, USA, 1992.

[3] W. B. Cross, L. Affleck, M. V. Kuznetsov, and I. P. Parkin, "Self-propagating high-temperature synthesis of ferrites MFe_2O4 (M = Mg, Ba, Co, Ni, Cu, Zn); reactions in an external magnetic field," *Journal of Materials Chemistry*, vol. 9, no. 10, p. 2545, 1999.

[4] K. Tomiyasu, J. Fukunaga, and H. Suzuki, "Magnetic short-range order and reentrant-spin-glass-like behavior in $CoCr_2O_4$ and $MnCr_2O_4$ by means of neutron scattering and magnetization measurements," *Physical Review B*, vol. 70, no. 21, Article ID 214434, 12 pages, 2004.

[5] V. T. Thanki, N. N. Jani, U. V. Chhaya, H. H. Joshi, and R. G. Kulkarni, "Magnetic properties of $CoFe_{2-x}Cr_xO_4$ synthesized by co-precipitation method," *AsianJournal of Physics*, vol. 6, no. 1-2, p. 222, 1997.

[6] G. A. Sawatzky, F. van der Woude, and A. H. Morrish, "Cation distributions in octahedral and tetrahedral sites of the ferrimagnetic spinel $CoFe_2O_4$," *Journal of Applied Physics*, vol. 39, no. 2, pp. 1204–1205, 1968.

[7] G. Lawes, B. Melot, K. Page et al., "Dielectric anomalies and spiral magnetic order in $CoCr_2O_4$," *Physical Review B*, vol. 74, no. 2, Article ID 024413, 6 pages, 2006.

[8] N. Menyuk, K. Dwight, and A. Wold, "Ferrimagnetic spiral configurations in cobalt chromite," *Journal de Physique*, vol. 25, pp. 528–536, 1964.

[9] A. Hauet, J. Teillet, B. Hannoyer, and M. Lenglet, "Mössbauer study of Co and Ni ferrichromites," *Physica Status Solidi (A)*, vol. 103, no. 1, pp. 257–261, 1987.

[10] H. Mohan, I. A. Shaikh, and R. G. Kulkarni, "Magnetic properties of the mixed spinel $CoFe_{2-x}CrxO_4$," *Physica B*, vol. 217, no. 3-4, pp. 292–298, 1996.

[11] B. G. Tolesha, S. E. Shrisath, M. L. Mane, S. M. Patange, S. S. Jadhav, and K. M. Jadhav, "Autocombustion high-temperature synthesis, structural, and magnetic properties of $CoCr_xFe_{2-x}O_4$ ($0 \le x \le 1.0$)," *The Journal of Physical Chemistry C*, vol. 115, no. 43, pp. 20905–20912, 2011.

[12] R. W. Siegel, "Positron annihilation spectroscopy," *Annual Review of Materials Science*, vol. 10, pp. 393–425, 1980.

[13] S. Biswas, S. Kar, S. Chaudhuri, and P. M. G. Nambissan, "Mn^{2+}-induced substitutional structural changes in ZnS nanoparticles as observed from positron annihilation studies," *Journal of Physics: Condensed Matter*, vol. 20, no. 23, Article ID 235226, 2008.

[14] P. Asoka-Kumar, M. Alatalo, V. J. Ghosh, A. C. Kruseman, B. Nielsen, and K. G. Lynn, "Increased elemental specificity of positron annihilation spectra," *Physical Review Letters*, vol. 77, no. 10, pp. 2097–2100, 1996.

[15] B. D. Cullity, *Elements of X-Ray Diffraction*, Addison-Wesley, Reading, Mass, USA, 1978.

[16] C. Dong, "*PowderX*: windows-95-based program for powder X-ray diffraction data processing," *Journal of Applied Crystallography*, vol. 32, p. 838, 1999.

[17] M. J. Buerger, *Crystal Structure Analysis*, Wiley, New York, NY, USA, 1960.

[18] H. Ohnish and T. Teranishi, "Crystal distortion in copper ferrite-chromite series," *Journal of the Physical Society of Japan*, vol. 16, pp. 35–43, 1961.

[19] V. T. Thanki, "Study on magnetic properties of oxide materials," *[Ph.D. thesis]*, Saurashtra University, Rajkot, India, 1996, (unpublished).

[20] J. V. Olsen, P. Kirkegaard, N. J. Pedersen, and M. Eldrup, "PALSfit: a new program for the evaluation of positron lifetime spectra," *Physica Status Solidi (C)*, vol. 4, no. 10, pp. 4004–4006, 2007.

[21] T. Koida, S. F. Chichibu, A. Uedono et al., "Correlation between the photoluminescence lifetime and defect density in bulk and epitaxial ZnO," *Applied Physics Letters*, vol. 82, no. 4, p. 532, 2003.

[22] A. Zubiaga, F. Tuomisto, F. Plazaola et al., "Zinc vacancies in the heteroepitaxy of ZnO on sapphire: influence of the substrate orientation and layer thickness," *Applied Physics Letters*, vol. 86, no. 4, Article ID 042103, 3 pages, 2005.

[23] P. Hautojarvi and C. Corbel, "For a detailed discussion on different cases of positron trapping in solids," in *Positron Spectroscopy of Solids*, pp. 491–532, IOS Press, Amsterdam, The Netherlands, 1995.

[24] P. M. G. Nambissan, C. Upadhyay, and H. C. Verma, "Positron lifetime spectroscopic studies of nanocrystalline $ZnFe_2O_4$," *Journal of Applied Physics*, vol. 93, no. 10, pp. 6320–6326, 2003.

[25] S. Chakraverty, S. Mitra, K. Mandal, P. M. G. Nambissan, and S. Chattopadhyay, "Positron annihiliation studies of some anomalous features of $NiFe_2O_4$ nanocrystals grown in SiO_2," *Physical Review B*, vol. 71, Article ID 024115, 8 pages, 2005.

[26] S. Chakrabarti, S. Chaudhuri, and P. M. G. Nambissan, "Positron annihilation lifetime changes across the structural phase transition in nanocrystalline Fe_2O_3," *Physical Review B*, vol. 71, no. 6, Article ID 064105, 6 pages, 2005.

[27] Y. Nagai, T. Nonaka, M. Hasegawa et al., "Direct evidence of positron trapping at polar groups in a polymer-blend system," *Physical Review B*, vol. 60, no. 17, pp. 11863–11866, 1999.

[28] T. Ghoshal, S. Biswas, S. Kar, S. Chaudhuri, and P. M. G. Nambissan, "Positron annihilation spectroscopic studies of solvothermally synthesized ZnO nanobipyramids and nanoparticles," *Journal of Chemical Physics*, vol. 128, no. 7, Article ID 074702, 2008, *Virtual Journal of Nanoscale Science and Technology*, vol. 17, no. 8, 2008.

[29] T. Ghoshal, S. Kar, S. Biswas, S. K. De, and P. M. G. Nambissan, "Vacancy-type defects and their evolution under Mn substitution in single crystalline ZnO nanocones studied by positron annihilation," *Journal of Physical Chemistry C*, vol. 113, no. 9, pp. 3419–3425, 2009.

[30] B. Roy, B. Karmakar, P. M. G. Nambissan, and M. Pal, "Mn substitution effects and associated defects in ZnO nanoparticles studied by positron annihilation," *Nano*, vol. 6, no. 2, p. 173, 2011.

[31] I. K. MacKenzie, *Positron Solid State Physics*, edited by W. Brandt and A. Dupasquier, North Holland, Amsterdam, The Netherlands, 1983.

[32] M. J. Puska and R. M. Nieminen, "Defect spectroscopy with positrons: a general calculational method," *Journal of Physics F*, vol. 13, no. 2, pp. 333–346, 1983.

[33] S. Mitra, K. Mandal, S. Sinha, P. M. G. Nambissan, and S. Kumar, "Size and temperature dependent cationic redistribution in $NiFe_2O_4(SiO_2)$ nanocomposites: positron annihilation and Mössbauer studies," *Journal of Physics D*, vol. 39, no. 19, pp. 4228–4235, 2006.

12

Vibrational Spectroscopy of Chemical Species in Silicon and Silicon-Rich Nitride Thin Films

Kirill O. Bugaev,[1,2] Anastasia A. Zelenina,[1,2] and Vladimir A. Volodin[1,2]

[1] *A.V. Rzhanov Institute of Semiconductor Physics, Russian Academy of Sciences, Lavrentieva Avenue 13, Novosibirsk 630090, Russia*
[2] *Physical Faculty, Novosibirsk State University, Pirogova Street 2, Novosibirsk 630090, Russia*

Correspondence should be addressed to Vladimir A. Volodin, volodin@isp.nsc.ru

Academic Editor: Alexander Milekhin

Vibrational properties of hydrogenated silicon-rich nitride (SiN_x:H) of various stoichiometry ($0.6 \leq x \leq 1.3$) and hydrogenated amorphous silicon (a-Si:H) films were studied using Raman spectroscopy and Fourier transform infrared spectroscopy. Furnace annealing during 5 hours in Ar ambient at 1130°C and pulse laser annealing were applied to modify the structure of films. Surprisingly, after annealing with such high-thermal budget, according to the FTIR data, the nearly stoichiometric silicon nitride film contains hydrogen in the form of Si–H bonds. From analysis of the FTIR data of the Si–N bond vibrations, one can conclude that silicon nitride is partly crystallized. According to the Raman data a-Si:H films with hydrogen concentration 15% and lower contain mainly Si–H chemical species, and films with hydrogen concentration 30–35% contain mainly Si–H_2 chemical species. Nanosecond pulse laser treatments lead to crystallization of the films and its dehydrogenization.

1. Introduction

The interest in amorphous, nano- and microcrystalline silicon films on nonrefractory inexpensive substrates and their crystallization is stimulated by demands of giant microelectronics. For example, the enlargement of sizes of flat-panel displays with active thin film transistor matrix can be described as "reverse Moor's law." Scientific interest in the silicon-rich nitride (SRN) films is growing due to their perspectives for application. The most important applications of SRN films are silicon-based optoelectronic devices [1], nonvolatile memory (flash-memory) devices [2], and new tandem solar cells based on silicon and silicon oxide (silicon nitride) heterostructures [3]. Silicon nanoclusters and SRN films can be applied in novel quantum-dots-based solar cells with enhanced efficiency [4].

Modern plasma-enhanced chemical vapor deposition (PECVD) technique allows the production of amorphous silicon films and SRN films with temperature of substrate lower than 100°C [5]. As reagents containing hydrogen are mainly used in PECVD, as-deposited films are hydrogenated. Hydrogen plays important role in electrical and optical properties of the films and also has influence on stability of the films. So, it is important to develop express, nondestructive spectroscopic approaches for studies of hydrogen in amorphous hydrogenated silicon (a-Si:H) films and hydrogenated SRN films. The present paper is devoted to studies of as-deposited films as well as furnace- and pulse-laser-treated films with the use of Raman spectroscopy and Fourier Transform Infrared (FTIR) spectroscopy.

2. Experiment Details

Hydrogenated SRN films (SiN_x:H) of different stoichiometry were deposited on the Si (100) substrates with the use of low-frequency PECVD method. The frequency of discharge was 55 kHz, and the films were deposited from the gas mixture of ammonia (NH_3) and monosilane (SiH_4). The stoichiometry parameter x depends on the ratio of ammonia/monosilane flows. The growth technology is described elsewhere in more details [6]. All SRN films were deposited at substrate temperature 100°C, the reactor pressure during deposition was 150 Pa, the deposition time was 16 minutes, and the plasma power was 400 W. Parameters of the films are shown in Table 1.

Amorphous silicon films were deposited on glass substrates using two methods. The first method is evaporation

TABLE 1: Growth parameters, thickness, and composition of hydrogenated SRN films.

Number of samples	Flow ratio NH_3/SiH_4	Thickness, nm	Growth temperature, °C	Stoichiometric parameter x
611	5	240	100	1.3
613	1.5	400	100	0.95
615	0.5	450	100	0.6

TABLE 2: Growth parameters, thickness, and hydrogen concentration of amorphous silicon films.

Number of samples	Thickness, nm	Growth temperature, °C	Hydrogen concentration, atomic per cents
1	100	20	0
2	90	280	10
3	110	225	15
4	75	200	30–35

of monocrystalline silicon target by electron beam in high vacuum and deposition onto unheated substrate. The films deposited by this method do not contain hydrogen. The second method is PECVD at various temperatures using different PECVD reactors. The parameters of the films are shown in Table 2. It is known that optical gap of a-Si:H depends on hydrogen concentration [7]. According to optical gap measurements from light transmission and reflection data, the films grown at temperature 280°C contain ≈10 atomic % of hydrogen (optical gap ≈ 1.4 eV); the films grown at temperature 225°C contain ≈15% of hydrogen (optical gap ≈ 1.6 eV); the films grown at temperature 200°C contain 30–35% of hydrogen (optical gap ≈ 2 eV).

Laser ellipsometry (λ = 632.8 nm) was applied to measure the thickness of the films. To study the states of hydrogen and vibration of the Si–N bonds, the FTIR spectroscopy was applied. The FTIR spectrometer FT-801, produced by scientific enterprise "SIMEX" (Novosibirsk, Russia), was used. The spectral range of the spectrometer is from 550 to 5500 cm^{-1}, and the spectral resolution can be optionally changed from 4 to 0.5 cm^{-1}. The structural properties of as-deposited and annealed films were studied using Raman spectroscopy technique. All Raman spectra were recorded in the back-scattering geometry. The 514.5 nm Ar^+ laser line was used as a light source. Triple spectrometer T64000 (Horiba Jobin Yvon) with micro-Raman setup and liquid-nitrogen-cooled CCD matrix detector was used. The spectral resolution was not worse than 1 cm^{-1}. All spectra were measured at room temperature. To avoid heating of the films, the laser beam was slightly defocused; the spot diameter was equal to about 4–6 μm; the laser power reaching the sample was within 2-3 mW. The spectra were registered in the $Z(XX)\overline{Z}$ polarization geometry. Furnace annealing of the samples in Ar ambient for 5 hours at temperature 1130°C was performed at the Institute of Electron Technology (Warsaw, Poland). The triple harmonic of YAG:Nd laser

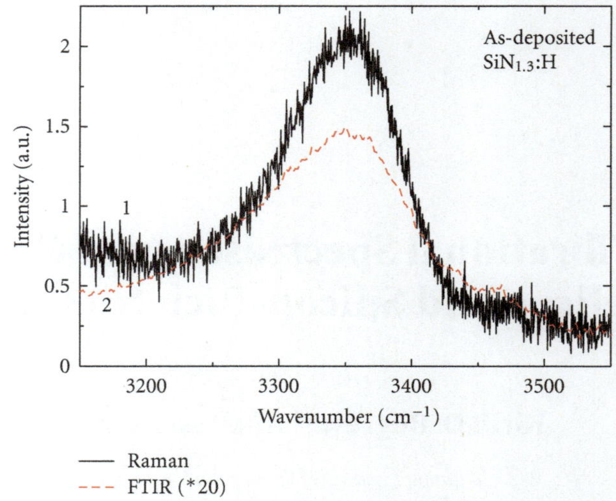

FIGURE 1: Raman and FTIR spectra of $SiN_{1.3}$:H film in frequency range of valence vibration of N–H bonds.

(λ = 335 nm, pulse duration of 10 ns) was used for laser treatments of the as-deposited films.

3. Results and Discussion

In Figure 1 one can see Raman spectroscopy and FTIR data concerning N–H valence (stretching) vibrations in sample 611 (stoichiometry is estimated as $SiN_{1.3}$:H). For convenience, the FTIR data was multiplied by 20 times. According to both Raman data and FTIR data, the position of N–H valence vibration peak is 3350 cm^{-1}. An advantage of Raman spectroscopy method is the possibility to use substrates that are not transparent in IR region. When the substrate is not transparent, only reflection setup can be used. Using multiple internal transmission, the sensitivity of IR spectroscopy can be notably improved [8]. Figure 2 shows shift of the position of N–H valence vibration peak versus NH_3 to SiH_4 ratio during deposition of SiN_x:H films. With changing of stoichiometry parameter x (1.3, 0.95, and 0.6) the frequency also changes (3350, 3342, and 3335 cm^{-1}) accordingly. The frequency of Si–H valence vibrations changes from 2170 to 2100 cm^{-1} with decrease of stoichiometry parameter x from 1.3 to 0.6 [6]. This is because of influence of neighboring atoms on frequencies of valence modes of Si–H and N–H bonds. It should be noted that Ali with coworkers [9] have also observed the shift of the position of N–H valence vibration peak versus stoichiometry of SiN_x:H films.

In Figure 3 one can see Raman spectroscopy and FTIR data concerning Si–H valence vibrations for $SiN_{1.3}$:H film. According to the FTIR data the position of Si–H valence vibration peak is 2170 cm^{-1}, but according to the Raman spectroscopy data the position of peak is 2187 cm^{-1}. A mercury lamp was used for back lighting of LCD monitor of experimental setup, and parasitic light can reach input slit of spectrometer. So, one can see in the Raman spectra a doublet (leakage from mercury lamp). One can see also a peak

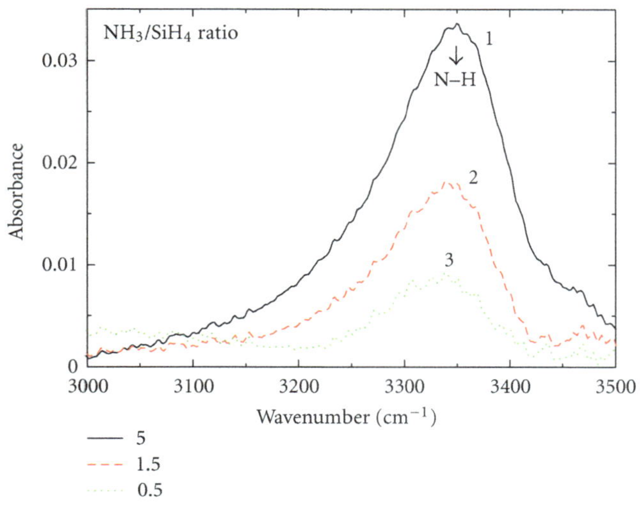

FIGURE 2: FTIR spectra of as-deposited SRN films in frequency range of valence vibration of N–H bonds.

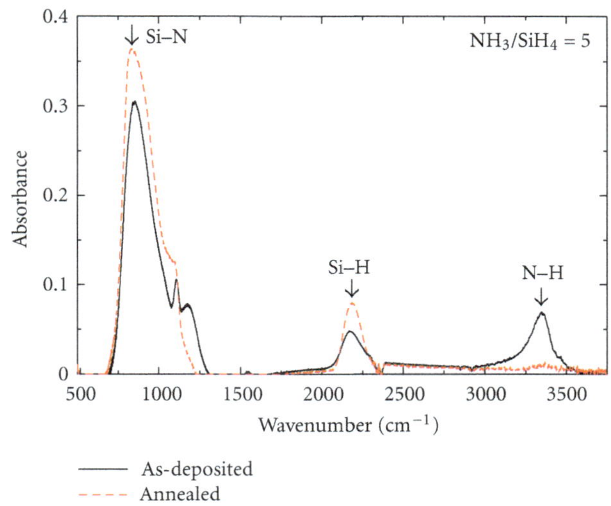

FIGURE 4: FTIR spectra of as-deposited and furnace-annealed $SiN_{1.3}$:H film.

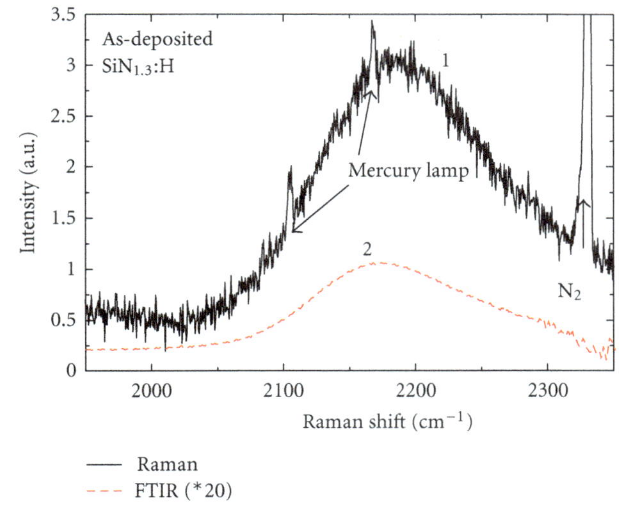

FIGURE 3: Raman and FTIR spectra of $SiN_{1.3}$:H film in frequency range of valence vibration of Si–H bonds.

due to Raman scattering on N_2 molecules in atmosphere. Both Raman and FTIR peaks are asymmetric, so one can assume presence of Si–H and SiH_2 bonds in this film. From fitting of FTIR spectrum by two Gaussians one can assume that frequency of Si–H valence vibration is 2170 cm^{-1} and frequency of Si–H_2 valence vibration is 2260 cm^{-1}. The difference in positions of the Raman scattering and the FTIR peaks may be due to different contribution of Si–H and SiH_2 bond valence vibration in dipole moment and polarisability.

Knowing the thickness of a film and effective absorption cross-sections for Si–N, Si–H, and N–H bonds, one can calculate the concentration of chemical bonds. The effective absorption cross-section S_{eff} for valence vibration of Si–N bonds is $1.6 \cdot 10^{-19}$ cm^2 [10], and S_{eff} for valence vibration of Si–H bonds is $0.57 \cdot 10^{-19}$ cm^2 [11]. According to different data, S_{eff} for valence vibration of N–H bonds is $0.48 \cdot 10^{-19}$ cm^2 [11] or $0.8 \cdot 10^{-19}$ cm^2 [12]. Concentration of chemical species can be calculated as $\text{N}(\text{cm}^{-3}) = A/(d(\text{cm}) \cdot S_{\text{eff}}(\text{cm}^2))$, where A is the height of absorbance peak (dimensionless parameter) and d is thickness. The results are shown in Table 3. The concentration of N–H species is calculated using S_{eff} according to Lanford [11] and Stein [12]. It should be noted that Stein and Wegener [12] have used multiple internal transmissions but Lanford and Rand [11] used normal transmission geometry. We also have used normal transmission geometry. So, the S_{eff} obtained by Lanford is more preferable for analysis of our data. From Table 3 one can see that total concentration of Si–N, Si–H, and N–H bonds decreases with decreasing of stoichiometry parameter x. This is because vibrations of Si–Si bonds do not change dipole momentum and are not active in FTIR. But they are active in Raman scattering. And Raman signal from amorphous silicon clusters was detected for sample number 615 [6]. The total concentration of Si–N bonds in stoichiometric Si_3N_4 is about $15.5 \cdot 10^{22}$ cm^{-3} [13]. So, one can assume that annealed near stoichiometric sample number 611 (Table 3) contains not only Si–N and Si–H bonds but also some concentration of Si–Si bonds. But concentration of such bonds is not high enough to be detected by Raman spectroscopy.

The FTIR spectra of as-deposited and annealed $SiN_{1.3}$:H films are shown in Figure 4 in range from 500 to 3900 cm^{-1}. Surprisingly, this nearly stoichiometric silicon nitride film still contains hydrogen after annealing with high-thermal budget, mainly in the form of the Si–H bonds. In this film, the peak, corresponding to absorption on stretching vibrations of the Si–H bonds, becomes even higher. The peak at 1175 cm^{-1} in spectrum of as-deposited film can be related to absorbance by bending vibrations of the N–H bonds [14]. Peculiarities at 850–900 cm^{-1} in spectrum of annealed film can be related to absorbance by stretching vibrations of the Si–N bonds in the α-modification of crystalline silicon

Table 3: Concentration of chemical species in SRN films according to FTIR data.

Number of samples	Si–N 10^{22} cm^{-3}	Si–H 10^{22} cm^{-3}	N–H (for S_{eff} according to Lanford) 10^{22} cm^{-3}	N–H (for S_{eff} according to Stein) 10^{22} cm^{-3}
611	8	3.2	5	3
611 annealed	10.5	6	bellow 0.1	bellow 0.06
613	5.58	3.64	0.9	0.54
615	3.75	2.92	0.32	0.19

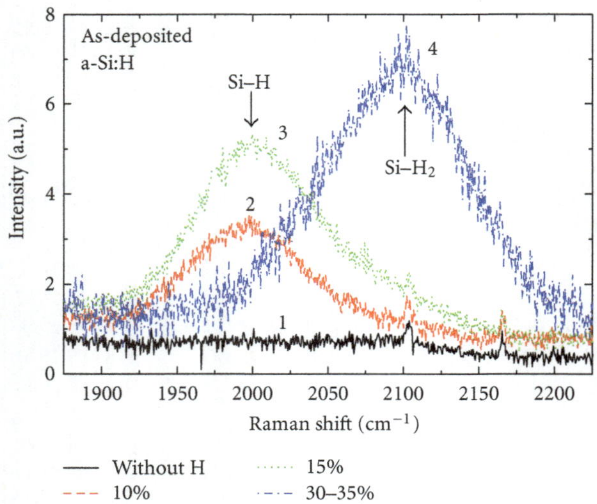

Figure 5: Raman spectra of a-Si:H films in frequency range of valence vibration of Si–H bonds.

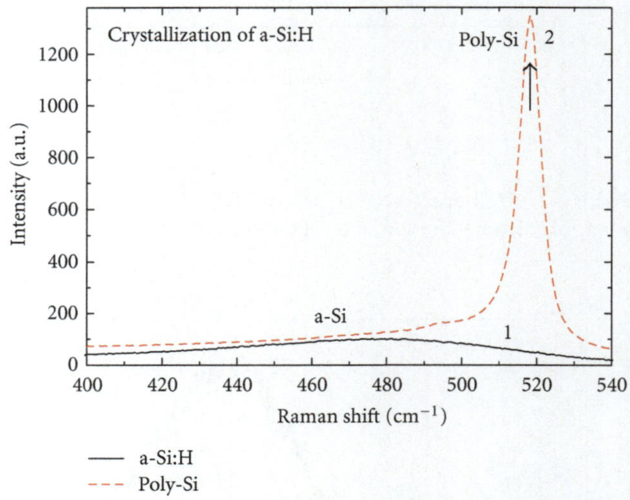

Figure 6: Raman spectra of as-deposited a-Si:H film and crystallized by laser pulse annealing film in frequency range of valence vibration of Si–H bonds.

nitride [15, 16]. Maybe partial crystallization of silicon nitride retards the dehydrogenization of the film. It has been detected earlier [17] that the closer is the stoichiometry parameter of the SRN films to that of stoichiometric Si_3N_4, the higher temperature is needed for effusion of hydrogen (dehydrogenization) from these films. However, short-time annealing at temperature 850°C [17] and annealing at 800°C for 2 hours [18] lead to total dehydrogenization of the SRN films, with the stoichiometry parameter x below 1. So, as it was abovementioned, one can assume that, for the case of nearly stoichiometric silicon nitride film, furnace annealing leads to so-called hardening of silicon nitride inhibiting hydrogen outdiffusion.

Figure 5 shows Raman spectra of amorphous silicon films with different hydrogen concentrations. All films were deposited on glass substrates that are not transparent in IR region, so it was impossible to analyze them using FTIR in normal geometry. One can see peaks due to scattering on valence vibrations of Si–H bonds. In Raman spectrum of the only film deposited in vacuum (without hydrogen) this peak is absent. As one can see from Raman spectra, film deposited at temperature 200°C contains mostly Si–H_2 bonds, because the position of Si–H_2 peak is about 2100 cm^{-1}. The frequency of Si–H peak is about 2000 cm^{-1} [19]. The films deposited at temperatures higher than 200°C contain mostly Si–H bonds. It should be noted that frequencies of stretch vibrations of Si–H bonds in silicon nitride and in a-Si:H are notably different: about 2200 cm^{-1} for the first case and 2000–2100 cm^{-1} for the second case. As it was shown by Hien et al. [20], the frequency of stretch vibrations of Si–H bonds in the case of neighbouring N and H atoms can reach 2122 cm^{-1}. In the case of only Si neighbouring atoms this frequency is 2012 cm^{-1} [20]. FTIR method can be used for quantitative hydrogen measurements [21], but for transmission experiments the substrate should be transparent in IR range.

Figure 6 shows the Raman spectra of as-deposited a-Si:H film and nanosecond-pulse-annealed film. The Raman spectrum of amorphous silicon contains broad peaks at approximately 480 cm^{-1} appearing as a result of effective density of trafnsverse optical (TO) vibration modes [19]. Due to scattering on optical phonon modes localized in the nanocrystals, the Raman spectrum of nanocrystals is characterized by narrow peak at position between 500 and 520 cm^{-1}. The position and the width of the peak strongly depend on size and structure of the nanocrystals according to dispersion of the localized modes [22]. The peak width is mainly determined by size dispersion of the nanocrystals. The intensity of the "nanocrystal" peak depends on the volume of nanocrystal phase. So, one can determine the volume part of nanocrystal phase: $V_{\text{nano-Si}}/V_{\text{full}} = I_{NC}/(I_{NC} + \sigma I_A)$, where I_{NC} and I_A are integral intensities of crystalline and amorphous peaks and σ is relation between the integral Raman cross-sections of nanocrystal Si and a-Si. According to the literature data, σ can change from 0.1 for large grain polycrystalline silicon up to 1.1 for nanocrystalline Si [23]. As one can see, the used nanosecond treatment leads to almost total crystallization of as-deposited film. According to peak position, the size of nanocrystals in our case should be about 10 nm [22]. It should be noted that no Si–H peaks were detected in nanosecond-pulse-annealed film. So, the pulse annealing can be used not only for crystallization but also for dehydrogenization.

4. Conclusion

In conclusion, the FTIR method was applied for determination of concentrations of chemical species in silicon-rich nitride films containing amorphous silicon nanoclusters. It was observed that nearly stoichiometric hydrogenated silicon nitride film contains hydrogen mainly in the form of the Si–H bonds even after 5 hours of annealing at temperature 1130°C. According to Raman spectroscopy data, nanosecond pulse laser treatment leads to crystallization and dehydrogenization of a-Si:H films with initial concentration of hydrogen up to 35%.

Acknowledgments

The authors are thankful to Dr. A. A. Popov (Yaroslavl Department of FTI RAS) and A. Kh. Antonenko for providing of as-deposited films, Professor A. Misiuk for furnace annealing, and Dr. S. A. Kochubei for help in laser annealing. They are pleased to thank directorate of Scientific and Educational Center "Nanosystems and Advanced Materials," Novosibirsk State University.

References

[1] N. M. Park, C. J. Choi, T. Y. Seong, and S. J. Park, "Quantum confinement in amorphous silicon quantum dots embedded in silicon nitride," *Physical Review Letters*, vol. 86, no. 7, pp. 1355–1357, 2001.

[2] M. Dai, K. Chen, X. Huang et al., "Formation and charging effect of Si nanocrystals in a-SiN_x/a-Si/a-SiN_x structures," *Journal of Applied Physics*, vol. 95, no. 2, pp. 640–645, 2004.

[3] C. W. Jiang and M. A. Green, "Silicon quantum dot superlattices: modeling of energy bands, densities of states, and mobilities for silicon tandem solar cell applications," *Journal of Applied Physics*, vol. 99, no. 11, Article ID 114902, pp. 1–7, 2006.

[4] A. J. Nozik, "Quantum dot solar cells," *Physica E*, vol. 14, no. 1-2, pp. 115–120, 2002.

[5] C. Doughty, D. C. Knick, J. B. Bailey, and J. E. Spencer, "Silicon nitride films deposited at substrate temperatures < 100°C in a permanent magnet electron cyclotron resonance plasma," *Journal of Vacuum Science and Technology A*, vol. 17, no. 5, pp. 2612–2618, 1999.

[6] T. T. Korchagina, D. V. Marin, V. A. Volodin, A. A. Popov, and M. Vergnat, "Structure and optical properties of SiNx : H films with Si nanoclusters produced by low-frequency plasma-enhanced chemical vapor deposition," *Semiconductors*, vol. 43, no. 11, pp. 1514–1519, 2009.

[7] J. D. Joannopoulos and G. Lucovsky, *The Physics of Hydrogenated Amorphous Silicon*, vol. I, Springer, Berlin, Germany, 1984.

[8] A. Milekhin, M. Friedrich, K. Hiller, M. Wiemer, T. Gessner, and D. R. T. Zahn, "Characterization of low-temperature wafer bonding by infrared spectroscopy," *Journal of Vacuum Science and Technology B*, vol. 18, no. 3, pp. 1392–1396, 2000.

[9] S. Ali, M. Gharghi, S. Sivoththaman, and K. Zeaiter, "Properties and characterization of low-temperature amorphous PECVD silicon nitride films for solar cell passivation," *Journal of Materials Science*, vol. 40, no. 6, pp. 1469–1473, 2005.

[10] V. I. Belyi, L. L. Vasilyeva, A. S. Ginkover et al., *Silicon Nitride in Electronics*, Elsevier, New York, NY, USA, 1988.

[11] W. A. Lanford and M. J. Rand, "The hydrogen content of plasma-deposited silicon nitride," *Journal of Applied Physics*, vol. 49, no. 4, pp. 2473–2477, 1978.

[12] H. J. Stein and H. A. R. Wegener, "Chemically bound hydrogen in CVD Si_3N_4: dependence on NH /SiH ratio and on annealing," *Journal of the Electrochemical Society*, vol. 124, no. 6, pp. 908–912, 1977.

[13] V. A. Gritsenko, "Atomic structure of the amorphous nonstoichiometric silicon oxides and nitrides," *Physics-Uspekhi*, vol. 51, no. 7, pp. 699–711, 2008.

[14] D. V. Tsu, G. Lucovsky, and M. J. Mantini, "Local atomic structure in thin films of silicon nitride and silicon diimide produced by remote plasma-enhanced chemical-vapor deposition," *Physical Review B*, vol. 33, no. 10, pp. 7069–7081, 1986.

[15] Y. N. Volgin and Y. I. Ukhanov, "Vibration spectra of silicon nitride," *Optics and Spectroscopy*, vol. 38, pp. 412–415, 1975.

[16] N. Wada, S. A. Solin, J. Wong, and S. Prochazka, "Raman and IR absorption spectroscopic studies on α, β, and amorphous Si_3N_4," *Journal of Non-Crystalline Solids*, vol. 43, no. 1, pp. 7–11, 1981.

[17] M. Molinari, H. Rinnert, and M. Vergnat, "Evolution with the annealing treatments of the photoluminescence mechanisms in a-SiN_x : H alloys prepared by reactive evaporation," *Journal of Applied Physics*, vol. 101, no. 12, Article ID 123532, pp. 1–8, 2007.

[18] G. Scardera, T. Puzzer, G. Conibeer, and M. A. Green, "Fourier transform infrared spectroscopy of annealed silicon-rich silicon nitride thin films," *Journal of Applied Physics*, vol. 104, no. 10, Article ID 104310, pp. 1–5, 2008.

[19] M. H. Brodsky, M. Cardona, and J. J. Cuomo, "Infrared and Raman spectra of the silicon-hydrogen bonds in amorphous silicon prepared by glow discharge and sputtering," *Physical Review B*, vol. 16, no. 8, pp. 3556–3571, 1977.

[20] T. T. T. Hien, C. Ishizaki, and K. Ishizaki, "SiH_x absorbance bands in Si_3N_4 powder surfaces analyzed by diffuse reflectance infrared fourier transform spectroscopy," *Journal of the Ceramic Society of Japan*, vol. 112, no. 1301, pp. 1–5, 2004.

[21] D. M. Goldie and S. K. Persheyev, "Quantitative hydrogen measurements in PECVD and HWCVD a-Si : H using FTIR spectroscopy," *Journal of Materials Science*, vol. 41, no. 16, pp. 5287–5291, 2006.

[22] V. Paillard, P. Puech, M. A. Laguna, R. Carles, B. Kohn, and F. Huisken, "Improved one-phonon confinement model for an accurate size determination of silicon nanocrystals," *Journal of Applied Physics*, vol. 86, no. 4, pp. 1921–1925, 1999.

[23] E. Bustarret, M. A. Hachicha, and M. Brunel, "Experimental determination of the nanocrystalline volume fraction in silicon thin films from Raman spectroscopy," *Applied Physics Letters*, vol. 52, no. 20, pp. 1675–1677, 1988.

Study on the Interaction of Bovine Serum Albumin with Ceftriaxone and the Inhibition Effect of Zinc (II)

Qiaoli Yue,[1] Tongfei Shen,[1] Changna Wang,[2] Chaohui Gao,[1] and Jifeng Liu[1]

[1] *Shandong Provincial Key Laboratory of Chemical Energy Storage and Novel Cell Technology, Department of Chemistry, Liaocheng University, Liaocheng 252059, Shandong, China*
[2] *Department of Chemistry and Biochemistry, Brigham Young University, Provo, UT 84602, USA*

Correspondence should be addressed to Qiaoli Yue, yueqiaoli@gmail.com

Academic Editor: Craig J. Eckhardt

The mechanism of the interaction between bovine serum albumin (BSA) and ceftriaxone with and without zinc (II) (Zn^{2+}) was studied employing fluorescence, ultraviolet (UV) absorption, circular dichroism (CD), and synchronous fluorescence spectral methods. The intrinsic fluorescence of BSA was quenched by ceftriaxone in a static quenching mode, which was authenticated by Stern-Volmer calculations. The binding constant, the number of binding sites, and the thermodynamic parameters were obtained, which indicated a spontaneous and hydrophobic interaction between BSA and ceftriaxone regardless of Zn^{2+}. Changes in UV absorption, CD, and synchronous fluorescence spectral data are due to the microenvironment of amide moieties in BSA molecules. In the BSA-ceftriaxone-Zn^{2+} system, Zn^{2+} must first interact with ceftriaxone forming a complex, which inhibits BSA binding to ceftriaxone. The present work uses spectroscopy to elucidate the mechanism behind the interaction between BSA and ceftriaxone in the presence and absence of Zn^{2+}. The BSA and ceftriaxone complex provides a model for studying drug-protein interactions and thus may further facilitate the study of drug metabolism and transportation.

1. Introduction

The interaction between biomacromolecules, especially between plasma proteins and drugs, has been an interesting research field in life sciences, chemistry, and clinical medicine [1]. Drug-albumin complexes may be considered as models for gaining fundamental insights into drug-protein interactions.

Serum albumin is a major soluble protein constituent of the circulatory system and has many physiological functions such as acting as a plasma carrier by nonspecifically binding to several hydrophobic steroid hormones and as a transport protein for hemin and fatty acids [2]. Albumins are characterized by a low content of tryptophan and methionine and a high content of cystine and charged amino acids [3–5]. Bovine serum albumin (BSA), an example of a mammalian albumin, has been studied extensively because of its stability, neutrality in many biochemical reactions, and low cost [6, 7]. Brown elucidated the 607 amino acid residue, primary structure of BSA in 1975, twenty one of which are tyrosine (Tyr) residues and two of which are tryptophan (Trp) residues located at positions 134 and 212, respectively [3, 6]. These two Trp residues cause BSA to have intrinsic fluorescence.

Ceftriaxone is a third-generation cephalosporin antibiotic. Cephalosporins are semisynthetic antibiotics produced by fungi *Cephalosporium* and, like penicillins, are β-lactam antibiotics, which have broad-spectrum activity against Gram-positive and Gram-negative bacteria. Ceftriaxone is often used to chelate metal ions, and we believe this improves its antibacterial activity to study the effect of metal ions on its biological activity [8, 9]. The structure of ceftriaxone contains –NH_2, –COOH, –CO, and N–C functional groups as shown in Scheme 1(a), which are all electron donors. The construction of ceftriaxone molecular model indicates that it is suitable for chelating transition metal ions such as Zn^{2+}, Cd^{2+} [10], Mn^{2+}, Fe^{3+}, Co^{2+}, Cu^{2+} [11, 12], and Pd^{2+} [13]. Ceftriaxone chelates metal ions as a bidentate monoanion

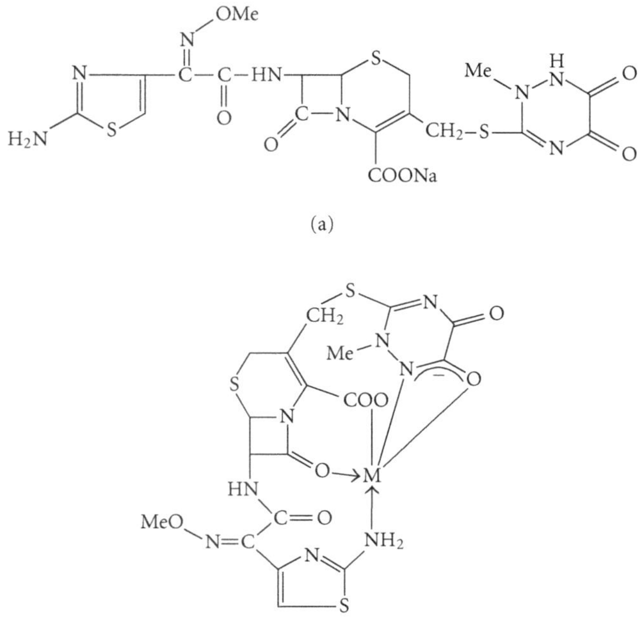

(a)

(b)

Scheme 1: Schematic drawings of the ceftriaxone (a) and ceftriaxone-M ((b), referred to in [11], Copyright 2005 Springer) molecule structure.

ligand through the β-lactam carbonyl and carboxylate group, as illustrated in Scheme 1(b). Furthermore, complexation with metal ions can make the drug more active and less toxic.

Common metal ions, especially transition metal ions, participate in many biochemical processes. Therefore, it is necessary to investigate protein-drug interactions in the presence of metal ions owing to the inevitable effects of metal ions in a ternary protein-drug-metal ion system [14]. Due to the intrinsic fluorescence of BSA from Trp-134 and Trp-212, the interaction between BSA and ligands such as drugs is frequently investigated by fluorescence methods [7, 15–18]. The effect of metal ions such as Cu^{2+} [14], Co^{2+} [19], and Fe^{3+} [20] in the interaction of serum albumin and drugs has been studied. The goal of this work is to reveal the interaction mechanism of BSA and ceftriaxone and the effect of Zn^{2+} on the present system by spectroscopic methods. This study may facilitate the study of metabolism and transportation of antibiotics as well as the relationship between protein and antibiotics in the presence of metal ions.

2. Materials and Methods

2.1. Instruments. All fluorescence measurements were carried out on an LS-55 fluorophotometer (Perkin Elmer, USA) equipped with a computer for data collection and a 1.0 cm quartz cell, with the slit's width of Ex/Em 5.0/5.0 nm and the PMT voltage at 700 V. Circular dichroism (CD) spectra were obtained on a J-810 circular dichroism chiroptical spectrometer (JASCO Co., Ltd, Japan). The UV spectra were obtained from a Lambda-40 spectrophotometer (Perkin Elmer, USA) coupled with a 1.0 cm quartz cell. All pH measurements were made with a pHs-3C digital pH meter (Shanghai Leici Device Works, China) with a combined glass-calomel electrode. The temperature was controlled by a water bath, and temperatures were kept in a certain range ($T \pm 0.1°C$) throughout the experiment.

2.2. Materials. In this work, all reagents used were of analytical-reagent grade unless specified. Doubly distilled deionized water was purified in a Milli-Q system (Millipore, Bedford, MA, USA). BSA (Sigma) was purchased from a local market without further purification to prepare a stock solution (50 μmol L^{-1}), which was kept in a brown flask at 4°C. A 1.0 mmol L^{-1} stock solution of ceftriaxone was prepared in a calibrated flask. The phosphate buffered saline (PBS) was prepared using Na_2HPO_4 and NaH_2PO_4 for controlling pH of the system at 7.4. Other reagents were purchased from a local market.

2.3. Procedures. A 3 mL solution, containing appropriate concentration of BSA, was titrated by successive additions of a 1.0 mM ceftriaxone solution. Titrations were done manually by using micropipettors.

In a typical fluorescence measurement, the fluorescence emission spectra were recorded in the presence and absence of Zn^{2+} ions ranging from 300 to 450 nm upon the excitation wavelength at 295 nm using the excitation and emission slit widths both of 5 nm. The experiments for discussing binding mechanism were conducted at three temperatures, 298, 308, and 318 K, as maintained by water bath.

Synchronous fluorescence spectra of BSA titrated with various concentrations of ceftriaxone were recorded from 250 to 320 nm ($\Delta\lambda$ = 60 nm) and from 260 to 320 nm ($\Delta\lambda$ = 15 nm), at which the spectrum only showed the spectroscopic behavior of Trp and tyrosine (Tyr) residues of BSA, respectively. The excitation and emission slit widths were set at 5/5 nm and 10/10 nm for $\Delta\lambda$ = 60 nm and $\Delta\lambda$ = 15 nm, respectively.

The UV-vis absorbance spectra of BSA-ceftriaxone system were recorded at 293 K. CD spectra were recorded from 200 to 280 nm at 0.2 nm with a scan of 50 points at 310 K in a thermostated cell holder, with three scans averaged for each CD spectrum. The results were expressed as ellipticity (mdeg), which was obtained in mdeg directly from the instrument. The molar ratio of BSA, ceftriaxone, and Zn^{2+} was varied as 1 : 0 : 0, 1 : 1 : 0, 1 : 0 : 1, and 1 : 1 : 1.

Binding studies of BSA and ceftriaxone in the presence or absence of Zn^{2+} were performed at two modes. First, the interaction between BSA and ceftriaxone without Zn^{2+} was studied. BSA concentration was fixed at 0.5 μM, and a series of ceftriaxone standard solutions was added. Second, the interaction of BSA-Zn^{2+} in the presence of various concentrations of ceftriaxone was investigated. The concentration for BSA and Zn^{2+} was kept at 0.5 μM and 1.0 μM, respectively, whereas ceftriaxone was then gradually added to the BSA-Zn^{2+} mixture. Fluorescence spectra of BSA were recorded from 300 to 500 nm upon excitation at 295 nm for both of the two modes.

3. Results and Discussion

3.1. Fluorescence Quenching Study of BSA by Ceftriaxone in the Presence or Absence of Zn^{2+}. In order to investigate the effect of ceftriaxone for BSA and BSA-Zn^{2+}, the fluorescence spectra were drawn in PBS (pH 7.4). In the liquid fluorescence, the inner filter effect (IFE) could not be ignored. Due to the increasing concentrations of a fluorescent substrate and quencher, the increasing absorbance of excitation and/or emission radiation introduces IFE that decreases the fluorescence intensity and results in a nonlinear relationship between the observed fluorescence intensity and the concentration of the fluorophore. The fluorescence intensity was corrected by multiplying appropriate correction factors according to the experiment results reported byGu and Kenny [21], as follows:

$$F_{\text{ideal}}(\lambda_{\text{ex}}, \lambda_{\text{em}}) = F_{\text{obs}}(\lambda_{\text{ex}}, \lambda_{\text{em}}) \text{CF}_p(\lambda_{\text{ex}}) \text{CF}_s(\lambda_{\text{em}}) \approx F_{\text{obs}}(\lambda_{\text{ex}}, \lambda_{\text{em}}) 10^{(a_{\text{em}}+a_{\text{ex}})/2}, \tag{1}$$

where $F_{\text{ideal}}(\lambda_{\text{ex}}, \lambda_{\text{em}})$ is fluorescence intensity observed at emission wavelength λ_{em} when excited at excitation wavelength λ_{ex}, and CF_p is the correction factor for the primary IFE (pIFE), which depends on the total absorbance of the sample at λ_{ex}, where CF_s is the correction factor for secondary IFE (sIFE), which depends on the total absorbance of the sample at λ_{em}. a_{ex} and a_{em} are the values of absorbance of the solution per cm at λ_{ex} and λ_{em}, respectively. F_{obs} is the observed fluorescence.

Both Trp and Tyr amino acid residues give fluorescence emissions when excited at 280 nm, while Trp alone has a fluorescence emission at an excitation wavelength of 295 nm [14]. We obtained the fluorescence spectra for the system of BSA with ceftriaxone in the presence or absence of Zn^{2+} by fixing an excitation wavelength of 295 nm. In addition, the fluorescence intensity in the whole procedure was corrected by exploiting (1), though the effect of the IFE was not obvious. BSA exhibits a strong fluorescence emission with band peak at 348 nm. The fluorescence intensity of BSA decreased gradually with the addition of ceftriaxone regardless of Zn^{2+} as shown in Figure 1.

3.2. Quenching Mechanism. Fluorescence quenching can be processed via different mechanisms, usually classified as dynamic quenching and static quenching. Dynamic and static quenching can be distinguished by their response to temperature [22]. Increased temperatures will quicken diffusion, produce large amount of collisional quenching, dissociate weakly bound complexes, and diminish static quenching. Based on this, the Stern-Volmer plots at different temperatures were drawn. As shown in Figure 2, the curves were linear. The quenching type may be static or dynamic, since the characteristic Stern-Volmer plot of combined quenching (both static and dynamic) has an upward curvature. Furthermore, the slope of the Stern-Volmer plots increased with the temperature. In this case, the quenching mode of ceftriaxone on the fluorescence of BSA was due to static quenching.

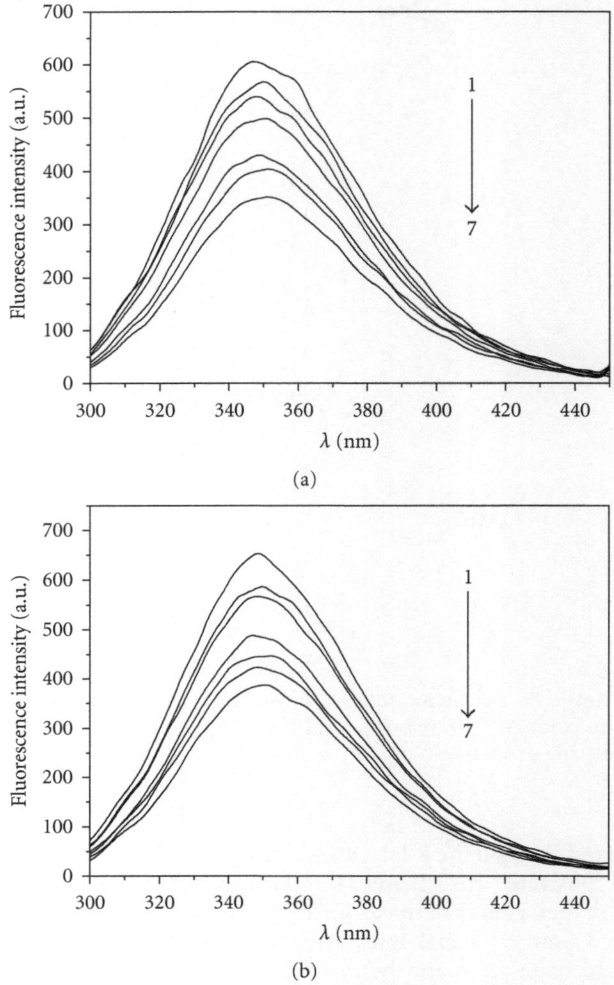

FIGURE 1: Fluorescence spectra of BSA in the absence (a) and presence (b) of Zn^{2+} with different ceftriaxone concentrations: 0, 1, 2, 4, 6, 8, and 10 μmol l^{-1} from (1) to (7), respectively. [BSA] = [Zn^{2+}] = 0.5 μmol l^{-1}.

The Stern-Volmer equation was utilized to further elaborate the fluorescence quenching mechanism. The relationship between quenching efficiency (F_0/F) and the concentration of quencher ([Q]) for the complete static or dynamic quenching mode should satisfy the following equation [23]:

$$\frac{F_0}{F} = 1 + K_{\text{sv}}[Q] = 1 + k_q \tau_0 [Q], \tag{2}$$

where F_0 and F are the fluorescence intensity of fluorophore in the absence and in the presence of quencher, respectively, [Q] is the concentration of quencher (ceftriaxone), K_{sv} is the Stern-Volmer constant corresponding to the slope of the plot for F_0/F versus [Q] in l mol^{-1}, k_q is the quenching rate constant for biomolecular quenching in l mol^{-1}s^{-1}, and τ_0 is the average fluorescence lifetime of fluorophore without quencher evaluated at about 5 ns [24–26]. The results listed in Table 1 show that the Stern-Volmer quenching constant K_{sv} is inversely correlated with temperature, and the value of k_q is larger than the limiting diffusion constant of the

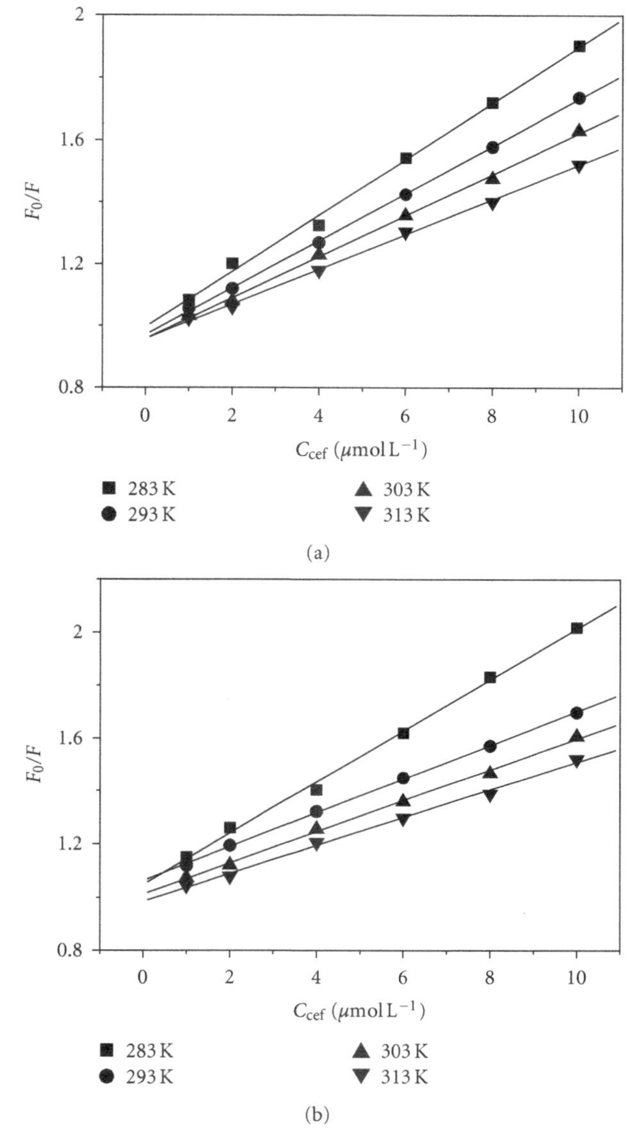

FIGURE 2: Stern-Volmer plots for the system of BSA without (a) and with (b) Zn^{2+} in the presence of ceftriaxone (with the concentrations from 0 to 10 μmol l^{-1}), at different temperatures. [BSA] = [Zn^{2+}] = 0.5 μmol l^{-1}.

biomolecule (2.0×10^{10} l mol^{-1}s^{-1}) [27]. This suggests that the fluorescence quenching was caused by a specific interaction between BSA and ceftriaxone regardless of Zn^{2+}. This evidence indicated that a static quenching was dominant in the system. From the results, it is also given that K_{sv} in the presence of Zn^{2+} was smaller than that in the absence of Zn^{2+}, indicating that quenching efficiency of ceftriaxone decreased by introducing Zn^{2+}. To confirm the static quenching mode, the lifetime of BSA and BSA-ceftriaxone was investigated with the results shown in Figure 3. It can be observed that the decay time of BSA fluorescence emission in the presence and absence of ceftriaxone is almost the same with the average lifetime at 5.1 ± 0.2 ns and 5.0 ± 0.1 ns, respectively. According to the literature, such slight differences indicated that the

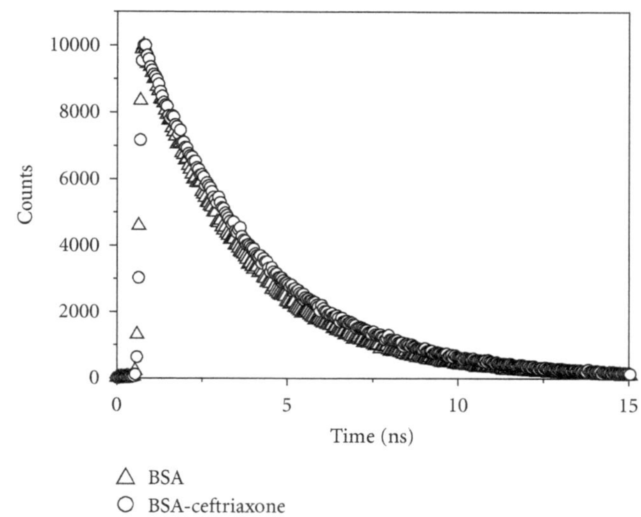

FIGURE 3: Fluorescence decay profiles of BSA in the presence and absence of ceftriaxone in 0.1 M PBS of pH 7.4. BSA 0.5 μmol l^{-1}, and ceftriaxone 2 μmol l^{-1}.

fluorescence quenching of BSA caused by ceftriaxone was attributed to the static quenching [22]. Thus, it is proposed that the complex formation takes a major role rather than dynamic collision for the interaction between BSA and ceftriaxone.

3.3. Influence of Zn^{2+} on the Binding Constant. Several equations can be used for the calculation of binding constant of the interaction between a protein and drug. One of the most frequently used equations is the Scatchard procedure equation since it can provide the actual stoichiometry for the binding sites of protein and ligands [28]

$$\frac{r}{D_f} = nK - rK, \qquad (3)$$

where r is the moles of ceftriaxone bound per mole of BSA, D_f is the molar concentration of free ceftriaxone, n is the binding stoichiometry per class of binding sites, and K is the equilibrium binding constant.

The linearity of Scatchard plots for the ceftriaxone-BSA system was obtained as shown in Figure 4. Table 2 summarizes the binding constant, K, values and the number of binding sites per BSA at four temperatures acquired in the absence and presence of Zn^{2+}. The values of n suggest that ceftriaxone binds to a single class of binding sites on BSA. From the values of K, it was illustrated that there was a moderate binding force between ceftriaxone and BSA. Serum albumin has a limited number of binding sites for endogenous and exogenous ligands. These ligands are typically reversibly bound and have binding constants ranging from 10^4 to 10^8 [29, 30]. A K value larger than 10^4 indicates a significant interaction between ligand and protein. In addition, one category of binding sites of BSA for ceftriaxone in the absence and presence of Zn^{2+} was obtained.

TABLE 1: Dynamic quenching constant between ceftriaxone and BSA without and with Zn^{2+}.

System	Temperature K	Stern-Volmer equation	K_q L mol^{-1} s^{-1}	r
Without Zn^{2+}	283	$y = 0.0903[Q] + 0.9969$	1.81×10^{13}	0.9982
	293	$y = 0.0671[Q] + 1.0039$	1.52×10^{13}	0.9966
	303	$y = 0.0592[Q] + 1.0074$	1.33×10^{13}	0.9991
	313	$y = 0.0517[Q] + 0.9965$	1.12×10^{13}	0.9981
With Zn^{2+}	283	$y = 0.0966[Q] + 1.0483$	1.93×10^{13}	0.9987
	293	$y = 0.0638[Q] + 1.0631$	1.28×10^{13}	0.9998
	303	$y = 0.0586[Q] + 1.0120$	1.17×10^{13}	0.9989
	313	$y = 0.0525[Q] + 0.9985$	1.05×10^{13}	0.9979

TABLE 2: Binding constant between ceftriaxone and BSA.

Temperature K	$K \times 10^5$ mol L^{-1}	n	ΔG kJ mol^{-1}	ΔH kJ mol^{-1}	ΔS J mol^{-1} K^{-1}
283	1.56	0.91	−13.28	2.49	55.72
293	1.99	0.92	−13.84		
303	2.37	0.92	−14.39		
313	2.93	0.92	−14.95		

In addition, it also can be observed from Table 2 that the presence of Zn^{2+} decreased the binding constants and binding sites of ceftriaxone-BSA complex, indicating that Zn^{2+} would affect the binding capacity of ceftriaxone binding to BSA. The smaller binding constant could result from two aspects: (1) competitive binding to BSA between Zn^{2+} and ceftriaxone and (2) Zn^{2+} binding to ceftriaxone inducing the conformational changes of ceftriaxone.

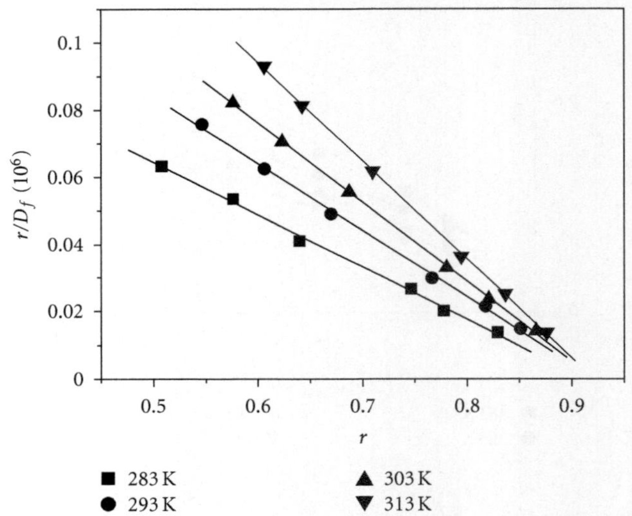

FIGURE 4: Scatchard plots for the BSA-ceftriaxone system in PBS (pH 7.4). BSA concentration: 0.5 μmol l^{-1}, λ_{ex} = 295 nm.

3.4. Thermodynamic Analysis. In general, intermolecular interacting forces between a small molecule and a biomacromolecule include hydrogen bonding, van der Waals force, and electrostatic and hydrophobic interactions. The thermodynamic parameters enthalpy change (ΔH) and entropy change (ΔS) of the reaction are important for confirming the binding mode. The temperature-dependent thermodynamic parameters for the ceftriaxone-BSA system in the absence and presence of Zn^{2+} are used to characterize the intermolecular forces between ceftriaxone and BSA. $\Delta H°$ and $\Delta S°$ for a binding reaction can be derived from the following van't Hoff equation:

$$\ln k = \frac{-\Delta H°}{RT} + \frac{\Delta S°}{R}, \quad (4)$$

$$\Delta G° = \Delta H° - T\Delta S°, \quad (5)$$

where K is the binding constant at the corresponding temperature, and R is the gas constant. The results of K, $\Delta H°$, $\Delta S°$, and the corresponding values of Gibbs free energy ($\Delta G°$) for the system of BSA-ceftriaxone and BSA-ceftriaxone-Zn^{2+} were presented in Table 2. The sign and magnitude of the thermodynamic parameters were associated with various interactions which might take place in the protein association processes. As shown in Table 2, the negative values of ΔG along with the positive values of ΔH and ΔS are obtained for the ceftriaxone-BSA interaction. $\Delta G < 0$ reveals that the binding process is spontaneous. $\Delta S > 0$ is evidence of hydrophobic interactions [31]. Moreover, $\Delta H \approx 0$ and $\Delta S > 0$ are characteristics of electrostatic interactions in aqueous solution. Therefore, ceftriaxone bound to BSA displayed mainly hydrophobic and electrostatic interactions.

3.5. UV-Vis Absorption Studies. UV-Vis absorption measurements are used to study structural changes and complex formation [20]. UV absorption spectra of ceftriaxone-BSA with or without Zn^{2+} systems were recorded and shown in Figure 5. The UV absorption spectrum of BSA shows a strong band with a maximum at 208 nm and a weak band with a maximum at 279 nm. The absorbance (279 nm) intensity of ceftriaxone-BSA system in the presence and absence of Zn^{2+}

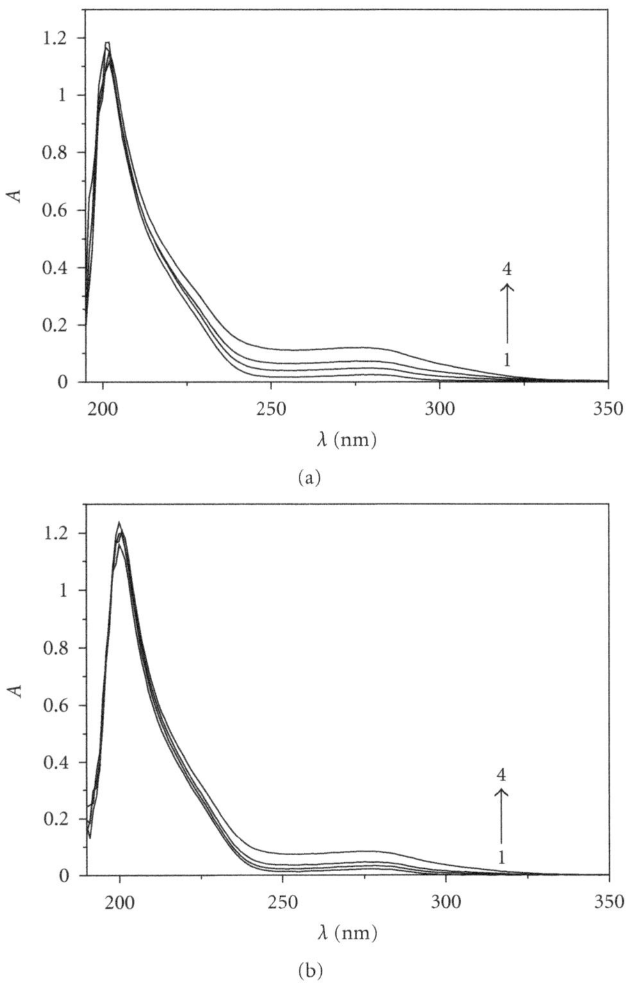

FIGURE 5: UV absorption spectra of BSA (A) and BSA-Zn^{2+} (B) in the presence of ceftriaxone. [BSA] = [Zn^{2+}] = 0.5 μmol l^{-1}, the concentration of ceftriaxone, 1–4, 0, 1, 2, and 5 μmol l^{-1}.

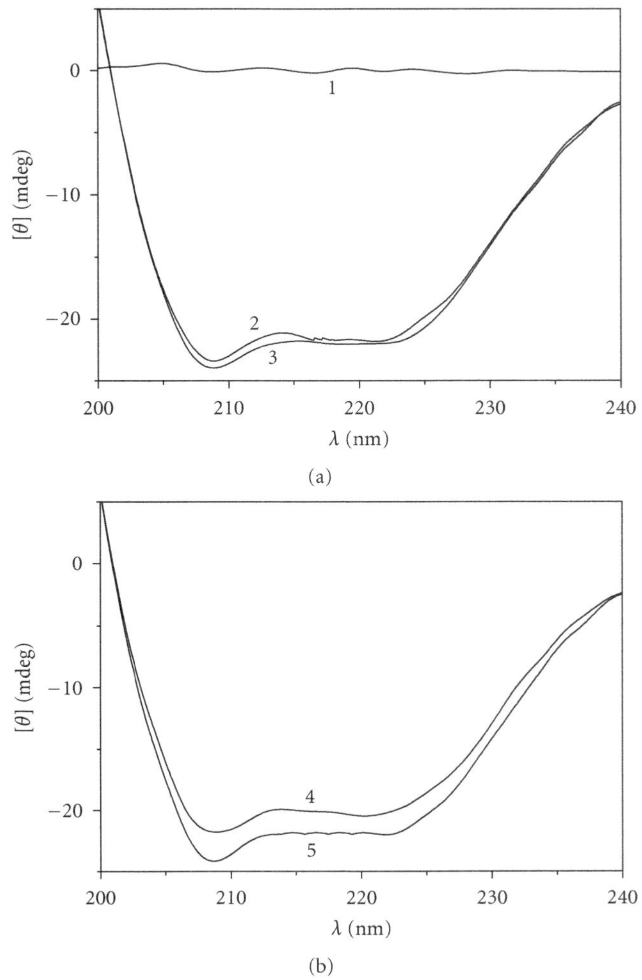

FIGURE 6: CD spectra of BSA (a) and BSA-Zn^{2+} (b) in the presence of ceftriaxone. 1 ceftriaxone, 2 BSA, 3 BSA-ceftriaxone, 4 BSA-Zn^{2+}-ceftriaxone, and 5 BSA-Zn^{2+}.

increased with increasing concentration of ceftriaxone, and the peak showed a slight blue shift (from 279 to 275 nm). However, the absorption at 208 nm did not significantly change. It was reported that the difference in the spectral peak of 208 nm is due to changes in the conformation of the peptide backbone associated with helix-coil transformation [32, 33]. In addition, the peak at the 279 nm region is related to the polarity of the microenvironment around Trp and Tyr residues of BSA. The interaction between ceftriaxone and BSA with or without Zn^{2+} may change the polarity of microenvironment around Trp and Tyr residues of BSA but does not cause a conformational change of BSA.

3.6. CD Studies. Due to the influence of Zn^{2+} on the spectra of BSA-ceftriaxone, it is important to determine whether the structure of BSA changes. CD is a sensitive technique to monitor conformational changes in protein structure [16]. CD spectra of BSA, BSA-ceftriaxone, BSA-Zn^{2+}, and BSA-ceftriaxone-Zn^{2+} are shown in Figure 6. In BSA spectrum, there are two minima in the ultraviolet region, one at 208 nm and the other at 222 nm, which are characteristic of the α-helical structure of a protein. Trynda-Lemiesz et al. explained that both of the negative peaks between 208-209 and 222-223 nm contribute to the transfer for the peptide bond of the α-helix [34]. Figure 6 shows that the bands of BSA are more pronounced in the presence of ceftriaxone with and without Zn^{2+}. The α-helix content in the secondary structure of BSA was determined (as listed in Table 3). From the data, it can be seen that the relative α-helix content of BSA increased in the presence of ceftriaxone, which suggested that the conformation of BSA changed by the addition of ceftriaxone. By comparison, the CD spectra for the participation of Zn^{2+} are similar to the native protein with the α-helix content of BSA decreasing slightly, which differs from previous results [14, 19]. Thus, the effect of Zn^{2+} on the BSA conformation may be ignored, and the binding of ceftriaxone is the main reason of conformational transition of BSA. In this case, Zn^{2+} may bind first to ceftriaxone forming complexes, thereby inhibiting BSA-ceftriaxone interactions. This may be the

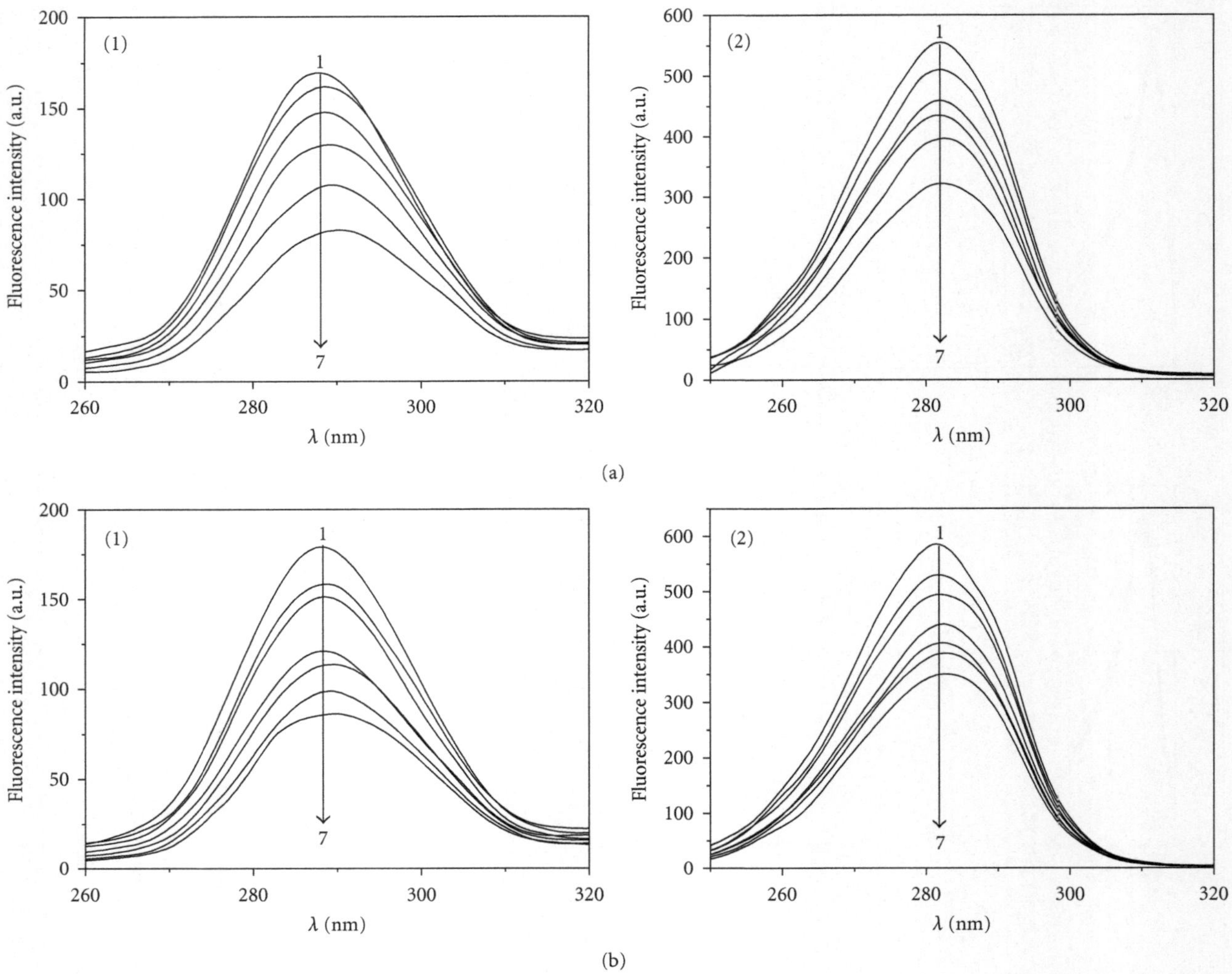

Figure 7: Synchronous fluorescence measurement for BSA without (a) and with (b) Zn^{2+} in the presence of ceftriaxone from (1) to (7) 0, 1, 2, 4, 6, 8, and 10 μmol l^{-1}, with $\Delta\lambda = 15$ nm (1) and $\Delta\lambda = 60$ nm (2), respectively.

Table 3: The contents of α-helix in the BSA in the presence of ceftriaxone without or with Zn^{2+}.

System	The content of α-helix (%)
2 μM BSA	49.6
2 μM BSA-2 μM ceftriaxone	52.4
2 μM BSA-2 μM Zn^{2+}	48.0
2 μM BSA-2 μM ceftriaxone-2 μM Zn^{2+}	48.7

reason why there is a distinct decrease of K in ceftriaxone-BSA system with Zn^{2+}.

3.7. Synchronous Fluorescence of BSA with Ceftriaxone in the Absence and Presence of Zn^{2+}. Synchronous mode fluorescence spectroscopy was applied to study the conformational changes of the protein due to this binding reaction by measuring the emission wavelength shift [1]. According to Miller [35], characteristic Trp residue information is obtained when $\Delta\lambda$ value between excitation and emission wavelength is maintained at 60 nm, while $\Delta\lambda$ value for Tyr at 15 nm. Figure 7 showed the effect of ceftriaxone on the fluorescence emission for Tyr (1) and Trp (2) in BSA structure without (A) or with Zn^{2+} (B). From Figure 7, it can be seen that the fluorescence for $\Delta\lambda = 15$ nm was much weaker than that of $\Delta\lambda = 60$ nm, and the position of the emission peak for $\Delta\lambda = 60$ nm shifted to blue side compared with that for $\Delta\lambda = 15$ nm in the presence and absence of Zn^{2+}. These observations that the fluorescence intensity of both Trp and Tyr decreases and that there is a notable red shift at maximum emission upon addition of ceftriaxone indicate that the conformation of protein has changed. It is likely due to the hydrophobic amino acid structure surrounding tryptophan and tyrosine residues in BSA, which tends to collapse slightly, so tryptophan and tyrosine residues are exposed more to the aqueous phase. According to He et al.'s explanations, the interaction of ceftriaxone with BSA affects the polarity around Trp and

Tyr residue microregions regardless of Zn^{2+} [36], which was in accordance with the result obtained from UV-Vis spectroscopy method mentioned above.

When Zn^{2+} was added to the BSA-ceftriaxone system, Zn^{2+} preferentially combined with ceftriaxone over BSA. In other words, when Zn^{2+} and ceftriaxone coexist in the blood, Zn^{2+} will react first with ceftriaxone molecules, which prevents BSA from effectively enwrapping ceftriaxone. The study of the binding phenomena is important in providing basic information on the pharmacological actions, biotransformation, and biodistribution.

Acknowledgments

The present work is supported by National Natural Science Foundation of China (Grants 21005036, 2107508, and 20875042), Key Project of Natural Science Foundation (ZR2010BX004, 2010GJC20808-15), and Tai-Shan Scholar Research Fund of Shandong Province.

References

[1] Y. Lu, Q. Q. Feng, F. L. Cui, W. W. Xing, G. S. Zhang, and X. J. Yao, "Interaction of 3'-azido-3'-deamino daunorubicin with human serum albumin: investigation by fluorescence spectroscopy and molecular modeling methods Bioorg," *Medicinal Chemistry Letters*, vol. 20, no. 23, pp. 6899–6904, 2010.

[2] P. A. Zunszain, J. Ghuman, T. Komatsu, E. Tsuchida, and S. Curry, "Crystal structural analysis of human serum albumin complexed with hemin and fatty acid," *BMC Structural Biology*, vol. 3, article 6, 2003.

[3] J. R. Brown, "Structure of bovine serum albumin," *Federation Proceedings*, vol. 34, p. 591, 1975.

[4] J. E. Patterson and and D. M. Geller, "Bovinemicrosomal albumin: aminoterminalsequence of bovineproalbumin," *Biochemical and Biophysical Research Communications*, vol. 74, no. 3, pp. 1220–1226, 1977.

[5] K. Hirayama, S. Akashi, M. Furuya, and K. Fukuhara, "Rapid confirmation and revision of the primary structure of bovine serum albumin by ESIMS and Frit-FAB LC/MS," *Biochemical and Biophysical Research Communications*, vol. 173, no. 2, pp. 639–646, 1990.

[6] T. J. Peters, "Serum albumin," *Advances in Protein Chemistry*, vol. 37, pp. 161–245, 1985.

[7] A. Tarushi, C. P. Raptopoulou, V. Psycharis, A. Terzis, G. Psomas, and D. P. Kessissoglou, "Zinc(II) complexes of the second-generation quinolone antibacterial drug enrofloxacin: structure and DNA or albumin interaction," *Bioorganic and Medicinal Chemistry*, vol. 18, no. 7, pp. 2678–2685, 2010.

[8] H. Gurdal, S. Usanmaz, and F. C. Tulunay, "The effects of ions on antibacterial activity of ofloxacin and ceftriaxone," *Chemotherapy*, vol. 37, no. 4, pp. 251–255, 1991.

[9] S. H. Auda, Y. Mrestani, D. H. Nies, C. Groâe, and R. H. H. Neubert, "Preparation, physicochemical characterization and biological evaluation of cefodizime metal ion complexes," *Journal of Pharmacy and Pharmacology*, vol. 61, pp. 753–758, 2009.

[10] A. I. El-Said, A. A. M. Aly, M. S. El-Meligy, and M. A. Ibrahim, "Mixed ligand Zinc(II) and Cadmium(II) complexes containing Ceftriaxone or Cephradine antibiotics and different donors," *Journal of the Argentine Chemical Society*, vol. 97, no. 2, pp. 149–165, 2009.

[11] J. R. Anacona and A. Rodriguez, "Synthesis and antibacterial activity of ceftriaxone metal complexes," *Transition Metal Chemistry*, vol. 30, no. 7, pp. 897–901, 2005.

[12] A. E. Ali, "Synthesis, spectral, thermal and antimicrobial studies of some new tri metallic biologically active ceftriaxone complexes," *Spectrochimica Acta A*, vol. 78, no. 1, pp. 224–230, 2011.

[13] S. Fu, Z. Liu, S. Liu, and A. Yi, "Study on the resonance Rayleigh scattering, second-order scattering and frequency doubling scattering spectra of the interactions of palladium(II)-ceftriaxone chelate with anionic surfactants and their analytical applications," *Talanta*, vol. 75, no. 2, pp. 528–535, 2008.

[14] M.-Y. Tian, X.-F. Zhang, L. Xie, J.-F. Xiang, Y.-L. Tang, and C.-Q. Zhao, "The effect of Cu^{2+} on the interaction between an antitumor drug-mitoxantrone and human serum albumin," *Journal of Molecular Structure*, vol. 892, no. 1–3, pp. 20–24, 2008.

[15] P. N. Naik, S. A. Chimatadar, and S. T. Nandibewoor, "Interaction between a potent corticosteroid drug—dexamethasone with bovine serum albumin and human serum albumin: A fluorescence quenching and fourier transformation infrared spectroscopy study," *Journal of Photochemistry and Photobiology B*, vol. 100, no. 3, pp. 147–159, 2010.

[16] X. Zhao, R. Liu, Z. Chi, Y. Teng, and P. Qin, "New insights into the behavior of bovine serum albumin adsorbed onto carbon nanotubes: comprehensive spectroscopic studies," *Journal of Physical Chemistry B*, vol. 114, no. 16, pp. 5625–5631, 2010.

[17] A. D. Bani-Yaseen, "Spectrofluorimetric study on the interaction between antimicrobial drug sulfamethazine and bovine serum albumin," *Journal of Luminescence*, vol. 131, no. 5, pp. 1042–1047, 2011.

[18] P. N. Naik, S. A. Chimatadar, and S. T. Nandibewoor, "Study on the interaction between antibacterial drug and bovine serum albumin: a spectroscopic approach," *Spectrochimica Acta A*, vol. 73, no. 5, pp. 841–845, 2009.

[19] H. Liang, J. Huang, C.-Q. Tu, M. Zhang, Y.-Q. Zhou, and P.-W. Shen, "The subsequent effect of interaction between Co^{2+} and human serum albumin or bovine serum albumin," *Journal of Inorganic Biochemistry*, vol. 85, no. 2-3, pp. 167–171, 2001.

[20] D. Li, M. Zhu, C. Xu, and B. Ji, "Characterization of the baicalein-bovine serum albumin complex without or with Cu $^{2+}$or Fe $^{3+}$ by spectroscopic approaches," *European Journal of Medicinal Chemistry*, vol. 46, no. 2, pp. 588–599, 2011.

[21] Q. Gu and J. E. Kenny, "Improvement of inner filter effect correction based on determination of effective geometric parameters using a conventional fluorimeter," *Analytical Chemistry*, vol. 81, no. 1, pp. 420–426, 2009.

[22] J. R. Lackowicz, *Principles of Fluorescence Spectroscopy*, Kluwer Academic/Plenum Publishers, New York, NY, USA, 2nd edn edition, 1999.

[23] M. R. Eftink, *Biophysical and Biochemical Aspects of Fluorescence Spectroscopy*, Plenum Press, New York, NY, USA, 1991.

[24] M.-X. Xie, M. Long, Y. Liu, C. Qin, and Y.-D. Wang, "Characterization of the interaction between human serum albumin and morin," *Biochimica et Biophysica Acta*, vol. 1760, no. 8, pp. 1184–1191, 2006.

[25] D. Li, J. Zhu, and J. Jin, "Spectrophotometric studies on the interaction between nevadensin and lysozyme," *Journal of Photochemistry and Photobiology A*, vol. 189, no. 1, pp. 114–120, 2007.

[26] D. Li, B. Ji, and J. Jin, "Spectrophotometric studies on the binding of vitamin C to lysozyme and bovine liver catalase," *Journal of Luminescence*, vol. 128, no. 9, pp. 1399–1406, 2008.

[27] J. R. Lakowicz and G. Weber, "Quenching of fluorescence by oxygen. A probe for structural fluctuations in macromolecules," *Biochemistry*, vol. 12, no. 21, pp. 4161–4170, 1973.
[28] G. Scatchard, "The attractions of proteins for small molecules and ions," *Annals of the New York Academy of Sciences*, vol. 51, pp. 660–672, 1949.
[29] X. M. He and D. C. Carter, "Atomic structure and chemistry of human serum albumin," *Nature*, vol. 358, no. 6383, pp. 209–215, 1992.
[30] C.-C. Lin and D.-E. Shieh, "The anti-inflammatory activity of *Scutellaria rivularis* extracts and its active components, baicalin, baicalein and wogonin," *American Journal of Chinese Medicine*, vol. 24, no. 1, pp. 31–36, 1996.
[31] P. D. Ross and S. Subramanian, "Thermodynamics of protein association reactions: forces contributing to stability," *Biochemistry*, vol. 20, no. 11, pp. 3096–3102, 1981.
[32] A. N. Glazer and E. L. Smith, "Studies on the ultraviolet difference spectra of proteins and polypeptides," *The Journal of biological chemistry*, vol. 236, pp. 2942–2947, 1961.
[33] H. Polet and J. Steinhardt, "Binding-induced alterations in ultraviolet absorption of native serum albumin," *Biochemistry*, vol. 7, no. 4, pp. 1348–1356, 1968.
[34] L. Trynda-Lemiesz, A. Karaczyn, B. K. Keppler, and H. Kozlowski, "Studies on the interactions between human serum albumin and trans-indazolium (bisindazole) tetrachlororuthenate(III)," *Journal of Inorganic Biochemistry*, vol. 78, no. 4, pp. 341–346, 2000.
[35] J. N. Miller, "Recent advances in molecular luminescence analysis," *Proceedings of the Analytical Division of the Chemical Society*, vol. 16, no. 7, pp. 203–208, 1979.
[36] W. He, Y. Li, J. Tian, H. Liu, Z. Hu, and X. Chen, "Spectroscopic studies on binding of shikonin to human serum albumin," *Journal of Photochemistry and Photobiology A*, vol. 174, no. 1, pp. 53–61, 2005.

14

Demonstrated Wavelength Portability of Raman Reference Data for Explosives and Chemical Detection

Timothy J. Johnson,[1] Yin-Fong Su,[1] Kristin H. Jarman,[1] Brenda M. Kunkel,[1] Jerome C. Birnbaum,[1] Alan G. Joly,[1] Eric G. Stephan,[1] Russell G. Tonkyn,[1] Robert G. Ewing,[1] and Glen C. Dunham[2]

[1] *Pacific Northwest National Laboratory, P.O. Box 999, Richland, WA 99354, USA*
[2] *Dunham Research, 3251 Kemptown Church Road, Monrovia, MD 21770, USA*

Correspondence should be addressed to Timothy J. Johnson, timothy.johnson@pnnl.gov

Academic Editor: Augustus Way Fountain

As Raman spectroscopy continues to evolve, questions arise as to the portability of Raman data: dispersive versus Fourier transform, wavelength calibration, intensity calibration, and in particular the frequency of the excitation laser. While concerns about fluorescence arise in the visible or ultraviolet, most modern (portable) systems use near-infrared excitation lasers, and many of these are relatively close in wavelength. We have investigated the possibility of porting reference data sets from one NIR wavelength system to another: We have constructed a reference library consisting of 145 spectra, including 20 explosives, as well as sundry other compounds and materials using a 1064 nm spectrometer. These data were used as a reference library to evaluate the same 145 compounds whose experimental spectra were recorded using a second 785 nm spectrometer. In 128 cases of 145 (or 88.3% including 20/20 for the explosives), the compounds were correctly identified with a mean "hit score" of 954 of 1000. Adding in criteria for when to declare a correct match versus when to declare uncertainty, the approach was able to correctly categorize 134 out of 145 spectra, giving a 92.4% accuracy. For the few that were incorrectly identified, either the matched spectra were spectroscopically similar to the target or the 785 nm signal was degraded due to fluorescence. The results indicate that imported data recorded at a different NIR wavelength can be successfully used as reference libraries, but key issues must be addressed: the reference data must be of equal or higher resolution than the resolution of the current sensor, the systems require rigorous wavelength calibration, and wavelength-dependent intensity response should be accounted for in the different systems.

1. Introduction

There are many spectroscopic methods that have decreased the size and cost of the instruments and simultaneously increased their detection potential so as to become routine methods for chemical detection: infrared spectroscopy [1–3], visible-NIR reflectance [4], terahertz methods [5, 6], and others. Recently Raman spectroscopy has followed this trend and evolved from a cumbersome method of chemical physics to a common forensic technique [7]. There are currently tens of thousands of deployed spectrometers, many on-line or at-line, and in recent years there has been an advent of portable, even handheld, systems [8, 9], as well as remote sensing devices [10, 11]. Such capabilities make Raman an especially attractive method in the security arena, where substances such as radioactive materials, explosives, and chemical-/biological-warfare agents are a danger not only on the battlefield, but now also to civilian targets. For explosives, detection at several tens of meters has now been demonstrated [12–17] as has Raman chemical imaging of trace explosives, for example, in fingerprints [18]. One implementation of the technique that has really stood out in recent years is Raman spectroscopy using near-infrared lasers [19, 20]. Such systems have become less cumbersome, inexpensive and still retain the specificity of Raman or IR spectroscopy; namely, they have the ability to detect thousands of different chemicals and substances [21–23].

Contemporary red and near-infrared lasers, particularly the diode lasers, offer great advantages in terms of economy, size, and ruggedness. Should more laser systems be realized at wavelengths longer than ca. 1400 nm, enormous advantages could also be realized in terms of eye safety, as the human cornea blocks wavelengths longer than 1400 nm [24]. At the near-infrared (NIR) wavelengths used in most modern portable systems, most species do not undergo an electronic transition, and thus, empirically, most molecules do not fluoresce or undergo resonance enhancement. The possibility thus exists to port reference data with good success, especially if one considers only NIR excitation wavelengths. But questions arise as to the transferability of Raman data: dispersive versus Fourier transform, wavelength calibration, intensity calibration, and in particular the frequency of the excitation laser [25–27]. Developing such databases is resource-intensive and typically the generated data are not transferred to other instruments. If the reference data were truly portable; however, it obviates the time-consuming and expensive process of re-recording instrument-specific spectral libraries. It may be that data will not be portable to all systems, for example, surface-probing systems or those which use ultraviolet lasers, since different Raman modes can be enhanced differently. However, with red or near-infrared excitation there is a greatly reduced likelihood of enhancement, fluorescence, or other perturbing phenomena; we have therefore investigated the possibility of porting reference data from one NIR system to another.

While not all aspects can be addressed, this paper considers the possibility of porting an entire database of spectra (i.e., a spectral library of a wide range of substances) recorded at one wavelength (1064 nm) and employing them for analysis at another wavelength (785 nm). A companion paper has been submitted [8] that discusses the specifics of adapting such high-resolution spectra to a modern handheld dispersive Raman system. We assume that such a demonstrated portability would be reflected amongst other systems such as those that use extant near-IR lasers (e.g., 633, 785, 830, 980, 1047, and 1064 nm), but could also be potentially useful for still other systems that may arise in the future, particularly those with laser wavelengths longer than 1400 nm, such as the lasers used in telecommunications (e.g., 1550 nm). At these longer wavelengths not only is there greater eye safety, but also less dispersion from atmospheric particles (aerosol Mie scattering), resulting in better transmission for obvious standoff applications such as explosives [13, 28]. We demonstrate this portability by testing a Raman spectra database of 145 chemicals, including 20 known explosives. To accomplish this, key efforts are made to calibrate on both the wavelength- and intensity-axes such that, in theory, the spectra from any one system will simply overlay those from another.

2. Experimental and Calibration

2.1. General Experimental. For these experiments, the spectra were measured using two different Fourier-transform (FT) Raman spectrometers, one using 1064 nm excitation, the other using 785 nm excitation. In many instances researchers wish to port data between systems, including between FT and dispersive systems. The merits of FT-spectrometers versus dispersive-instruments for specific applications can be deliberated elsewhere; this paper focuses only on the portability of the data. For both FT instruments [29] and for the dispersive instruments (especially smaller units) frequent and accurate wavelength calibration is paramount [30, 31]. For this study, it is important to note that while both are FT instruments from the same manufacturer, the instruments are of different wavelength (785 and 1064 nm) and have different étendues (numerical aperture), different interferometers (linear air-bearing versus mechanical flex-pivot), different detectors (though both are of the same type, power detection as opposed to photon counting), and differing optical components [7, 22]. Both instruments have a dedicated sample compartment, with either 90- or 180-degree backscatter spectral light collection; for the present work, the 180° configuration was used including a mirror behind the quartz cuvette for liquid samples. For solids, typically ~5 mg of samples was pressed into an Al sample cup for use with the 180° collection geometry. While wavelength calibration is critical for any Raman system, the absolute calibration in general can follow more easily for an FT instrument (the Connes advantage) [32]. The intensity calibration for most Raman systems, including the present ones, is not an absolute photometric calibration, but only removes the wavelength-dependent response of the optical system, which arises largely from the detector.

For the 1064 nm data, the spectra of the neat liquids and solids were recorded using a Bruker FRA 106 Raman module coupled to the interferometer of a Bruker IFS 66v/S vacuum Fourier infrared spectrometer used in purged mode [33, 34]. The 66v interferometer [35] used a germanium on calcium fluoride beamsplitter and a white light source for initial interferometer alignment. The FRA 106 Raman accessory consists of a CW 1064 nm Nd:YAG laser with a focused or unfocused laser spot [19, 36] and a notch filter after the F/1 collection lens that removes the Rayleigh scattered light. A 6.0 mm manual aperture [37, 38] was added before the interferometer to achieve the specified 2.0 cm^{-1} spectral resolution [37, 39]. The Raman-scattered light is modulated by the FTIR interferometer and focused onto a liquid-nitrogen cooled Ge detector inside the 106 accessory.

For the 785 nm data, a dual wavelength dedicated FT-Raman instrument, the Bruker MultiRam 27, was used in the 785 nm configuration. The MultiRam is a dedicated FT-Raman instrument, but instead of a linear air-bearing interferometer, it has a flex-pivot mechanical bearing system. The MultiRam 27 interferometer is a faster optic f/2.9 versus the f/4.5 of the IFS 66v, so the MultiRam was used with an internal aperture setting of 3.5 mm. The MultiRam beamsplitter is silicon coated onto quartz, and the Raman detector for the 785 channel is a thermo-electrically cooled avalanche pin-diode detector. The MultiRam also uses a proprietary notch filter to remove the Rayleigh line before the interferometer [36]. Both the 1064 and the 785 nm systems were equipped with a (custom) Stokes/anti-Stokes filter that passes light on both the low- and high-frequency side of the notch frequency.

All data were recorded at 2.0 cm^{-1} resolution, collecting the 180°-backscattered light. The excitation laser power was varied as needed, but seldom exceeded 500 mW. Typically between 300 and 2000 interferograms were averaged, depending on signal quality. Data were collected using forward-backward double-sided interferograms and (due to the broad interferograms) phase-corrected using the power method. To achieve the data portability it is clear that the data must be calibrated in two dimensions, namely (1) the wavelength and (2) the intensity axes. For any given sample, the spectrally corrected plots should overlay one another if done correctly and in absence of any wavelength-dependent phenomena. Extensive calibration for wavelength and intensity proceeded as described below. Most chemicals were purchased as the highest purity available from common chemical suppliers, mostly Sigma-Aldrich. The mineral specimens were from the "Rocks and Minerals of Canada collection" from American Educational Products. The data were only collected after the instrument was wavelength calibrated. The intensity calibration scales the relative response on the y-axis as a function of wavelength (described below) and arbitrarily sets the strongest peak to a value of 2.0 with the minimum set to 0.0.

2.2. Wavelength Calibration. For a Fourier-transform Raman system, frequency calibration is not trivial as it requires an absolute calibration of both (1) the scattered light recording system and (2) the instantaneous (e.g., shot to shot) or averaged frequency of the Raman excitation laser, typically after some stabilization period. To calibrate the interferometer there are several options such as using the narrow absorption lines of a gas such as CO or N_2O at low pressure [34, 37]. For both the present spectrometers the interferometer's wavelengths are referenced to those of the HeNe laser, as is common in commercial FT systems. Originally these were calibrated in the near-IR by using a white light source in the sample compartment and observing the 2 ← 0 overtone transition of neat carbon monoxide [34, 40] in a 10 cm cell. The transition frequencies were fit by adjusting the HeNe reference frequency. Using unapodized spectra, a series of CO lines near 4300 cm^{-1} in the first overtone band were used to match the reported frequencies of Maki and Wells [40]. This proved to be somewhat cumbersome as to where to locate the gas cell, as the interferometer path has few unobstructed beam paths > 10 cm. Recognizing other alternatives, such as an atomic lamp calibration standard, [29] it was found more expedient for reliable calibration of the interferometer to use an array of Hg lines [41] by bleeding in small amounts of light from the overhead fluorescent lamps. A typical emission spectrum of such lines can be seen in Figure 1. The method is preferred to the CO absorption technique as the lamp emission light follows exactly the same optical train as the Raman scattered light, and because the Hg-lines method is easier to implement. Using the five Hg lines depicted in Figure 1 along with the NIST (air) frequencies, the interferometer frequency is calibrated such that the sum of the deviations (obs.-ref.) for all five lines is ≤ 0.1 cm^{-1} on a daily basis. Typically, the average deviation for any given line is < ±0.1 cm^{-1}.

For an FT-Raman system, calibration of the interferometer frequencies alone is not sufficient. The absolute wavelength of the continuous wave excitation laser must also be set. There are methods to achieve this including using absolute Raman wavelength standards such as cyclohexane C_6H_{12} frequency shifts [25, 31, 42–44]. If using just one or two such transitions, however, it is possible as a happenstance that a small frequency offset of the Raman laser could be equally offset by a small miscalibration of the interferometer frequency in the opposite direction. This can be especially true for certain NIR solid state lasers with strong room temperature frequency fluctuation. In the present system we have opted to use a laser line notch filter that passes both the Stokes and anti-Stokes lines. In this method the calibration of the excitation laser is accomplished by setting the software value of the Raman laser frequency such that the absolute values of the Stokes and anti-Stokes frequency shifts for a given vibrational mode are made equivalent. If the input value of the excitation laser frequency were incorrect, the nominal Raman "zero" frequency position would be shifted, resulting in differing Stokes and anti-Stokes frequencies. For the Raman Nd:YAG 1064 nm laser systems, a daily Stokes versus anti-Stokes calibration is performed using a strong scatterer, typically elemental sulfur [19, 38]. Low frequency modes are advantageous here as the detector responsivity is limited to approximately −1,000 cm^{-1} anti-Stokes shift (detector's bandgap or work function cutoff). The 785 nm laser wavelength was calibrated using 785 nm $KClO_3$ Raman spectra against the $KClO_3$ spectra obtained from Stokes and anti-Stokes calibrated 1064 nm system. For most species, however, the sample itself can be employed to calibrate the Raman laser frequency as seen in Figure 2 for $KMnO_4$. The $KMnO_4$ provided several low frequency lines with which one can obtain calibration accuracies of the Stokes and anti-Stokes Raman shifts of better than 0.2 cm^{-1} using a few of the strongest lines. For both the 1064 nm Nd:YAG laser and the 785 nm GaAlAs laser, the nominal frequency was adjusted until the agreement was within ±0.20 cm^{-1}. Due to the inherent Raman laser linewidths, typically 1.0 to 1.5 cm^{-1}, greater precision is not warranted.

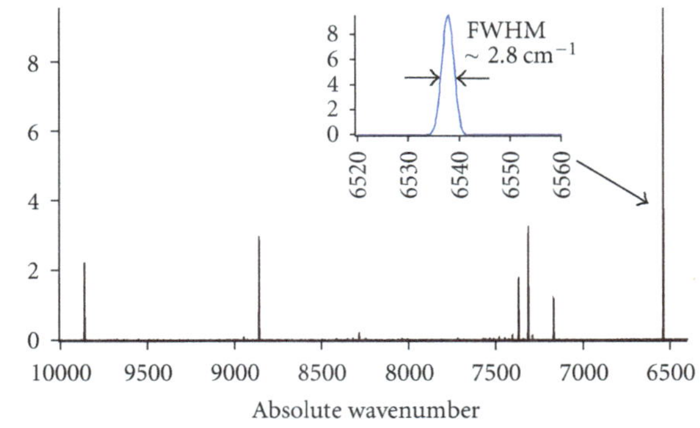

FIGURE 1: Hg-line spectrum used for interferometer calibration. Several lines are used to calibrate the absolute wavelength response of the interferometer. The Raman laser frequency ~9397 cm^{-1} is bracketed by the Hg lines. The inset shows the linewidth (FWHM) of one of the Hg lines.

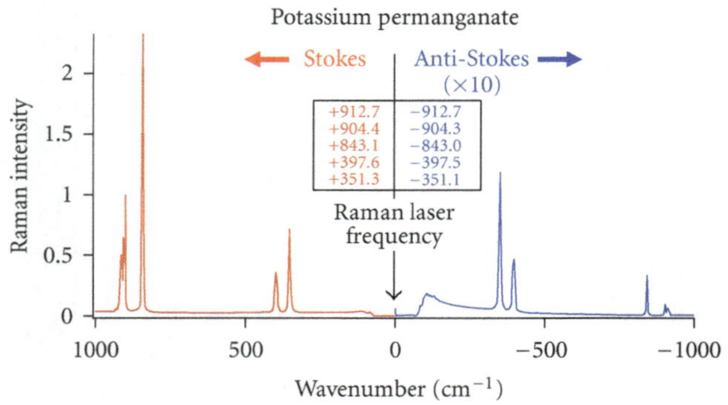

FIGURE 2: Calibration scheme demonstrating how Stokes and anti-Stokes lines are used to calibrate the laser frequency such that the Stokes lines are at the same shift with the opposite sign of the anti-Stokes lines. The present data show Raman lines of potassium permanganate. Typically the values are ≤ 0.20 cm^{-1} absolute value of each other. The excitation laser line is nominally at 9397.58 cm^{-1} (air) absolute frequency, but has typical daily variations of ±0.50 cm^{-1} or more.

2.3. Intensity Calibration. In some instances, intensity calibration plays a lesser role as chemical identification is accomplished only via peak look-up tables. In many cases, however, the relative amplitude of the peaks plays a key role as peak heights as well as positions are used for identification. This is more crucial when methods such as partial least squares are used for chemometric analysis [45]. In some cases absolute scattering coefficients are obtained, and in most cases this is done by comparison to an absolute standard [7]. As Ray and McCreery have pointed out, [27] obtaining Raman intensity values that are corrected even on a relative scale, that is, that properly ratio out the instrumental response, is more challenging than first supposed. The challenge lies in the fact that the Raman experiment is a single-beam experiment; that is, it is not a self-ratioing measurement such as an infrared transmittance measurement.

In our earlier measurements, the calibration of the scattering intensity for its wavelength dependence was accomplished using a broadband source of known brightness [46]. Other workers have accomplished this using calibrated blackbodies whose absolute emittance was known as a function of drive current. However, the FRA 106 Raman module has a 3000 K white light source whose intensity is scattered into the spectrometer off a NIST-traceable Spectralon surface mounted at the sample position. The instrument response function as a function of wavelength was generated by ratioing this spectrum relative to the intensity predicted by Planck's law for a 3000 K source. The quality of the intensity correction was assessed by measuring the intensities in the corrected Stokes and anti-Stokes spectra; the relative intensities were within a few percent of the expected value [21] and the agreement was good.

While a useful method, this is not especially practical for field instruments. Recently, the National Institute of Standards and Technology [25] has developed standard calibrated targets SRM 2241 and SRM2244, for use with either 785 or 1064 nm excitation lasers, respectively. By measuring the spectrum of such a target and comparing it to the NIST-calibrated spectrum we are able to determine a calibrated system response curve from 3500 to 100 cm^{-1}. After determining the system response in this manner all that is required for intensity calibration is a simple spectral multiplication. Subsequent spectra are then multiplied by this frequency-dependent y-axis scaling factor.

An example of the relative response factor for our 1064 nm system between +3500 to +100 cm^{-1} can be seen in Figure 3, as well as a Raman spectrum of *ortho*-xylene before and after the correction factor has been applied. It is seen that the intensity correction is changing the most rapidly at the ends of the spectrum near the system notch filter cutoff and at the red end of the spectrum ($\Delta\nu = 3500\ cm^{-1}$, $\nu \sim 5900\ cm^{-1}$) where the Ge detector is approaching its bandgap cutoff. The intensity correction results using the two different methods were found to give comparable results.

2.4. Molecule Selection. To test the library's portability a list of species was derived. The list of the composite library contains a total of 145 molecules. Of these, 20 are known explosive materials specifically selected and these are summarized with some of their properties in Table 1. For the nonexplosive materials we intentionally chose several different categories and colors of molecules and materials so as to attain a good cross-representation. While a more extensive list is always desirable for statistical purposes, time and budget constraints would not allow this. Many of the reference materials are pure chemicals commonly found in a modern laboratory, but care was taken such that the list represented a good variety of chemical classes, rather than just a few classes such as, for example, alkanes, alkenes, and alcohols. The composite list was designed to contain at least three species from each of the several chemical categories found in Table 2: aldehydes, alkenes, amines, amino acids, aromatics, carboxylic acids, halogenated organics, ketones, thiols, mineral acids, ionic simple salts, and organometallics. Table 2 characterizes the chemical breakdown of these 125 non-explosive materials. For those pure chemical species with two or more functional groups, the compound could be counted in two different categories; for example, 1, 3-dichlorobenzene would be tallied both as an aromatic and as a halogenated organic. Besides pure chemicals found in a chemical laboratory, the list was further expanded to contain common household and industrial materials, and also materials such as mineral acids, common minerals and some biological samples. Some of the common materials included items such as wood, concrete, minerals, and dirt; plastics such as nylon, Teflon, and a water bottle; other fabrics such as wool, cotton, leather; and household materials.

The list was thus completed and expanded to include a total spectral database of 145 compounds and materials including the 20 explosives as described above—with a significant effort to derive a representative potpourri of different chemical classes. What is only partially indicated is that several of the non-explosive chemical samples were

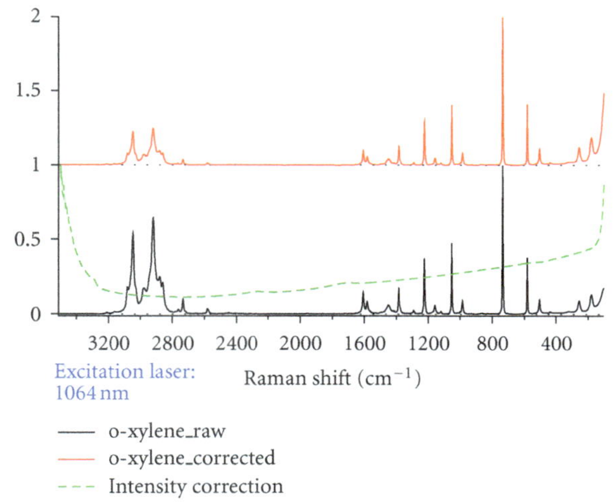

FIGURE 3: Response of the 1064 nm FT-Raman system before and after correcting for the spectrometer intensity response function using the NIST 2244 standard plotted as a function of Raman shift. The x-axis represents 5897 to 9327 cm^{-1} absolute wavenumbers, with the intensity correction factor generated by the standard plotted as the dashed green line. The "raw" spectrum is plotted in black as the lower trace, the corrected spectrum in red as the upper trace. These spectra are arbitrarily scaled 0 to 1 on the ordinate and have been vertically offset for clarity.

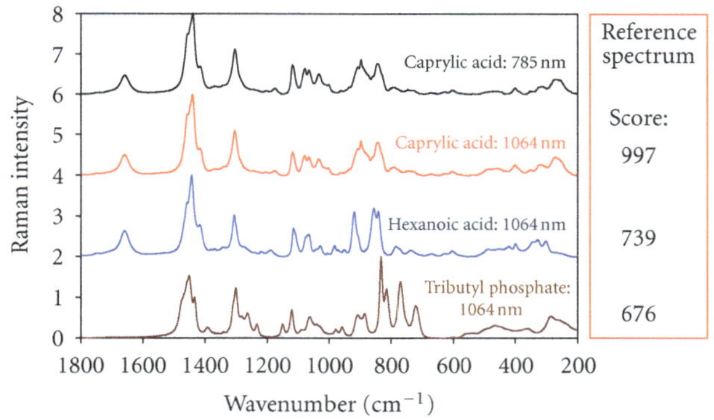

FIGURE 4: Example of a successful identification for comparing the 785 nm data versus the 1064 nm library. The result is typical; for most species the 1064 nm spectrum of the same compound is identified as the top candidate "hit" for the 785 spectrum. Scores for the top three hits are at right. Spectra are scaled 0 to 2 and have been vertically offset for clarity.

intentionally selected due to their mixture of light and dark materials, that is, their physical colors. While there was a long list of clear liquids and white solids (the "unknown white powder" challenge), there were also many deeply colored materials, which are more susceptible to fluorescence at 785 nm. To further ensure a representative cross-section, specific examples were also *intentionally* chosen because they are known to fluoresce, in some cases more strongly at 785 than 1064 nm. These include species such as calcium hypochlorite, but also some ink dyes as suggested by the work of Geiman et al. [47], namely, acid red 52 dye and crystal violet dye. The complete list of chemical species and breakdown by chemical category is available as Supplementary material available online at doi:10.1155/2012/297056, with the classes summarized in Table 2.

3. Results and Discussion: Spectroscopy

As discussed above, the intensity calibration scales only the relative (wavelength-dependent) response as seen in Figure 3, with the spectrum maximum arbitrarily set to a value of 2.0. Absolute cross-sections are not needed for this study as the goal is only for identification not quantification. That is to say, the present results do not explicitly account for the "ν^4" factor associated with excitation wavelength. Strictly speaking, if a power-detecting element is used (common in NIR systems), the $(\nu_0 - \nu_k)^4$ expression is appropriate whereas if a photon-counting device (common in visible Raman systems) is used, the relative response term is $\nu_0(\nu_0 - \nu_k)^3$, where ν_0 is the laser frequency and ν_k the vibrational mode frequency [22]. As has been noted before [7, 22] any measured intensity of a Raman band in fact has contributions from 3 sources, namely, (1) the Raman activity itself that is intrinsic to the scattering species; this is the term that is typically calculated by theoretical models, (2) a statistical mechanical term accounting for temperature (which affects the population of the scattering vibrational state), and (3) the well-known "ν^4" factor that depends on the laser excitation frequency and the type of photometric detection in the Raman spectrometer as discussed above.

The quality of the derived spectra is very high due to the laboratory nature of our experiments with adequate averaging time. After proper wavelength and (relative) intensity calibration, and barring any substantial fluorescence or resonance effects, the spectra from the differing wavelength systems should nearly superimpose on one another. In almost all cases our results have shown this to be the case. The 1064 reference data can readily serve as reference spectra for 785 systems, as shown for the two FT-Raman systems shown here, as well as for dispersive 785 handheld Raman systems [8]. This is seen for a typical example in Figure 4 and was the case for almost all measured species. The search algorithm and criteria for declaring a match are described below, but Figure 4 illustrates the results of this process for caprylic acid. In particular, caprylic acid (octanoic acid) is a "785 clean spectrum" so the search clearly identifies the correct compound at 1064 as the top hit. The traces of the "search" 785 spectrum and the "reference" 1064 spectrum are seen as the top two traces in the plot and are identical to the eye. The second hit, with a relatively high score, is the chemically and spectroscopically similar hexanoic acid, with tributyl phosphate detected as the third hit.

For the explosives, the commercial grade materials such as those used for mining and demolition are sold by trade names and generally consist of a mixture of energetic materials. The variety and relative percent concentrations of the materials used vary widely and depend upon required performance and cost. Many of the commercial products, while sold under different names, have some strong commonalities

Table 1: Summary of the 20 explosives measured by both the 785 and 1064 nm Raman systems. In several cases the materials are named by their primary chemical component(s), followed by the common or commercial name, since the primary components define the spectrum. AN: ammonium nitrate, EGDN: ethylene glycol dinitrate, PETN: Pentaerythritol tetranitrate, NC: nitrocellulose, NG: nitroglycerine.

Sample	Name	Comments/chemical composition
1	AN/$NaNO_3$-Coalmex	Often contains AN, $NaNO_3$, EGDN
2	AN/$NaNO_3$-Dyno AP	Emulsion explosive for mining, AN, $NaNO_3$
3	AN/$NaNO_3$-Dynomax Pro	Dynamite blasting gelatine, NG, EGDN, NC, AN, $NaNO_3$
4	AN/$NaNO_3$-Powerfrac	Often contains AN, $NaNO_3$, EGDN, S, NG, NC
5	AN/$NaNO_3$-Red-D Gel-B	Dynamite- often contains NG, EGDN, NC, AN, $NaNO_3$
6	HMX	Octahydro-1,3,5,7-tetranitro-1,3,5,7-tetrazocine
7	Nitroglycerine	NG, Liquid, 1,2,3-Trinitroxypropane
8	Nitroguanidine	$CH_4N_4O_2$, used in powders to reduce barrel erosion
9	PETN	Pentaerythritol tetranitrate
10	PETN-Primasheet 1000	PETN-65% based rubberized explosive
11	PETN-Semtex 10	Largely contains ~85% PETN
12	PETN-Semtex A	Largely contains ~84% PETN
13	PETN-Semtex H	Often contains >25% PETN, <65% RDX
14	Picric acid	2,4,6-Trinitrophenol, forms sensitive metal salts
15	RDX	1,3,5-Trinitroperhydro-1,3,5-triazine
16	RDX-C4	Contains ~91% RDX plus plasticizers and binders
17	RDX-PE4	Similar to C-4, an RDX-based explosive (white)
18	RDX-PW4	Largely RDX-based plastic explosive
19	TNB	Trinitrobenzene, commercial mining high explosive
20	TNT	2,4,6-Trinitrotoluene

Table 2: Chemical and material classes of the 125 nonexplosive chemicals and materials that are amongst the species that were tested. For the pure chemicals, species with multiple functional groups were tabulated in each class.

Class/type of compound	Number of compounds in class	Class/type of compound	Number of compounds in class
Organic classes		Inorganic classes	
Alcohol	14	Borane	1
Aldehyde	3	Inorganic or rare earth salt	17
Alkane	6	Mineral, dirt	7
Alkene, alkyne	7	Mineral acid	3
Amides, carbamides	5	Organometallic	3
Amine	11	Peroxides, water	2
Amino acid	2		
Aromatic	31	Other classes	
Carbohydrate	4	Building material	2
Carboxylic acid	13	Household product	4
Ester	11	Synthetic or natural fabric	2
Ether	11		
Halogenated organic	6	Phenomenological	
Ketone	5	Fluorescent material (known)	>10
Nitrile, nitro	3		
Phosphines, phosphonates, and so forth.	4		
Sulfides, thiols	2		
Sulfates, sulfones, and so forth.	3		

in the material components. For example, many mining materials contain various levels of ammonium nitrate and sodium nitrate along with other materials. Due to the strong Raman activity of the NO_3^- moiety, both NH_4NO_3 (AN) and $NaNO_3$ provide features that dominate the spectrum, while the smaller features help aide slightly in the identification of the specific formulation. Thus, the explosive materials listed in Table 1 are segregated and categorized based upon their primary chemical components or compound groups. Subtle differences due to noise and baseline correction can lead to slight changes in relative peak intensity and even peak position. Due to the similar makeup of several of the explosives, such differences were likely the cause of the top hit of an analyte not being an exact match to that formulation but the top hit was for a formulation containing nearly identical chemical components to those in the analyte. When the library generated at 1064 nm was used for identifying samples measured at 785 nm and was based upon selecting the correct compound group (e.g., if any of the AN/$NaNO_3$ formulations were identified), all 20 explosive formulations in Table 1 were correctly identified.

There were a few outliers in the statistical data, a few of which merit additional explanation: for a few species, the fluorescence phenomenon stimulated by the excitation laser light generated a signal that is more intense than the light scattered by the weaker Raman effect; this was true for those species selected due to their 785 nm fluorescence [47]. In fluorescence, the excitation light is absorbed and re-emitted via an electronic transition that does not involve a change in the spin quantum number; at room temperature the fluorescence bands typically manifest themselves as a broad and featureless background. In some cases, this fluorescence background signal can be removed via baseline correction, but many researchers have shown that shifting to longer wavelength excitation is even more advantageous, as Lewis et al. have shown for certain explosives [48, 49]. In the present studies a broad fluorescence was observed in ~10 out of 145 cases, but identification was still possible for a few of these materials. Typically the method used in most sensors is to simply filter the analysis via baseline correction such as with a rolling circle filter or some other baseline correction algorithm. While the phenomena are real optical phenomena, they are usually featureless and are simply removed as backgrounds via a software algorithm, leaving only the Raman peaks.

Interestingly, there were also a few anomalous situations where there was significant spectral change (not just a fluorescence background) depending upon the excitation wavelength. One example, illustrated in Figure 5, was for neat Er_2O_3. While a few peaks are common and are likely Raman bands, it appears that either the 1064 nm or 785 nm excitation, or both, has excited transitions into low-lying electronic states. The resulting fluorescence then dominates the emission spectrum. In this particular case the fluorescence(s) are not broad and featureless as in the other cases, but are structured and are likely associated with the pseudoatomic transitions of the Er^{3+} cation. The translanthanides and trans-actinides typically have multiple low-lying electronic states that give the species their colors. In their studies of Er_2O_3 nanocolloids, Patel et al. [50] discuss an upconversion signal observed at 740 nm when pumping with a 980 nm multimode diode laser that is resonant with $^4I_{11/2}$ state. This was ascribed to a two photon excitation processes of the type $^4I_{15/2} \rightarrow {}^4I_{11/2}$ and $^4I_{13/2} \rightarrow {}^4F_{9/2}$ that resulted (after internal conversion) to a $^4I_{9/2} \rightarrow {}^4I_{15/2}$ emitted. The structured bands we see in the 785 nm Raman spectrum near 2500 cm^{-1} shift correspond to 10,235 cm^{-1} absolute or about 977 nm and are likely emission from the original state into which their 980 nm laser pumped, that is,

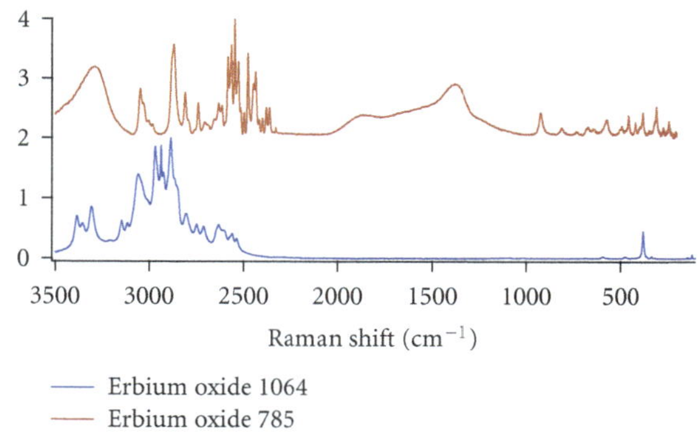

FIGURE 5: Results of the Er_2O_3 spectra at the two different excitation wavelengths. The spectra have been vertically offset for clarity with *y*-axis scaled 0–2. The results are discussed in the text.

an $^4I_{11/2} \rightarrow {}^4I_{15/2}$ emission, though further measurements would be needed to confirm this [51].

4. Results and Discussion: Statistical Analysis

For the analysis, our overarching goal was to determine whether a spectrum recorded using 785 nm excitation can be successfully identified using a library created from 1064 nm reference data. To test the hypothesis that the 785 nm and 1064 nm spectra are interchangeable after wavelength and intensity corrections, a comparison study was conducted, where the Bruker OPUS commercial spectral searching software was used and the 785 nm intensity-corrected spectra were treated as test cases and searched against a 1064 nm intensity-corrected library. The software has several different search algorithms, each with several possible variations to it. However, our intent was not to perform a comprehensive study of different algorithms, but rather to measure the similarity of the 785 nm and 1064 nm spectra under ordinary laboratory conditions. Therefore, the OPUS basic search algorithm was used.

The OPUS basic search algorithm works by matching peaks without regard to their intensities. In particular, for any given spectrum, a peak table is constructed. This peak table is used to create a *signature spectrum* consisting of Lorentz curves for every peak, each of which has a height of one. To compare a test case to a library compound, the algorithm looks up all of the test signature peaks in the library signature. Those test peaks that appear in the library signature are assigned a value of one, whereas those that are different or do not appear in the library signature are assigned a lesser value, with a minimum of zero. A *hit quality* is then calculated as the average of these values times 1000. More simply, when all test case peaks are found in given library signature spectrum, the hit quality is 1000. When none of them are found, the hit quality is zero. For example, when three of four test signature peaks are found in a given library entry, the hit quality is $3/4*1000 = 750$. For a given test case, this comparison process is repeated for every entry of the spectral library, and the three library compounds with the highest hit quality—the top three hits—are reported.

OPUS allows for some parameter adjustment in its basic search algorithm. In particular, there is an adjustable "search sensitivity" parameter. While specific details about this parameter are unavailable, its value, an integer between 1 and 20, relates to the sensitivity with which OPUS identifies peaks, 20 dictating a much closer match. Loosely interpreted, a higher search sensitivity means the peak locations must be more precise for the algorithm to give a higher hit score, to the point that Bruker did not recommend a setting of 15 or higher for this application. The search sensitivity also affects the number of peaks created in the temporary peak table for the spectrum being searched. Thus, a higher search sensitivity implies a stronger match and implies more peaks may have been created and matched than a lower sensitivity value. OPUS suggests testing different values to determine which is best for a particular application. For our application, a search sensitivity of 7 was chosen because preliminary results showed this value tended to generate favorable search results in terms of good discrimination.

A 1064 nm library was created in OPUS using all 145 compound spectra. The 785 nm spectra were then individually loaded and searched against the 1064 nm library via the basic search algorithm. In terms of the explosives, using the chemical formulation grouping, all 20 explosives were correctly identified. It should also be noted that for many of these formulations the top 3 or 4 hits were for similar compounds and all had very similar high scores. Thus the explosive materials listed in Table 1 are segregated and categorized based upon primary chemical components or compound groups. When the library generated at 1064 nm was used for identifying samples measured at 785 nm and was based upon selecting the correct compound group, all 20 explosive formulations were correctly identified. We also note that, for the explosives, the first hit minus second hit refers to the first hit minus the first hit outside of its compound group. For example, NH_4NO_3-$NaNO_3$-Coalmex has a top hit of 1000. Its second, third, fourth, and fifth hits are also 1000 with each hit belonging to the same compound group. Its first hit outside of the group is potassium nitrate with a hit score of 468. Therefore the distance between first and second hits for AN-NaNO3 is 532, the difference of 1000 and 468.

While the explosives were all easily identified with the 1064 imported library, for the more general case of relying on an algorithm for identifications, it is important to first establish objective criteria for declaring a match. For example, when the top hit score is near 1000, there is strong evidence for a match, but not so when the top hit score is near 400. Even in the case of a high score, if the top two hit scores are so close as to be virtually identical (say 985 and 988), then there is ambiguity in declaring one or the other a match. A concrete example of this is provided in Figure 6. In the figure, hair gel had petroleum jelly as its top hit with hair gel as its second hit. This implies the search algorithm cannot, for certain, determine which spectrum is the correct match. In this particular case, because the hair gel was determined to be made of almost exclusively petroleum jelly with a few

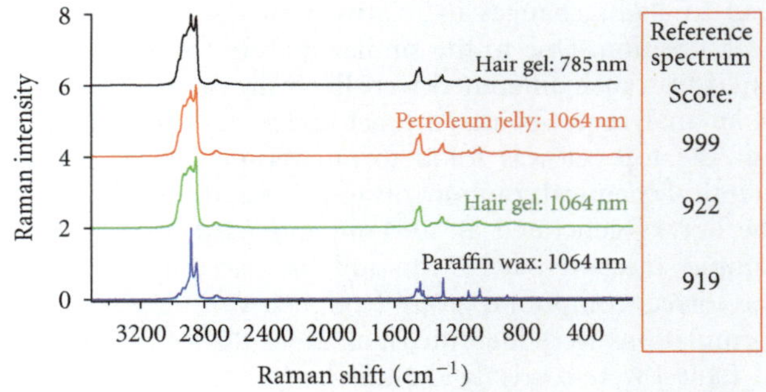

Figure 6: Spectral search example of one of the few false hits showing hair gel at 785 nm with petroleum jelly (1064) as its top hit and hair gel (1064) as its second hit. The distance between the top two hits is more than 60 (with a distance of 77) and the top hit is above 775 so it is declared a match. These spectra are all scaled 0–2 on the y-axis and have been vertically offset for clarity.

trace additives (namely a coloring agent) the major peaks were therefore in the same positions. Hair gel is an example of a "false hit"; however, in the more general case, a "false miss" is usually caused by peaks in (nearly) the same x-axis location but which have slightly different peak intensities or profiles. In such cases a more definitive criterion is needed to declare a match.

Based on the above considerations, two criteria were thus established for matching the 785 nm spectra to the 1064 nm library. The first criterion, referred to as the naïve criterion, selects the top hit as a match regardless of the hit score. Using this criterion, 128 out of the 145 spectra at 785 nm (88.3%) were correctly identified. The second method, referred to here as the dual criteria, is designed to take into account not only the top hit but also the difference between the first and second hits. In this case, the top hit must be above a minimum threshold to be declared a match, and the distance between the first and second hit must be larger than a specified value. These dual criteria ensure that (1) the top hit closely matches the test spectrum, and (2) there are no other library spectra similar enough to the test spectrum to cause ambiguity. After initial experimentation, the dual criteria were set so that the minimum hit threshold was 775 and the minimum required distance between the first and second hit was at least 60. Those test spectra meeting both criteria were declared a match. All others were declared "uncertain."

Figure 7 contains the results of matching the 785 nm spectra to the 1064 nm library using the dual criteria. The y-axis contains the top hit score and the x-axis is the distance between the first two hit scores. The points inside the shaded region are the ones that are declared a match using the dual criteria. The markers indicate whether the corresponding spectrum was a (1) correct match (true hit), (2) an incorrect match (false hit), (3) an incorrect elimination (false miss), or (4) a correct elimination (mismatch excluded).

Table 3 provides a comparison between the results using the naïve criteria and the dual criteria. As can be seen, the number of false matches using the naïve criteria is 17, and this number drops to 4 using the dual criteria. This large

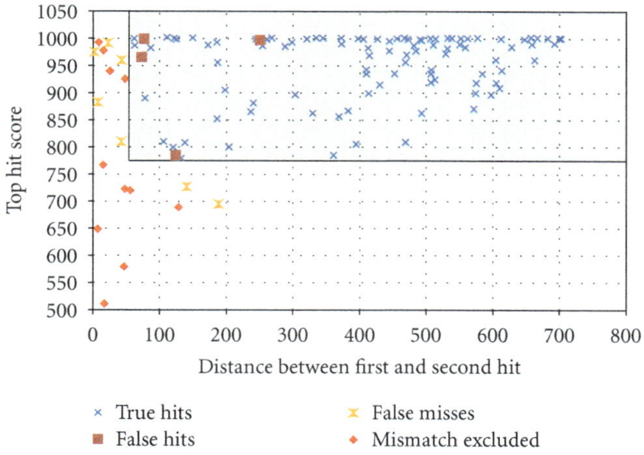

FIGURE 7: 785 nm-to-1064 nm comparison results using the dual criteria with all possible outcomes indicated. The true hits occur when the top hit is the correct spectrum and is declared a match, the false hits occur when a top hit is an incorrect spectrum but is declared a match, the false misses occur when the top hit is the correct spectrum but is declared not a match, and the mismatch excluded occurs when the top hit is an incorrect spectrum and is declared not a match.

decrease in false matches is due to the addition of a new classification category, "uncertain," for those test spectra that do not meet both of the dual criteria. Eleven of the original false matches are moved into the "uncertain" category when the dual criteria are used. Also because of the "uncertain" category, the number of false misses increases when going from the naïve to the dual criteria. In particular, the naïve criterion always declares a match and so the number of false misses must be zero. On the other hand, the dual criteria do not always find a match, and the number of mismatches is 7. Despite the increase in the number of false misses, we note the overall error, the sum of the false matches and false misses, decreases from 17 to 11 when going from the naïve criterion to the dual criteria, suggesting the dual criteria improve accuracy of the identification process.

5. Summary and Conclusions

We measured Raman spectra of 145 sundry chemicals, including 20 explosives, representing multiple chemical classes using both 785 nm and 1064 nm excitation. This cross-section represented many different chemical classes (e.g., alkanes, alcohols, nitrates, etc.) along with common materials (concrete, sand, wood, etc.). From the measured data, a 1064 nm library was constructed and the 785 nm spectra were individually searched against this library to determine the success of searching against reference spectra recorded at a different wavelength. All 20 explosives tested were correctly and readily identified using the 1064 to 785 nm ported data. Moreover, the initial results indicated that after all searches were complete 88% of the 785 nm spectra were correctly identified using the naïve criteria. These results clearly show good discrimination and utility

TABLE 3: Results of spectral searching using the naïve and dual criteria. See text for details.

Search criteria	No. of false matches (%)	No. of false misses (%)
Naïve	17 (11.7%)	0 (0%)
Dual	4 (2.8%)	7 (4.8%)

of reference data from a different NIR wavelength system. The use of the "dual" criteria reduced the total number of errors even further, resulting in a 92.4% correct identification rate. It is again noted that for those 785 nm cases where fluorescence prevented identification, this would likely have happened regardless of the excitation wavelength of the reference data.

In the near-infrared there are now several dispersive and also Fourier transform spectrometers available. With but a few caveats our results have shown that the data can in fact be ported from one system to another. The data can ideally be used with many common near infrared lasers: 785, 830, 960, 1047, 1064 or 1550 nm. This is especially true for algorithms that use only relative peak heights or peak positions for identification. For visible or UV systems, it is more important to consider those phenomena that can change relative peak heights of different vibration modes, resonance Raman, fluorescence, and so forth. It is important to note that the reference data must be of equal or higher resolution compared with the target spectra and the spectrometers require rigorous wavelength calibration. Ideally wavelength-dependent intensity response should be accounted for as well.

Acknowledgments

The research described in this paper was supported by the Combating Terrorism Technical Support Office (CTTSO) and the Joint Improvised Explosive Device Defeat Organization (JIEDDO). The authors thank sponsors for their support. The Pacific Northwest National Laboratory is a multiprogram national laboratory operated by Battelle for the U.S. Department of Energy under Contract DE-AC05-76RL01830. Portions of the research were performed using EMSL, a national scientific user facility sponsored by the Department of Energy's Office of Biological and Environmental Research and located at PNNL. We thank Drs. Arno Simon and Juergen Sawatzki of Bruker Optik for discussions about the spectrometers and also Dr. Martin Pirzer for discussions about OPUS software specifics. We also thank Dr. Trish McDaniel, Dr. Lou Wasserzug, and Dr. Peter Hotchkiss for their involvement and support of this research.

References

[1] I. R. Burling, R. J. Yokelson, S. K. Akagi et al., "Airborne and ground-based measurements of the trace gases and particles emitted by prescribed fires in the United States," *Atmospheric Chemistry and Physics*, vol. 11, no. 23, pp. 12197–12216, 2011.

[2] B. E. Bernacki and M. C. Phillips, "Standoff hyperspectral imaging of explosives residues using broadly tunable external cavity quantum cascade laser illumination," in *Chemical, Biological, Radiological, Nuclear, and Explosives (CBRNE) Sensing XI*, vol. 7665, article 76650I of *Proceedings of SPIE*, Orlando, Fla , USA, April 2010.

[3] T. J. Johnson, R. L. Sams, S. D. Burton, and T. A. Blake, "Absolute integrated intensities of vapor-phase hydrogen peroxide (H_2O_2) in the mid-infrared at atmospheric pressure," *Analytical and Bioanalytical Chemistry*, vol. 395, no. 2, pp. 377–386, 2009.

[4] W. M. Wendlandt and H. G. Hecht, *Reflectance Spectroscopy*, Interscience Publishers, London, UK, 1966.

[5] T. J. Johnson, N. B. Valentine, and S. W. Sharpe, "Mid-infrared versus far-infrared (THz) relative intensities of room-temperature Bacillus spores," *Chemical Physics Letters*, vol. 403, no. 1–3, pp. 152–157, 2005.

[6] D. M. Sheen, D. L. McMakin, and T. E. Hall, "Detection of Explosives by Millimeter-wave Imaging," in *Counterterrorist Detection Techniques of Explosives*, J. Yinon, Ed., pp. 237–278, Elsevier, Amsterdam, The Netherlands, 2007.

[7] R. L. McCreery, "Photometric standards for Raman spectroscopy," in *Handbook of Vibrational Spectroscopy*, J. M. Chalmers and P. R. Griffiths, Eds., Wiley, New York, NY, USA, 2002.

[8] J. C. Weatherall, J. Barber, C. S. Brauer et al., "Adapting Raman spectra from laboratory spectrometers to portable detection systems," *Applied Spectroscopy.*

[9] P. Vandenabeele, K. Castro, M. Hargreaves, L. Moens, J. M. Madariaga, and H. G. M. Edwards, "Comparative study of mobile Raman instrumentation for art analysis," *Analytica Chimica Acta*, vol. 588, no. 1, pp. 108–116, 2007.

[10] N. R. Butt, M. Nilsson, A. Jakobsson et al., "Classification of Raman spectra to detect hidden explosives," *IEEE Geoscience and Remote Sensing Letters*, vol. 8, no. 3, pp. 517–521, 2011.

[11] A. Ehlerding, I. Johansson, S. Wallin, and H. Östmark, "Electro-optical remote sensing, photonic technologies, and applications IV," vol. 7835, article 783507 of *Proceedings of SPIE*, Toulouse, France, September 2010.

[12] S. Wallin, A. Pettersson, H. Östmark, and A. Hobro, "Laser-based standoff detection of explosives: a critical review," *Analytical and Bioanalytical Chemistry*, vol. 395, no. 2, pp. 259–274, 2009.

[13] A. Pettersson, I. Johansson, S. Wallin, M. Nordberg, and H. Östmark, "Near real-time standoff detection of explosives in a realistic outdoor environment at 55 m distance," *Propellants, Explosives, Pyrotechnics*, vol. 34, no. 4, pp. 297–306, 2009.

[14] M. D. Ray, A. J. Sedlacek, and M. Wu, "Ultraviolet mini-Raman lidar for stand-off, in situ identification of chemical surface contaminants," *Review of Scientific Instruments*, vol. 71, no. 9, pp. 3485–3489, 2000.

[15] M. Wu, M. Ray, K. H. Fung, M. W. Ruckman, D. Harder, and A. J. Sedlacek, "Stand-off detection of chemicals by UV Raman spectroscopy," *Applied Spectroscopy*, vol. 54, no. 6, pp. 800–806, 2000.

[16] B. Zachhuber, G. Ramer, A. Hobro, E. T. H. Chrysostom, and B. Lendl, "Stand-off Raman spectroscopy: a powerful technique for qualitative and quantitative analysis of inorganic and organic compounds including explosives," *Analytical and Bioanalytical Chemistry*, vol. 400, no. 8, pp. 2439–2447, 2011.

[17] B. Zachhuber, C. Gasser, E. T. H. Chrysostom, and B. Lendl, "Stand-off Raman spectroscopy: a powerful stand-off spatial offset Raman spectroscopy for the detection of concealed content in distant objects," *Analytical Chemistry*, vol. 83, no. 24, pp. 9438–9442, 2011.

[18] E. D. Emmons, A. Tripathi, J. A. Guicheteau, S. D. Christesen, and A. W. Fountain, "Raman chemical imaging of explosive-contaminated fingerprints," *Applied spectroscopy*, vol. 63, no. 11, pp. 1197–1203, 2009.

[19] B. Schrader, A. Hoffmann, A. Simon, and J. Sawatzki, "Can a Raman renaissance be expected via the near-infrared Fourier transform technique?" *Vibrational Spectroscopy*, vol. 1, no. 3, pp. 239–250, 1991.

[20] D. B. Chase, "Fourier transform Raman spectroscopy," *Journal of the American Chemical Society*, vol. 108, no. 24, pp. 7485–7488, 1986.

[21] D. A. Long, *Raman Spectroscopy*, McGraw-Hill, New York, NY, USA, 1977.

[22] S. D. Williams, T. J. Johnson, T. P. Gibbons, and C. L. Kitchens, "Relative Raman intensities in C_6H_6, C_6D_6, and C_6F_6: a comparison of different computational methods," *Theoretical Chemistry Accounts*, vol. 117, no. 2, pp. 283–290, 2007.

[23] T. A. Blake, J. F. Kelly, N. B. Gallagher, P. L. Gassman, and T. J. Johnson, "Passive standoff detection of RDX residues on metal surfaces via infrared hyperspectral imaging," *Analytical and Bioanalytical Chemistry*, vol. 395, no. 2, pp. 337–348, 2009.

[24] T. Okuno, "Thermal effect of visible light and infra-red radiation (IR-A, IR-B AND IR-C) on the eye—a study of infra-red cataract based on a model," *Annals of Occupational Hygiene*, vol. 38, no. 4, pp. 351–359, 1994.

[25] S. J. Choquette, E. S. Etz, W. S. Hurst, D. H. Blackburn, and S. D. Leigh, "Relative intensity correction of Raman spectrometers: NIST SRMs 2241 through 2243 for 785 nm, 532 nm, and 488 nm/514.5 nm excitation," *Applied Spectroscopy*, vol. 61, no. 2, pp. 117–129, 2007.

[26] M. Fryling, C. J. Frank, and R. L. McCreery, "Intensity calibration and sensitivity comparisons for CCD/Raman spectrometers," *Applied Spectroscopy* , vol. 47, pp. 1965–1974, 1993.

[27] K. G. Ray and R. L. McCreery, "Simplified calibration of instrument response function for Raman spectrometers based on luminescent intensity standards," *Applied Spectroscopy*, vol. 51, no. 1, pp. 108–116, 1997.

[28] R. Martini, C. Glazowski, E. A. Whittaker et al., "Optical free-space communications at middle infra-red wavelengths," in *Quantum Sensing and Nanophotonic Devices*, M. Razeghi and G. J. Brown, Eds., vol. 5359 of *Proceedings of SPIE*, pp. 196–202, San Jose, Calif, USA, January 2004.

[29] A. W. Fountain III, C. K. Mann, and T. J. Vickers, "Routine wavenumber calibration of an FT-Raman spectrometer," *Applied Spectroscopy*, vol. 49, no. 7, pp. 1048–1053, 1995.

[30] W. S. Hurst, S. J. Choquette, and E. S. Etz, "Requirements for relative intensity correction of Raman spectra obtained by column-summing charge-coupled device data," *Applied Spectroscopy*, vol. 61, no. 7, pp. 694–700, 2007.

[31] P. Vandenabeele, "Evaluation of a spectral searching algorithm for the comparison of Raman band positions," *Spectrochimica Acta A*, vol. 80, no. 1, pp. 27–31, 2011.

[32] R. J. Bell, *Introductory Fourier Transform Spectroscopy*, Academic Press, New York, NY, USA, 1972.

[33] S. W. Sharpe, R. L. Sams, T. J. Johnson, P. M. Chu, G. C. Rhoderick, and F. R. Guenther, "Creation of 0.10 cm^{-1} resolution, quantitative, infrared spectral libraries for gas samples," in *Vibrational Spectroscopy-based Sensor Systems*, vol. 4577 of *Proceedings of SPIE*, pp. 12–24, Boston, Mass, USA, 2002.

[34] S. W. Sharpe, T. J. Johnson, R. L. Sams, P. M. Chu, G. C. Rhoderick, and P. A. Johnson, "Gas-phase databases for quantitative infrared spectroscopy," *Applied Spectroscopy*, vol. 58, no. 12, pp. 1452–1461, 2004.

[35] J. E. Bertie, "Specification of components, methods, and parameters in Fourier transform spectroscopy by Michelson and related interferometers: IUPAC recommendations," *Pure and Applied Chemistry*, vol. 70, no. 10, pp. 2039–2045, 1998.

[36] B. Schrader, A. Hoffmann, A. Simon, R. Podschadlowski, and M. Tischer, "NIR-FT-Raman-spectroscopy, state of the art," *Journal of Molecular Structure*, vol. 217, pp. 207–220, 1990.

[37] T. J. Johnson, R. L. Sams, T. A. Blake, S. W. Sharpe, and P. M. Chu, "Removing aperture-induced artifacts from Fourier transform infrared intensity values," *Applied Optics*, vol. 41, no. 15, pp. 2831–2839, 2002.

[38] D. Bonen, T. J. Johnson, and S. L. Sarkar, "Characterization of principal clinker minerals by FT-Raman microspectroscopy," *Cement and Concrete Research*, vol. 24, no. 5, pp. 959–965, 1994.

[39] J. W. C. Johns, "High resolution and the accurate measurement of intensities," *Mikrochimica Acta*, vol. 93, no. 1–6, pp. 171–188, 1987.

[40] A. G. Maki and J. S. Wells, "Wavenumber calibration tables from heterodyne frequency measurements," National Institute of Standards and Technology, Special Publication 821, U.S. Government Printing Office, Washington, DC, 1991, http://physics.nist.gov/PhysRefData.

[41] D. Hutsebaut, P. Vandenabeele, and L. Moens, "Evaluation of an accurate calibration and spectral standardization procedure for Raman spectroscopy," *Analyst*, vol. 130, no. 8, pp. 1204–1214, 2005.

[42] A. W. Fountain III, T. J. Vickers, and C. K. Mann, "Factors that affect the accuracy of Raman shift measurements on multichannel spectrometers," *Applied Spectroscopy*, vol. 52, no. 3, pp. 462–468, 1998.

[43] M. Fryling and C. J. Frank, "Intensity calibration and sensitivity comparisons for CCD/Raman spectrometers,"*Applied Spectroscopy*, vol. 47, no. 12, pp. 1965–1974, 1993.

[44] "Standard Guide for Testing the Resolution of a Raman Spectrometer," ASTM E2529-06, ASTM Publication, www.astm .org/.

[45] J. B. Forrester, N. B. Valentine, Y.-F. Su, and T. J. Johnson, "Chemometric analysis of multiple species of Bacillus bacterial endospores using infrared spectroscopy: discrimination to the strain level," *Analytica Chimica Acta*, vol. 651, no. 1, pp. 24–30, 2009.

[46] B. M. Kunkel, Y.-F Su, and R. G. Tonkyn, "Raman database considerations for near-infrared systems," in *Optics and Photonics for Counterterrorism and Crime Fighting VII*, vol. 8189, article 818905 of *Proceedings of SPIE*, Prague, Czech Republic, September 2011.

[47] I. Geiman, M. Leona, and J. R. Lombardi, "Application of Raman spectroscopy and surface-enhanced Raman scattering to the analysis of synthetic dyes found in ballpoint pen inks," *Journal of Forensic Sciences*, vol. 54, no. 4, pp. 947–952, 2009.

[48] M. L. Lewis, I. R. Lewis, and P. R. Griffiths, "Anti-Stokes Raman spectrometry with 1064-nm excitation: an effective instrumental approach for field detection of explosives," *Applied Spectroscopy*, vol. 58, no. 4, pp. 420–427, 2004.

[49] N. W. Daniel Jr., I. R. Lewis, and P. R. Griffiths, "Interpretation of Raman spectra of nitro-containing explosive materials. Part II: the implementation of neural, fuzzy, and statistical models for unsupervised pattern recognition," *Applied Spectroscopy*, vol. 51, no. 12, pp. 1868–1879, 1997.

[50] D. Patel, C. Vance, N. King, M. Jessup, and S. Sarkisov, "Preparation, fluorescence spectroscopy and AFM analysis of erbium oxide nanocolloid," in *Optical Components and Materials VI*, vol. 7212, article 72121F of *Proceedings of SPIE*, 2009.

[51] S. Heer, K. Kömpe, H.-U. Güdel, and M. Haase, "Highly efficient multicolour upconversion emission in transparent colloids of lanthanide-doped $NaYF_4$ nanocrystals," *Advanced Materials*, vol. 16, no. 23-24, pp. 2102–2105, 2004.

Recent Applications of Ion Mobility Spectrometry in Diagnosis of Vaginal Infections

Zeev Karpas, Gabi Cohen, Elias Atweh, Geoff Barnard, and Moshe Golan

3QBD Ltd., 8 Haplada Street, Arad 89106, Israel

Correspondence should be addressed to Zeev Karpas, karpas4@netvision.net.il

Academic Editor: Dafang Zhong

Vaginal infections (vaginosis) globally affect more than 15% of the female population of reproductive age. However, diagnosis of vaginosis and differentiating between the three common types: bacterial vaginosis (BV), vulvovaginal candidiasis (VVC), and trichomoniasis are challenging. Elevated levels of the biogenic amines, trimethylamine (TMA), putrescine, and cadaverine have been found in vaginal discharge fluid of women with vaginosis. Ion mobility spectrometry (IMS) is particularly suitable for measurement of amines even in complex biological matrices due to their high proton affinity and has been shown to be suitable for the diagnosis of vaginal infections. Recent developments that have increased the accuracy of the technique for diagnosis of BV and simplified sample introduction are described here.

1. Introduction

There are three common types of vaginal infections (vaginosis): bacterial vaginosis (BV), vulvovaginal candidiasis (VVC or yeast infection), and trichomoniasis that affect women worldwide [1]. According to many studies the prevalence of these infections varies considerably from country to country, and between different populations within each country but according to a conservative estimate more than 15% of women of reproductive age are affected by one or more of these infections at any given time [2]. Symptoms may include copious discharge of vaginal fluid, fishy odor, pain, itching, or burning. However, in many cases the symptoms are not noticed, that is, an asymptomatic infection [1]. The cause of BV is generally attributed to an overgrowth of anaerobic microorganisms and it has been associated with several types like Gardnerella, Mobiluncus, Bacteroides, Mycoplasma hominis, Ureaplasma urealyticum, and so forth. These microorganisms may normally be present in vaginal discharge fluid, but their growth is inhibited by Lactobacilli that maintain a low pH and produce hydrogen peroxide. Vulvovaginal candidiasis is caused by some types of candida (fungi) while trichomoniasis is caused by trichomonads that are parasites. When the balance between the pathological microorganisms and the Lactobacilli is disturbed a vaginal infection may erupt. This could be due to the use of antibiotic medication that deleteriously affects the Lactobacilli and thus allows the fungi to thrive leading to a yeast infection. An infection can also result from practices like frequent douching or unhygienic habits, allergy, or even sensitivity to semen.

Correctly diagnosing the vaginal infection could be complicated due to the similarity of symptoms of the three common infections or when women are asymptomatic. Several laboratory methods have been proposed to overcome this problem. The Amsel test includes a visual inspection of the discharge, measuring its pH, a "whiff test" performed by adding a drop of alkaline solution, and a "wet mount" microscopic examination [3]. The Gram stain method involves staining the vaginal discharge sample with a dye and using a microscope to quantify three morphotypes (Lactobacilli, Gram variable rods and curved rods) and analyzing the results according to the Nugent score [4–6]. Cultures may be used to diagnose candidiasis and trichomoniasis but they are not effective for determination of bacterial vaginosis as some anaerobic microorganisms are normally present in the discharge fluid. Other methods involve Antigen-detecting immunoassays, the OSOM Trichomonas Rapid Test, DNA probes, and polymerase chain reaction (PCR). These are useful for detecting trichomonads as concisely summarized recently [1, 7]. However, each of these methods may either have a low accuracy (low specificity or sensitivity), high cost, long response time or place demands of time, and skill on the physician.

The need for a rapid, reliable, inexpensive diagnostic test that can differentiate between BV and VVC has been delineated [8]. The fact that biogenic amines are produced by the action of the microorganisms has been well known and has served as the basis for several commercial kits that are used to diagnose these vaginal infections [9–11]. However, a simple analytical procedure that can differentiate between the biogenic amines and thus serve to accurately diagnose the vaginal infections and determine BV was only reported less than a decade ago [12, 13].

Classical, or linear, ion mobility spectrometry (IMS) is an analytical technique used for separation of ions on the basis of their mobility, or drift velocity, under the influence of a weak electric field ($E/N < 2Td$). The main factors that affect the drift velocity, in addition to the mass of the ion, are the composition, pressure and temperature of the gas through which the ion drifts, and the magnitude of the electric field, as explained in detail previously [14]. These parameters are accounted for in calculation of the reduced mobility of the ion. The ion shape, size, and internal charge distribution also affect the drift velocity and are important mainly for differentiating between isomers and isobaric ions [14]. In most practical applications of IMS, as in the present study, the drift gas is air at ambient pressure. The fact that amines in general, and biogenic amines in particular, have high proton affinities [15] and are thus readily ionized through atmospheric pressure chemical ionization (APCI) conditions enables their sensitive and specific detection even in a complex biological sample like vaginal discharge fluid or food products [16–19]. The vapors of volatile biogenic amines like trimethylamine (TMA) or semi-volatile diamines like putrescine (PUT) and cadaverine (CAD) can be measured directly as they emanate from the sample. The emission of these vapors from biological samples can be enhanced chemically by application of an alkaline solution and physically by heating the sample, as shown previously [12, 13]. Two schematic representations of the structure of the three biogenic amines and triethylphosphate (TEP) are shown in Figure 1. On the left, the space-filling model is shown (blue-N atoms, black-C atoms, gray-H atoms, red-O atoms, and orange-P atoms). The schematics on the right show important features of the amines regarding the protonation site of the ions formed: the nitrogen lone-pair of electrons shown for TMA and a proton-bridged cyclic structure for the diamines where the proton is between the two amine groups [20].

Since the first publication of the method for diagnosing vaginal infections by measurement of biogenic amines with IMS [12, 13] there have been several developments that have helped increase the reliability and accuracy of the method, as well as making it simpler to perform. These will be described and discussed in the following article.

2. Experimental

Commercially available chemicals were used including Putrescine (1,4 Diamino-butane, Aldrich 99%), Cadaverine (1,5 Diaminopentane, Sigma > 97%), and trimethylamine (50% aqueous solution, Acros, New Jersey). Triethylphosphate (TEP) (Aldrich 99.8%) was used as the dopant. and 2, 4-Lutidine (Acros 99%) was used to calibrate the reduced mobility scale. Reduced mobility values were calculated relative to 2,4-lutidine (1.90 $cm^2 V^{-1} sec^{-1}$ at 90°C) [21].

A linear ion mobility spectrometer with a corona discharge ionization source (VG-Test, 3QBD, Arad, Israel) was operated at 90°C as described in detail previously [22].

Ambient air, purified by passing over an activated charcoal filter, was used as the drift gas and carrier gas. The drift cell consisted of a stack of metal rings separated from one another by Teflon rings in order to form a linear electric field gradient of 270 V cm^{-1}. The dopant (TEP) was introduced continuously into the ionization region from a homemade permeation tube emanating vapors at a rate of about 1 μg min^{-1}. The samples were dripped onto a cotton swab (50 μL by calibrated pipette), and one drop of a 15% KOH alkaline solution was added. The swab was then placed in a sample holder that was inserted into the sample block of the IMS where the analysis was triggered by a microswitch. For determination of volatile TMA a halogen lamp was turned on for the duration of the measurement (25 secs). For determination of the semivolatile biogenic amines, putrescine, and cadaverine, a longer measurement time was used in order to enhance their emanation. The mobility spectrum was measured at a rate of 80 kHz, and every second the averaged spectrum is recorded and stored. The nominal resolution was 25, as calculated by dividing the drift time of the ion by the full width at half maximum (FWHM).

In some cases, particularly when samples are sent to a central laboratory for diagnosis, the swab with the sample of vaginal discharge fluid was stored in a gel test tube (Amies gel, Copan, Italy). Extended storage may lead to migration of biogenic amines from the swab to the gel, and this has been tested with samples of TMA that were deposited on a swab that was placed in the gel and stored for different periods at room temperature, refrigerator, and deep freeze (25°C, 4°C and −18°C, resp.).

3. Results and Discussion

3.1. Calibration. Needless to say that for ethical and practical reasons calibration could not be done with real vaginal discharge fluid so alternative approaches were deployed. First a dilute aqueous solution of TMA was used, and calibration of the system was carried out by adding a fixed volume (50 μL) of an aqueous solution containing between 125 and 2000 ng of TMA to a cotton swab and measuring the mobility spectrum as described above. The intrarun reproducibility was 15% which is quite typical for measurement of volatile samples by IMS. The interrun reproducibility was 30% so that by performing a daily calibration procedure a clear

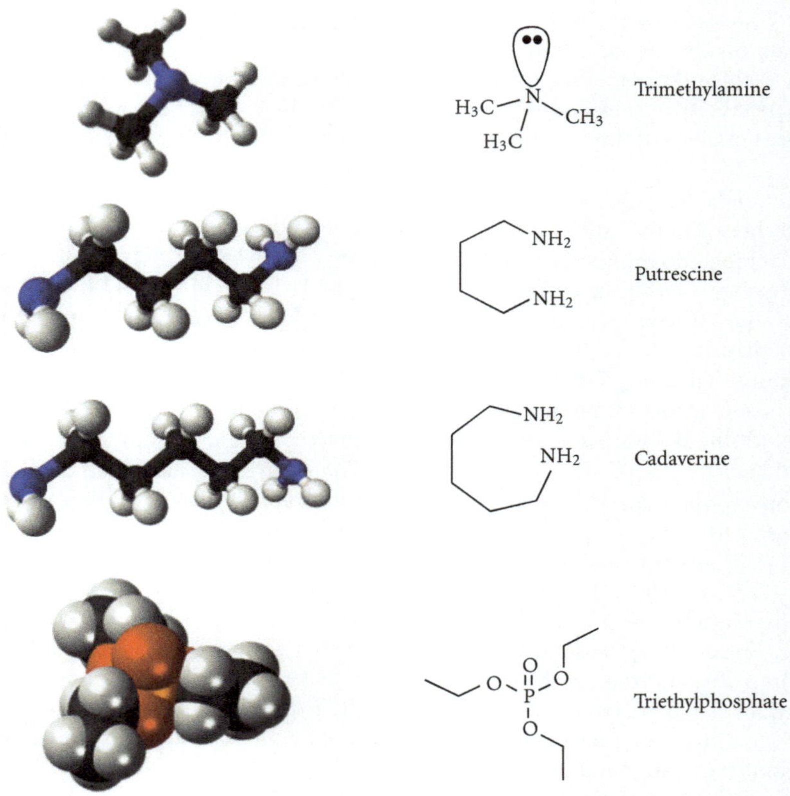

FIGURE 1: The structure of trimethylamine (TMA), putrescine, cadaverine and triethylphosphate (TEP). Note the lone pair of electrons on the nitrogen atom of TMA and the cyclic representation of the structure of putrescine and cadaverine.

distinction was found between samples collected from BV-negative and BV-positive patients.

One of the main features of IMS is that when product ions are formed the population of the reactant ions is depleted. The use of the ratio TMA/(TMA + TEP) somewhat provides a correction for this and also allows all the quantitative results to be displayed conveniently on an axis between zero and one. In the present study the level of the other biogenic amines (putrescine and cadaverine) was low, so they had little influence on the BV diagnosis. In cases where their level is significant the approach is to calculate and display the ratio between each amine and the total intensity of the ions, for example, TMA/(TMA + PUT + CAD + TEP).

Figure 2(a) shows the calibration curve obtained by measuring the ratio of the areas of the product ion peak (TMA) divided by the sum of the product ion and reactant ion (TEP) as a function of the amount of TMA placed on the swab. Each point represents the average of three measurements. The second approach was adopted in order to better simulate a biological sample by the use of the solution in canned tuna (in water not in oil) that contained about 0.4 mg mL^{-1} of TMA [23]. The tuna water was diluted so that the actual TMA concentrations were equivalent to the concentrations used in aqueous solution calibration curve. Figure 2(b) depicts the calibration curve thus obtained for TMA in water from canned tuna.

The calibration curves in Figures 2(a) and 2(b) are typical of the quantitative response of IMS instruments, that is, the response approaches saturation at high concentrations or large amounts of the analytes. In this case, after about 2500 ng of TMA the calibration curves start to level out. This happens when the intensity ratio of the analyte TMA/(TMA + TEP) reaches about 0.6 and is indicative of depletion of the TEP reactant ions.

3.2. New Instrumental and Sample Introduction Techniques. The VG-Test differs in several aspects from the PT-IMS (Rotem Industries, Israel) that was used in the early tests for diagnosing vaginal infections [12, 13]. The radioactive ^{63}Ni source that was used in the PT-IMS has been replaced by a corona discharge ionization source. The operating temperature of the drift tube was reduced from 130°C in the PT-IMS to 90°C in the VG-Test without deleterious consequences. The pump that was used to generate the flow of the carrier and drift gas streams in the PT-IMS was reinforced by a second pump that shortened clearance times of the system and reduced "memory effects". Nonylamine that was the dopant in the PT-IMS that formed the reactant ion was replaced by triethylphosphate (TEP) that is safer and easier to handle as it is not degraded by reaction with atmospheric moisture. However, the most radical change was in the sample introduction technique: in the PT-IMS the

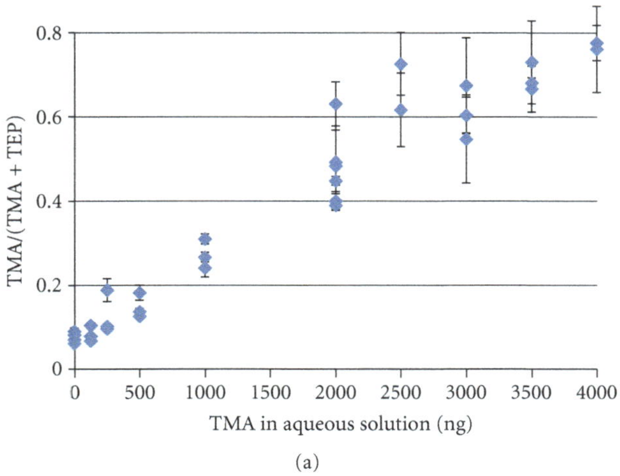

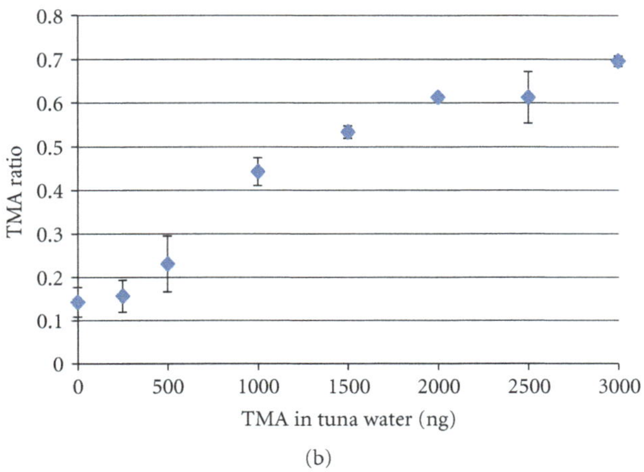

Figure 2: (a) The ratio TMA/(TMA + TEP) as a function of the amount of TMA in aqueous solution that was deposited on a cotton swab. (b) The ratio TMA/(TMA + TEP) as a function of the amount of TMA in water from canned tuna that was deposited on a cotton swab.

swab with the sample of vaginal discharge was stirred in a vial with distilled water, followed by addition of alkaline solution, and then the vial was connected to the carrier flow stream. For enhanced emanation of the semivolatile amines the vial had to be placed in a beaker with hot water. The inlet system of the VG-Test simplified the sample introduction considerably: the swab was placed in a module (housing), and a drop of alkaline solution was added directly to the swab followed by insertion of the housing into the inlet where heating by a halogen lamp started automatically according to the analytical procedure that was selected. This also diminished the "memory effects". Typical mobility spectra obtained for a blank swab and a swab upon which 1 μg of TMA for calibration was deposited are shown in Figure 3. Note the TEP peak at a drift time of 7.2 ms and the TMA peak at 5.9 ms.

3.3. Sensitivity and Limit of Detection. The limit of detection (LOD) and limit of quantification (LOQ) of the method were calculated from the uncertainty in the spectrum of

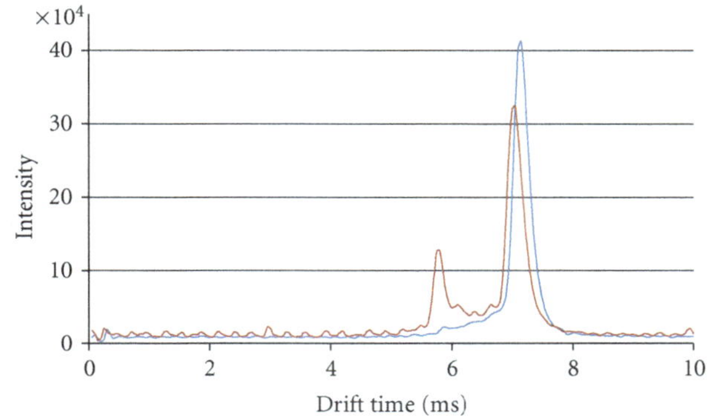

Figure 3: Typical mobility spectra of a blank swab (blue trace) and a swab upon which 1 μg of TMA was deposited for calibration (red trace). Note the TEP peak at a drift time of 7.2 ms and the TMA peak at 5.9 ms.

blank samples. The uncertainty was 10.3 mV, and based on the sensitivity calibration (0.86 mV ng^{-1}) the LOD and LOQ were calculated as 39 and 120 ng, respectively. Thus, the linear range for quantification of TMA in aqueous solution or in tuna water (that represents a biological matrix) is between 120 to 2500 ng of TMA deposited on cotton swab. This was certainly sufficiently sensitive to readily distinguish between samples collected from a patient with a BV infection where the TMA level in a sample of vaginal discharge fluid exceeded 2000 ng and a sample collected from a healthy woman where the TMA level was typically well below 1000 ng. The mobility spectra shown in Figure 4 were obtained from actual swabs with vaginal discharge fluid collected from a patient without a vaginal infection (blue trace) and a patient with bacterial vaginosis (red trace).

3.4. Effect of Storage in Gel. As expected, the storage temperature strongly affects the rate of TMA migration from the swab into the gel. After three days at room temperature (25°C) about 90% of the TMA was lost from the swab but could be detected in the gel. A similar period in a refrigerator (4°C) led to loss of about 84% of the TMA while storage in a deep freeze (−18°C) resulted in loss of about 50% of the TMA originally deposited on the swab. Figure 5 shows the calculated peak area of the TMA retained on the swab after deposition of 1 μg of TMA as a function of the storage time in a refrigerator. Initially, after one day about 75% of the deposited TMA migrated from the swab to the gel, but after one week the amount of TMA retained on the swab remained almost constant.

Therefore, a calibration curve for swabs stored in gel would differ significantly from that obtained from a fresh swab, and corrections must be made in consideration of the storage time and temperature. In case the swab contains elevated levels of TMA then a BV-positive diagnosis is unmistakable. However, in order to confirm a BV-negative diagnosis, particularly for swabs that have been stored for several days in gel, the TMA level in the gel should also be tested.

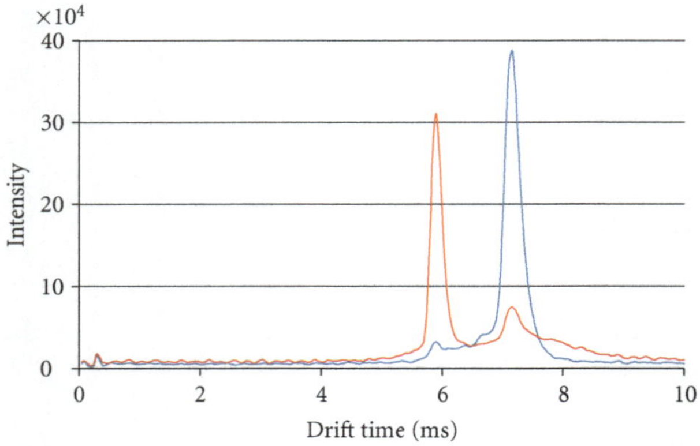

FIGURE 4: Typical mobility spectra of a swab collected from a patient without a vaginal infection (blue trace) and a swab from a patient with BV (red trace).

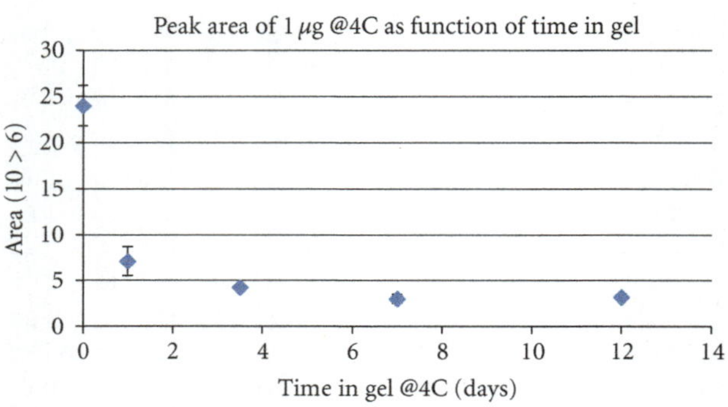

FIGURE 5: The calculated peak area of the TMA retained on the swab after deposition of 1 μg of TMA as a function of the storage time in a refrigerator.

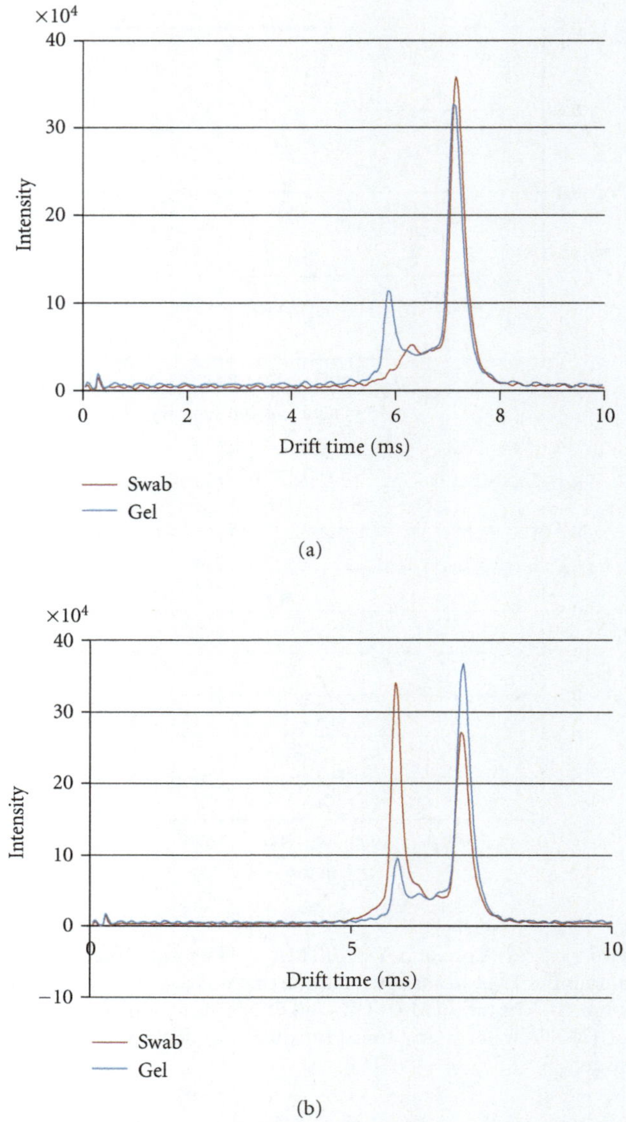

FIGURE 6: The signal intensity as a function of drift time for the swab (red trace) and gel (blue trace) of a vaginal discharge fluid that was stored in gel. (a) After storage in gel for one week; (b) after storage in gel for less than two days.

3.5. Preliminary Tests of Vaginal Samples Stored in Gel. Samples that were stored in gel were diagnosed according to the Nugent score either as positive for BV or as negative for BV and were subsequently analyzed by the VG-Test. Of 18 samples that had Nugent scores of 7–10 (i.e., BV-positive samples) 15 were observed to have elevated TMA levels, two had slightly elevated levels of cadaverine, and one was stored for over a week which probably led to migration of almost all the TMA from the swab to the gel. Of 21 samples that were with Nugent scores of 1–3 (i.e., BV-negative samples) according to the VG-Test analysis 19 had little or no TMA and the remaining two had elevated cadaverine levels indicating a vaginal infection but not BV. As mentioned above the calibration for a swab stored in gel differs from a fresh swab, but nevertheless the accuracy of the BV diagnosis was impressive, especially as other biogenic amines could be seen in the few samples where there was a disagreement between the laboratory diagnosis and IMS analysis.

Figure 6 shows the mobility spectra recorded from a swab with vaginal discharge fluid and from a fraction of the gel in which the swab was stored. Figure 6(a) depicts the mobility spectra of the swab and a sample of the gel after storage in the gel of one week while Figure 6(b) shows the spectra after storage of less than two days. Evidently, the spectra show that the long storage time resulted in migration of the TMA from the swab to the gel, while after a short storage period only a small fraction of TMA was transferred to the gel.

These studies made it possible to extend the diagnosis of bacterial vaginosis that had previously been demonstrated for fresh swabs of vaginal discharge fluid [12, 13] to the diagnosis of vaginitis in swabs that have been stored in gel. Thus, the method can be of use to central laboratories that use this storage method to ship the samples from the clinic to the laboratory.

4. Summary

Recent developments in the application of ion mobility spectrometry (IMS) for diagnosing vaginal infections, or specifically bacterial vaginosis (BV), were described. The main features are improvements in the sample introduction methodology that make the system more user friendly, replacement of the radioactive ionization source by a corona discharge source, and changing the n-nonylamine dopant by triethylphosphate (TEP) that is more stable and easy to use. In addition, calibration techniques have been improved, and the use of water from canned tuna chunks as a biological matrix provides a better simulant for body fluids. The linear range is typical of the response of IMS instruments spanning a range of 120 ng to 2500 ng, but is suitable for diagnosis of BV on the basis of the TMA level in a sample vaginal discharge fluid. Finally, the study of the effects of storage in gel of vaginal discharge fluid samples makes it possible to extend the diagnostic method to samples that are tested in central laboratories.

References

[1] STD Facts—Bacterial Vaginosis, http://www.cdc.gov/std/bv/.

[2] E. H. Koumans, M. Sternberg, C. Bruce et al., "The prevalence of bacterial vaginosis in the United States, 2001–2004; associations with symptoms, sexual behaviors, and reproductive health," *Sexually Transmitted Diseases*, vol. 34, no. 11, pp. 864–869, 2007.

[3] R. Amsel, P. A. Totten, and C. A. Spiegel, "Nonspecific vaginitis. Diagnostic criteria and microbial and epidemiologic associations," *American Journal of Medicine*, vol. 74, no. 1, pp. 14–22, 1983.

[4] R. P. Nugent, M. A. Krohn, and S. L. Hillier, "Reliability of diagnosing bacterial vaginosis is improved by a standardized method of gram stain interpretation," *Journal of Clinical Microbiology*, vol. 29, no. 2, pp. 297–301, 1991.

[5] K. Chaijareenont, K. Sirimai, D. Boriboonhirunsarn, and O. Kiriwat, "Accuracy of Nugent's score and each Amsel's criteria in the diagnosis of bacterial vaginosis," *Journal of the Medical Association of Thailand*, vol. 87, no. 11, pp. 1270–1274, 2004.

[6] E. D. Hapsari, M. Hayashi, and H. Matsuo, "Clinical characteristics of vaginal discharge in bacterial vaginosis diagnosed by Nugent's criteria," *Clinical and Experimental Obstetrics and Gynecology*, vol. 33, no. 1, pp. 5–9, 2006.

[7] http://emedicine.medscape.com/article/257141-workup/.

[8] J. D. Sobel and P. Hay, "Diagnostic techniques for bacterial vaginosis and vulvovaginal candidiasis requirement for a simple differential test," *Expert Opinion on Medical Diagnostics*, vol. 4, no. 4, pp. 333–341, 2010.

[9] V. K. Hogan, J. F. Culhane, J. Hitti, V. A. Rauh, K. F. McCollum, and K. J. Agnew, "Relative performance of three methods for diagnosing bacterial vaginosis during pregnancy," *Maternal and Child Health Journal*, vol. 11, no. 6, pp. 532–539, 2007.

[10] N. C. Kampan, S. S. Suffian, N. S. Ithnin, M. Muhammad, S. Z. S. Zakaria, and M. A. Jamil, "Evaluation of BV blue test kit for the diagnosis of bacterial vaginosis," *Sexual and Reproductive Healthcare*, vol. 2, no. 1, pp. 1–5, 2011.

[11] H. Wolrath, U. Forsum, P. G. Larsson, and H. Borén, "Analysis of bacterial vaginosis-related amines in vaginal fluid by gas chromatography and mass spectrometry," *Journal of Clinical Microbiology*, vol. 39, no. 11, pp. 4026–4031, 2001.

[12] Z. Karpas, W. Chaim, R. Gdalevsky, B. Tilman, and A. Lorber, "Novel application for ion mobility spectrometry: diagnosing vaginal infections through measurement of biogenic amines," *Analytica Chimica Acta*, vol. 474, no. 1-2, pp. 115–123, 2002.

[13] W. Chaim, Z. Karpas, and A. Lorber, "New technology for diagnosis of bacterial vaginosis," *European Journal of Obstetrics Gynecology and Reproductive Biology*, vol. 111, no. 1, pp. 83–87, 2003.

[14] G. A. Eiceman and Z. Karpas, *Ion Mobility Spectrometry—Second Edition*, CRC Press, Boca Raton, Fla, USA, 2005.

[15] S. G. Lias, J. F. Liebman, and R. D. Levin, "Evaluated gas phase basicities and proton affinities of molecules; heats of formation of protonated molecules," *Journal of Physical and Chemical Reference Data*, vol. 13, pp. 695–808, 1984.

[16] G. M. Bota and P. B. Harrington, "Direct detection of trimethylamine in meat food products using ion mobility spectrometry," *Talanta*, vol. 68, no. 3, pp. 629–635, 2006.

[17] Z. Karpas, B. Tilman, R. Gdalevsky, and A. Lorber, "Determination of volatile biogenic amines in muscle food products by ion mobility spectrometry," *Analytica Chimica Acta*, vol. 463, no. 2, pp. 155–163, 2002.

[18] Z. Hashemian, A. Mardihallaj, and T. Khayamian, "Analysis of biogenic amines using corona discharge ion mobility spectrometry," *Talanta*, vol. 81, no. 3, pp. 1081–1087, 2010.

[19] M. A. Awan, I. Fleet, and C. L.P. Thomas, "Optimising cell temperature and dispersion field strength for the screening for putrescine and cadaverine with thermal desorption-gas chromatography-differential mobility spectrometry," *Analytica Chimica Acta*, vol. 611, no. 2, pp. 226–232, 2008.

[20] Z. Karpas, "Evidence of proton-induced cyclization of α,ω-diamines from ion mobility measurements," *International Journal of Mass Spectrometry and Ion Processes*, vol. 93, no. 2, pp. 237–242, 1989.

[21] G. A. Eiceman, E. G. Nazarov, and J. A. Stone, "Chemical standards in ion mobility spectrometry," *Analytica Chimica Acta*, vol. 493, no. 2, pp. 185–194, 2003.

[22] Z. Karpas, O. Litvin, G. Cohen, J. Mishin, E. Atweh, and A. Burlakov, "The reduced mobility of the biogenic amines: trimethylamine, putrescine, cadaverine, spermidine and spermine," *International Journal for Ion Mobility Spectrometry*, vol. 14, no. 1, pp. 3–6, 2011.

[23] G. Barnard, G. Cohen, E. Atweh, and Z. Karpas, "Measurement of biogenic amines in biological fluids by ion mobility spectrometry," *International Journal for Ion Mobility Spectrometry*, vol. 14, pp. 207–211, 2011.

16

A Review of Optical Nondestructive Visual and Near-Infrared Methods for Food Quality and Safety

Jarmo T. Alander,[1] Vladimir Bochko,[1] Birgitta Martinkauppi,[1] Sirinnapa Saranwong,[2] and Timo Mantere[1]

[1] Department of Electrical and Energy Engineering, University of Vaasa, 65101 Vaasa, Finland
[2] Bruker Optics KK, 1-4-1 Shinkawa, Chuo-ku, Tokyo 104-0033, Japan

Correspondence should be addressed to Jarmo T. Alander; jal@uwasa.fi

Academic Editor: Jaan Laane

This paper is a review of optical methods for online nondestructive food quality monitoring. The key spectral areas are the visual and near-infrared wavelengths. We have collected the information of over 260 papers published mainly during the last 20 years. Many of them use an analysis method called chemometrics which is shortly described in the paper. The main goal of this paper is to provide a general view of work done according to different FAO food classes. Hopefully using optical VIS/NIR spectroscopy gives an idea of how to better meet market and consumer needs for high-quality food stuff.

1. Introduction

Consumers are the driving force in the food market. They have become more health conscious, demanding and willing to pay for the "good quality." The consumers' trust for food industry has been diminished due to food scandals thus making it important to improve the safety monitoring.

The "quality" and "safety" have different meanings and aspects which depend on the food class, target market, criteria, and application. In some sense, the safety is part of the quality but here we separate these two because we like to emphasize the difference between how a human perceive the food and how we can evaluate threats of health. Some unsafe food cannot be detected by manual inspection. The food quality and safety evaluation has become more important and the need for more comprehensive assessment for all food batches is adequate.

Even though a lot of research and development work has been done on food safety and quality, more needs to be done to find economic ways of monitoring food safety. In this paper, our purpose is to go through the subject according to Food and Agriculture Organizations of the United Nations (FAO) categories (http://www.fao.org/) and publications related to quality and safety. These two factors can be evaluated in many different methods and techniques.

The traditional methods are sensory evaluation, chemical analysis, and microbiological analysis. A more recent approach in technical sense is optical techniques, especially the multispectral methods in visual and near-infrared wavelengths. The optical methods can be used for nondestructive, fast, real-time, online monitoring of all samples (Huang et al. [1]). Table 1 gives a general comparison of these four approaches.

The sensory evaluation is the oldest method and still used every day. The evaluation is done by both consumers and professional food observers. It is relatively fast but hardly suitable for a large amount of samples due to many reasons like observer's fatiguing. The reliability varies with food item group and by person. The results are typically very subjective, depending on many non-food-related factors, and difficult to evaluate quantitatively.

The chemical analysis is an objective technique which often provides reliable results if the monitored batch is represented well by the samples measured. It has some

TABLE 1: Comparison of method properties.

Method	Subjective/objective	Sample/all
Sensory	Subjective	All
Chemical	Objective	Sample
Microbiological	Objective	Sample
NIR	Objective	All

limitations—the processes may not work correctly due to some substances, the need of accurate and correct calibrations, the need of complex multiphased processing, and the need of experts for the analysis. It might not be suitable if the distribution of the substance content is nonuniform. It cannot provide a point measurement of the sample, only an average over the whole sample. Since the sample is destroyed, chemical analysis cannot be used for monitoring all food units. In addition, the price of the analysis per measurement may be quite high and it produces toxic waste; thus, the number of samples for monitoring must be limited due to economic and ecological reasons. It is generally not very well suited for fast, online product management.

The microbiological analysis is usually applied to detect biological organism like fungi or bacteria but not so much for identifying other content. It gives often objective and reliable results provided that, for example, the culture do grow, the growth is detectable, or the sample is not contaminated by some other sources. The analysis is typically slow and more tedious and expensive than the previous two methods. The microbiological analysis is suitable only for low speed, offline monitoring using a representative sample set of the batch. The main problem is that it allows defective products to be manufactured even though the cheapest solution might be to immediately stop processing and clean up the manufacturing unit.

To overcome the disadvantages and limitations of the previous methods, new economic, fast, and environment friendly techniques are sought after. As a result, optical multispectral measurements and imaging have recently become more and more popular in food measurements because they lend themselves easily to online monitoring of all samples even without touching the samples. The price of the measurement instruments has also decreased so they provide a viable option for the traditional methods. However, analysis of spectral measurements is often not easy and requires expertise. The mathematical and statistical models created might not be general and need to be adjusted to new conditions and products. Nowadays, a typical optical system employs ultraviolet (fluorescence) and visible wavelengths (VIS), near-infrared (NIR) or infrared (IR) areas or a combinations of them. The near-infrared spectrum is especially interesting since it is related to overtones and combinations of such chemical bonds as C–H, O–H, and N–H which have influence on many properties of food.

1.1. History. Food and agriculture have been subjects for near-infrared (NIR) applications from the beginning of NIR measurement history. The earliest work which involved nondestructive moisture content determination of grain and seeds is published in [2] of the United State Department of Agriculture (USDA). Later, the success of NIR spectroscopy is marked by the establishment of wheat protein analysis throughout Canada by Phil Williams of the Canadian Grain Commission in 1975 [3]. In this area of application, at first, NIR was restricted to quality evaluation of low-moisture products such as cereal grains including wheat, corn, and soybean due to interference of water absorption in the long-wavelength region, especially in the combination area of 1900–2500 nm. In the latter stages of NIR applications, the development of silicon sensors allowed acquisition in the short-wavelength region of 700–1100 nm, leading to the analysis of thicker and higher-moisture samples such as fruit, fish, meat, and intact kernels. The use of the short-wavelength region allows nondestructive measurement of relatively thick samples since the electromagnetic wave in this region has better penetration because the absorption of water in this third overtone band becomes weak.

According to Osborne [4], the major limitation of using NIR spectroscopy for food analysis is its dependence on less precise reference methods. Since the chemical information of sample is analyzed from the changes in spectra, in most food samples NIR spectroscopy needs calibration against a reference method with the ingredient of interest.

Other optical methods have been studied for food quality lately by many researchers, for example, Farkas and Dalmadi [5], Kaffka [6], and Huang et al. [7]. The optical methods could alleviate these problems although they are not so accurate. Another advantage of optics is that it can be imaging and thus characterizing the sample better. A popular spectral area for food research has been near-infrared (NIR) in which many publications can be found like Huang et al. [7], Pallav et al. [8], Ozaki et al. [9], and Osborne [4].

Berrueta et al. [10] have recently made a review of supervised pattern recognition methods in food analysis. Reid et al. [11] have made a review of different food authentication methods including NIRS. Cen and He [12] give a review of NIR reflectance spectroscopy in determination of food quality.

Du and Sun [13] give a review of machine learning techniques such as neural networks and evolutionary optimization in computer vision for food quality.

1.2. Optical NIR Measurements for Food

1.2.1. NIR Measurement Modes. There are four standard measurement modes for the acquisition of NIR spectra from a sample: transmission, reflection, transflection, and interaction (interactance). Illustrations of these are shown in Figure 1. The selection of measurement mode depends on many factors like sample type or installment location. Generally, the design and implementation of measurement system and the selection of measurement mode require at least some expertise.

In the transmission measurement mode (Figure 1(a)), incident light propagates through the sample and is measured as it exits the sample at a point directly opposite to the light

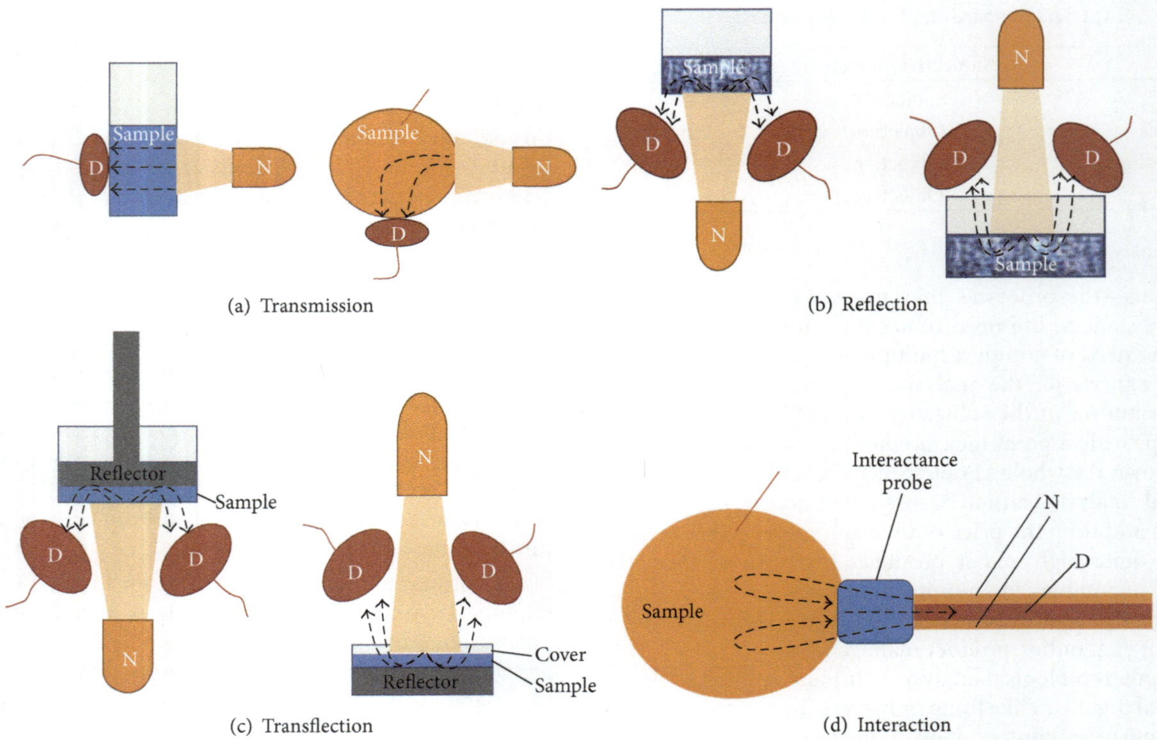

FIGURE 1: Sample presentation for near-infrared (NIR) spectroscopy. N: NIR light source, D: detector.

source. Occasionally, in the case of transmission for solid samples, the detector is placed at an angle of 90 degrees to the source.

For the reflection mode (Figure 1(b)), the light source and the detector are placed on the same side of sample. To avoid surface reflection, detectors are usually placed at an angle of 45 degrees to the sample plane. In some cases, an integrating sphere is used to concentrate light reflected from the sample before it reaches the detector by capturing and integrating light from various directions.

Transflection (Figure 1(c)) is the combination of transmission and reflection modes. The sample is positioned as if for reflection mode measurement, but a reflector is situated opposite to the light source on the far side of the sample, reflecting light that is transmitted through the sample back through the sample again to the detector. The samples, generally in liquid form, are placed in a holder of 0.1–0.3 mm width along the direction of light propagation. Therefore, the physical path length of transflection mode is double the sample thickness. Transflection is often used to reduce path length as the thinnest cuvette (standard sample holder for transmission measurements) is generally 1.0 mm thick. In some cases, transflection is employed for NIR instruments without transmission capability.

The fourth configuration, the interactance mode (Figure 1(d)), is also a combination of transmission and reflection but for solid samples. In this mode, the sample is illuminated by a fiber-optic cable(s) in direct contact with the sample surface or the sample holder. Light propagates inside the sample and interacts with the interior through reflection, transmission, and/or absorption. Fibre bundles that are placed on the same side as the source subsequently return the light that propagates inside the sample to the detector. The direct contact between the fibre bundles (for both the source and the detector) and the sample eliminates the effect of surface reflection and maximizes the penetration depth.

1.2.2. Types of Food Samples. From the product viewpoint, NIR applications in food and agriculture could be classified into three groups by the state of the sample: (i) liquid samples, (ii) ground and relatively small solid samples, and (iii) relatively large samples that require nondestructive or noninvasive measurement.

A benefit in working with samples from the first group (liquid samples) is the handling of optical path length. In the case of a clear liquid such as soy sauce, filtrated fruit juice, or vegetable oil, samples can be contained in a quartz cuvette, a disposable glass vial, or a transflectance sample holder, all of which would provide a relatively high consistency of optical path length due to the low scattering characteristics of the samples. The measurements used are transmission or transflection. The sample thickness can be adjusted based on the absorptivity of major composition(s) and the constituent of interest.

Figure 2 shows NIR spectra in the long-wavelength region of water and oil acquired in various physical path lengths. According to the figure, if the sample analyzed is water-rich and the constituent of interest has absorption

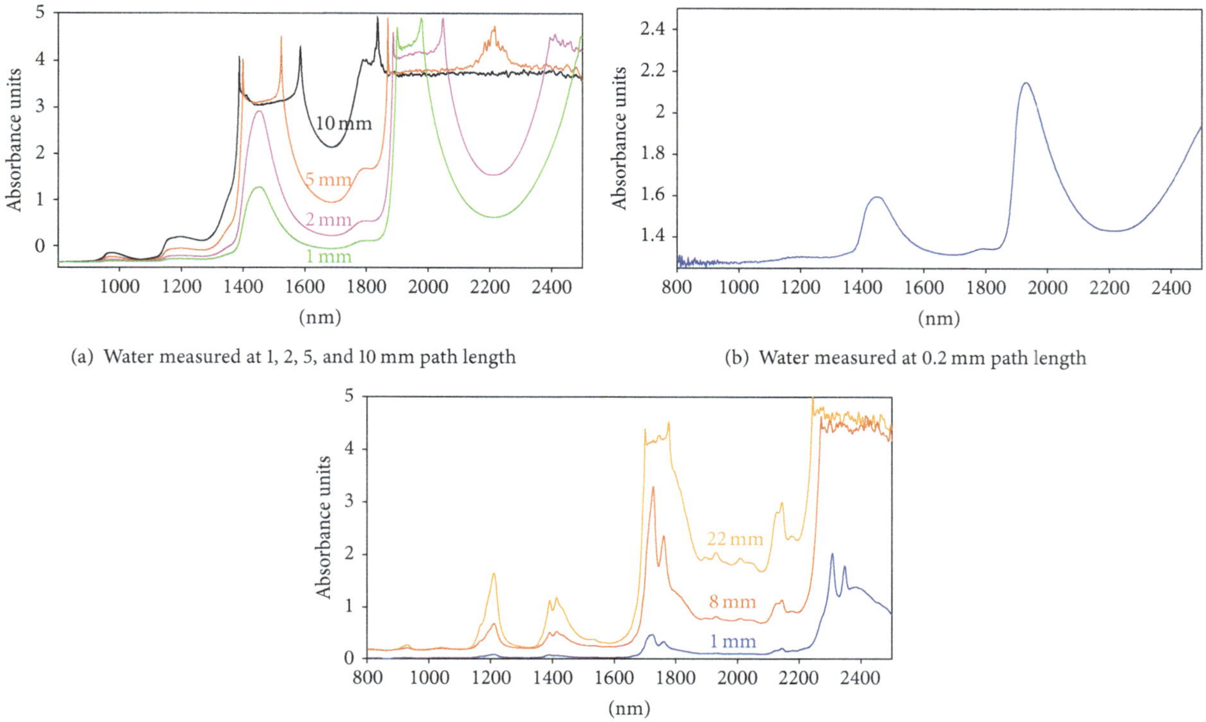

(a) Water measured at 1, 2, 5, and 10 mm path length

(b) Water measured at 0.2 mm path length

(c) Vegetable oil measured at 1, 8, and 22 mm path length

FIGURE 2: NIR spectra of water and oil measured at different path lengths. (a) Water spectra when the path length is 1, 2, 5, or 10 mm. (b) Water spectra when the path length is 0.2 mm. (c) Vegetable oil spectra when the path length is 1, 8, or 22 mm. The shape and level of measured spectra depend on the path length.

bands overlapped with water absorption, the appropriate physical path length would be 0.2 mm for the combination region measurement (1800–2100 nm), 1.0 mm for the first overtone region (1400–1500 nm), and 10 mm for the second overtone region (900–1000 nm). However, as a forementioned, the decision for optical path length is also dependent on the molar absorptivity, that is, the ability of a particular substance to absorb light at a given wavelength, and the concentration of that constituent of interest. In many cases, the constituent of interest has absorption bands on the shoulder of water bands such as in the region of 2100–2300 nm. In such cases, sample cells with longer physical path length, say 1 mm, could be more appropriate than the 0.2 mm sample cell due to lower absorptivity of substances in this region.

On the other hand, if the sample is oil-based, sample cells with significant longer path length could be used due to the lower absorptivity of oil compared to water. In the case where the constituent of interest has very low absorptivity and/or concentration, for example, acid value or peroxide value in fresh vegetable oil, increasing physical path length to obtain adequate information is necessary.

However, one should keep in mind that the analysis of clear samples typically differs from that of the analysis of scattering samples. When colloids or suspensions are analyzed, the scattering characteristics of the samples are stronger, and thus the optical path length, that is, the actual distance that NIR wave travels, is much longer than that of clear samples. Due to this, the path length would have to be reduced. Figure 3 shows the NIR spectra of homogenized milk (colloid) and vegetable-mix juice (suspension) using a 4 mm path length. In this case, milk has a stronger absorption due to its composition and the scattering characteristics of the constituent small fat globules. Vegetable-mix juice has less absorbance compared to milk but stronger than that of water at 5 mm path length (for water, see Figure 2). This is due to the scattering characteristics of the constituent suspended particles.

1.2.3. Some Important Considerations for Measurements. Another point to be aware of for liquid samples with water as a major constituent is the importance of temperature control. It is well documented [14–17] that water absorption bands in the NIR region shift in both peak position and amplitude according to sample temperature, as the number of hydrogen bonds in the water molecules is temperature dependent. Increasing temperature results in a higher percentage of water molecules with fewer hydrogen bonds due to the higher kinetic energy. Therefore, in many cases, temperature control is necessary to perform precise NIR measurements. Since many oil-based samples are in solid or opaque form at room temperature, temperature control is required not mainly to maintain spectral consistency but to increase sample transparency.

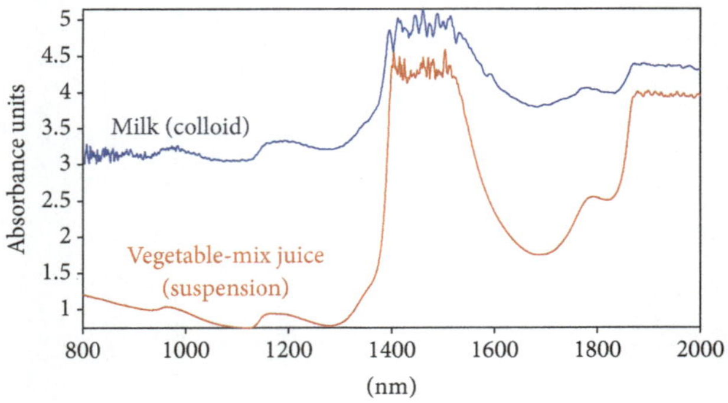

FIGURE 3: NIR of milk (colloid) and vegetable-mix juice (suspension) measured with 4 mm physical path length. Note: the preamplifier gain used for milk was stronger than that of vegetable-mix juice due to very weak transmitted light.

For ground or small-particle samples, NIR spectra are usually acquired in reflection mode due to strong scattering characteristics. Such samples are generally placed in sample holders with quartz or glass windows. Quartz is more expensive but has lower OH absorption compared to glass resulting in better precision in some constituents, especially those with OH functional groups. However, in the case where the absorption of the constituent of interest occurs at different wavelength region than the OH absorption in glass and there is enough concentration of those constituents, it is possible to use glass-window sample holder or glass Petri dish. For many food manufacturers, the use of glass or quartz is prohibited and plastic such as a polystyrene dish is used as a sample holder. In that case, the absorption band of the constituent would fall within the absorption bands for polystyrene, especially in the combination region beyond 2000 nm. If the absorption bands of the constituent can be precisely measured in the shorter-wavelength region by arranging an adequate path length and detector, the use of plastic dish would be applicable. For samples with a relatively coarse particle size, there is the possibility that light could pass through the space between samples and be lost. One way to reduce this problem is to place a reflector on the back side of the sample cell. During the measurement, spinning the sample cell will increase the acquisition area, thus reducing error due to sample nonuniformity especially in relatively coarse or large-particle samples. Examples of samples belonging to this group include ground and intact cereal grains, cereal meals such as soy bean meals or corn gluten meals, crushed processed foods/snacks, and powder milk or drink.

Finally, we consider samples that require nondestructive measurement. Samples in this group include meat, fish, intact fruit, and intact nut kernels. For the most part, these samples are sold intact and the constituent of interest is not evenly distributed between the surface and the sample core. To obtain information from deep inside the sample, two modes of measurement are used for this group: the interaction mode and the transmission mode. As mentioned earlier, most of the applications in this group use silicon detectors that allow spectral acquisition in the short-wavelength region of 700–1100 nm. The measurement parameters could be moisture and sugar contents for intact fruit and fat, protein, and moisture content in meat and kernels, as well as detection of various internal disorders or pest in agricultural products.

To employ NIR spectroscopy for routine quality control, it is not enough to develop calibration equations. A system developer needs to consider various factors affecting calibration results including instrument status (source strength, detector responses, and wavelength selection mechanism), atmospheric effects (temperature and humidity), and physical conditions (dust, vibration, and stray light) as well as transferability of the equations developed between various instrument in the same period of time and the transferability to the newer version of instrument and software upon the end of instrument life span.

1.3. Chemometrics. Optical measurement results connected with chemical analysis and statistics are called chemometrics. Chemometrics has become an established technique for handling chemical data. Good reviews of chemometrics are provided by Geladi and Grahn [18], and Workman and Springsteen [19], and Koljonen et al. [20], which present a review of genetic algorithm optimization in chemometry. The next text is a short introduction for the subject, and for a deeper insight, the previous reviews are recommended.

The first step of chemometrics is to collect the data. This includes sample selection, measurements, and chemical analysis. The measurement geometry, measurement conditions, number of samples, and ingredients need to be very carefully designed and determined. The samples need to be representative of the food batches because interpolation or generalization of the model most probably does not work. The samples are first subjected to optical measurement and after that chemical analysis. One should also notice that the chemical analysis might sometimes provide unreliable results. Sometimes even the samples may change during time. The data should be divided into a training and test sets.

After the measurements, the equations to transform optical data to chemical data should be set up. Since many functions and equations are available, it might be useful to test a couple of them to select the best results. After the selection, the equations need to be solved and coefficients to be determined with the training set. A typical approach for modeling is regression. The training of the model can be understood as a "calibration." The optical methods are calibrated to produce the amount of a given substance or substances.

The training data should be subjected to detection of residuals, outliers, and nonlinearity. The outliers should be excluded and nonlinearity corrected if present. Otherwise the results might be distorted. Typically the number of wavelengths from the optical measurements needs to be reduced due to their high amount and interdependence. Many techniques are available for wavelength reduction (e.g., splines or Savitzky-Golay filtering).

Maybe the most often used regression models are multiple linear regression (MLR) or "ordinary" least squares

(LS), ridge regression (RR), principal component regression (PCR), and partial least squares (PLS). The general equation for the regression is

$$y = b_0 + b_1x_1 + b_2x_2 + \cdots + b_kx_k + f = b_0 + \sum_{i=1}^{k} b_ix_i + f = b_0 + f + BX, \quad (1)$$

where y is the response (from chemical analysis), x_k are the predictors (from optical measurements, k pieces), b_k are the regression coefficient to be determined, b_0 is the offset and a constant factor, and f is the residual. If X and y are mean-centered, then $b_0 = 0$.

The solution for regression coefficient for the LS is

$$b = \left(X^TX\right)^{-1}X^Ty, \quad (2)$$

where T means transpose of the matrix. The LS may not work since the inverse might not exist or may be unstable. It is also notoriously sensitive to noise. The results should be also diagnosed using the following method: coefficient of determination and root mean squared error of calibration.

A better solution for problems related to inversion can be RR:

$$b = \left(X^TX + rI\right)^{-1}X^Ty, \quad (3)$$

where I is an identity matrix and r is a ridge constant. The ridge constant should be as small as possible. However, there is no established method for the selection of the ridge constant or diagnostic of results.

PCR also provides an easy inversion of matrices. It requires applying the principal component analysis (PCA) to reduce dimensions of X and to obtain the eigenvectors. The eigenvectors are orthogonal. The PCA also reduces the noise present in X. The equation for regression can be formulated as

$$y = Vd + f, \quad d = \left(V^TV\right)^{-1}V^Ty, \quad (4)$$

where V is the principal components and d contains coefficients. The number of components needs to be determined by testing and checking.

Assume that y matrix consists of several measurements of y. PLS decomposes both y and X matrices. It also removes noise in both of these matrices. Its formulae are

$$y = UQ^T + F, \qquad X = TP^T + E, \quad (5)$$

where T and U are projections (principal components) of X and y on eigenvectors and P and Q are eigenvectors of X and y. The projections are obtained by decomposing the corresponding matrix. Error terms are marked as E and F.

Both PLS and PCR calculate one component at a time using regression between t and u. Then a residual is calculated and the calculation of the next component is based on this residual.

The obtained results for equations need to be evaluated by using the test set. The predicted values are calculated using test values from the optical measurement set and parameters obtained from the training phase. The difference between actual and predicted values in the test set can be evaluated using the root mean squared error of prediction and the coefficient of determination.

The visualization gives a good view of how the model is working. This could be done, for example, by plotting root mean squared error of calibration, root mean squared error of prediction, calculated loading plot, residual versus sample, and residual versus predicted test value.

1.4. Food Quality and Safety

1.4.1. Safety Factors. We consider here food safety to factors which may cause danger to health either directly or indirectly.

(1) *Foreign Bodies.* A foreign body can be anything unwanted in a food product. It can be from the raw material for the product or external material added indeliberately; for example, the raw material can contain bones which are shattered to sharp bone fraction pieces during the manufacturing process. The product may be littered with machine parts due machine blistering or residues from other sources like pieces of plastic.

(2) *Incorrect, Contaminated, or Phony Ingredients in Product.* Incorrect food ingredients refer here to any material which should not be in the food. They can be residues from previous production like peanut remnants or accidently added wrong material. The food can be contaminated from external sources like manure or too much pesticide. The phony ingredients are deliberately added to generate more income from consumers or manufacturers. The phony items can be rip-offs like cheap wine sold as a more expensive version or even very dangerous ones like melanin milk.

(3) *Spoilage.* The spoilage is considered in this paper as a decomposition of the food. It is often a byproduct of the activity of harmful bioagents like fungi or bacteria. Other causes are wrong handling, like overheating, freezing or wrong harvesting time, wrong environment, like too hot environment or environment that has gases like ethene which mature fruits too fast, or age.

(4) *Bioagents: Fungi, Yeast, and Molds; Microbes, Bacteria, and Viruses.* Fungi, yeast, or molds can be used in the production of certain food, but usually they are serious quality problems. This paper concentrates on articles on harmful effects of the bioorganisms like food poisoning or infections. The detection of smaller bioorganisms like bacteria or viruses is usually very hard due to their small size.

1.4.2. Quality Factors. Food quality consists of several factors. The validity or importance of a factor depends on the food group and on the evaluator. For example, the fat content is irrelevant for apples but very relevant for meat. The factors for this paper are appearance and taste.

(1) *Appearance.* The appearance of food is related to the visual experience of the product. It is affected by the perceived colors

and textures as well as personal preferences and opinions. The environment in which the food is presented has its own effect. Since the color perceived depends on the prevailing illumination, for example, there are already available special lamps to emphasize the color of meat. Good texture can be achieved with proper processing.

(2) *Taste.* Taste is composed of several ingredients and is affected greatly by smell and also appearance. In this paper, we limit the taste to the actual food ingredients or state-like amount of sugar or acidity.

Smell. The smell can be used to separate bad food from good food. It is also essential to make food more delicious.

Juiciness. One attribute especially for fruit is juiciness. It is related to the water content and distribution of fruits.

Noise. Consumers often expect certain noise when eating food like cereals.

Freshness. Freshness is important especially for fish and other sea products.

Authentication. Some food products have been falsified. An example of this kind of products is wine. Spectral methods can be used to separate authentic wine from nonauthentic wine. The NIR spectra is used for food authentication by Sun [21].

Essential Ingredients. Food can contain many elements which have positive effects on health. These elements include, for example, flavonoids and vitamins in plants flavonoids and vitamins or omega-6 fatty acid in fish. It is generally thought that the higher the amount obtained from food the better the effect on health. Ingredients can be correlated using chemometrics.

Unwanted Ingredients. Food contains also ingredients which consumers want to avoid for real or imaginary reasons. These ingredients can be preservatives which can in fact make the food safer. Other harmful ingredients include excessive amount of fat or salt.

Unwanted Objects. Although food can be perfectly safe and edible, it can still contain unwanted objects. This kind of objects can range from machine parts to stones and insect parts.

In their book, Workman and Weyer [22] explain theoretical backgrounds on how different qualities can be analyzed and found from NIR spectra; these qualities include alkanes and cycloalkanes, methylene groups, alkenes and alkynes, aromatic compounds, hydroxyls, water, carbonyls, amines and amides, carbohydrates, amino acids, peptides, and proteins.

High Quality. Affordable price consistently.

Chemometric Models. None known.

1.5. Optimization. Optimization with NIR spectroscopy is mostly connected with the selection of optimal wavelengths that correlate best with the object of interest. Usually we want to use as few wavelengths as possible, since the less the data needs to be analyzed the faster it is. However, we need to be sure that the selected wavelengths are adequate for reliable results.

A special type of genetic algorithms is called genetic programming (GP). It uses genetic algorithms to find a function or a program to solve a given problem. GP has been demonstrated to find solutions to problems that have been found by the most skilled engineers as patents and other inventions [23]. Another view of GP is to say that it is symbolic regression, that is, a method to find a function that best satisfies the given constraints. Yet another usage of GP is to find rules to explain data. GP is an extremely flexible method to solve computational intelligence problems. The main drawback of GP is its slow processing which prevents its usage in most online applications. However, in the systems design or similar phase, GP can be a nice tool to find good solutions automatically after the proper fitness function is implemented.

While spectroscopy produces huge amounts of raw data, there is a strong need to automatically process this data to compress it and to find the features that are relevant or correlate to those that are looked for. Norris and Williams [24] used wavelength selection so that they selected 12 wavelengths that produced the closest prediction with the required constituent in wheat measurements, particle size, and protein content. Goodacre and Kell [25] have demonstrated spectroscopical imaging GP applications related directly or indirectly to food quality. More specifically the application was to find the algebraic formula of the wavelengths that give the best discrimination between the bacterial samples examined. More generally the authors claim that GP is one potential tool to decrypt metabolomic data for biomedical applications.

With several materials, the VIS/NIR reflectance works best if the wavelengths used are optimized between 400 and 1800 nm [26]. This was also true with wholegrain measurement, where they found wavelengths between 1100 and 1400 nm giving the highest correlations with most of the constituents of interest with wholegrain measurements. However, the protein content was best determined from the 1700 to 2500 nm wavelength band.

1.5.1. Illustrative Example of Tomato Classification. The spectral methods lend themselves very well to fruit quality inspection. In this example, the transmittance spectra of tomatoes were measured using a spectrophotometer (Ocean optics). The spectra were further handled: they were smoothed to reduce noise and then downsampled by splines. An example of tomato spectra is shown in Figure 4. The transmittance is measured from 400 nm to 1000 nm. The spectra differ in both visible and NIR areas. The difference is caused by the ripening process.

The spectral data itself contains a lot of information, part of which is in some cases irrelevant or redundant. The dimensionality of data can therefore be reduced. Here the locally linear embedding was used for this purpose in a nonlinear manner. The visualization of results can be seen in Figure 5, in which the feature space is spanned by the first two embedded components.

The kernel Fisher discriminant (KFD) was used to classify tomatoes into good and bad ones. As shown in Figure 5, the tomatoes are classified 100% correctly into these two groups.

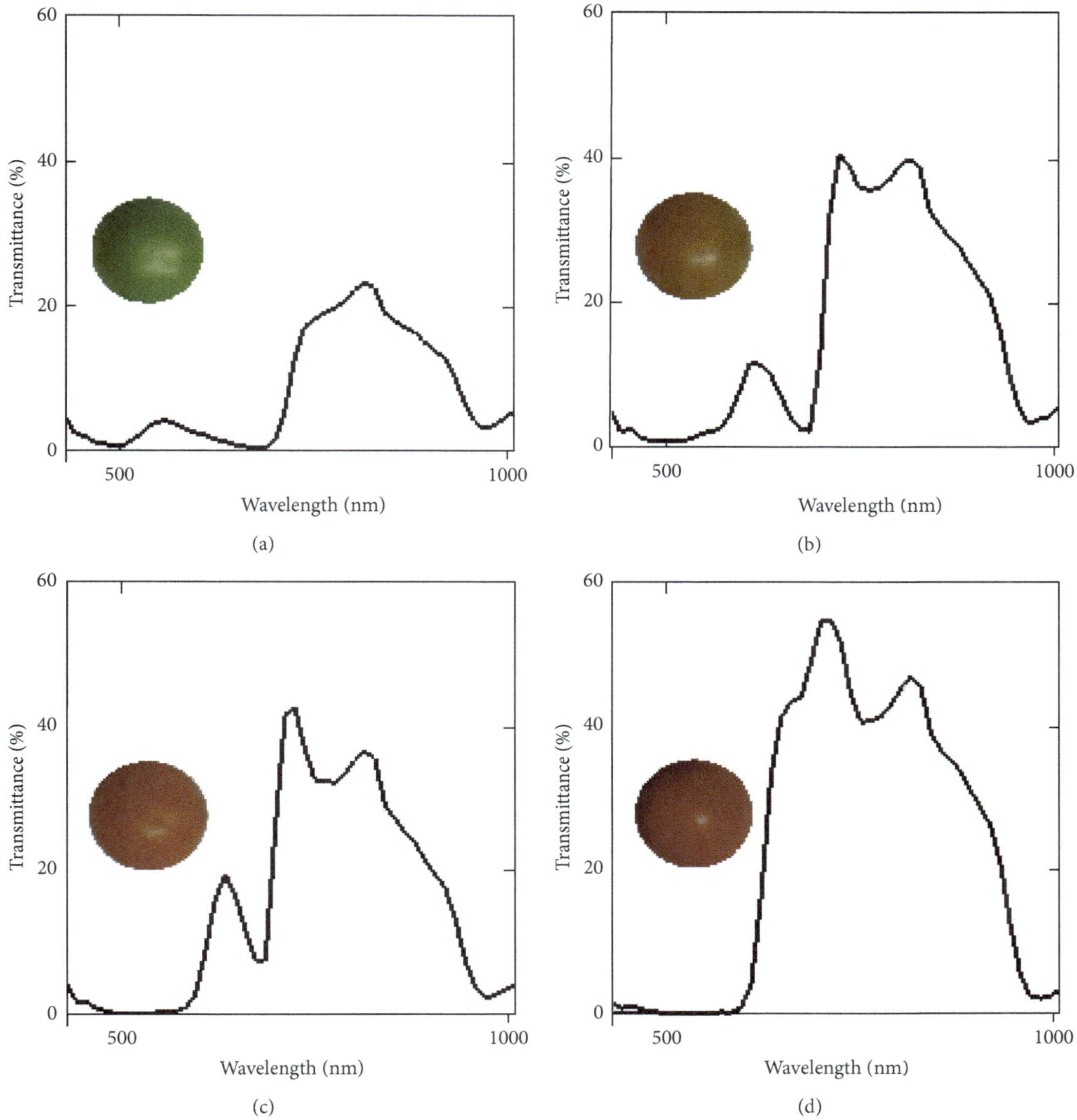

FIGURE 4: Transmittance spectra for different tomatoes of the same cultivar from 400 nm to 1000 nm. (a) A raw tomato, (b) a partially raw tomato, (c) an acceptably ripe tomato, and (d) a ripe tomato.

The red-colored tomatoes are ripe while the yellow and green are raw. The boundary between the groups is visualized with red curve (Figure 5).

2. Review of Hyperspectral/Multispectral Methods for Different Food Groups

This paper follows the FAO classification in grouping food which divides the food and animal feed into five classes:

(i) class A—primary food commodities of plant origin,

(ii) class B—primary food commodities of animal origin,

(iii) class C—primary animal feed commodities,

(iv) class D—processed foods of plant origin,

(v) class E—processed foods of animal origin.

Each of these classes has subclasses. Class A consists of five subclasses: fruits, vegetables, grasses, nuts and seeds, and herbs and spices. Class B has the following classes: mammalian products, poultry products, aquatic animal products, amphibians and reptilians, and invertebrate animals. Class C has only one class, primary feed commodities of plant origin. Subclasses, which belong to class D, are secondary food commodities of plant origin, derived products of plant origin, manufactured foods (single ingredient) of plant origin, and manufactured foods (multi-ingredient) of plant origin. Class E is made of the following subclasses: secondary food commodities of animal origin, derived edible products of animal origin, manufactured food (single ingredient) of

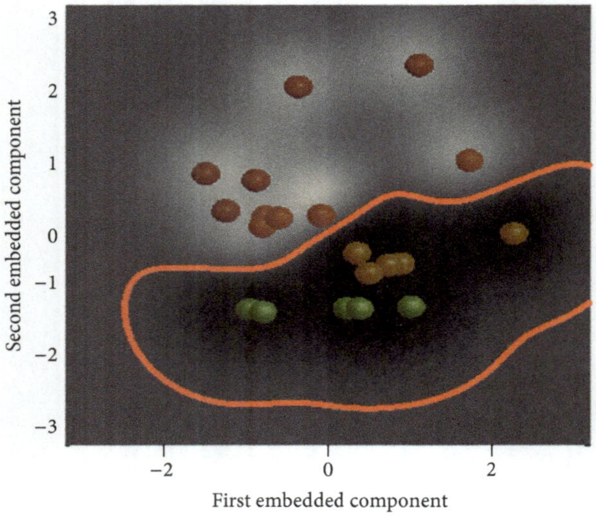

Figure 5: The feature space is spanned by the first two components. The red curve indicates the boundary between good and bad tomatoes. The good tomatoes are outside the red curve.

animal origin, and manufactured food (multi-ingredient) of animal origin.

More detailed reviews of the selected papers for each of the classes can be found at the Appendices A (for class A), B (for class B), C (for class D), and D (for class E). The papers in class C are not considered in this paper.

2.1. Class A—Primary Food Commodities of Plant Origin. Class A has five groups: fruits, vegetables, grasses, nuts and seeds, and herbs and spices.

The fruit group consists of 6 subgroups: citrus, pome and stone fruits, berries and other small fruits, and assorted tropical and subtropical fruits with edible or inedible peels. Optical and spectroscopic methods [27] and near-infrared spectroscopy [28] present a review of promising methods for fruit quality.

Vegetables include bulb vegetables, cabbage (*Brassica* (cole or cabbage) vegetables, Head cabbages, Flowerhead cabbages), fruiting vegetables either cucurbits or other than cucurbits, leafy vegetables (including Brassica leafy vegetables), legume vegetables, pulses, root and tuber vegetables. Mushrooms and fungi belong also to this group. Reviews of vegetable quality have been made by Butz et al. [27] and Nicolaï et al. [28].

The grasses have two groups: cereal grains and grasses used for sugar or syrup production. The nuts and seeds' group consists of tree nuts, oilseed, and seed for beverages, and sweets. The herbs and spices' group consists of herbs and spices as the name indicates.

Table 2 provides a short description for each of the subgroups of class A. We list parameters which have successfully been evaluated for the group and methods applied.

2.2. Class B—Primary Food Commodities of Animal Origin. Class B consists of the following groups: mammalian products, poultry products, aquatic animals, amphibians and reptilians, and invertebrate animals. Examples of publications can be found at Table 3.

Mammalian products consist of nonmarine meat, mammalian fats, edible offals, and milks. It seems that NIR technologies are more acceptable, for example, to the AOAC (Association of Official Agricultural Chemists).

The poultry can be studied by other optical techniques than just reflectance or transmittance. Autofluorescence has been suggested for food by Andersen et al. [29] and for poultry by Wold and Mielnik [30].

FAO includes in aquatic animals fish (freshwater, marine and diadromous, i.e., migrating between fresh and saltwater), fish roe and milt, fish offals, marine mammals, and crustaceans. Seafood and its products are an important source of nutrition [31]. Their special challenge is the changes in properties from the moment of catch or harvest due to metabolic processes, microbiological processes, and microbiological fauna change. This change is described as a function of spoilage and freshness parameters.

The invertebrate animals' group consists of molluscs (including *cephalopods*) and other invertebrate animals.

2.3. Class D—Processed Foods of Plant Origin. Class D has four groups: secondary food commodities of plant origin, derived products of plant origin, manufactured foods (single ingredient) of plant origin, and manufactured foods (multi-ingredient) of plant origin.

Secondary food commodities of plant origin are dried fruits, vegetables and herbs, secondary food commodities of plant origin, and miscellaneous secondary food commodities of plant origin.

Derived products of plants for human consumption include cereal grain milling fractions, teas, vegetable oils: crude, edible, or refined, origin including miscellaneous derived edible products of plant origin, and fruit juices.

Manufactured foods (single ingredient) of the plant origin group naturally consist of fruits, vegetables, or miscellaneous manufactured single-ingredient foods from plants. Manufactured foods (multi-ingredient) of the plant origin group are all manufactured, multi-ingredient cereal products. Table 4 shows examples of published works in this area.

2.4. Class E—Processed Foods of Animal Origin. Class E has the following subgroups: secondary food commodities of animal origin, derived edible products of animal origin, manufactured food (single-ingredient) of animal origin, and manufactured food (multi-ingredient) of animal origin.

Secondary food commodities of animal origin consist of dried meat and fish products and secondary milk products. We include here also products using dried meat and fish. An example of this is ham.

Derived edible products of the animal origin group include processed crustaceans and animals, milk, and derived fats. Manufactured food (single ingredient) of the animal origin group is the manufactured milk products using only single ingredient.

Table 2: Different food groups, their parameters and methods applied for class A.

Food group/appendix	Examples of parameters evaluated and one of related research articles	Methods applied
Fruit/A.1	*Quality* Taste (Mehinagic et al. [46]) Content (i) Sugar (Walsh et al. [47]) (ii) Soluble solid (Carlini et al. [48]) (iii) acid content (León et al. [51]) (iv) water/moisture (Blakey et al. [50]) (v) titratable acidity and ph (Ying and Liu [56]) (vi) vitamin C and polyphenol (Pissard et al. [57]) Translucency (Terdwongworakul et al. [53]) Internal quality (Magwaza et al. [54]) Maturity (Tanaka and Kojima [58]) Defects (Blasco et al. [262]) *Safety* Internal insect detection (Xing et al. [60]) Fecal and other contaminations (Kim et al. [61] and Mehl et al. [263]) Storage (Veraverbeke et al. [63])	MLR, PCR, PLS, PCA Genetic algorithms for wavelength and calibration model selection
Vegetables/A.2	*Quality* Appearance (Flores Rojas [65] and Gowen et al. [84]) Content (i) Cyst nematode (Fernández Pierna et al. [64]) (ii) Sugar (Tsuta et al. [67]) (iii) Soluble solid content, lycopene, β-carotene, and carotenoid (Pedro and Ferreira [68]) (Iv) Protein (Liu et al. [69]) (v) Phytic acid (Delwiche et al. [73]) (vi) Dry matter (Van Dijk et al. [70]) (vii) Protein (Pazdernik et al. [75]) (viii) Freshness (Esquerre et al. [81]) (ix) Moisture (Lamb and Hurburgh Jr. [78]) Structural characteristics (Siska and Hurburgh Jr. [83]) Damage (Esquerre et al. [81]) *Safety* Incorrect ingredients (Roggo et al. [72]) Bioagents (Pearson and Wicklow [85])	Texture metrics, wavelength selection, PLS, MLR, PCA
Grasses/A.3	*Quality* Content (i) Moisture, gluten, alveograph parameters, and protein (Miralbés [90]) (ii) Scab, vomitoxin, and ergosterol (Dowell et al. [89]) (iii) Rutin and D-chiro-inositol (Yang and Ren [91]) (iv) Carotenoids (Brenna and Berardo [92]) (v) Phenolics and antioxidant capacity (Zhang et al. [94]) (vi) Neutral acid detergent fibers and acid detergent lignin (Stubbs et al. [95]) (vii) Nitrogen and protein (Wiley et al. [96]) structural characteristics (i) Softness/hardness (Manley et al. [98]) (ii) Density (Siska and Hurburgh Jr. [83]) Taste (Kawamura et al. [99]) *Safety* many impurities like plastics or faeces (Fernández Pierna et al. [64])	PLS, MLR, PCA

Table 2: Continued.

Food group/appendix	Examples of parameters evaluated and one of related research articles	Methods applied
Nuts and seeds/A.4	*Quality* Content (i) Moisture, fat, stearic acid, and oleic acids (Davrieux et al. [100]) (ii) Sinigrin, gluconapin, and glucosinolate (Font et al. [101]) (iii) Lipids, water, protein, and chlorophyll (Font et al. [102]) (iv) Oil and fatty acid content (Kim et al. [103]) (v) Starch (Hacisalihoglu et al. [104]) (vi) Amino acids (Wang et al. [105]) Authentication (Downey et al. [107])	PLS
Herbs and spices/A.5	*Quality:* Content (i) Kava pyrones (Gautz et al. [110]) (ii) Moisture, volatile content, and coloring strength (Zalacain et al. [111] (iii) Curcuminoids (Tanaka et al. [112]) (iv) Ginsenosides (Ren and Chen [113]) (v) Lignans (Kwan et al. [114]) Origin (Zalacain et al. [111])	PLS

Manufactured milk products using multi-ingredient belong to manufactured food (single ingredient) of the animal origin group. Here we consider dairy products like cheese or yogurts. Some publications has been listed in Table 5.

2.5. Other Foods Not Included in FAO Criteria. Some foods are not included in FAO classification although they are commonly used. For example, honey is not included. There is no clear category for processed foods like hamburgers or sausages. Some of the research done for these foods is introduced here.

2.5.1. Authentication. There have been cases in which the ingredients of manufactured food have been replaced by a cheaper version. For example, beef hamburger can be adulterated with mutton, pork, skim milk powder, or wheat flour. Ding and Xu [32] used near-infrared for detecting and evaluating the adulteration level in the adulterated hamburgers with 5–25% phony ingredients. The accuracy increases as a function of adulteration rate since there were more ingredients to be detected. The highest detection rate was 92.7%.

ElMasry et al. [33] have applied hyperspectral imaging (Imspector N17E, Specim, Spectral Imaging Ltd., Oulu, Finland) in the region 900–1700 nm for predicting colour, pH, and tenderness of fresh beef. With more improvement in terms of speed and processing, their approach has potential for nondestructive quality measurements.

Liao et al. [34] have used VIS/NIR reflectance spectroscopy for the online assessment of the pH value in fresh pork. Their result show that VIS/NIR spectroscopy can be used for the online prediction of pH and that wavelength selection can provide a cost-effective calibration model. For wavelength selection, they compared variable importance in projection (VIP), uninformative varible elimination (UVE), genetic algorithms, and success project algorithm (SPA).

Collell et al. [35] have used NIR reflectance spectroscopy to predict moisture, water activity, and salt content on the surface of dry-cured ham during the process. Their results show that the proposed method could be used in the dry-curing process of ham. However, the optical properties of the meat surface seem to strongly affect the measured spectra.

Authenticity and origin are also important issues for honey especially when it is claimed to be collected by bees from certain flowers. One example is avocado honey. Dvash et al. [36] applied near-infrared reflectance spectroscopy to obtain the concentration of specific sugar (perseitol) for this honey.

Downey and Kelly [37] studied sulfited strawberry and raspberry purées which have been adulterated with cooked apples (10–75% w/w). They applied VIS and NIR transreflectance measurement on 0.1 mm thick samples. The best wavelength bands were 400–1098 nm with 75% correct classification rate for strawberry and 750–1098 nm with 95% rate for raspberry. PLS regression was used for the prediction of apple content with 400–1880 nm range for raspberry and with 1100–1880 nm range for strawberry. They concluded that apple adulteration can be detected if its content exceeds 25% for raspberry and 20% for strawberry purées.

Fügel et al. [38] made a review about authenticity control of fruit purées, fruit preparations, and jams.

2.5.2. Content. Qiu et al. [39] applied near-infrared transmittance to quantify the chemical content of honey. They found that 1 mm optical path length was the best for this purpose. Modified partial least squares were applied for the determination of content except for moisture. Moisture was determined using just partial least squares. The models were successful for fructose, glucose, sucrose, and maltose

TABLE 3: Class B groups.

Food group/appendix	Examples of parameters evaluated and one of related research articles	Methods applied
Mammalian products/B.1	*Quality* Taste (Byrne et al. [120]) Color (Van den Oord and Wesdorp [124]) Diet (Prieto et al. [130]) Content (i) pH (Andrés et al. [128]) (ii) Moisture (Chan et al. [135]) (iii) Fat (Sierra et al. [136]) (iv) Iron (Hong and Yasumoto [139]) (v) Protein and dry content (Prieto et al. [140]) Maturity (Prieto et al. [154]) Shear force (Yancey et al. [155]) Tenderness (Shackelford et al. [126]) Texture (Byrne et al. [[120]) Sensory characteristics (Prieto et al. [159]) Structural properties (Shackelford et al. [126]) Drip loss (Geesink et al. [160]) Authentication (Dian et al. [168] and Arvanitoyannis and Houwelingen-Koukaliaroglou [264]) Freshness (Downey and Beauchêne [173]) *Safety* Spoilage (Horváth et al. [116]) Contamination (Windham et al. [118]) Bioagents (Ellis et al. [117]) Adulteration of milk (Sato et al. [181])	Color, geometrical features, texture Multivariate data analysis, ratios, MLR, PCA, PLS
Poultry products/B.2	*Quality* Content (fat, protein, iron) Authenticity/origin (Hong and Yasumoto [139]) Storage and handling (Liu et al. [192]) *Safety* Safety in general (Chen et al. [182]) Fecal contamination (Park et al. [188]) Spoilage (Lin et al. [190])	Wavelength selection PLS Correlation
Aquatic animal products/B.3	*Quality* Temperature Appearance (olsen [195]) Identification/authentication (Cozzolino et al. [197]) Freshness (Uddin et al. [198]) Content (i) Water, fat, and protein (Gjerde and Martens [204]) (ii) Free fatty acids (Zhang and Lee [203]) (iii) Salt and moisture (Huang et al. [7]) (iv) Crude lipid (Mathias et al. [209]) Defects like bruises (Lin et al. [211]) *Safety* Incorrect ingredients (Font et al. [212]) Spoilage (Lin et al. [202])	Wavelength PCA, linear discriminant analysis PLS
Invertebrate animals and spices/B.5	*Quality* Authenticity (Gayo and Hale [213]) Content (Brown [215])	PLC, PCA

Table 4: Class D food groups, their parameters, and methods applied.

Food group/appendix	Examples of parameters evaluated and one of related research articles	Methods applied
Secondary food commodities of plant origin/D.1	Content (i) Starch, sugar, nitrogen, and ash (Lebot et al. [216]) (ii) Hollow heart (Dacal-Nieto et al. [217]) (iii) Bioactive components (Jourbert et al. [218]) (iv) Glucan and protein (Bellato et al. [219]) Adulteration (Laasonen et al. [224])	Wavelength selection PLS PCA
Derived products of plant origin/D.2	Content (i) Dry extract and sugars (Li et al. [225]) (ii) Acids (Chen et al. [226]) (iii) Alcohol (González-Sáiz et al. [227]) (iv) Soluble solids and pH (Liu and He [229]) (v) Oils (Ng et al. [230])	Multivariate calibration PLS, MLR
Manufactured foods (single ingredient) of plant origin aquatic animal products/D.3	Content (i) Carbohydrate (Chen and Lei [232]) (ii) Soluble solids and pH (Liu et al. [238]) (iii) Methanol (Dambergs et al. [233]) (iv) Acids (Fan et al. [236]) Protein-lipid-carbohydrate interactions (Bruun [237]) Authentication (Cozzolino et al. [239]) Age	Wavelength selection PLS, LS, PCA
manufactured foods (multi-ingredient) of plant origin/D.4	Content Fat (Vines et al. [243]) Moisture (Kays et al. [245]) Impurity (Fernández Pierna et al. [64]) Total fiber amount (Kays et al. [246]) Structural properties (de Alencar Figueiredo et al. [253]) Handling and fermentation (Bruun et al. [254])	PLS Wavelength selection

Table 5: Different food groups, their parameters, and methods applied for class E.

Food group/appendix	Examples of parameters evaluated and one of related research articles	Methods applied
Secondary food commodities of animal origin/E.1	Appearance (García-Rey et al. [257])	Wavelength
Manufactured food (multi-ingredient) of animal origin/E.5	Appearance (Lucas et al. [258]) Smell and taste (Çelik [259]) Content (Lucas et al. [258], Arvanitoyannis and Van Houwelingen-Koukaliaroglou [264]) pH (Navrátil et al. [260]) Origin (Cozzi et al. [261]) Authenticity (Woodcock et al. [265])	PLS, wavelength discriminant analysis

content but poor and unreliable for free acid, lactone, and hydroxymethylfurfural.

NIR spectroscopy is applicable also for wholemeals. Fontaine et al. [40] used near-infrared for detecting essential amino acids for different meals. They concluded that the NIRS calibration model was good for evaluating the content of protein-rich feed ingredient for amino acids methionine, cystine, lysine, and threonine, tryptophan with 85–98% variation explained and was better than crude protein regression. The protein and moisture contents were also evaluated.

Kim et al. [41] used NIR reflectance (400–2498 nm) and partial least-square modeling to detect the total dietary fiber in meals. Meals have different handling alternatives: homogenization, homogenization and drying, and homogenization, drying, and defatting. The reference method for total dietary fiber evaluation was AOAC 991.43. The predictions were good. The main advantage over the NIR-based method was faster evaluation. The traditional method takes 4 days.

Zeng et al. [42] developed a NIR transmission system for determining the amount of starch in liquid gravies with low concentration. The best wavelengths were 605, 623, 919, 1017, and 1031 nm for the regression model. The starch content makes possibility to evaluate the viscosity of liquid gravies.

2.5.3. Authenticity. Sausages can be made with different fats. In some cases, sausages can be identified based on their fat substitutes [43]. The quality of them has been studied, for example, by Gaitán-Jurado et al. [44] and Ortiz-Somovilla et al. [45].

3. Conclusions

Nondestructive methods for food quality measurement are warmly welcomed by the food industry. Optical engineering provides a large variety of measurement techniques, some of which, like optical and near-infrared spectroscopy and imaging, have especially high potential for various food-quality-related measurements. And indeed, during the last few decades, it has seen an increasing number of applications, not only in food quality laboratories, but also in factories, and even in farm fields and orchards, where the food quality arises. In this paper, we have shortly described the most prominent application examples in this rapidly growing area of food quality techniques. Unfortunately there are already so many important works in this area that we were able to consider only a small fraction of the total literature existing, which means that too many important contributions had to be omitted from this paper. However, the authors hope that even in its current imperfect state this paper could help those who are trying to find solutions or model approaches to their food quality problems from optics, imaging, and spectroscopy.

Appendices

A.

A.1. Class A, Type 01: Fruits

A.1.1. Taste. Taste is an elemental parameter for fruits. Mehinagic et al. [46] demonstrated promising results in combining taste and spectra for apples.

A.1.2. Content and Factors Related to Content. A popular ingredient sought after for a fruit is sugar, for example, Walsh et al. [47]. Another important content is soluble solid content. Carlini et al. [48] evaluated the soluble solid content using VIS/NIR spectra. They modeled spectra using partial least-square regression.

The time-of-flight technique with NIR was applied to detect sugar and acid contents of apples [49]. The modeling used multiple linear regression, principal component regression, and partial least squares. The prediction precision was high and especially superior for acid content.

NIR has been used to predict water content of several different fruits. The water content of avocados was used by Blakey et al. [50] to predict ripening of the fruit. The water content was predicted using PLS (partial least squares) model with 14 factors in the region of 1100–2000 nm.

León et al. [51] studied the intact olive VIS/NIR spectra (400–1700 nm) for evaluation of oil, moisture, oleic acid, and linoleic acid contents. They compared spectra from different female parent trees and harvest year. The partial least-square calibration models were not transferable across year.

Zhang et al. [52] have recently used partial least-square (PLS) calibration model for the measurement of soluble solids content (SSC) of fresh jujube (*Ziziphus zizyphus* L.). The device used was FieldSpec3 (Analytical Spectral Device, USA). The correlation found was over 0.88 indicating the feasibility of the method for SSC measurement.

Terdwongworakul et al. [53] have used NIR transmittance spectroscopy for the quality assessment of mangosteen fruits. The goal was to develop a method for nondestructive detection of translucency in mangosteen fruit. The water peak absorbance at 955 nm seems to be an important factor for discriminating translucent and normal fruits.

Magwaza et al. [54] give a recent review of NIRS in external and internal quality analysis of citrus fruits.

Shi et al. [55] have used the genetic algorithm (GA) for wavelength selection for the determination of SSC in Fuji apples. They used NIRS in the range from 1065 to 1625 nm and were able to reduce the number of wavelengths from 88 to 17 with their GA approach.

Ying and Liu [56] have used FT-NIR spectroscopy for the internal quality assessment of pears. For wavelength selection, they also used genetic algorithms (GAs). The quality factors tested were sugar content (SC), titratable acidity (TA), and pH. Despite the complexity of the spectral data, the GA found calibration models that significantly outperformed those based on full-spectrum analyses.

Pissard et al. [57] suggested the use of NIR spectroscopy and multivariate calibration to determine the levels of vitamin C, total polyphenol, or sugar content. They stated that the prediction performance was very good.

A.1.3. Maturity. Maturity or fruit ripeness affects the quality and taste of the fruit. Tanaka and Kojima [58] studied the growth stages and maturity of Japanese pear fruit using the sugar concentration as an indicator. Pear juice from different maturity stages produced different spectra, the absorption bands of which were related to carbohydrate. Multiple linear regression models were applied to determine the content of sucrose, fructose, sorbitol, and glucose sugars with 3–7 wavelength variables. Sirisomboon et al. [59] used NIR with principal component analysis (PCA) to classify tomatoes. The model provided a prediction for maturity.

A.1.4. Unwanted Objects. Insects cause damage to fruits in many ways. Some of them can grow inside the fruit like in tart cherry. Xing et al. [60] have used genetic algorithms to find the optimal wavelengths for detection of internal insect (plum curculio (Coleoptera: Curculionidae)) infestation that in tart cherry. Based on the GA wavelength selection on the reflectance spectra (580–980 nm), three to four wavelength regions were selected. The results were checked using partial least-squared discriminant analysis (PLSDA) and they showed that the wavelengths selected resulted in classification efficiency similar to that of the full spectral range. Different GA runs resulted in somewhat different results, which are typical for the wavelength selection problem.

A.1.5. Contamination. Fruit skin can be contaminated by fecal material which can have pathogens like *E. coli*. Kim et al. [61] applied multispectral imaging 450–851 nm to detect fecal with thin and thick layer on apples with different green and red colorations. This was somehow successful for thick patches. Apple coloration caused problems with thin layer. Kim et al. [62] studied also fluorescence spectral information. They identified four optimal bands (450, 530, 685, and 735 nm) for fecal detection and suggested the use of a simple ratio.

A.1.6. Storage. Veraverbeke et al. [63] utilized NIR transmission spectra to solutions of single wax components and extracted apple wax. The wax was collected from apples with different parameters: apple cultivar (Jonagold, Jonagored, and Elshof), picking data (early, commercial, and late), storage time (0, 4, or 8 months in controlled atmosphere), and shelf life periods (0, 1, and 2 weeks). They applied canonical discriminant analysis on the first derivative of spectra. The proposed NIR technique could identify cultivar and storage duration but not very well the picking date or shelf life period. It seems that storage time and cultivars are related to differences in the number of aliphatic chains, like alkanes and esters, and the presence of α-farnesene.

A.2. Class A, Type 002: Vegetables. Fernández Pierna et al. [64] have used NIR hyperspectral imaging and chemometrics for the detection of cyst nematode (*Heterodera schachtii*) infected sugar beets. They used a pushbroom system (SWIR XEVA CL 2.5 320 TE4 camera from XENICS using an Imspector N25E spectrograph with a cooled, temperature-stabilised mercury-cadmium-telluride (MCT) detector from Specim Ltd., Oulu, Finland) combined with HyperPro and HyperSee software (BurgerMetrics SIA, Riga, Latvia). The samples were moved with a conveyor belt at 2 mm/s speed. The result of the study was that NIR imaging has potential for, for example, plant breeders to select cyst-nematode-resistant traits.

A.2.1. Appearance. The quality of a vegetable is sometimes evaluated by its texture. One of these kinds of vegetables is *Asparagus*, the texture of which has been evaluated by Flores Rojas [65]. The textural quality parameters considered in their research were maximum shear force and cutting energy. The evaluation was based on results from a scanning monochromator of 400–2500 nm and a combination of diode array and scanning monochromator of 350–2500 nm.

A.2.2. Content. Sugar concentration is important for many vegetables because it affects their sensory evaluation and storage ability. Sugar content of melon was evaluated and visualized using NIR absorption spectra by the autor in [66]. His paper indicated a strong inverse correlation for a melon between sugar content and specific wavelength (676 nm) near to the chlorophyll absorption band. The visualization was implemented by taking an image through a filter with band around 676 nm and converting the pixel values to another unit (absorbance) to assign them with a color. However, there was still a bias caused by melon color. Later, they developed another visualization method using deeper NIR region to prevent bias from melon color [67]. They noticed that the two 2nd derivative absorbances in 874 nm and 902 correlated strongly to the sugar content. For this visualization, two band-pass filters were employed.

Pedro and Ferreira [68] applied NIR and multivariate calibration to evaluate soluble solid, lycopene, β-carotene, and carotenoid contents of tomato. The PLS-1 was used for calibration. Prediction accuracy r values were over 0.99.

Protein content of edible fungi (*Auricularia auricular*) was studied by Liu et al. [69] They applied partial least-square, multiple linear regression, and least square support vector machine on data and compared performances of different preprocessing including Savitzky-Golay smoothing, standard normal variate, multiplicative scatter correction, first derivative, second derivative, and direct orthogonal signal correction. Wavelength selection was based on successive projections algorithm. Liu et al. [69] concluded that the best full spectrum partial least-square model was obtained using "raw" data whereas optimal successive projection algorithm with multiple linear regression, successive projection algorithm with partial least squares, and successive projection algorithm with least-square support vector machine models were achieved with spectra preprocessed multiplicative scatter correction. The optimal model was produced by successive projection algorithm with least-square support vector machine combination.

For cooked potatoes, dry matter content is important. Van Dijk et al. [70] studied three potato cultivars, cv. Nicola (a firm cooking potato), cv. Irene (a mealy cooking potato), and cv. Bintje (intermediate between firm and mealy cooking potato). They related NIR spectra to dry mater content and sensory perceived texture of steam cooked potatoes. The dry mater content affected most on the texture while the effect of chosen cultivar is minimal. Storage did not have any effect on the relationships.

Dacal-Nieto et al. [71] have used genetic algorithm (GA) for feature selection in potato classification done by the help of computer vision.

Roggo et al. [72] applied NIR for sucrose content with very small sucrose ester of fatty acid (SEF, fresh beet) and got satisfactory results for brix, marc, sugar in molasses, and juice purity but lower accuracy results for glucose, nitrogen, sodium, and potassium.

Soybean is an important vegetable in many countries. Soybeans have phytic acid which has nutritional and environmental importance. To evaluate the levels of the phytic acid, Delwiche et al. [73] measured its complement, inorganic phosphorus, using diffuse reflectance of near-infrared (1100–2500 nm) for ground meal and transmittance (600–1900 nm) for single beans. The results were not very good with partial least-square regression but give an indication. Kovalenko et al. [74] developed calibration equations for estimating amino acid composition of whole soybeans from near-infrared spectra. They also concluded that cysteine and tryptophan could not be estimated from spectra while leucine and lysine have a high correlation.

Pazdernik et al. [75] applied near-infrared reflectance spectroscopy to evaluate crude glycinin fraction of soybean protein. They find that the fraction was positively correlated to crude protein concentration and negatively to oil concentration. The high crude protein and fraction content was related to higher growing temperatures while the lower growing temperature produced higher oil concentration. Pazdernik et al. [76] studied also amino and fatty acid compositions. Hollung et al. [77] applied NIR reflectance on soybeans to find samples with a low content of oligosaccharides and nonstarch polysaccharides. They were able to predict that total content of nonstarch polysaccharides could be predicted using partial least-square regression.

The moisture content of whole soybean in the range of 5 to 22% can be determined with great accuracy using near-infrared transmittance (800 to 1100 nm) and difference in absorbance, stepwise multiple linear regression, or partial least-square modeling [78].

The classification based on soybean position can be useful, too. Bennett et al. [79] analyzed field-grown soybean seeds with different nodal positions with NIR. They found that seed protein content varied as a function of place: those from plant basal region have higher oil content and lower protein content while the opposite was true for the apex. Later in [80], they studied the seed quality traits (protein, oil) utilizing NIR to determine percentage of oil and protein. They noticed that early-planted crop has increased the amount of oil (3–8%) while there was not much change in protein content.

A.2.3. Freshness. Quality of mushroom can decrease if too long time elapsed from harvest. Esquerre et al. [81] used VIS and NIR absorption spectra to detect bruises and evaluate the freshness. The partial least-square model was employed and it was mostly influenced by changes in water matrix and enzymatic browning.

A.2.4. Authentication. Li et al. [82] have recently tested classification methods for the authentication of pure camellia oil NIR spectra. Sample authenticities were confirmed by gas chromatography. The classification methods were unsupervised principal component analysis (PCA), hierarchical cluster analysis (HCA), supervised classification techniques based on discriminant analysis (DA), and radical basis function neural network (RBFNN). The classification rate of all methods reached 98.3% when smoothing, first derivative, and autoscaling were used. The good performance showed that NIR spectroscopy could be successfully used as a rapid, simple, and nondestructive method to discriminate pure camellia oil.

A.2.5. Structural Characteristics. The physical and chemical properties of soybean and corn are correlated with size and density. However, the change on size and density has no effect in accuracy of determination of soybean protein and oil content. The effect is very small for determining protein, oil, and starch contents for corn [83].

A.2.6. Damage. Vegetables, mushrooms, and fungi can suffer physical damages for too rough handling. VIS/NIR absorption seems to be useful for quantifying bruise damages for mushrooms [81, 84].

A.2.7. Incorrect Ingredients. Roggo et al. [72] suggested the use of NIR spectroscopy for decreasing pollution of heavy metals, lead acetate, caused by the legal methods, polarimetric measurement for sucrose content, and wet chemical analysis for quality control.

A.2.8. Bioagents. The detection of fungi from food assumes those the spectra of fungi differ from that of the food itself. This assumption has been shown to be valid, for example, by Pearson and Wicklow [85] who detected the fungi in corn kernels, Gordon et al. [86] for food grain, and Berardo et al. [87] for maize.

Jin et al. [88] have used GAs to find the optimal wavelengths when trying to discriminate aflatoxin-producing *Aspergillus flavus* (AF13) fungi and non-toxin-producing *Aspergillus* strains (AF2038, AF283, and AF38). Their hyperspectral image was taken of several strains cultured 7 days on potato dextrose agar using both broad-spectrum UV light source and tungsten halogen bulbs. GA was used to find the most important principal components (PCs) among the first 40 PCs. The selected PCs were given to a support vector machine (SVM) for training for classification between toxigenic and atoxigenic fungi species. The discrimination accuracy varied between the strains and the light source used ranging from 70% to 95% pixel-wise.

Berardo et al. [87] applied NIR for the detection and identification of mycotoxigenic fungi and their toxic metabolites. They analyzed the samples for fungi infection, ergosterol, and fumonisin B_1 content. They concluded that NIR is suitable for monitoring mold contamination and separating the contaminated lots from clean lots.

A.3. Class A, Type 003: Grasses

A.3.1. Content. Cereal grains have many important ingredients which are searched in many studies. Also moisture is an important parameter. Dowell et al. [89] have applied NIRS for scab, vomitoxin, and ergosterol determination in single-wheat kernels. The results obtained show that the NIRS method was able to find all scab-damaged kernels that were identified by official inspectors.

Miralbés [90] measured NIR transmittance 850–1048 nm of whole wheat from different countries and evaluated their moisture, protein, wet gluten, dry gluten, and alveograph parameters (W, P, and P/L) using partial least squares. The alveograph parameters required division of wheat samples into two groups according to their deformation energy values.

Tartary buckwheat has beneficial ingredients, rutin and D-*chiro*-inositol, against hemorrhagic diseases and insulin-resistant diseases. Yang and Ren [91] applied NIR reflectance to evaluate their content. They developed a partial least-square method and used the results from HPLC as a reference. A good chemometric correlation existed.

Brenna and Berardo [92] studied maize whose antioxidant compounds like carotenoids are useful against reactive oxygen substances. They combined NIR and HPLC results for determining carotenoid concentration of grinded maize using modified partial least squares.

Rice is another very important food plant. Kawamura et al. [93] used NIR transmittance and PLS for successful evaluation of moisture and protein contents of undried rough rice. However, they found that their method was not applicable to determine amylase content. Zhang et al. [94] suggested use of NIR to predict total phenolics, flavonoid content and antioxidant capacity of dehulled grain of brown rice. The phenolic content, and antioxidant capacity were successfully determined using mPLS and PLS while this failed for flavonoids.

Stubbs et al. [95] measured residue from cultivars of spring wheat (*Triticum aestivum* L.), winter wheat, and spring barley (*Hordeum vulgare*) with NIR and made models to estimate neutral detergent fiber (NDF), acid detergent fiber (ADF), acid detergent lignin (ADL), carbon (C), sulfur (S), nitrogen (N), and C : N. The models were successful for neutral acid detergent fibers and acceptable for acid detergent lignin, for others less successful.

Wiley et al. [96] studied barley mutants. They evaluated grain nitrogen and protein composition. The grains were measured with NIR reflectance spectra, which were resolutionally enhanced by derivative spectroscopy. The advantage of derivative spectroscopy is to allow wavelengths to be identified with clear differences in contribution from the associated chemical bonds. They compare the 4th derivative of wholemeal flour with high and low *N* containing grains.

A.3.2. Structural Characteristics. Structural characteristics like hardness have effect on quality. For maize, the hardness was evaluated using NIR spectra in [97]. Manley et al. [98] used hyperspectral images of 1000 to 2498 nm wavelength range to classify yellow maize kernels. They found clusters for different endosperms (vitreous, floury, and 3rd type) which present varying ratios of soft and hard kernels. Cereal composition reveals the properties of inner structure. Siska and Hurburgh Jr. [83] applied NIR transmittance and PLS to obtain corn density which is related to starch and protein.

A.3.3. Taste. Taste from cereal food is affected by the grains themselves. The study of Kawamura et al. [99] indicates that VIS/NIR reflectance could be useful for rough classification of wholegrain milled rice using multiple linear regression (MLR) analysis into three groups: poor, better, and best taste.

A.3.4. Safety. Fernández Pierna et al. [64] applied near-infrared hyperspectral images for separating impurities from cereal grains. The cereal grains were wheat, spelt, and barley while the contaminants were a variety of different matters (straw, broken grains, grains from other crops, weed seeds, insects, plastic, stones, pieces of wood and paintings, and animal feces) classified into different categories. For each category, a discriminant model was developed. The method was based on PCA and support vector machine.

A.4. Class A, Type 004: Nuts and Seeds

A.4.1. Content. Nuts of shea tree can be used as a source of nut butter. The moisture, fat, stearic acid, and oleic acid can be determined using NIRS [100].

Front et al. [101] studied the sinigrin, gluconapin, 4-hydroxyglucobrassicin, and total glucosinolate of Indian mustard seed. They modeled the NIR spectra with modified partial least-square regression for chemometrics. The coefficients of determination were over 0.8 for all other except for 4-hydroxyglucobrassicin which was only 0.33. Font et al. [102] measured intact and ground seed of Indian mustard, Ethiopian mustard, and rapeseed and applied modified partial least-square regression to combine near-infrared spectra and acid detergent fiber values. The sample type did not have any significance on the evaluation. According to them, the most important factors for modeling were C–H and O–H groups of lipids, water, protein, and chlorophyll.

Kwan et al. [103] determined the oil and fatty acid contents of intact perilla seed using NIR reflectance. They concluded that mPLS produced good predictions for oil, oleic acid, and linolenic acid contents whereas the predictions for palmitic, stearic, and linoleic acid were unsuitable for screening purposes.

The content of intact common bean was subjected to NIR reflectance in order to determine seed weight, protein, and starch for 91 genotypes [104]. PLS model was able to predict protein content with high accuracy (SEP) = 1.6% and reasonable accuracy for weight and starch. Seed color had a correlation to spectra; black seed has very different spectra. Color did not affect model accuracy.

Wang et al. [105] used near-infrared and PLS to study proteins and amino acids in peanuts. They found the results acceptable. Yang and Sun [106] demonstrated that the hardness of licorice seeds can be done with NIR.

A.4.2. Authentication. Downey et al. [107] applied near- and mid-infrared spectra for authentication of lyophilized coffee samples of Arabica and Robusta cultivars. They used factorial discriminant analysis and partial least-square models.

Downey et al. [108] applied VIS and NIR transflectance spectra to detect adulteration of pure extra virgin olive oils.

The adulteration rates for the olive oil were 1% (w/w) and 5% (w/w) sunflower oil. The best classification was obtained using 1st derivative of 400–2498 nm spectra. The adulterant content was predicted with 0.8% standard error accuracy.

Woodcock et al. [109] studied European olive oil samples to specify their geographical origins with *near-infrared* transflectance spectroscopy (1100–2498 nm). The samples were collected from harvests of 2005, 2006, and 2007. The detection rate for Ligurian ones was 92.8% and non-Ligurian ones 81.5% when using the first-derivative data pretreatment and discriminant partial least-square regression.

A.5. Class A, Type 005: Herbs and Spices

A.5.1. Content. The herbs and spices contain many characteristic ingredients, the amount of which is often a concern. Often the goal is to find an alternative method for HPLC. The herbs and spices can come in many forms like leaves, roots, or powders. Chemometrics can be applied to all these forms.

Another example is kava which is an herb with anxiolytic properties caused by active compounds called kavalactones or kava pyrones. As a natural product, the amounts of active compounds vary between specimens. The NIR reflectance and partial least squares can be used to detect these compounds reliablly [110]. The results from the developed method are reproducible and have repeatability comparable to HPLC methods.

NIR reflectance can be used to determine several factors of the chemical composition of saffron according to Zalacain et al. [111]: moisture and volatile content, coloring strength, $E^{1\%}_{1\,\mathrm{cm}}$(250 nm), and $E^{1\%}_{1\,\mathrm{cm}}$(330 nm), established on the ISO 3632 Technical Specification Normative and used to certify saffron quality in the international market. They also analysed five main crocetin glycosides (compounds causing saffron color). The best correlations obtained for *trans*-crocetin di (β-d-gentibiosyl) ester was $R^2 = 0.93$, for *trans*-crocetin (β-D-glucosyl)-(β-D-gentibiosyl) $R^2 = 0.94$, and for picrocrocin (saffron bitterness) $R^2 = 0.92$.

Tanaka et al. [112] quantified curcuminoids (total curcuminoid content, curcumin, demethoxycurcumin, and bisdemethoxycurcumin) of turmeric. They applied 2nd derivative preprocessing and standard normal variate to NIR data and after that partial least-square regression. Characteristic absorption of curcuminoids was detected in approximately 1700 nm and 2300 nm–2320 nm. The results obtained have a high correlation with HPLC.

Many herbs and spices are often milled to powder-like Ginseng. Chemical methods like HPLC are expensive and sometimes they can be replaced by NIR/VIS spectroscopy 400–2500 nm. The authors in [113] studied it for American Ginseng root powder to detect ginsenosides. They found that it is applicable for major ginsenosides, Rb_1, Re, and m-Rb and total ginsenosides.

Sesame has some antioxidative-activity-related ingredients like lignans (including lipid-soluble lignans sesamin and sesamolin) and water-soluble lignin glycosides (sesaminol triglucoside and sesaminol diglucoside). Kim et al. [114] developed a NIR reflectance method for evaluating the lignin content of intact sesame seeds. The modeling was done using mPLS regression using results from high-performance liquid chromatography analysis as reference values. The results were good ingredients except for sesaminol diglucoside owning to its small amount.

Xin et al. [115] suggested a method to separate raw and processed herbs. They measured the herbs in the NIR area and used support vector machine, random forests, and linear discriminant analysis in analysing and classification. The obtained results were excellent.

A.5.2. Origin. Sometimes the origin is important for consumers. Zalacain et al. [111] suggested the use of near-infrared to determine the origin of saffron (Iran, Greece, and Spain).

B.

B.1. Class B, Type 06: Mammalian Products

B.1.1. Spoilage. Meat spoils very easily and thus produces a health risk not to mention unpleasantness. So excluding certain delicacies, the spoilage should be avoided or at least the spoilt meat should be removed before going to consumer. The spoilage for different meats has been studied for pork meat [116] and beef [117]. These papers considered also bacteria and microbiological factors.

B.1.2. Contamination. The fecal material is sometimes found on animal carcasses. Windham et al. [118] suggested the use of four optimal wavelengths: 434, 517, 565, and 628 nm. These wavelengths were selected using multivariate data analysis and VIS/NIR reflectance data. Different spectral ratios were tested to find optimal classification. The ratios 565 nm/517 nm provide 100% detection rate in carcasses of broilers fed a corn or soybean meal diet. Another ratio, 574 nm/588 nm was shown to be optimal for a larger variety of diets.

B.1.3. Taste (including Juiciness). One of the earliest references to taste of meat was already published in 1969 [119]. Byrne et al. [120] used NIR reflectance (750–1098 nm) to examine the taste of heifer cow. The samples from the *longissimus dorsi* (2.5 cm thick) were taken at 2, 7, and 14 days *post mortem*. They were vacuum-packaged in plastic bags and stored at −20°C.

The results from taste panel and chemical processes for lamb meat were evaluated against NIR reflectance by Andrés et al. [121]. They evaluated a set of 232 muscle samples from Texel and Scottish Blackface lambs. They concluded that absorbance is correlated to juiciness, flavour, abnormal flavor, and overall liking of lamb meat, but the correlation does not provide very clear classification. It was applicable for separating extreme samples.

B.1.4. Appearance. The appearance consists of color, geometrical features, and texture of meat. Texture of meat here is the general surface appearance.

Color of meat depends on water content and concentrations of myoglobin and also the relative proportions of myoglobin's different forms, oxymyoglobin, myoglobin, and ferric myoglobin [122, 123].

Meat color has been studied for decades and several methods have been suggested. For example, Van den Oord and Wesdorp [124] suggested the use of reflectance differences at 630 and 580 nm. Later, beef color prediction has been studied by Leroy et al. [125] and Shackelford et al. [126]. Shackelford et al. [126] showed that the best color prediction with NIR was obtained in reflectance mode. Leroy et al. [125] studied also beef, *longissimus thoracis*, and obtained similar results. For pork, color is an especially important parameter for customers which make their buy/not buy partially based on it. Cozzolino et al. [127] applied VIS/NIR for prediction.

Andrés et al. [128] applied VIS/NIR (400–2500 nm) for determining $L^*a^*b^*$ colour parameters. The results for luminosity at 0 and 60 min had a good predictability.

B.1.5. Diet. The diet affects the composition of beef: their subcutaneous fat and its fatty acid composition [129]. Prieto et al. [130] report that NIR can be used to separate whether or not the beef cows has been fed with flaxseed.

B.1.6. pH. pH is affected by light scattering [131–133] study produced over 0.8 values for coefficient of correlation when using first derivative and multiplicative scatter correction for calibration and validation sets. The obtained precision is good; only 4.27% of predictions had difference more than 0.2 pH. Andrés et al. [128] predicted pH of beef with NIR with good accuracy.

B.1.7. Content. Important factors for meat are lean-to-fat ratio, meat percentage, intramuscular fat, marbling, protein, water contents and muscle area. Certain ingredient, or their amount can be used as a guarantee for quality and used as a mark of quality.

Moisture of meat has been evaluated by Liao et al. [133] with coefficient of correlation near 0.8 for validation set using the first derivative and multiplicative scatter correction. Tornberg et al. [134] with no processing showed very poor predictability (0.21) for both VIS (400–800 nm) and NIR (800–2500). Chan et al. [135] used first derivate and preprocessing of spectra (400–1700 nm) and obtained results over 0.7 for the validation set.

Liao et al. [133] showed over 0.75 values for the calibration, and the validation sets for protein content. Values around 0.7 were obtained by Chan et al. [135].

The intramuscular fat is important quality factor which has been studied for beef [130, 136, 137] and pork [133, 137]. Sierra et al. [136] studied the composition and content of intramuscular fat has been studied by using NIR transmittance 850 to 1050 nm. They evaluated the fatty acid profile for samples of ground beef from the *longissimus thoracis* of yearling bulls. The study indicates that it is possible with relatively good accuracy to evaluate saturated, branched, and monounsaturated fatty acids. NIR transmittance also provides information about conjugated linoleic acids. Prieto et al. [130] used NIR reflectance 400 to 2498 nm for determining concentration of polyunsaturated fatty acids and their biohydrogenation products in beef cows which have been fed on special diet (flaxseed).

Studies of intramuscular fat for pork [132–135, 137, 138] have dissimilar results. The best results were from Prevolnik et al. [137] where the coefficient of determination value was over 0.9. Very poor results were obtained by Tornberg et al. [134] when no preprocessing was used with VIS (400–800 nm) whereas NIR (800–2500 nm) provided almost twice as good results. The first derivative of spectra with preprocessing provided further improvement [135]. One reason for difference could be different wavelength areas: Tornberg et al. [134] used visible range 400–800 nm, Barlocco et al. [138] used NIR (800–2500 nm), Savenije et al. [132] used VIS/NIR (400–2500 nm), and Liao et al. [133] used VIS/NIR (350–1100 nm). The homogeneity of samples varies [133, 137, 138]; for example, Liao et al. [133] and Savenije et al. [132] used intact pork samples. Minced pork samples were used, for example, in [137, 138]. Liao et al. [133] studied in addition to intramuscular fat protein and water content for pork in 350–1100 nm.

The intramuscular fat and water for lamb meat were predicted at a higher accuracy and they seem to correlate to the sensory parameters [121].

Iron-related substances can also be evaluated from meat. Hong and Yasumoto [139] used NIR diffuse reflectance from 1100 to 2500 nm to study heme and nonheme iron in raw meats: beef, pork, chicken, rabbit, duck, horse, lamb, and mutton. The second derivate was taken from data and after that multiple regression equations were applied. The result indicates that the multiple correlation coefficients are over 0.9 for both hemes and total iron. Hong and Yasumoto [139] obtained the best results with multiple regression equation at wavelengths where NH groups in protein molecules and C–C and CH in the porphyrin ring of heme have absorption.

The same NIR wavelength range (1100–2500 nm) has been tested for evaluating crude protein (CP), myoglobin, collagen, ether extract (EE), gross energy (GE), dry matter (DM), and ash content of oxen meat [140]. Prieto et al. [140] found that myoglobin, collagen, and ash contents could be evaluated with poor predictability. They speculate that this was due to lack of correlation between these parameters and ether extract content. However, ether extract evaluation was successful and important since this parameter is highly related to the organoleptic characteristics (like juiciness, flavor, and texture) in oxen meat samples. Also other parameters like crude protein, dry matter, and gross energy, which have a high correlation with ether extract content, were predicted with an acceptable degree of accuracy.

Pérez-Marín et al. [141] applied NIR for all stages of pigs: the live animal, carcass in slaughterhouse, subcutaneous fat sample with skin, skin-free subcutaneous fat sample, and transverse section. They develop prediction model for the four main fatty acids in Iberian pig fat, obtaining for palmitic acid SECV values of 1.24% for in vivo analysis and 0.82% for carcass analysis, for stearic acid 0.67% and 0.94%, for oleic acid 1.42% and 1.48%, and for linoleic 0.36% and 0.55%. The model proved the feasibility of NIR because it accounted for between 60% and 74% of the variation of spectra of live animals and between 31% and 87% of variation of carcasses.

For Iberian pigs, also intact fat or adipose tissues are studied [142–144].

Lamb muscles appear to have special absorption bands at 424 nm and 550 nm. These bands are related to the Soret band and the oxymyoglobin absorption band [145, 146]. The absorption bands (at 540 nm and at 580 nm) correlate with both myoglobin and oxymyoglobin [145, 147, 148], but myoglobin has also another absorption band at 555 nm [145–150] with the absorption band at 574 nm with oxyhaemoglobin absorption.

According to Murray [151], an absorption band at 762 nm is caused by the OH third overtone or according to Swatland [145] and Delpy and Cope [149], an absorption band produced by the oxidation of the myoglobin (deoxymyoglobin). Pigments and other characteristics (intramuscular fat, fatty-acids, and moisture) cause most of the spectral information used for the VIS region (400–700 nm) and the NIR region around 1400–2300 nm.

Similar results are reported by others like Downey et al. [152]; Lomiwes et al. [153] predicted glycogen concentration of prerigor muscle.

B.1.8. Maturity. The age of animal can affect the meat quality since there can be differences in intramuscular fat and water contents. Prieto et al. [154] applied successfully NIR reflectance 1100–2500 nm to separate adult, over 4 years old castrated steers and young, under 14 months old cattle. The samples were taken from *longissimus thoracis* muscle.

B.1.9. Shear Force. Shear force value indicates how much force is needed to cut the meat and it is related to tenderness. In fact, Yancey et al. [155] studied the connection between shear force (Warner-Bratzler shear force (WBSF), and Meullenet-Owens razor shear (MORS)) and tenderness (consumer evaluation, also with overall impression) for beef ribeye rolls taken from Select, low Choice, top (upper 2/3) Choice, and Prime quality grade carcasses with visible and near-infrared spectroscopy reflectance. Steaks were cut 2.54 cm thick and put to aging for 14 or 28 days. The longer aging of samples was related to increased tenderness. The 2nd derivatives of VIS/NIR measurements were found to be capable for predicting tenderness and overall impression than shear force measures.

Many studies for shear force have been made, for example, Byrne et al. [120] 1998, Park et al. [156] Rodbøtten [157] and Venel et al. [158]. Byrne et al. [120] evaluated Warner-Bratzler shear force using NIR 750–1098 nm for Heifer cows. They obtained over 0.7 value of correlation coefficient for both groups, for samples taken at 2, 7 and 14 days *post mortem*, vacuum-packaged in plastic bags and stored at −20°C. However, the results seem to vary. Prieto et al. [159] studied beef and cooked pork and suggest significant variation as a function of position. Shackelford et al. [126] applied shear force detection online for beef. The NIR transmission measurement provided better results than reflectance measurements [125]. The ability was limited *longissimus thoracis* at 2 or 8 days *post mortem*. Venel et al. [158] suggest using additional features (age, grade, pH, and sex) to improve results.

The results for pork meat have different conclusions. Study of Geesink et al. [160] indicates poor predictability. Liao et al. [133] obtained relatively good prediction for the calibration set but over 2.5 times poorer for the validation set. It has been discussed that the problem is the significant variations at different positions of the samples.

B.1.10. Tenderness. Tenderness of a beef describes the resistance to bite or shear meat and is related to the hardness. It should be noted that the characteristics can correlate to each other, like tenderness of meat that has a positive correlation to flavor and juiciness [161].

Generally the customers prefer beef with low effort needed and thus tenderness is an important factor related to quality. NIR models have predictions rating fair to good for the tenderness of postrigor *longissimus* muscle in many publications [120, 148, 156, 162], and [157]. Common measurement ranges were 1100–2500 nm and 750–1100 nm. It seems that absorption increases when the meat becomes tougher. Tenderness predictions for prerigor longissimus muscle were not so good [134, 157] which may indicate that the change from before rigor to after rigor and parameters related to it might be critical for tenderness. The applicability of techniques is different: extracted part of a muscle sample [156, 162–164] or a very small area ($4\,cm^2$) [120]. Shackelford et al. [165] have extended the technique for exposed longissimus cross-section of ribbed beef. Later, an online detection of tenderness has been tested by Shackelford et al. [126].

B.1.11. Texture of Meat. Byrne et al.'s [120] results showed correlation coefficient values of 0.5 and 0.7 for heifer meat texture by NIR. Texture of lamb meat has been evaluated by Andres et al. [121], but although absorbance does correlate to texture, it was not suitable for classification.

B.1.12. Acceptability. The acceptability of heifer meat seems to be very poorly predicted by NIR according to Byrne et al. [120]. The correlation coefficient values were under 0.5.

B.1.13. Sensory Characteristics other than Appearance. It would be advantageous to classify the beef based on its quality directly in the abattoir. Prieto et al. [159] applied VIS/NIR fiber optics measurements (350 to 1800 nm) to very fresh beef muscle to correlate it with lab color, cooking loss, instrumental texture (Volodkevitch, 10-day aged meat; slice shear force, 3 and 14-day aged meat), and sensory characteristics. However, the results indicated that only the sensory characteristics correlation was useful.

B.1.14. Structural Properties. Structural properties are related to functional quality. They can be water holding capacity, isometric tension, muscle fiber shortening, pH, and cooking loss.

Shackelford et al. [126] evaluated NIR for water-holding capacity and cooking loss for beef. The average reflectance showed higher absorbance for samples with highest cooking loss with meat 2 days *post mortem*. The difference between spectra diminished when the meat was 8 days *post mortem*. The transmission spectra improved the results. The drip loss prediction was not very high. It was best for reflectance at day 8 with a correlation coefficient of 0.54. Tornberg et al. [134] studied the drip loss for porcine meat using wavelength 802–2500 nm. They obtained a correlation coefficient of 0.64.

The study of Prieto et al. [159] suggest that VIS/NIR may not be very useful for determining cooking loss. Drip loss and shear force for pork were studied by Geesink et al. [160]. The data (reflectance spectra (1000–2500 nm), water-holding capacity, shear force, ultimate pH, and colour (L*, a*, b*-value) of 96 pork *longissimus*) were obtained two days *post mortem* and subjected to stepwise multiple linear regression (SMLR) and partial least-square regression (PLSR). The drip loss was predicted successfully while shear force was not.

B.1.15. Other Subjects. Shackelford et al. [165] studied optimal measurement for NIR and VIS. They recommended for online VIS/NIR studies of *longissimus* quality traits of ribbed beef carcasses: a high-intensity probe (with a 50 mm diameter hole in the limiter plate), data acquirement was done with averaging 20 spectra per observation, measurement place *longissimus* from one side per carcass, and at a standardized bloom time for spectra taking. Offline measurement application to lamb and pork and smaller muscles may require a reduction in the size of the opening in the limiter plate.

Information about sample characteristics, process, and slaughter conditions can be found, for example, in Cozzolino et al. [166], Cozzolino et al. [146], Cozzolino et al. [167], Mitsumoto et al. [148] 1991, Swatland [145], and Cozzolino and Murray [150].

B.1.16. Authentication. Authenticity of meet has several viewpoints. One viewpoint is the origin, another adulteration, and the last one ecological point. The common factor is that consumers should get what they intend to buy, even when the food bought is otherwise acceptable. Some people prefer organic food for their perceived superiority. The composition of meat, flavor, color, and fatty acid composition [168] of cattle and sheep depend heavily on diet.

Dian et al. [168] used VIS 400–700 nm and VIS/NIR 400–2500 nm with laboratory monochromator for separating pasture-fed and concentrate-fed lamb carcasses. They evaluate for this purpose three methods: (1) composition based on carotenoid absorption and perirenal fat reflectance at 450–510 nm, (2) composition based on perirenal fat reflectance at 400–700 nm, and (3) full spectrum of 400–2500 nm. The measured target was especially the perirenal fat because it is always present and can be removed without affecting carcass quality, and it provides better results than subcutaneous caudal fat [169]. The methods assessed the pasture-fed lambs by classification rates of 89.1%, 90.8%, and 97.5% and for concentrate-fed lambs by 98.6%, 98.6%, and 97.8%, respectively.

Cozzolino and Murray [170] applied VIS/NIR 400–2500 nm spectra for classifying the meat from beef, lamb, pork, and chicken. The achieved accuracy was over 80% with PCA and PLS.

Ding and Xu [171] obtained ideal results while separating beef and kangaroo samples. All kangaroo samples were identified with 100% accuracy.

B.1.17. Drip Loss/Water Holding Capacity. The water-holding capacity is a major pig meat technological property. Prevolnik et al. [172] used artificial neural network to predict protein drip loss from NIR with error of 2.5–2.6%.

B.1.18. Meat Freshness. The meat freshness can be thought to be related to its butchery date and possible freezing/thawing. Downey and Beauchêne [173] successfully applied VIS/NIR for separating the fresh meat from frozen/thawed beef using factorial discriminant procedure. Another study to separate frozen and unfrozen beef was done by Thyoldt and Isaksson [174].

B.1.19. Spoilage. Spoilage of beef has been studied by Ellis et al. [117]; Horváth et al. [116] applied NIR spectroscopy to pork meet.

B.1.20. Bioagents. Hobson et al. [175] suggested the use of spectral methods for detecting the metabolic state of microorganisms.

B.1.21. Milks

General Quality. Milk quality measured using NIR has been suggested by Kawasaki et al. [176]. Al-Qadiri et al. [177] studied the quality with visible (starting at 600 nm) and short infrared NI Karoui and De Baerdemaeker [178] give a review of analytical methods including NIRS coupled with chemometric tools for the quality determination and identity of dairy products.

Content. Laporte and Paquin [179] used partial and overall PLS calibration of near-infrared transmission data to evaluate its fat, crude protein, true protein, and casein content. The results indicated that NIR provides reliable results for the evaluation purposes and overall PLS calibration can improve the results.

Šašić and Ozaki [180] subjected raw milk to SW-NIR (800–1100 nm) measurements. They predicted the fat, protein, and lactose contents using PLS regression and band assignment. The correlation of spectra and milk protein was sufficiently high. Fat content prediction needed pretreatment. Prediction of lactose content was limited.

Adulteration. Milks are also subject to adulteration. Some cases of the adulteration can be subtle like adding foreign fat. Sato et al. [181] used NIR analysis to detect these fats.

B.2. Class B, Type 07: Poultry Products. Chicken meat products need to be controlled because of their safety risk and quality. These issues have been considered in many papers such as Chen et al. [182], Chen and Marks [183], Chen and

Marks [184], Chen et al. [185], McElhinney et al. [186], and Rannou and Downey et al. [187].

Park et al. [188] have recently constructed a prototype multispectral imaging system for online poultry carcass fecal contamination detection. They used a line-scan imaging system using 517, 565, 600, and 802 nm. The longest wavelength was used for removing artifacts caused by cuticle.

B.2.1. Incorrect Ingredients. The poultry meat can be contaminated by many things. One of the most dangerous things is the manure which contains many disease-causing microorganisms.

B.2.2. Content. The poultry meat is evaluated for ash, dry matter (DM), crude protein (CP), and fat content. The meat can be in different formats-like fresh or dried. Freeze dried ostrich meat tenderloin (*M. ambiens*), big drum (*M. iliofibularis*), and fan fillet (*M. gastrocnemius*) were evaluated by Viljoen et al. (2005) using NIR 1100–2500 nm and PLS 0. The results were accurate for crude protein and fat, but the prediction of ash and dry matter was not very successful.

Iron-related substances can also be evaluated from meat. Hong and Yasumoto [139] studied heme and nonheme iron in raw meats (e.g., chicken and duck) using NIR diffuse reflectance. They applied multiple regression equations, the second derivative data (1100 to 2500 nm) with multiple correlation coefficients over 0.9 for both hemes and total iron. The best MREs were obtained with the NIR data recorded at wavelengths where NH groups in protein molecules and C–C and CH in the porphyrin ring of heme give absorption.

B.2.3. Authenticity/Origin. The origin of chicken meat affects its value. Ding et al. [189] applied VIS/NIR (750–1100 nm) to separate broilers from Chinese local chickens using pieces of chicken meat cut.

B.2.4. Spoilage. Lin et al. [190] used chicken as a material for investigating spoilage. Poultry can be spoilt with too much heating. It seems that the intensity of wavelength bands around 445 nm and 560 nm decreases with cooking time when using correlation analysis [191]. After this decrease, the intensity begins to vary at 475, 520, and 585 nm.

B.2.5. Storage and Handling. It has been indicated that thawing may swallow the reabsorption of certain fluids like soluble proteins, vitamins, and salts by the meat. However, there is a risk of meat discoloration and lipid oxidation if the time for thawing is too long. Liu et al. [192] studied the effect of chilled and frozen storages (temperature, time) and shear force on chicken breasts using generalized 2D correlation analysis of VIS and NIR spectra. They noticed that the temperature of frozen storage changed intensities at 445 and 560 nm. The NIR spectra decreased. Tender and tough muscles caused difference in NIR spectral intensity. Liu et al. [193] examined the thawing behavior of frozen chicken meat with 2D VIS/NIR correlation.

B.3. Class B, Type 08: Aquatic Animal Products. Visible spectroscopy may sometimes be more useful than NIRS as found by Stormo et al. [194] who used visible spectroscopy for assessing endpoint temperature of heat-treated surimi. Their method shows potential for use in the food processing industry.

B.3.1. Appearance. Blood in fish is sometimes an unwanted characteristic. Olsen [195] and Olsen et al. [196] suggested that it might be possible to develop a method using VIS/NIR to access residual blood in intact cod muscle.

B.3.2. Identification and Authentication. Fishmeal batches can be made from different fish species. Cozzolino et al. [197] used NIR reflectance (1100 nm–2500 nm) and chemometrics to identify and authenticate the patches. They concluded that dummy partial least-square regression achieves 80–82% correct classification. Linear discriminant analysis based on principal component analysis scored achieved over 80% correct rate.

B.3.3. Freshness. Fish quality is largely determined by freshness. The freshness parameter can be linked to the time of catch or harvest or freezing. It can refer to instrumental analysis or sensory evaluation.

Uddin et al. [198] suggest using of VIS/NIR spectroscopy to determine whether the fish were frozen-thawed or fresh. Bøknaes et al. [199] used NIR spectroscopy and sensory quality evaluation to rate the freshness of thawed and chilled cod fillets packed in modified atmosphere. They studied the estimation of storage temperature, storage period, and chill storage period for fish. The storage time for cod and salmon was also investigated with VIS/NIR spectra by Nilsen et al. [200]. Several techniques like visible spectroscopy, texture, and image analysis can be combined to improve results [201]. The spoilage of rainbow trout has been studied by Lin et al. [202].

Zhang and Lee [203] used near-infrared for determining free fatty acids content of fish oil and mackerel. Since the change in free fatty acids is related to hypoxanthine, they suggested it to be used as an indicator of freshness. They found that the partial least-square model produced better results than multiple linear regression.

B.3.4. Content. Earliest fish study seems to be by Gjerde and Martens [204], they demonstrated that NIR can be used to predict water, fat, and protein contents in rainbow trout. They used reflectance spectroscopy to minced and freeze-dried samples. Since then, a lot of fish studies have been published and some at-line and online food quality controls. However, the technique is not yet a standard everyday instrumental tool. This might be due to high instrumentation costs and the need of models between spectroscopic reading and quality parameters in order to analyze them [31].

The fat content of fish has and still is under research. It has been applied on different species and different forms of fish: frozen and thawed glazed bigeye tuna [205], albacore [206], frozen skipjack [207, 208], raw and frozen horse mackerel

[206], frozen and thawed mackerel. Fish are sometimes ranked by their fat using visual inspection which seems to be an inaccurate method [205].

Mathias et al. [209] suggested the use of NIR for protein content analysis of freeze-dried freshwater fish. The standard deviation between results from NIR (mean lipid content) and chemical analysis was 3.7%. Mathias et al. [209] suggested also the use of NIR for lipid content analysis of freeze-dried freshwater fish. The standard deviation between results from NIR (mean lipid content) and chemical analysis was 10%.

Application of short-wavelength near-infrared spectra (600–1100 nm) can be used to determine the salt and moisture contents of salmon roe [7].

Lee et al. [210] applied diffusive reflectance of short-wavelength near-infrared in the range of 700–1000 for predicting the amount of crude lipid of intact, whole rainbow trout. The best correlation between measured and reference values was obtained from measurements at a midposition between dorsal and adipose fins above the lateral line. The modeling was done using multivariate calibration models, multiple linear regression, and partial least squares.

B.3.5. Defects. Lin et al. [211] used VIS and SW-NIR (600–1100 nm) to find bruises in intact, whole Pacific pink salmon. The measurements were done in diffuse reflectance mode and without removing skin or scales. The spectral results were compared against digital images of filleted fish with partial least-square model. The correct classification rate for nonbruised fish spectra was 84%, while for bruised it was 81%.

B.3.6. Incorrect Ingredients. Inorganic arsenic is an unwanted and unhealthy ingredient. Font et al. [212] subjected freeze-dried samples of red crayfish to inorganic arsenic content analysis. They applied a modified partial least-square regression for modeling with acid-digestion solvent-extracted hydride generation atomic absorption spectrometry measurements. The data was standardized by applying standard normal variate and detrending algorithms and preprocessed with 2nd derivative.

B.3.7. Bioagents. Fish can contain parasites which range from plain ugly appearance to dangerous for human health. Some food-borne illnesses caused by microorganisms can be prevented by proper handling like cooking.

B.3.8. Spoilage. Fish is spoiled easily and the spoiled fish is usually unsuitable for human consumption. Lin et al. [202] studied the spoilage for rainbow trout. They used 600–1100 nm region which may be applicable to detect various fundamental molecular vibrations of several functional groups like CH, O–H, N–H, and C=O. The absorption, reflectance, and transmittance of biochemical components help to determine the chemical composition and thus spoilage.

B.4. Class B, Type 009: Amphibians and Reptilians. The amphibians and reptilians group includes frogs, lizards, snakes, and turtles. This area seems to be lacking research papers.

B.5. Class B, Type 010: Invertebrate Animals. The invertebrate animals group consists of molluscs (including cephalopods) and other invertebrate animals.

B.5.1. Authenticity. Some edible invertebrate animals are more valuable and sought after than others. The processing of seafood causes appearance changes which make it very hard for a consumer to separate between authentic and adulterated products. This increases the risk of adulteration. Gayo and Hale [213] applied VIS and NIR spectra to find adulteration of crab meat. The crab meats (Atlantic blue and blue swimmer) were mixed with surimi-based adulterant in 10% increase. The adulterant increase caused decreasing water absorbance. They concluded that both PLC and principal component analysis can detect successfully adulteration. In another study, by Gayo et al. [214], VIS/NIR was applied to separate the Atlantic blue crabmeat adulterated with blue swimmer crabmeat (10% increment). Adulteration of meat changed the water absorption bands which could be used for adulteration detection.

B.5.2. Content. Brown [215] applied VIS and NIR (350–2500 nm) for homogenized oyster meat and analyzed the data quantitatively and qualitatively. The results for moisture (r^2 = 0.92, SEP = 0.53% wet weight), protein (r^2 = 0.97, SEP = 0.18% wet weight), fat (r^2 = 0.97, SEP = 0.11% wet weight), and glycogen (r^2 = 0.94, SEP = 0.24% weight) indicate that VIS/NIR is useful for oyster evaluations.

C.

C.1. Class D, Type 12: Secondary Food Commodities of Plant Origin

C.1.1. Content. Lebot et al. [216] applied NIR spectra (350–2500 nm) for flours from tropical root and tuber crops (cassava, sweet potato, taro, and yam). They evaluated starch, total sugars, cellulose, total nitrogen, and ash (total minerals) contents of the flours. The prediction accuracy achieved over 85% confidence for starch, sugar, and nitrogen, while for ash (minerals) it was 71%. The method was unsuitable for cellulose. The advantage of NIR is its cheapness.

Dacal-Nieto et al. [217] have recently used NIR multi-spectral imaging to detect the hollow heart of potato tubers. The camera used is sensitive from 900 nm to 1700 nm and used Specim Imspector N17E (Specim Oy, Oulu, Finland) spectrograph. Several wavelength selection methods were used, out of which the genetic algorithm was marginally the best, that is, gave the best classification accuracy in the classification step.

Green honeybush contains mangiferin and hesperidin which are bioactive components. Jourbert et al. [218] predicted content of these ingredients in dried plant material using diffuse NIR reflectance with results suitable for screening purposes.

Bellato et al. [219] studied wholegrains and flour of oats. The purpose was to determine β-glucan and protein content. They applied NIR reflectance for flour and grains and NIR transmittance for flours. The model for chemometric relationship was a partial robust M-regression, a version of partial least squares. The models using reflectance values stood out best.

For example, Schimleck et al. [220] evaluated successfully air-dried leaves from *Melaleuca cajuputi* tree for their 1.8-cineole (Eucalyptol) content and oil concentrations using NIR.

Evaluating roots provides an attractive possibility to select the individual containing the highest amount of ingredients or classify the roots in different groups according to some preset content classes. It is also possible to compare different root processing methods. Joubert et al. [221] studied Devil's claw root for evaluating iridoids (harpagoside and 8-ρ-coumaroyl harpagide (8ρCHG)) and moisture content from different drying conditions. The NIR study indicates that the harpagoside retention was lower in sun-dried samples than tunnel dried or freeze-dried samples. They concluded that the best retention was obtained from tunnel-drying at 50°C. NIRS was very suitable for predicting moisture and suitable for semiquantitative classification using the iridoids.

Leaves of green tea were subjected to NIR reflectance measurement and PLS modeling to predict polyphenol and alkaloid compounds (Schulz et al. [222]). The prediction of parameters of green tea was very successful for gallic acid, epicatechin, epigallocatechin, epicatechin gallate, epigallocatechin gallate, caffeine, and theobromine, but a lower accuracy was obtained for total polyphenol content probably due to the colorimetric method (lack of specificity).

The dry matter of leaves can be evaluated very precisely (Schulz et al. [222]) for green tea $R^2 = 0.94$; SD/SECV = 4.12. Furthermore, it is possible to discriminate tea leaves of different age by principal component analysis on the basis of the received NIR spectra.

Hernández-Hierro et al. [223] examined visible-NIR spectroscopy for freeze-dried broccoli. They concluded that it is potentially usable for screening of almost all individual glucosinolates and total glucosinolates.

C.1.2. Adulteration. Laasonen et al. [224] used PLS and pretreated NIR reflectance spectra to identify fast *Echinacea purpurea* dried and milled root from samples adulterated/substituted by other species (*Echinacea angustifolia*, *Echinacea pallida*, or *Parthenium integrifolium*). Minimum adulteration rate for detection was 10%.

C.2. Class D, Type 13: Derived Products of Plant Origin

C.2.1. Content. Juices contain many important ingredients whose concentration needs to be evaluated. Li et al. [225] measured transmission spectra of orange juice dry extract in the range of 1100–2500 nm. They used multivariate calibration to evaluate the amount of sugars (glucose, fructose, and sucrose) and acids (citric and malic acids). The reference values were obtained using enzymatic assays. The best model was obtained using partial least-square regression. Fruit juice from raw Japanese apricot which is also known as Japanese plum was also subjected to evaluation of citric and malic acid contents. Chen et al. [226] used NIR spectra of 1100–1850 nm and partial least-square regression for producing acceptable prediction results for the evaluation.

Sometimes the concentration of juice is important and one needs to get a high yield. For example, González-Sáiz et al. [227] applied successfully NIR and multivariate calibration for monitoring alcoholic fermentation of onion juice. The goal was the valorization of worthless onions.

Cen et al. [228] used VIS/NIR (325–1075 nm) for determining the soluble solids contents and pH in orange juice. The data was preprocessed with Wavelet packet transform, standard normal variate transformation (SNV), and Savitzky-Golay first-derivative transformation. Partial least-square (PLS) regression was applied and it produced best results with SNV.

We include in this juice group also cola beverages. For a cola beverage, important parameters are, for example, pH. Liu and He [229] applied chemometrics to VIS/NIR spectra to get the soluble solid content and pH of cola beverage with excellent prediction precision. Least squares with support vector machine provided better results than partial least squares.

Different oils have also ingredients which need to be evaluated. Ng et al. [230] used NIR (1100–2500 nm) for the evaluation of degradation products (including total polar material and free fatty acids) of frying oils. According to their results, partial least-square regression is better in 700–1100 nm area than forward stepwise multiple linear regression, while in the whole range, they are equal. The best correlations were over 0.98.

C.3. Class D, Type 014: Manufactured Foods (Single Ingredient) of Plant Origin

C.3.1. Content. Cozzolino et al. [231] did a review about near-infrared spectroscopy applied to grapes and wine.

Chen and Lei [232] applied VIS and NIR spectra to obtain the amount of carbohydrate in soy milk powder. Their goal was to find a high speed method for this. They searched optimal wavelength band in spectra using simulated annealing and wavelet packet transform tree. After this, different partial least-square models were tested using 20 variables.

Dambergs et al. [233] determined the methanol content of wine-fortifying spirit. They developed partial least-square model for NIR spectra with results from gas chromatography used as a reference.

Sáiz-Abajo et al. [234] used NIR to classify the samples into red/white wine vinegar and alcohol vinegar samples. They used orthogonal signal correction (OSC) method to remove information related to tartaric acid. Sáiz-Abajo et al. [235] studied 14 parameters which are important for the quality of wine vinegar.

Fan et al. [236] have recently used NIRS with several wavelength selection methods to predict the total acid of vinegar. Their main conclusion was that wavelength selection is essential for the creation of better regression models for acid prediction.

Bruun [237] has studied protein-lipid-carbohydrate interactions and their importance to the food quality.

Feng Liu et al. [238] have recently used VIS/NIR spectroscopy and least-square support vector machine (LS-SVM) for the soluble solid content (SSC) and pH determination of fruit vinegar. The results indicate that the method could be used in the online monitoring of vinegar fermentation.

C.3.2. Authenticity. Cozzoline et al. [239] used VIS and NIR spectra to classify Chardonnay and Riesling white wines according to their Australian varietal origins. Discriminant partial least-square (DPLS) regression was the most successful model tested; achieving 100% correct labeling for Riesling. Later, Liu et al. [240] classified Tempranillo wines from Australia and Spain according to their geographical origins. They accompanied both VIS and NIR data (400–2500 nm) for multivariate analysis PCA, DPLS, PLS-DA, and LDA from which PLS-DA was the best with high correct classification rate.

Buratti et al. [241] have tested NIR and IR spectroscopy combined with electronic nose and tongue for the offline monitoring of wine fermentation. The results are promising for the development of an online fermentation monitoring method based on these techniques.

C.3.3. Age. Yu et al. [242] applied LS-support vector machine (SUM) to NIR spectra to determine age and oenological parameters of rice wine. The inputs to LS-SVM applied were scores of ten principal components (PCs) and radial basis functions. LS-SVM was a little better than PLS for alcohol content, titratable acidity, and pH.

C.4. Class D, Type 015: Manufactured Foods (Multi-Ingredient) of Plant Origin

C.4.1. Content. Information about fat is often searched by consumer, for example, in cereals. However, the information is typically an estimate because the methods used like calorimetry destroy samples. Vines et al. [243] predicted the total amount of fat using near-infrared and modified partial least-square regression with good results when compared against reference values from the AOAC 996.01 method. Windham et al. [244] tested the effect of moisture to the calibration models and found that it has a significant effect. However, they developed a new partial least-square regression model to be robust enough to provide good prediction in a wide range of residual moistures. Kays et al. [245] used NIR reflectance (1100 to 2498 nm) to determine the total dietary fiber in cereal and grain products. AOAC (991.43) enzymatic-gravimetric method produced the reference samples.

Fernández Pierna et al. [64] have recently used NIR hyperspectral imaging and chemometrics for the detection of impurities in cereals. They used a whiskbroom system MatrixNIR Chemical Imaging instrument (Malvern instruments, Analytical Imaging, Columbia, MD, USA)) with an InGaAs FPA along with two liquid crystal tuneable filters (LCTF) for wavelength selection. The results show potential for NIR imaging in impurity detection. However, the whiskbroom system might not be as good in this task as a pushbroom system that they used in another case study with sugar beets.

C.4.2. Total Fiber Amount. Kays et al. [246] evaluated the total fiber amount of cereals using NIR spectra. The samples with high sugar and crystalline sugar were selected because a unique spectrum of crystalline sugar affects NIR spectra. The developed partial least-square model predicted the content well. Kays et al. [247] extended the NIR model to cereals with high fat or sugar and fat. The extension was necessarily because high fat or sugar may differ substantially from spectral properties.

Kays and Barton II [248] used NIR reflectance (1104–2494 nm) to determine the chemometric model for predicting gross energy and available energy of cereals. The reference values were obtained from bomb calorimetry. In another study, Kays and Barton II [249] determined accurately soluble and insoluble dietary fiber using NIR reflectance spectra (NIRSystems 6500 monochromator) and reference values from the AOAC Method 991.43. The modeling was done using modified PLS models.

Another aspect is how much energy comes from fat of ground cereals. Kays and Barton II [250] applied successfully near-infrared in the range of 1104–2494 nm to determine this using a version of PLS regression model. The sample reference was obtained using calorimetry. Archibald and Kays [251] analyzed whole intact breakfast cereals, snack foods, wholegrains, and milled products. They found that total dietary fiber could be evaluated from NIR reflectance spectra and moisture has not any effect on the results. Transmission and reflectance 700–1100 nm NIR spectra of cookies were measured by Ozanich et al. [252] to determine moisture and oil contents. Moisture content has highest correlation (and absorbance) with 965 nm wavelength (a combination band of H_2O) while oil content with 930 nm (a 3rd overtone CH stretch of methylene) in the normalized spectra.

C.4.3. Structural Properties. de Alencar Figueiredo et al. [253] studied a wide range of sorghum cores with high genetic diversity. They used NIR reflectance to determine amylase, protein, and lipid contents, endosperm texture, and hardness.

C.4.4. Handling/Fermentation. Bruun et al. [254] used NIR for analyzing gluten powder with different handling (heat, moisture). The NIR data was processed with second-derivative transformation and extended multiplicative signal correction. The moisture caused shifts and intensity changes in the protein bands. Bruun et al. [254] assumed this to be due to changes in water binding and microenvironments of the amino acid side chains. Heat caused an increase at 2209 nm and decrease at 2167–2182 nm which could be due to an α-helix to β-sheet transformation. Next, in [255] different salts were added to the hydrated gluten. The same preprocessing was used and PCA was applied on this data. The results showed a correlation between spectra and salt type and concentration.

Sinelli et al. [256] used NIR (750–1100 nm) and MIR (4000–600 cm^{-1}) with a fiberoptic surface interactance for dough proofing. Dough proofing is a phase during which the dough starts to ferment after the mixture of ingredients, it affects the quality of bread, and it was also noticed that NIR spectra changed.

D.

D.1. Class E, Type 016: Secondary Food Commodities of Animal Origin. Secondary food commodities of animal origin consist of dried meat and fish products and secondary milk products. We include here also products using dried meat and fish. An example of this is ham.

D.1.1. Appearance. García-Rey et al. [257] studied appearance (texture and color) of sliced samples from deboned and cross-sectioned dry-cured in VIS/NIR 400–2200 nm with a fiber-optic probe. The classification accuracy for pastiness was 88.5% and 79.7% when compared against sensory evaluation. They classified into two classes: defect class and no-defect class.

D.2. Class E, Type 017: Derived Edible Products of Animal Origin. Derived edible products of the animal origin group include processed crustaceans and animals, milk, and derived fats.

D.3. Class E, Type 018: Manufactured Food (Single Ingredient) of Animal Origin. Manufactured food (single ingredient) of the animal origin group is the manufactured milk products using only single ingredient.

D.4. Class E, Type 019: Manufactured Food (Multi-Ingredient) of Animal Origin. Manufactured milk products using multi-ingredient belong to manufactured food (single ingredient) of the animal origin group. Here we consider dairy products like cheese or yogurts.

D.4.1. Appearance. Lucas et al. [258] applied VIS and NIR for estimating color parameters (brightness, redness, and yellowness) with good results.

D.4.2. Smell. NIR seems to be suitable. Çelik [259] used NIR to evaluate yogurts' aroma and its fermentation [260].

D.4.3. Taste. Taste depends on the goodness of the fermentation process. NIR can be combined with electric nose as a method for online fermentation monitoring and visualizing the state of the fermentation [260].

D.4.4. Content. The titratable acidity of the cultures for yogurt and sour milk can be predicted by forming a model using PLS with NIRS signals [260].

Lucas et al. [258] applied VIS/NIR (400–2500 nm) for 4 cheese varieties. They estimated the content of cheese for the following items: dry matter, fat, pH, retinol, α-tocopherol, β-carotene, xanthophylls, sodium chloride, calcium, potassium, magnesium, zinc, and total antioxidant capacity. The data was handled with modified partial least squares. The coefficient of determination and residual predictive deviation indicated that the spectra provided a good estimate for dry matter, fat, β-carotene, sodium chloride, calcium, and zinc in fresh cheeses while it was poor for the rest (pH, retinol, α-tocopherol, xanthophylls, potassium, magnesium, and total antioxidant capacity).

D.4.5. pH. pH of the dairy product is important. Navrátil et al. [260] predicted its value using NIR.

D.4.6. Origin. Cozzi et al. [261] developed a NIR- (1100–2500 nm) based method using discriminant analysis to discriminate Asiago d'Allevo cheese from different production chains (alpine farms, mountain, and lowland factories). Alpine products were discriminable from factory products due to their different fatty acids profile and color. The results were confirmed by chemical data.

Abbreviations

ADF: Acid detergent fiber
ADL: Acid detergent lignin
AM: Amylose content
AOAC: Association of Official Agricultural Chemists
CH: Carbon-hydrogen
CP: Crude protein
DA: Discriminant analysis
DM: Dry matter
DPLS: Discriminant partial least squares
EE: Ether extract
ET: Endosperm texture
FAO: Food and Agriculture Organization
FPA: Focal plane array
GA: Genetic algorithms
GE: Gross energy
GP: Genetic programming
HCA: Hierarchical cluster analysis
HD: Hardness
HPLC: High-performance liquid chromatography
IR: Infrared
KDF: Kernel Fisher discriminant
L^*, a^*, b^*: Lab color space
LCTF: Liquid crystal tuneable filter
LDA: Linear discriminant analysis
LI: Lipid content
LS: Least squares
MCT: Mercury-cadmium-telluride
MIR: Mid-infrared
MORS: Meullenet-Owens razor shear
mPLS: Modified partial least squares

MRE: Multiple regression equations
NDF: Neutral detergent fiber
NIR: Near-infrared
NIRS: Near-infrared spectroscopy
OSC: Orthogonal signal correction
PC: Principal components
PCA: Principal component analysis
pH: Potential hydrogen
PLS: Partial least squares
PLSDA: Partial least-squared discriminant analysis
PLSR: Partial least-square regression
PR: Protein content
R^2: Coefficient of determination of a linear regression
RPFNN: Radical basis function neural network
SC: Sugar content
SMLR: Stepwise multiple linear regression
SEF: Sucrose ester of fatty acid
SPA: Successiveprojection algorithm
SSC: Soluble solid content
SVM: Support vector machines
SW-NIR: Short wavelength near-infrared
TA: Titratable acidity
UV: Ultraviolet
UVE: Uninformative variable elimination
VIP: Variable importance in projection
VIS: Visible wavelengths
VIS/NIR: Visual and near-infrared
WBSF: Warner-Bratzler shear force.

Acknowledgments

The authors want to acknowledge Tekes (the Finnish Funding Agency for Technology and Innovation) and industrial partners Riitan Herkku, Oy Ari Green Plant Ltd., Jepuan Peruna Oy, Etnovia Oy, and Specim, Spectral Imaging Ltd., for funding this research in a project called TarhaNIRSo (FruitForce) in Sapuska Research Program of Tekes. They also like to acknowledge Professor Paul Geladi for his many advices on NIRS applications.

References

[1] H. Huang, H. Yu, H. Xu, and Y. Ying, "Near infrared spectroscopy for on/in-line monitoring of quality in foods and beverages: a review," *Journal of Food Engineering*, vol. 87, no. 3, pp. 303–313, 2008.

[2] K. H. Norris and J. R. Hart, "Direct spectrophotometric determination of moisture content of grain and seeds," in *Proceedings of the International Symposium on Humidity and Moisture*, 1965.

[3] P. Williams, "Grains and seeds," in *Near-Infrared Spectroscopy in Food Science and Technology*, Y. Ozaki, W. F. McClure, and A. A. Christy, Eds., chapter 7.1, pp. 165–217, John Wiley & Sons, 2007.

[4] B. G. Osborne, "Near-infrared spectroscopy in food analysis," in *Encyclopedia of Analytical Chemistry*, John Wiley & Sons, 2006.

[5] J. Farkas and I. Dalmadi, "Near infrared and fluorescence spectroscopic methods and electronic nose technology for monitoring foods," *Progress in Agricultural Engineering Sciences*, vol. 5, no. 1, pp. 1–29, 2009.

[6] K. Kaffka, "How the NIR technology came to and spread in Europe for quality assessment and control in the food industry," *Acta Alimentaria*, vol. 37, no. 2, pp. 141–145, 2008.

[7] Y. Huang, T. M. Rogers, M. A. Wenz et al., "Detection of sodium chloride in cured salmon roe by SW-NIR spectroscopy," *Journal of Agricultural and Food Chemistry*, vol. 49, no. 9, pp. 4161–4167, 2001.

[8] P. Pallav, G. G. Diamond, D. A. Hutchins, R. J. Green, and T. H. Gan, "A near-infrared (NIR) technique for imaging food materials," *Journal of Food Science*, vol. 74, no. 1, pp. E23–E33, 2009.

[9] Y. Ozaki, W. F. McClure, and A. A. Christy, Eds., *Near-Infrared Spectroscopy in Food Science and Technology*, Wiley-Interscience, 2006.

[10] L. A. Berrueta, R. M. Alonso-Salces, and K. Héberger, "Supervised pattern recognition in food analysis," *Journal of Chromatography A*, vol. 1158, no. 1-2, pp. 196–214, 2007.

[11] L. M. Reid, C. P. O'Donnell, and G. Downey, "Recent technological advances for the determination of food authenticity," *Trends in Food Science and Technology*, vol. 17, no. 7, pp. 344–353, 2006.

[12] H. Cen and Y. He, "Theory and application of near infrared reflectance spectroscopy in determination of food quality," *Trends in Food Science and Technology*, vol. 18, no. 2, pp. 72–83, 2007.

[13] C.-J. Du and D.-W. Sun, "Learning techniques used in computer vision for food quality evaluation: a review," *Journal of Food Engineering*, vol. 72, no. 1, pp. 39–55, 2006.

[14] M. Iwamoto, J. Uozumi, and K. Nishinari, "Preliminary investigation of the state of water in foods by near infrared spectroscopy," in *NIR/NIT Conference*, Budapest, Hungary, 1987.

[15] M. Iwamoto, S. Kawano, and H. Abe, "Water in foods. Analysis of hydrogen bonding related to water in foods," *NIR News*, vol. 6, no. 3, pp. 10–12, 1995.

[16] H. Maeda, Y. Ozaki, M. Tanaka, N. Hayashi, and T. Kojima, "Near infrared spectroscopy and chemometrics studies of temperature-dependent spectral variations of water: relationship between spectral changes and hydrogen bonds," *Journal of Near Infrared Spectroscopy*, vol. 3, pp. 191–201, 1995.

[17] H. Büning-Pfaue, "Analysis of water in food by near infrared spectroscopy," *Food Chemistry*, vol. 82, no. 1, pp. 107–115, 2003.

[18] P. Geladi and H. Grahn, *Multivariate Image Analysis*, John Wiley & Sons, 1996.

[19] J. Workman Jr. and A. Springsteen, *Applied Spectroscopy: A Compact Reference for Practitioners*, Academic Press, 1998.

[20] J. Koljonen, T. E. M. Nordling, and J. T. Alander, "A review of genetic algorithms in near infrared spectroscopy and chemometrics: past and future," *Journal of Near Infrared Spectroscopy*, vol. 16, no. 3, pp. 189–197, 2008.

[21] D.-W. Sun, *Modern Techniques for Food Authentication*, Elsevier/Academic Press, 2008.

[22] J. Workman Jr. and L. Weyer, *Practical Guide to Interpretive Near-Infrared Spectroscopy*, CRC Press, Boca Raton, Fla, USA, 2008.

[23] J. R. Koza, "Genetic programming as a means for programming computers by natural selection," *Statistics and Computing*, vol. 4, no. 2, pp. 87–112, 1994.

[24] K. Norris and P. Williams, "Optimization of mathematical treatment of raw near-infrared signal in the measurement of

protein in hard red spring wheat. I. Influence of particle size," *Cereal Chemistry*, vol. 61, pp. 158–165, 1983.

[25] R. Goodacre and D. B. Kell, "Evolutionary computing for the characterisation and qualification of microbial systems from hyperspectral data," in *FT-IR Meeting in Berlin*, 2000.

[26] P. Williams and D. Sobering, "Comparison of commercial near infrared transmittance and reflectance instruments for analysis of whole grains and seeds," *Journal Near Infrared Spectroscopy*, vol. 1, no. 1, pp. 25–32, 1993.

[27] P. Butz, C. Hofmann, and B. Tauscher, "Recent developments in noninvasive techniques for fresh fruit and vegetable internal quality analysis," *Journal of Food Science*, vol. 70, no. 9, pp. R131–R141, 2005.

[28] B. M. Nicolaï, K. Beullens, E. Bobelyn et al., "Nondestructive measurement of fruit and vegetable quality by means of NIR spectroscopy: a review," *Postharvest Biology and Technology*, vol. 46, no. 2, pp. 99–118, 2007.

[29] C. M. Andersen, J. P. Wold, and S. B. Engelsen, "Autofluorescence spectroscopy in food analysis," in *Handbook of Food Analysis Instruments*, chapter 15, pp. 347–364, CRC Press, 2009.

[30] J. P. Wold and M. Mielnik, "Nondestructive assessment of lipid oxidation in minced poultry meat by autofluorescence spectroscopy," *Journal of Food Science*, vol. 65, no. 1, pp. 87–95, 2000.

[31] L. Nollet and F. Toldra, *Handbook of Seafood and Seafood Products Analysis*, CRC Press, Boca Raton, Fla, USA, 2010.

[32] H. B. Ding and R. J. Xu, "Near-infrared spectroscopic technique for detection of beef hamburger adulteration," *Journal of Agricultural and Food Chemistry*, vol. 48, no. 6, pp. 2193–2198, 2000.

[33] G. ElMasry, D.-W. Sun, and P. Allen, "Near-infrared hyperspectral imaging for predicting colour, pH and tenderness of fresh beef," *Journal of Food Engineering*, vol. 110, no. 1, pp. 127–140, 2012.

[34] Y. Liao, Y. Fan, and F. Cheng, "On-line prediction of pH values in fresh pork using visible/near-infrared spectroscopy with wavelet de-noising and variable selection methods," *Journal of Food Engineering*, vol. 109, no. 4, pp. 668–675, 2012.

[35] C. Collell, P. Gou, J. Arnau, and J. Comaposada, "Nondestructive estimation of moisture, water activity and NaCl at ham surface during resting and drying using NIR spectroscopy," *Food Chemistry*, vol. 129, no. 2, pp. 601–607, 2011.

[36] L. Dvash, O. Afik, S. Shafir et al., "Determination by near-infrared spectroscopy of perseitol used as a marker for the botanical origin of avocado (*Persea americana* Mill.) honey," *Journal of Agricultural and Food Chemistry*, vol. 50, no. 19, pp. 5283-5287, 2002.

[37] G. Downey and J. D. Kelly, "Detection and quantification of apple adulteration in diluted and sulfited strawberry and raspberry purées using visible and near-infrared spectroscopy," *Journal of Agricultural and Food Chemistry*, vol. 52, no. 2, pp. 204–209, 2004.

[38] R. Fügel, R. Carle, and A. Schieber, "Quality and authenticity control of fruit purées, fruit preparations and jams-a review," *Trends in Food Science & Technology*, vol. 16, no. 10, pp. 433–441, 2005.

[39] P. Y. Qiu, H. B. Ding, Y. K. Tang, and R. J. Xu, "Determination of chemical composition of commercial honey by near-infrared spectroscopy," *Journal of Agricultural and Food Chemistry*, vol. 47, no. 7, pp. 2760–2765, 1999.

[40] J. Fontaine, J. Hörr, and B. Schirmer, "Near-infrared reflectance spectroscopy enables the fast and accurate prediction of the essential amino acid contents in soy, rapeseed meal, sunflower meal, peas, fishmeal, meat meal products, and poultry meal," *Journal of Agricultural and Food Chemistry*, vol. 49, no. 1, pp. 57–66, 2001.

[41] Y. Kim, M. Singh, and S. E. Kays, "Near-infrared spectroscopy for measurement of total dietary fiber in homogenized meals," *Journal of Agricultural and Food Chemistry*, vol. 54, no. 2, pp. 292–298, 2006.

[42] W. Zeng, H. Zhang, and T. Lee, "Direct determination of the starch content in gravy by near-infrared spectroscopy," *Journal of Agricultural and Food Chemistry*, vol. 44, no. 6, pp. 1460–1463, 1996.

[43] M. R. Ellekjaer, T. Naes, T. Isaksson, and R. Solheim, "Identification of sausages with fat-substitutes using near infrared spectroscopy," in *Near Infrared Spectroscopy: Bridging the Gap Between Data Analysis and NIR Applications*, pp. 320–326, Ellis Horwood, London, UK, 1992.

[44] A. J. Gaitán-Jurado, V. Ortiz-Somovilla, F. España-España, J. Pérez-Aparicio, and E. J. De Pedro-Sanz, "Quantitative analysis of pork dry-cured sausages to quality control by NIR spectroscopy," *Meat Science*, vol. 78, no. 4, pp. 391–399, 2008.

[45] V. Ortiz-Somovilla, F. España-España, A. J. Gaitán-Jurado, J. Pérez-Aparicio, and E. J. De Pedro-Sanz, "Proximate analysis of homogenized and minced mass of pork sausages by NIRS," *Food Chemistry*, vol. 101, no. 3, pp. 1031–1040, 2006.

[46] E. Mehinagic, G. Royer, D. Bertrand, R. Symoneaux, F. Laurens, and F. Jourjon, "Relationship between sensory analysis, penetrometry and visible-NIR spectroscopy of apples belonging to different cultivars," *Food Quality and Preference*, vol. 14, no. 5-6, pp. 473–484, 2003.

[47] Walsh, J. Guthrie, and J. Burney, "Application of commercially available, low-cost, miniaturised NIR spectrometers to the assessment of the sugar content of intact fruit," *Australian Journal of Plant Physiology*, vol. 27, no. 12, pp. 1175–1186, 2000.

[48] P. Carlini, R. Massantini, and F. Mencarelli, "Vis-NIR measurement of soluble solids in cherry and apricot by PLS regression and wavelength selection," *Journal of Agricultural and Food Chemistry*, vol. 48, no. 11, pp. 5236–5242, 2000.

[49] S. Tsuchikawa and T. Hamada, "Application of time-of-flight near infrared spectroscopy for detecting sugar and acid contents in apples," *Journal of Agricultural and Food Chemistry*, vol. 52, no. 9, pp. 2434–2439, 2004.

[50] R. J. Blakey, J. P. Bower, and I. Bertling, "Influence of water and ABA supply on the ripening pattern of avocado (*Persea americana* Mill.) fruit and the prediction of water content using Near Infrared Spectroscopy," *Postharvest Biology and Technology*, vol. 53, no. 1-2, pp. 72–76, 2009.

[51] L. León, A. Garrido-Varo, and G. Downey, "Parent and harvest year effects on near-infrared reflectance spectroscopic analysis of olive (*Olea europaea* L.) fruit traits," *Journal of Agricultural and Food Chemistry*, vol. 52, no. 16, pp. 4957–4962, 2004.

[52] S. Zhang, Y. Zhao, W. Guo et al., "Robust model of fresh jujube soluble solids content with near-infrared (NIR) spectroscopy," *African Journal of Biotechnology*, vol. 11, no. 32, pp. 8133–8140, 2012.

[53] A. Terdwongworakul, N. Nakawajana, S. Teerachaichayut, and A. Janhiran, "Determination of translucent content in mangosteen by means of near infrared transmittance," *Journal of Food Engineering*, vol. 109, no. 1, pp. 114–119, 2012.

[54] L. S. Magwaza, U. L. Opara, H. Nieuwoudt, P. J. Cronje, W. Saeys, and B. Nicolaï, "NIR spectroscopy applications for internal

and external quality analysis of citrus fruit-a review," *Food and Bioprocess Technology*, vol. 5, no. 2, pp. 425–444, 2012.

[55] B. Shi, B. Ji, D. Zhu, Z. Tu, and Z. Qing, "Study on genetic algorithms-based NIR wavelength selection for determination of soluble solids content in fuji apples," *Journal of Food Quality*, vol. 31, no. 2, pp. 232–249, 2008.

[56] Y. Ying and Y. Liu, "Nondestructive measurement of internal quality in pear using genetic algorithms and FT-NIR spectroscopy," *Journal of Food Engineering*, vol. 84, no. 2, pp. 206–213, 2008.

[57] A. Pissard, J. A. Fernández Pierna, V. Baeten et al., "Nondestructive measurement of vitamin C, total polyphenol and sugar content in apples using near-infrared spectroscopy," *Journal of the Science of Food and Agriculture*, vol. 93, no. 2, pp. 238–244, 2013.

[58] M. Tanaka and T. Kojima, "Near-infrared monitoring of the growth period of Japanese pear fruit based on constituent sugar concentrations," *Journal of Agricultural and Food Chemistry*, vol. 44, no. 8, pp. 2272–2277, 1996.

[59] P. Sirisomboon, M. Tanaka, T. Kojima, and P. Williams, "Nondestructive estimation of maturity and textural properties on tomato "Momotaro" by nearinfrared spectroscopy," *Journal of Food Engineering*, vol. 112, no. 3, pp. 218–226, 2012.

[60] J. Xing, D. Guyer, D. Ariana, and R. Lu, "Determining optimal wavebands using genetic algorithm for detection of internal insect infestation in tart cherry," *Sensing and Instrumentation for Food Quality and Safety*, vol. 2, no. 3, pp. 161–167, 2008.

[61] M. Kim, A. Lefcourt, K. Chao, Y. Chen, I. Kim, and D. Chan, "Multispectral detection of fecal contamination on apples based on hyperspectral imagery: part I application of visible and near-infrared reflectance imaging," *Transactions of the ASABE*, vol. 45, no. 6, pp. 2027–2037, 2002.

[62] M. Kim, A. Lefcourt, Y. Chen, I. Kim, D. Chan, and K. Chao, "Multispectral detection of fecal contamination on apples based on hyperspectral imagery: part II application of hyperspectral fluorescence imaging," *Transactions of the American Society of Agricultural Engineers*, vol. 45, no. 6, pp. 2039–2047, 2002.

[63] E. A. Veraverbeke, J. Lammertyn, B. M. Nicolaï, and J. Irudayaraj, "Spectroscopic evaluation of the surface quality of apple," *Journal of Agricultural and Food Chemistry*, vol. 53, no. 4, pp. 1046–1051, 2005.

[64] J. A. Fernández Pierna, P. Vermeulen, O. Amand, A. Tossens, P. Dardenne, and V. Baeten, "NIR hyperspectral imaging spectroscopy and chemometrics for the detection of undesirable substances in food and feed," *Chemometrics and Intelligent Laboratory Systems*, vol. 117, pp. 233–239, 2012.

[65] K. U. Flores Rojas, *Determinación no destructiva de parámetros de calidad de frutas y hortalizas mediante espectroscopía de reflectancia en el infrarrojo cercano [doctoral thesis]*, Universidad de Córdoba, Córdoba, Spain, 2009.

[66] J. Sugiyama, "Visualization of sugar content in the flesh of a melon by near-infrared imaging," *Journal of Agricultural and Food Chemistry*, vol. 47, no. 7, pp. 2715–2718, 1999.

[67] M. Tsuta, J. Sugiyama, and Y. Sagara, "Near-infrared imaging spectroscopy based on sugar absorption band for melons," *Journal of Agricultural and Food Chemistry*, vol. 50, no. 1, pp. 48–52, 2002.

[68] A. M. K. Pedro and M. M. C. Ferreira, "Nondestructive determination of solids and carotenoids in tomato products by near-infrared spectroscopy and multivariate calibration," *Analytical Chemistry*, vol. 77, no. 8, pp. 2505–2511, 2005.

[69] F. Liu, Y. He, and G. Sun, "Determination of protein content of *Auricularia auricula* using near infrared spectroscopy combined with linear and nonlinear calibrations," *Journal of Agricultural and Food Chemistry*, vol. 57, no. 11, pp. 4520–4527, 2009.

[70] C. Van Dijk, M. Fischer, J. Holm, J. Beekhuizen, T. Tolle-Smits, and C. Boeriu, "Texture of cooked potatoes (*Solanum tuberosum*). 1. Relationships between dry matter content, sensory-perceived texture, and near-infrared spectroscopy," *Journal of Agricultural and Food Chemistry*, vol. 50, no. 18, pp. 5082–5088, 2002.

[71] A. Dacal-Nieto, E. Vázquez-Fernández, A. Formella, F. Martin, S. Torres-Guijarro, and H. González-Jorge, "A genetic algorithm approach for feature selection in potatoes classification by computer vision," in *Proceedings of the 35th Annual Conference of the IEEE Industrial Electronics Society (IECON '09)*, pp. 1955–1960, November 2009.

[72] Y. Roggo, L. Duponchel, and J. Huvenne, "Quality evaluation of sugar beet (*Beta vulgaris*) by near-infrared spectroscopy," *Journal of Agricultural and Food Chemistry*, vol. 52, no. 5, pp. 1055–1061, 2004.

[73] S. R. Delwiche, L. O. Pordesimo, A. M. Scaboo, and V. R. Pantalone, "Measurement of inorganic phosphorus in soybeans with near-infrared spectroscopy," *Journal of Agricultural and Food Chemistry*, vol. 54, no. 19, pp. 6951–6956, 2006.

[74] I. V. Kovalenko, G. R. Rippke, and C. R. Hurburgh Jr., "Determination of amino acid composition of soybeans (*Glycine max*) by near-infrared spectroscopy," *Journal of Agricultural and Food Chemistry*, vol. 54, no. 10, pp. 3485–3491, 2006.

[75] D. L. Pazdernik, S. J. Plehn, J. L. Halgerson, and J. H. Orf, "Effect of temperature and genotype on the crude glycinin fraction (11s) of soybean and its analysis by near-infrared reflectance spectroscopy (near-IRS)," *Journal of Agricultural and Food Chemistry*, vol. 44, no. 8, pp. 2278–2281, 1996.

[76] D. L. Pazdernik, A. S. Killam, and J. H. Orf, "Analysis of amino and fatty acid composition in soybean seed, using near infrared reflectance spectroscopy," *Agronomy Journal*, vol. 89, no. 4, pp. 679–685, 1997.

[77] K. Hollung, M. Øverland, M. Hrustić et al., "Evaluation of non-starch polysaccharides and oligosaccharide content of different soybean varieties (*Glycine max*) by near-infrared spectroscopy and proteomics," *Journal of Agricultural and Food Chemistry*, vol. 53, no. 23, pp. 9112–9121, 2005.

[78] D. T. Lamb and C. R. Hurburgh Jr., "Moisture determination in single soybean seeds by near-infrared transmittance," *Transactions of the American Society of Agricultural Engineers*, vol. 34, no. 5, pp. 2123–2129, 1991.

[79] J. O. Bennett, A. H. Krishnan, W. J. Wiebold, and H. B. Krishnan, "Positional effect on protein and oil content and composition of soybeans," *Journal of Agricultural and Food Chemistry*, vol. 51, no. 23, pp. 6882–6886, 2003.

[80] J. O. Bennett, O. Yu, L. G. Heatherly, and H. B. Krishnan, "Accumulation of genistein and daidzein, soybean isoflavones implicated in promoting human health, is significantly elevated by irrigation," *Journal of Agricultural and Food Chemistry*, vol. 52, no. 25, pp. 7574–7579, 2004.

[81] C. Esquerre, A. A. Gowen, C. P. O'donnell, and G. Downey, "Initial studies on the quantitation of bruise damage and freshness in mushrooms using visible-near-infrared spectroscopy," *Journal of Agricultural and Food Chemistry*, vol. 57, no. 5, pp. 1903–1907, 2009.

[82] S. Li, X. Zhu, J. Zhang, G. Li, D. Su, and Y. Shan, "Authentication of pure camellia oil by using near infrared spectroscopy and pattern recognition techniques," *Journal of Food Science*, vol. 77, no. 4, pp. C374–C380, 2012.

[83] J. Siska and C. R. Hurburgh Jr., "Corn density measurement by near-infrared transmittance," *Transactions of the American Society of Agricultural Engineers*, vol. 38, no. 6, pp. 1821–1824, 1995.

[84] A. A. Gowen, C. P. O'Donnell, M. Taghizadeh, P. J. Cullen, J. M. Frias, and G. Downey, "Hyperspectral imaging combined with principal component analysis for bruise damage detection on white mushrooms (Agaricus bisporus)," *Journal of Chemometrics*, vol. 22, no. 3-4, pp. 259–267, 2008.

[85] T. C. Pearson and D. T. Wicklow, "Detection of corn kernels infected by fungi," *Transactions of the ASABE*, vol. 49, no. 4, pp. 1235–1245, 2006.

[86] S. H. Gordon, B. C. Wheeler, R. B. Schudy, D. T. Wicklow, and R. V. Greene, "Neural network pattern recognition of photoacoustic FTIR spectra and knowledge-based techniques for detection of mycotoxigenic fungi in food grains," *Journal of Food Protection*, vol. 61, no. 2, pp. 221–230, 1998.

[87] N. Berardo, V. Pisacane, P. Battilani, A. Scandolara, A. Pietri, and A. Marocco, "Rapid detection of kernel rots and mycotoxins in maize by near-infrared reflectance spectroscopy," *Journal of Agricultural and Food Chemistry*, vol. 53, no. 21, pp. 8128–8134, 2005.

[88] J. Jin, L. Tang, Z. Hruska, and H. Yao, "Classification of toxigenic and atoxigenic strains of *Aspergillus flavus* with hyperspectral imaging," *Computers and Electronics in Agriculture*, vol. 69, no. 2, pp. 158–164, 2009.

[89] F. E. Dowell, S. E. Ram, and L. M. Seitz, "Predicting scab, vomitoxin, and ergosterol in single wheat kernels using near-infrared spectroscopy," *Cereal Chemistry*, vol. 76, no. 4, pp. 573–576, 1999.

[90] C. Miralbés, "Prediction chemical composition and alveograph parameters on wheat by near-infrared transmittance spectroscopy," *Journal of Agricultural and Food Chemistry*, vol. 51, no. 21, pp. 6335–6339, 2003.

[91] N. Yang and G. Ren, "Application of near-infrared reflectance spectroscopy to the evaluation of rutin and D-chiro-inositol contents in tartary buckwheat," *Journal of Agricultural and Food Chemistry*, vol. 56, no. 3, pp. 761–764, 2008.

[92] O. V. Brenna and N. Berardo, "Application of near-infrared reflectance spectroscopy (NIRS) to the evaluation of carotenoids content in maize," *Journal of Agricultural and Food Chemistry*, vol. 52, no. 18, pp. 5577–5582, 2004.

[93] S. Kawamura, M. Natsuga, and K. Itoh, "Determination of undried rough rice constituent content using near-infrared transmission spectroscopy," *Transactions of the American Society of Agricultural Engineers*, vol. 42, no. 3, pp. 813–818, 1999.

[94] C. Zhang, Y. Shen, J. Chen, P. Xiao, and J. Bao, "Nondestructive prediction of total phenolics, flavonoid contents, and antioxidant capacity of rice grain using near-infrared spectroscopy," *Journal of Agricultural and Food Chemistry*, vol. 56, no. 18, pp. 8268–8272, 2008.

[95] T. L. Stubbs, A. C. Kennedy, and A. Fortuna, "Using NIRS to predict fiber and nutrient content of Dryland cereal cultivars," *Journal of Agricultural and Food Chemistry*, vol. 58, no. 1, pp. 398–403, 2010.

[96] P. R. Wiley, G. J. Tanner, P. M. Chandler, and R. S. Anderssen, "Molecular classification of barley (*Hordeum vulgare* L.) mutants using derivative NIR spectroscopy," *Journal of Agricultural and Food Chemistry*, vol. 57, no. 10, pp. 4042–4050, 2009.

[97] G. Fox and M. Manley, "Hardness methods for testing maize kernels," *Journal of Agricultural and Food Chemistry*, vol. 57, no. 13, pp. 5647–5657, 2009.

[98] M. Manley, P. Williams, D. Nilsson, and P. Geladi, "Near infrared hyperspectral imaging for the evaluation of endosperm texture in whole yellow maize (*Zea maize* L.) Kernels," *Journal of Agricultural and Food Chemistry*, vol. 57, no. 19, pp. 8761–8769, 2009.

[99] S. Kawamura, M. Natsuga, and K. Itoh, "Visual and near-infrared reflectance spectroscopy for rice taste evaluation," *Transactions of the American Society of Agricultural Engineers*, vol. 40, no. 6, pp. 1755–1759, 1997.

[100] F. Davrieux, F. Allal, G. Piombo et al., "Near infrared spectroscopy for high-throughput characterization of shea tree (*Vitellaria paradoxa*) nut fat profiles," *Journal of Agricultural and Food Chemistry*, vol. 58, no. 13, pp. 7811–7819, 2010.

[101] R. Font, M. del Río, J. M. Fernández-Martínez, and A. de Haro-Bailón, "Use of near-infrared spectroscopy for screening the individual and total glucosinolate contents in indian mustard seed (*Brassica juncea* L. Czern. & Coss.)," vol. 52, no. 11, pp. 3563–3569, 2004.

[102] R. Font, M. del Río, J. M. Fernández, and A. de Haro, "Acid detergent fiber analysis in oilseed brassicas by near-infrared spectroscopy," *Journal of Agricultural and Food Chemistry*, vol. 51, no. 10, pp. 2917–2922, 2003.

[103] S. Kim, H. Park, and G. Choung, "Nondestructive determination of oil content and fatty acid composition in perilla seeds by near-infrared spectroscopy," *Journal of Agricultural and Food Chemistry*, vol. 55, no. 5, pp. 1679–1685, 2007.

[104] G. Hacisalihoglu, B. Larbi, and Mark Settles, "Near-infrared reflectance spectroscopy predicts protein, starch, and seed weight in intact seeds of common bean (*Phaseolus vulgaris* L.)," *Journal of Agricultural and Food Chemistry*, vol. 58, no. 2, pp. 702–706, 2010.

[105] L. Wang, Q. Wang, H. Liu, L. Liu, and Y. Du, "Determining the contents of protein and amino acids in peanuts using near-infrared reflectance spectroscopy," *Journal of the Science of Food and Agriculture*, vol. 93, no. 1, pp. 118–124, 2013.

[106] L. Yang and Q. Sun, "Recognition of the hardness of licorice seeds using a semi-supervised learning method and near-infrared spectral data," *Chemometrics and Intelligent Laboratory Systems*, vol. 114, pp. 109–115, 2012.

[107] G. Downey, R. Briandet, R. H. Wilson, and E. K. Kemsley, "Near- and mid-infrared spectroscopies in food authentication: coffee varietal identification," *Journal of Agricultural and Food Chemistry*, vol. 45, no. 11, pp. 4357–4361, 1997.

[108] G. Downey, P. McIntyre, and A. N. Davies, "Detecting and quantifying sunflower oil adulteration in extra virgin olive oils from the Eastern Mediterranean by visible and near-infrared spectroscopy," *Journal of Agricultural and Food Chemistry*, vol. 50, no. 20, pp. 5520–5525, 2002.

[109] T. Woodcock, G. Downey, and C. P. O'Donnell, "Confirmation of declared provenance of European extra virgin olive oil samples by NIR spectroscopy," *Journal of Agricultural and Food Chemistry*, vol. 56, no. 23, pp. 11520–11525, 2008.

[110] L. D. Gautz, P. Kaufusi, M. C. Jackson, H. C. Bittenbender, and C. Tang, "Determination of kavalactones in dried kava (*Piper methysticum*) powder using near-infrared reflectance

spectroscopy and partial least-squares regression," *Journal of Agricultural and Food Chemistry*, vol. 54, no. 17, pp. 6147–6152, 2006.

[111] A. Zalacain, S. A. Ordoudi, E. M. Díaz-Plaza et al., "Near-infrared spectroscopy in saffron quality control: determination of chemical composition and geographical origin," *Journal of Agricultural and Food Chemistry*, vol. 53, no. 24, pp. 9337–9341, 2005.

[112] K. Tanaka, Y. Kuba, T. Sasaki, F. Hiwatashi, and K. Komatsu, "Quantitation of curcuminoids in curcuma rhizome by near-infrared spectroscopic analysis," *Journal of Agricultural and Food Chemistry*, vol. 56, no. 19, pp. 8787–8792, 2008.

[113] G. Ren and F. Chen, "Simultaneous quantification of ginsenosides in American ginseng (*Panax quinquefolium*) root powder by visible/near-infrared reflectance spectroscopy," *Journal of Agricultural and Food Chemistry*, vol. 47, no. 7, pp. 2771–2775, 1999.

[114] S. Kim, H. Park, and G. Choung, "Nondestructive determination of lignans and lignan glycosides in sesame seeds by near infrared reflectance spectroscopy," *Journal of Agricultural and Food Chemistry*, vol. 54, no. 13, pp. 4544–4550, 2006.

[115] N. Xin, X. Gu, H. Wu, Y. Hu, and Z. Yang, "Discrimination of raw and processed *Dipsacus asperoides* by near infrared spectroscopy combined with least squares-support vector machine and random forests," *Spectrochimica Acta—Part A*, vol. 89, pp. 18–24, 2012.

[116] K. Horváth, Z. Seregély, É. Andrássy, I. Dalmadi, and J. Farkas, "A preliminary study using near infrared spectroscopy to evaluate freshness and detect spoilage in sliced pork meat," *Acta Alimentaria*, vol. 37, no. 1, pp. 93–102, 2008.

[117] D. I. Ellis, D. Broadhurst, D. B. Kell, J. J. Rowland, and R. Goodacre, "Rapid and quantitative detection of the microbial spoilage of meat by fourier transform infrared spectroscopy and machine learning," *Applied and Environmental Microbiology*, vol. 68, no. 6, pp. 2822–2828, 2002.

[118] W. R. Windham, D. P. Smith, B. Park, K. C. Lawrence, and P. W. Feldner, "Algorithm development with visible/near-infrared spectra for detection of poultry feces and ingesta," *Transactions of the American Society of Agricultural Engineers*, vol. 46, no. 6, pp. 1733–1738, 2003.

[119] H. L. A. Tarr, "Flavor of flesh foods," *Advances in Chemistry*, vol. 56, pp. 190–202, 1969.

[120] C. E. Byrne, G. Downey, D. J. Troy, and D. J. Buckley, "Non-destructive prediction of selected quality attributes of beef by near-infrared reflectance spectroscopy between 750 and 1098 nm," *Meat Science*, vol. 49, no. 4, pp. 399–409, 1998.

[121] S. Andrés, I. Murray, E. A. Navajas, A. V. Fisher, N. R. Lambe, and L. Bünger, "Prediction of sensory characteristics of lamb meat samples by near infrared reflectance spectroscopy," *Meat Science*, vol. 76, no. 3, pp. 509–516, 2007.

[122] M. Renerre and R. Labas, "Biochemical factors influencing metmyoglobin formation in beef muscles," *Meat Science*, vol. 19, no. 2, pp. 151–165, 1987.

[123] M. Renerre, "Factors involved in the discoloration of beef meat," *International Journal of Food Science and Technology*, vol. 25, no. 6, pp. 613–630, 1990.

[124] A. H. A. Van den Oord and J. J. Wesdorp, "Colour rating and pigment composition of beef: relation between colour measurement by reflectance spectrophotometry, subjective colour evaluation, and the relative concentration of oxymyoglobin and ferric myoglobin in chilled retail-packed beef," *Journal of Food Technology*, vol. 6, no. 1, pp. 15–20, 1971.

[125] B. Leroy, S. Lambotte, O. Dotreppe, H. Lecocq, L. Istasse, and A. Clinquart, "Prediction of technological and organoleptic properties of beef *Longissimus thoracis* from near-infrared reflectance and transmission spectra," *Meat Science*, vol. 66, no. 1, pp. 45–54, 2004.

[126] S. D. Shackelford, T. L. Wheeler, and M. Koohmaraie, "On-line classification of US Select beef carcasses for longissimus tenderness using visible and near-infrared reflectance spectroscopy," *Meat Science*, vol. 69, no. 3, pp. 409–415, 2005.

[127] D. Cozzolino, N. Barlocco, A. Vadell, F. Ballesteros, and G. Gallieta, "The use of visible and near-infrared reflectance spectroscopy to predict colour on both intact and homogenised pork muscle," *LWT—Food Science and Technology*, vol. 36, no. 2, pp. 195–202, 2003.

[128] S. Andrés, A. Silva, S. L. Soares-Pereira, C. Martins, A. M. Bruno-Soares, and L. Murray, "The use of visible and near infrared reflectance spectroscopy to predict beef M. longissimus thoracis et lumborum quality attributes," *Meat Science*, vol. 78, no. 3, pp. 217–224, 2008.

[129] J. D. Wood, M. Enser, A. V. Fisher et al., "Fat deposition, fatty acid composition and meat quality: a review," *Meat Science*, vol. 78, no. 4, pp. 343–358, 2008.

[130] N. Prieto, M. E. R. Dugan, O. López-Campos, T. A. McAllister, J. L. Aalhus, and B. Uttaro, "Near infrared reflectance spectroscopy predicts the content of polyunsaturated fatty acids and biohydrogenation products in the subcutaneous fat of beef cows fed flaxseed," *Meat Science*, vol. 90, no. 1, pp. 43–51, 2012.

[131] J. R. Anderson, C. Borggaard, A. J. Rasmussen, and L. P. Houmoller, "Optical measurements of pH in meat," *Meat Science*, vol. 53, no. 2, pp. 135–141, 1999.

[132] B. Savenije, G. H. Geesink, J. G. P. van der Palen, and G. Hemke, "Prediction of pork quality using visible near infrared reflectance spectroscopy," *Meat Science*, vol. 73, no. 1, pp. 181–184, 2006.

[133] Y. Liao, Y. Fan, and F. Cheng, "On-line prediction of fresh pork quality using visible/near-infrared reflectance spectroscopy," *Meat Science*, vol. 86, no. 4, pp. 901–907, 2010.

[134] E. Tornberg, M. Wahlgren, J. Brøndum, and S. B. Engelsen, "Pre-rigor conditions in beef under varying temperature- and pH-falls studied with rigometer, NMR and NIR," *Food Chemistry*, vol. 69, no. 4, pp. 407–418, 2000.

[135] D. E. Chan, P. N. Walker, and E. W. Mills, "Prediction of pork quality characteristics using visible and near-infrared spectroscopy," *Transactions of the ASAE*, vol. 45, no. 5, pp. 1519–1527, 2002.

[136] V. Sierra, N. Aldai, P. Castro, P. Osoro, A. Coto-Montes, and M. Oliván, "Prediction of the fatty acid composition of beef by near infrared transmittance spectroscopy," *Meat Science*, vol. 78, no. 3, pp. 248–255, 2008.

[137] M. Prevolnik, M. Candek-Potokar, D. Skorjanc et al., "Predicting intramuscular fat content in pork and beef by near infrared spectroscopy," *Journal of Near Infrared Spectroscopy*, vol. 13, no. 2, pp. 77–85, 2005.

[138] N. Barlocco, A. Vadell, F. Ballesteros, G. Galietta, and D. Cozzolino, "Predicting intramuscular fat, moisture and Warner-Bratzler shear force in pork muscle using near infrared reflectance spectroscopy," *Animal Science*, vol. 82, no. 1, pp. 111–116, 2006.

[139] J. Hong and K. Yasumoto, "Near-infrared spectroscopic analysis of heme and nonheme iron in raw meats," *Journal of Food Composition and Analysis*, vol. 9, no. 2, pp. 127–134, 1996.

[140] N. Prieto, S. Andrés, F. J. Giráldez, A. R. Mantecón, and P. Lavín, "Potential use of near infrared reflectance spectroscopy (NIRS) for the estimation of chemical composition of oxen meat samples," *Meat Science*, vol. 74, no. 3, pp. 487–496, 2006.

[141] D. Pérez-Marín, E. De Pedro Sanz, J. E. Guerrero-Ginel, and A. Garrido-Varo, "A feasibility study on the use of near-infrared spectroscopy for prediction of the fatty acid profile in live Iberian pigs and carcasses," *Meat Science*, vol. 83, no. 4, pp. 627–633, 2009.

[142] E. De Pedro, N. Núñez, J. García et al., "Implementing near infrared spectroscopy in the Iberian pig industry," *OPtions Méditerranéennes, Series A*, no. 76, pp. 225–227, 2007.

[143] J. García-Olmo, A. Garrido, and E. De Pedro, "Methodological aspects on near infrared analysis of Iberian pig fat using interactance-reflectance fiber optic mode," *Journal of Near Infrared Spectroscopy*, vol. 6, no. 1–4, pp. A307–A312, 1998.

[144] I. González-Martín, C. González-Pérez, J. Hernández-Méndez, and N. Álvarez-García, "Determination of fatty acids in the subcutaneous fat of Iberian breed swine by near infrared spectroscopy (NIRS) with a fibre-optic probe," *Meat Science*, vol. 65, no. 2, pp. 713–719, 2003.

[145] H. J. Swatland, *On-Line Evaluation of Meat*, Technomic Publishing, Lancaster, UK, 1995.

[146] D. Cozzolino, I. Murray, J. R. Scaife, and R. Paterson, "Study of dissected lamb muscles by visible and near infrared reflectance spectroscopy for composition assessment," *Animal Science*, vol. 70, no. 3, pp. 417–423, 2000.

[147] R. A. Lawri, *Meat Science*, Pergamon Press, Oxford, UK, 1985.

[148] M. Mitsumoto, S. Maeda, T. Mitsuhashi, and S. Ozawa, "Near-infrared spectroscopy determination of physical and chemical characteristics in beef cuts," *Journal of Food Science*, vol. 56, no. 6, pp. 1493–1496, 1991.

[149] D. T. Delpy and M. Cope, "Quantification in tissue near-infrared spectroscopy," *Philosophical Transactions of the Royal Society B*, vol. 352, no. 1354, pp. 649–659, 1997.

[150] D. Cozzolino and I. Murray, "Effect of sample presentation and animal muscle species on the analysis of meat by near infrared reflectance spectroscopy," *Journal of Near Infrared Spectroscopy*, vol. 10, no. 1, pp. 37–44, 2002.

[151] I. Murray, "The NIR spectra of homologous series of organic compounds," in *Proceedings of the International NIR/NIT Conference*, Budapest, Hungary, 1986.

[152] G. Downey, J. McElhinney, and T. Fearn, "Species identification in selected raw homogenized meats by reflectance spectroscopy in the mid-infrared, near-infrared, and visible ranges," *Applied Spectroscopy*, vol. 54, no. 6, pp. 894–899, 2000.

[153] D. Lomiwes, M. M. Reis, E. Wiklund, O. A. Young, and M. North, "Near infrared spectroscopy as an on-line method to quantitatively determine glycogen and predict ultimate pH in pre rigor bovine *M. longissimus dorsi*," *Meat Science*, vol. 86, no. 4, pp. 999–1004, 2010.

[154] N. Prieto, S. Andrés, F. J. Giráldez, A. R. Mantecón, and P. Lavín, "Discrimination of adult steers (oxen) and young cattle ground meat samples by near infrared reflectance spectroscopy (NIRS)," *Meat Science*, vol. 79, no. 1, pp. 198–201, 2008.

[155] J. W. S. Yancey, J. K. Apple, J. F. Meullenet, and J. T. Sawyer, "Consumer responses for tenderness and overall impression can be predicted by visible and near-infrared spectroscopy, Meullenet-Owens razor shear, and Warner-Bratzler shear force," *Meat Science*, vol. 85, no. 3, pp. 487–492, 2010.

[156] B. Park, Y. R. Chen, W. R. Hruschka, S. D. Shackelford, and M. Koohmaraie, "Near-infrared reflectance analysis for predicting beef longissimus tenderness," *Journal of Animal Science*, vol. 76, no. 8, pp. 2115–2120, 1998.

[157] R. Rødbotten, *Prediction of beef tenderness by NIR spectroscopy, and examination of the influence of ante- and post-mortem factors on tenderness [Ph.D. thesis]*, Agricultural University of Norway, Bergen, Norway, 2001.

[158] C. Venel, A. M. Mullen, G. Downey, and Troy, "Prediction of tenderness and other quality attributes of beef by near infrared reflectance spectroscopy between 750 and 1100 nm; further studies," *Journal of Near Infrared Spectroscopy*, vol. 9, no. 3, pp. 185–198, 2001.

[159] N. Prieto, D. W. Ross, E. A. Navajas et al., "On-line application of visible and near infrared reflectance spectroscopy to predict chemical-physical and sensory characteristics of beef quality," *Meat Science*, vol. 83, no. 1, pp. 96–103, 2009.

[160] G. H. Geesink, F. H. Schreutelkamp, R. Frankhuizen et al., "Prediction of pork quality attributes from near infrared reflectance spectra," *Meat Science*, vol. 65, no. 1, pp. 661–668, 2003.

[161] R. J. Winger and C. J. Hagyard, "Juiciness—its importance and some contributing factors," in *Quality Attributes and Their Measurement in Meat, Poultry and Fish Products*, A. M. Pearson and T. R. Dutson, Eds., vol. 9 of *Advances in Meat Research*, chapter 4, pp. 94–124, Aspen Publisher, Frederick, Md, USA, 1994.

[162] K. I. Hildrum, B. N. Nilsen, M. Mielnik, and T. Næs, "Prediction of sensory characteristics of beef by near-infrared spectroscopy," *Meat Science*, vol. 38, no. 1, pp. 67–80, 1994.

[163] K. I. Hildrum, T. Isaksson, T. T. Naes, B. N. Nilsen, M. Rodbotten, and P. Lea, "Near infrared reflectance spectroscopy in the prediction of sensory properties of beef," *Journal of Near Infrared Spectroscopy*, vol. 3, no. 2, pp. 81–87, 1995.

[164] T. Næs and K. I. Hildrum, "Comparison of multivariate calibration and discriminant analysis in evaluating NIR spectroscopy for determination of meat tenderness," *Applied Spectroscopy*, vol. 51, no. 3, pp. 350–357, 1997.

[165] S. D. Shackelford, T. L. Wheeler, and M. Koohmaraie, "Development of optimal protocol for visible and near-infrared reflectance spectroscopic evaluation of meat quality," *Meat Science*, vol. 68, no. 3, pp. 371–381, 2004.

[166] D. Cozzolino, I. Murray, and R. Paterson, "Visible and near infrared reflectance spectroscopy for the determination of moisture, fat and protein in chicken breast and thigh muscle," *Journal of Near Infrared Spectroscopy*, vol. 4, no. 1–4, pp. 213–223, 1996.

[167] D. Cozzolino, D. De Mattos, and D. Vaz Martins, "Visible/near infrared reflectance spectroscopy for predicting composition and tracing system of production of beef muscle," *Animal Science*, vol. 74, no. 3, pp. 477–484, 2002.

[168] P. H. M. Dian, D. Andueza, M. Jestin, I. N. Prado, and S. Prache, "Comparison of visible and near infrared reflectance spectroscopy to discriminate between pasture-fed and concentrate-fed lamb carcasses," *Meat Science*, vol. 80, no. 4, pp. 1157–1164, 2008.

[169] P. H. M. Dian, D. Andueza, C. M. P. Barbosa, S. Amoureux, M. Jestin, and P. C. F. Carvalho, "Methodological developments in the use of visible reflectance spectroscopy for discriminating pasture-fed from concentrate-fed lamb carcasses," *Animal*, vol. 1, no. 8, pp. 1198–1208, 2007.

[170] D. Cozzolino and I. Murray, "Identification of animal meat muscles by visible and near infrared reflectance spectroscopy,"

Lebensmittel-Wissenschaft und-Technologie, vol. 37, no. 4, pp. 447–452, 2004.

[171] H. B. Ding and R. J. Xu, "Differentiation of beef and kangaroo meat by visible/near-infrared reflectance spectroscopy," *Journal of Food Science*, vol. 64, no. 5, pp. 814–817, 1999.

[172] M. Prevolnik, M. Čandek-Potokar, M. Novič, and D. Škorjanc, "An attempt to predict pork drip loss from pH and colour measurements or near infrared spectra using artificial neural networks," *Meat Science*, vol. 83, no. 3, pp. 405–411, 2009.

[173] G. Downey and D. Beauchêne, "Discrimination between fresh and frozen-then-thawed beef m. longissimus dorsi by combined visible-near infrared reflectance spectroscopy: a feasibility study," *Meat Science*, vol. 45, no. 3, pp. 353–363, 1997.

[174] K. Thyholt and T. Isaksson, "Differentiation of frozen and unfrozen beef using near-infrared spectroscopy," *Journal of the Science of Food and Agriculture*, vol. 73, no. 4, pp. 525–532, 1997.

[175] N. S. Hobson, I. Tothill, and A. P. Turner, "Microbial detection," *Biosensors and Bioelectronics*, vol. 11, no. 5, pp. 455–477, 1996.

[176] M. Kawasaki, S. Kawamura, H. Nakatsuji, and M. Natsuga, "Online real-time monitoring of milk quality during milking by near infrared spectroscopy," in *ASABE Meeting Presentation*, 2005.

[177] H. M. Al-Qadiri, M. Lin, M. A. Al-Holy, A. G. Cavinato, and B. A. Rasco, "Monitoring quality loss of pasteurized skim milk using visible and short wavelength near-infrared spectroscopy and multivariate analysis," *Journal of Dairy Science*, vol. 91, no. 3, pp. 950–958, 2008.

[178] R. Karoui and J. De Baerdemaeker, "A review of the analytical methods coupled with chemometric tools for the determination of the quality and identity of dairy products," *Food Chemistry*, vol. 102, no. 3, pp. 621–640, 2007.

[179] M. Laporte and P. Paquin, "Near-infrared analysis of fat, protein, and casein in cow's milk," *Journal of Agricultural and Food Chemistry*, vol. 47, no. 7, pp. 2600–2605, 1999.

[180] S. Šašić and Y. Ozaki, "Short-wave near-infrared spectroscopy of biological fluids. 1. Quantitative analysis of fat, protein, and lactose in raw milk by partial least-squares regression and band assignment," *Analytical Chemistry*, vol. 73, no. 1, pp. 64–71, 2001.

[181] T. Sato, S. Kawano, and M. Iwamoto, "Detection of foreign fat adulteration of milk by near infrared spectroscopic method," *Journal of Dairy Science*, vol. 73, no. 12, pp. 3408–3413, 1990.

[182] Y. R. Chen, R. W. Huffman, B. Park, and M. Nguyen, "Transportable spectrophotometer system for on-line classification of poultry carcasses," *Applied Spectroscopy*, vol. 50, no. 7, pp. 910–916, 1996.

[183] H. Chen and B. P. Marks, "Evaluating previous thermal treatment of chicken patties by visible/near-infrared spectroscopy," *Journal of Food Science*, vol. 62, no. 4, pp. 753–780, 1997.

[184] H. Chen and B. P. Marks, "Visible/near-infrared spectroscopy for physical characteristics of cooked chicken patties," *Journal of Food Science*, vol. 63, no. 2, pp. 279–282, 1998.

[185] Y. R. Chen, B. Park, R. W. Huffman, and M. Nguyen, "Classification of on-line poultry carcasses with backpropagation neural networks," *Journal of Food Process Engineering*, vol. 21, no. 1, pp. 33–48, 1998.

[186] J. McElhinney, G. Downey, and T. Fearn, "Chemometric processing of visible and near infrared reflectance spectra for species identification in selected raw homogenised meats," *Journal of Near Infrared Spectroscopy*, vol. 7, no. 3, pp. 145–154, 1999.

[187] H. Rannou and G. Downey, "Discrimination of raw pork, chicken and turkey meat by spectroscopy in the visible, near- and mid-infrared ranges," *Analytical Communications*, vol. 34, no. 12, pp. 401–404, 1997.

[188] B. Park, S. Yoon, W. R. Windham, K. C. Lawrence, M. S. Kim, and K. Chao, "Line-scan hyperspectral imaging for real-time in-line poultry fecal detection," *Sensing and Instrumentation for Food Quality and Safety*, vol. 5, no. 1, pp. 25–32, 2011.

[189] H. Ding, R.-J. Xu, and D. K. O. Chan, "Identification of broiler chicken meat using a visible/near-infrared spectroscopic technique," *Journal of the Science of Food and Agriculture*, vol. 79, no. 11, pp. 1382–1388, 1999.

[190] M. Lin, M. Al-Holy, M. Mousavi-Hesary, H. Al-Qadiri, A. G. Cavinato, and B. A. Rasco, "Rapid and quantitative detection of the microbial spoilage in chicken meat by diffuse reflectance spectroscopy (600-1100 nm)," *Letters in Applied Microbiology*, vol. 39, no. 2, pp. 148–155, 2004.

[191] Y. Liu and Y. Chen, "Two-dimensional correlation spectroscopy study of visible and near-infrared spectral variations of chicken meats in cold storage," *Applied Spectroscopy*, vol. 54, no. 10, pp. 1458–1470, 2000.

[192] Y. Liu, F. E. Barton II, B. G. Lyon, W. R. Windham, and C. E. Lyon, "Two-dimensional correlation analysis of visible/near-infrared spectral intensity variations of chicken breasts with various chilled and frozen storages," *Journal of Agricultural and Food Chemistry*, vol. 52, no. 3, pp. 505–510, 2004.

[193] Y. Liu, Y. Chen, and Y. Ozaki, "Two-dimensional visible/near-infrared correlation spectroscopy study of thermal treatment of chicken meats," *Journal of Agricultural and Food Chemistry*, vol. 48, no. 3, pp. 901–908, 2000.

[194] S. K. Stormo, A. H. Sivertsen, K. Heia, and D. Skipnes, "Endpoint temperature of heat-treated surimi can be measured by visible spectroscopy," *Food Control*, vol. 26, no. 1, pp. 92–97, 2012.

[195] S. H. Olsen, *Quantification and characterisation of residual blood in fish muscle—impact of slaughtering methods [Ph.D. thesis]*, University of Tromsø UIT, Tromsø, Norway, 2011.

[196] S. H. Olsen, N. K. Sørensen, R. Larsen, E. O. Elvevoll, and H. Nilsen, "Impact of pre-slaughter stress on residual blood in fillet portions of farmed Atlantic cod (Gadus morhua)—measured chemically and by Visible and Near-infrared spectroscopy," *Aquaculture*, vol. 284, no. 1–4, pp. 90–97, 2008.

[197] D. Cozzolino, A. Chree, J. R. Scaife, and I. Murray, "Usefulness of Near-infrared reflectance (NIR) spectroscopy and chemometrics to discriminate fishmeal batches made with different fish species," *Journal of Agricultural and Food Chemistry*, vol. 53, no. 11, pp. 4459–4463, 2005.

[198] M. Uddin, E. Okazaki, S. Turza, Y. Yumiko, M. Tanaka, and Y. Fukuda, "Non-destructive visible/NIR spectroscopy for differentiation of fresh and frozen-thawed fish," *Journal of Food Science*, vol. 70, no. 8, pp. C506–C510, 2005.

[199] N. Bøknaes, K. N. Jensen, C. M. Andersen, and H. Martens, "Freshness assessment of thawed and chilled cod fillets packed in modified atmosphere using near-infrared spectroscopy," *Lebensmittel-Wissenschaft und-Technologie*, vol. 35, no. 7, pp. 628–634, 2002.

[200] H. Nilsen, M. Esaiassen, K. Heia, and F. Sigernes, "Visible/near-infrared spectroscopy: a new tool for the evaluation of fish freshness?" *Journal of Food Science*, vol. 67, no. 5, pp. 1821–1826, 2002.

[201] G. Olafsdottir, P. Nesvadba, C. Di Natale et al., "Multisensor for fish quality determination," *Trends in Food Science and Technology*, vol. 15, no. 2, pp. 86–93, 2004.

[202] M. Lin, M. Mousavi, M. Al-Holy, A. G. Cavinato, and B. A. Rasco, "Rapid near infrared spectroscopic method for the detection of spoilage in rainbow trout (*Oncorhynchus mykiss*) fillet," *Journal of Food Science*, vol. 71, no. 1, pp. S18–S23, 2006.

[203] H. Zhang and T. Lee, "Rapid near-infrared spectroscopic method for the determination of free fatty acid in fish and its application in fish quality assessment," *Journal of Agricultural and Food Chemistry*, vol. 45, no. 9, pp. 3515–3521, 1997.

[204] B. Gjerde and H. Martens, "Predicting carcass composition of rainbow trout by near-infrared reflectance spectroscopy," *Journal of Animal Breeding and Genetics*, vol. 104, pp. 137–148, 1987.

[205] J. Shimamoto, K. Hasegawa, S. Hattori, Y. Hattori, and T. Mizuno, "Non-destructive determination of the fat content in glazed bigeye tuna by portable near infrared spectrophotometer," *Fisheries Science*, vol. 69, no. 6, pp. 1247–1256, 2003.

[206] J. Shimamoto, K. Hasegawa, K. Ide, and S. Kawano, "Nondestractive determination of the fat content in raw and frozen horse mackerel by near infrared spectroscopy," *Nippon Suisan Gakkaishi*, vol. 67, no. 4, pp. 717–722, 2001.

[207] S. Yamauchi, T. Sawada, and S. Kawano, "Nondestructive determination of fat content in frozen skipjack by near infrared (NIR) spectroscopy with fiber optics in interactance mode," *Nippon Suisan Gakkaishi*, vol. 65, no. 4, pp. 747–752, 1999.

[208] J. Shimamoto, S. Hiratsuka, K. Hasegawa, M. Sato, and S. Kawano, "Rapid non-destructive determination of fat content in frozen skipjack using a portable near infrared spectrophotometer," *Fisheries Science*, vol. 69, no. 4, pp. 856–860, 2003.

[209] J. A. Mathias, P. C. Williams, and D. C. Sobering, "The determination of lipid and protein in freshwater fish using near-infrared reflectance spectroscopy," *Aquaculture*, vol. 61, no. 3-4, pp. 303–311, 1987.

[210] M. H. Lee, A. G. Cavinato, D. M. Mayes, and B. A. Rasco, "Noninvasive short-wavelength near-infrared spectroscopic method to estimate the crude lipid content in the muscle of intact rainbow trout," *Journal of Agricultural and Food Chemistry*, vol. 40, no. 11, pp. 2176–2180, 1992.

[211] M. Lin, A. G. Cavinato, D. M. Mayes et al., "Bruise detection in Pacific pink salmon (*Oncorhynchus gorbuscha*) by visible and short-wavelength near-infrared (SW-NIR) spectroscopy (600-1100 nm)," *Journal of Agricultural and Food Chemistry*, vol. 51, no. 22, pp. 6404–6408, 2003.

[212] R. Font, M. Del Río-Celestino, D. Vélez, A. De Haro-Bailón, and R. Montoro, "Visible and near-infrared spectroscopy as a technique for screening the inorganic arsenic content in the red crayfish (*Procambarus clarkii* Girard)," *Analytical Chemistry*, vol. 76, no. 14, pp. 3893–3898, 2004.

[213] J. Gayo and S. A. Hale, "Detection and quantification of species authenticity and adulteration in crabmeat using visible and near-infrared spectroscopy," *Journal of Agricultural and Food Chemistry*, vol. 55, no. 3, pp. 585–592, 2007.

[214] J. Gayo, S. A. Hale, and S. M. Blanchard, "Quantitative analysis and detection of adulteration in crab meat using visible and near-infrared spectroscopy," *Journal of Agricultural and Food Chemistry*, vol. 54, no. 4, pp. 1130–1136, 2006.

[215] M. R. Brown, "Rapid compositional analysis of oysters using visible-near infrared reflectance spectroscopy," *Aquaculture*, vol. 317, no. 1–4, pp. 233–239, 2011.

[216] V. Lebot, A. Champagne, R. Malapa, and D. Shiley, "NIR determination of major constituents in tropical root and tuber crop flours," *Journal of Agricultural and Food Chemistry*, vol. 57, no. 22, pp. 10539–10547, 2009.

[217] A. Dacal-Nieto, A. Formella, P. Carrión, E. Vazquez-Fernandez, and M. Fernández-Delgado, "Non-destructive detection of hollow heart in potatoes using hyperspectral imaging," in *Computer Analysis of Images and Patterns: 14th International Conference (CAIP '11)*, P. Real, D. Díaz-Pernil, H. Molina-Abril, A. Berciano, and W. Kropatsch, Eds., vol. 6855 of *Lecture Notes in Computer Science*, pp. 180–187.

[218] E. Joubert, M. Manley, and M. Botha, "Use of NIRS for quantification of mangiferin and hesperidin contents of dried green honeybush (*Cyclopia genistoides*) plant material," *Journal of Agricultural and Food Chemistry*, vol. 54, no. 15, pp. 5279–5283, 2006.

[219] S. Bellato, V. Del Frate, R. Redaelli et al., "Use of near infrared reflectance and transmittance coupled to robust calibration for the evaluation of nutritional value in naked oats," *Journal of Agricultural and Food Chemistry*, vol. 59, no. 9, pp. 4349–4360, 2011.

[220] L. R. Schimleck, J. C. Doran, and A. Rimbawanto, "Near infrared spectroscopy for cost effective screening of foliar oil characteristics in a Melaleuca cajuputi breeding population," *Journal of Agricultural and Food Chemistry*, vol. 51, no. 9, pp. 2433–2437, 2003.

[221] E. Joubert, M. Manley, B. R. Gray, and H. Schulz, "Rapid measurement and evaluation of the effect of drying conditions on harpagoside content in Harpagophytum procumbens (devil's claw) root," *Journal of Agricultural and Food Chemistry*, vol. 53, no. 9, pp. 3493–3502, 2005.

[222] H. Schulz, U. H. Engelhardt, A. Wegent, H. Drews, and S. Lapczynski, "Application of near-infrared reflectance spectroscopy to the simultaneous prediction of alkaloids and phenolic substances in green tea leaves," *Journal of Agricultural and Food Chemistry*, vol. 47, no. 12, pp. 5064–5067, 1999.

[223] J. M. Hernández-Hierro, J. Valverde, S. Villacreces et al., "Feasibility study on the use of visible-near-infrared spectroscopy for the screening of individual and total glucosinolate contents in broccoli," *Journal of Agricultural and Food Chemistry*, vol. 60, no. 30, pp. 7352–7358, 2012.

[224] M. Laasonen, T. Harmia-Pulkkinen, C. L. Simard, E. Michiels, M. Räsänen, and H. Vuorela, "Fast identification of Echinacea purpurea dried roots using near-infrared spectroscopy," *Analytical Chemistry*, vol. 74, no. 11, pp. 2493–2499, 2002.

[225] W. Li, P. Goovaerts, and M. Meurens, "Quantitative analysis of individual sugars and acids in orange juices by near-infrared spectroscopy of dry extract," *Journal of Agricultural and Food Chemistry*, vol. 44, no. 8, pp. 2252–2259, 1996.

[226] Y. C. Chen, H. Zhang, and R. Matsunaga, "Rapid determination of the main organic acid composition of raw Japanese apricot fruit juices using near-infrared spectroscopy," *Journal of Agricultural and Food Chemistry*, vol. 54, no. 26, pp. 9652–9657, 2006.

[227] J. M. González-Sáiz, C. Pizarro, I. Esteban-Díez et al., "Monitoring of alcoholic fermentation of onion juice by NIR spectroscopy: valorization of worthless onions," *Journal of Agricultural and Food Chemistry*, vol. 55, no. 8, pp. 2930–2936, 2007.

[228] H. Cen, Y. He, and M. Huang, "Measurement of soluble solids contents and pH in orange juice using chemometrics and Vis–NIRS," *Journal of Agricultural and Food Chemistry*, vol. 54, no. 20, pp. 7437–7443, 2006.

[229] F. Liu and Y. He, "Use of visible and near infrared spectroscopy and least squares-support vector machine to determine soluble solids content and pH of cola beverage," *Journal of Agricultural and Food Chemistry*, vol. 55, no. 22, pp. 8883–8888, 2007.

[230] C. L. Ng, R. L. Wehling, and S. L. Cuppett, "Method for determining frying oil degradation by near-infrared spectroscopy," *Journal of Agricultural and Food Chemistry*, vol. 55, no. 3, pp. 593–597, 2007.

[231] D. Cozzolino, R. G. Dambergs, L. Janik, W. U. Cynkar, and M. Gishen, "Analysis of grapes and wine by near infrared spectroscopy," *Journal of Near Infrared Spectroscopy*, vol. 14, no. 5, pp. 279–289, 2006.

[232] X. Chen and X. Lei, "Application of a hybrid variable selection method for determination of carbohydrate content in soy milk powder using visible and near infrared spectroscopy," *Journal of Agricultural and Food Chemistry*, vol. 57, no. 2, pp. 334–340, 2009.

[233] R. G. Dambergs, A. Kambouris, L. Francis, and M. Gishen, "Rapid analysis of methanol in grape-derived distillation products using near-infrared transmission spectroscopy," *Journal of Agricultural and Food Chemistry*, vol. 50, no. 11, pp. 3079–3084, 2002.

[234] M.-J. Sáiz-Abajo, J.-M. González-Sáiz, and C. Pizarro, "Classification of wine and alcohol vinegar samples based on near-infrared spectroscopy. Feasibility study on the detection of adulterated vinegar samples," *Journal of Agricultural and Food Chemistry*, vol. 52, no. 25, pp. 7711–7719, 2004.

[235] M. J. Sáiz-Abajo, J. M. González-Sáiz, and C. Pizarro, "Prediction of organic acids and other quality parameters of wine vinegar by near-infrared spectroscopy. A feasibility study," *Food Chemistry*, vol. 99, no. 3, pp. 615–621, 2006.

[236] W. Fan, Y. Shan, G. Li, H. Lv, H. Li, and Y. Liang, "Application of competitive adaptive reweighted sampling method to determine effective wavelengths for prediction of total acid of vinegar," *Food Analytical Methods*, vol. 5, no. 3, pp. 585–590, 2012.

[237] S. W. Bruun, *Protein-lipid-carbohydrate interactions and importance to food quality—spectroscopic detection of macromolecular interactions [Ph.D. thesis]*, Technical University of Denmark, Lyngby, Denmark, 2006.

[238] F. Liu, W. Kong, and Y. He, "Quality determination of fruit vinegars using visible/near infrared spectroscopy and least squares-support vector machine," *African Journal of Agricultural Research*, vol. 7, no. 15, pp. 2395–2404, 2012.

[239] D. Cozzolino, H. E. Smyth, and M. Gishen, "Feasibility study on the use of visible and near-infrared spectroscopy together with chemometrics to discriminate between commercial white wines of different varietal origins," *Journal of Agricultural and Food Chemistry*, vol. 51, no. 26, pp. 7703–7708, 2003.

[240] L. Liu, D. Cozzolino, W. U. Cynkar, M. Gishen, and C. B. Colby, "Geographic classification of Spanish and Australian tempranillo red wines by visible and near-infrared spectroscopy combined with multivariate Analysis," *Journal of Agricultural and Food Chemistry*, vol. 54, no. 18, pp. 6754–6759, 2006.

[241] S. Buratti, D. Ballabio, G. Giovanelli et al., "Monitoring of alcoholic fermentation using near infrared and mid infrared spectroscopies combined with electronic nose and electronic tongue," *Analytica Chimica Acta*, vol. 697, no. 1-2, pp. 67–74, 2011.

[242] H. Yu, H. Lin, H. Xu, Y. Ying, B. Li, and X. Pan, "Prediction of enological parameters and discrimination of rice wine age using least-squares support vector machines and near infrared spectroscopy," *Journal of Agricultural and Food Chemistry*, vol. 56, no. 2, pp. 307–313, 2008.

[243] L. L. Vines, S. E. Kays, and P. E. Koehler, "Near-infrared reflectance model for the rapid prediction of total fat in cereal foods," *Journal of Agricultural and Food Chemistry*, vol. 53, no. 5, pp. 1550–1555, 2005.

[244] Windham, S. E. Kays, and F. E. Barton II, "Effect of cereal product residual moisture content on total dietary fiber determined by near-infrared reflectance spectroscopy," *Journal of Agricultural and Food Chemistry*, vol. 45, no. 1, pp. 140–144, 1997.

[245] S. E. Kays, Windham, and F. E. Barton II, "Prediction of total dietary fiber in cereal products using near-infrared reflectance spectroscopy," *Journal of Agricultural and Food Chemistry*, vol. 44, no. 8, pp. 2266–2271, 1996.

[246] S. E. Kays, F. E. Barton II, W. R. Windham, and D. S. Himmelsbach, "Prediction of total dietary fiber by near-infrared reflectance spectroscopy in cereal products containing high sugar and crystalline sugar," *Journal of Agricultural and Food Chemistry*, vol. 45, no. 10, pp. 3944–3951, 1997.

[247] S. E. Kays, Windham, and F. E. Barton II, "Prediction of total dietary fiber by near-infrared reflectance spectroscopy in high-fat- and high-sugar-containing cereal products," *Journal of Agricultural and Food Chemistry*, vol. 46, no. 3, pp. 854–861, 1998.

[248] S. E. Kays and F. E. Barton II, "Rapid prediction of gross energy, and utilizable energy in cereal food products using near-infrared reflectance spectroscopy," *Journal of Agricultural and Food Chemistry*, vol. 50, no. 5, pp. 1284–1289, 2002.

[249] S. E. Kays and F. E. Barton II, "Near-infrared analysis of soluble and insoluble dietary fiber fractions of cereal food products," *Journal of Agricultural and Food Chemistry*, vol. 50, no. 10, pp. 3024–3029, 2002.

[250] S. E. Kays and F. E. Barton II, "Energy from fat determined by near-infrared reflectance spectroscopy," *Journal of Agricultural and Food Chemistry*, vol. 52, no. 6, pp. 1669–1674, 2004.

[251] D. D. Archibald and S. E. Kays, "Determination of total dietary fiber of intact cereal food products by near-infrared reflectance," *Journal of Agricultural and Food Chemistry*, vol. 48, no. 10, pp. 4477–4486, 2000.

[252] R. M. Ozanich, M. L. Schrattenholzer, and J. B. Callis, "Noninvasive determination of moisture and oil content of wheat-flour cookies—near-infrared spectroscopy in the wavelength range 700-1100 nm," in *Biosensor Design and Application*, vol. 511 of *ACS Symposium Series*, chapter 12, pp. 137–164, 1992.

[253] L. F. de Alencar Figueiredo, F. Davrieux, G. Fliedel et al., "Development of NIRS equations for food grain quality traits through exploitation of a core collection of cultivated sorghum," *Journal of Agricultural and Food Chemistry*, vol. 54, no. 22, pp. 8501–8509, 2006.

[254] S. W. Bruun, I. Søndergaard, and S. Jacobsen, "Analysis of protein structures and interactions in complex food by near-infrared spectroscopy. 1. Gluten powder," *Journal of Agricultural and Food Chemistry*, vol. 55, no. 18, pp. 7234–7243, 2007.

[255] S. W. Bruun, I. Søndergaard, and S. Jacobsen, "Analysis of protein structures and interactions in complex food by near-infrared spectroscopy. 2. Hydrated gluten," *Journal of Agricultural and Food Chemistry*, vol. 55, no. 18, pp. 7244–7251, 2007.

[256] N. Sinelli, E. Casiraghi, and G. Downey, "Studies on proofing of yeasted bread dough using near- and mid-infrared spectroscopy," *Journal of Agricultural and Food Chemistry*, vol. 56, no. 3, pp. 922–931, 2008.

[257] R. M. García-Rey, J. García-Olmo, E. De Pedro, R. Quiles-Zafra, and M. D. Luque de Castro, "Prediction of texture and colour of dry-cured ham by visible and near infrared spectroscopy using a fiber optic probe," *Meat Science*, vol. 70, no. 2, pp. 357–363, 2005.

[258] A. Lucas, D. Andueza, E. Rock, and B. Martin, "Prediction of dry matter, fat, pH, vitamins, minerals, carotenoids, total antioxidant capacity, and color in fresh and freeze-dried cheeses by visible-near-infrared reflectance spectroscopy," *Journal of Agricultural and Food Chemistry*, vol. 56, no. 16, pp. 6801–6808, 2008.

[259] E. S. Çelik, *Determination of aroma compounds and exopolysaccharides formation by lactic acid bacteria isolated from traditional yogurts [M.S. thesis]*, İzmir Institute of Technology, İzmir, Turkey, 2007.

[260] M. Navrátil, C. Cimander, and C. Mandenius, "On-line multisensor monitoring of yogurt and filmjölk fermentations on production scale," *Journal of Agricultural and Food Chemistry*, vol. 52, no. 3, pp. 415–420, 2004.

[261] G. Cozzi, J. Ferlito, G. Pasini, B. Contiero, and F. Gottardo, "Application of near-infrared spectroscopy as an alternative to chemical and color analysis to discriminate the production chains of Asiago d'Allevo cheese," *Journal of Agricultural and Food Chemistry*, vol. 57, no. 24, pp. 11449–11454, 2009.

[262] J. Blasco, N. Aleixos, J. Gómez, and E. Moltó, "Citrus sorting by identification of the most common defects using multispectral computer vision," *Journal of Food Engineering*, vol. 83, no. 3, pp. 384–393, 2007.

[263] P. M. Mehl, K. Chao, M. Kim, and Y. R. Chen, "Detection of defects on selected apple cultivars using hyperspectral and multispectral image analysis," *Applied Engineering in Agriculture*, vol. 18, no. 2, pp. 219–226, 2002.

[264] I. S. Arvanitoyannis and M. Van Houwelingen-Koukaliaroglou, "Implementation of chemometrics for quality control and authentication of meat and meat products," *Critical Reviews in Food Science and Nutrition*, vol. 43, no. 2, pp. 173–218, 2003.

[265] T. Woodcock, C. C. Fagan, C. P. O'Donnell, and G. Downey, "Application of near and mid-infrared spectroscopy to determine cheese quality and authenticity," *Food and Bioprocess Technology*, vol. 1, no. 2, pp. 117–129, 2008.

17

The Nanofabrication and Application of Substrates for Surface-Enhanced Raman Scattering

Xian Zhang,[1] Qin Zhou,[1,2] Yu Huang,[1] Zhengcao Li,[1] and Zhengjun Zhang[1]

[1] *Advanced Materials Laboratory, Department of Materials Science and Engineering, Tsinghua University, Beijing 100084, China*
[2] *Institute of Nuclear and New Energy Technology, Tsinghua University, Beijing 100084, China*

Correspondence should be addressed to Zhengjun Zhang, zjzhang@tsinghua.edu.cn

Academic Editor: Rolf W. Berg

Surface-enhanced Raman scattering (SERS) was discovered in 1974 and impacted Raman spectroscopy and surface science. Although SERS has not been developed to be an applicable detection tool so far, nanotechnology has promoted its development in recent decades. The traditional SERS substrates, such as silver electrode, metal island film, and silver colloid, cannot be applied because of their enhancement factor or stability, but newly developed substrates, such as electrochemical deposition surface, Ag porous film, and surface-confined colloids, have better sensitivity and stability. Surface enhanced Raman scattering is applied in other fields such as detection of chemical pollutant, biomolecules, DNA, bacteria, and so forth. In this paper, the development of nanofabrication and application of surface-enhanced Ramans scattering substrate are discussed.

1. Introduction

Surface enhanced Raman Scattering was discovered by Fleischmann et al. in 1974 as a phenomenon where the intensity of Raman spectrum from pyridine adsorbed on a silver electrode is extrodinarily enhanced [1, 2]. Then this effect was verified in many kinds of metals such as Ag, Au, Cu, Li, Na, K, and so forth [2, 3], while high enhancement is just found in Ag, Au, and Cu. But the phenomenon was not correctly recognized at that time, then Jeanmaire and Van Duyne reported in 1977 that the enhancement of intensity of Raman spectrum from pyridine on silver electrode is attributed to the enhancement of the scattering cross-section of pyridine [4]. Then the phenomenon was accepted as surface enhanced Raman scattering (SERS).

Cross-sections of Raman scattering are typically 14 orders of magnitude smaller than those of fluorescene. So the weak intensity of Raman spectrum, the limits from the intensity of excitation sources, and the sensitivity of detector hindered the application of Raman spectrum for many years. The discovery of surface enhanced Raman scattering made Raman scattering more applicable in a wide range of research fields, while the interest and attention focused on the mechanism of SERS promotes the development of the theory on enhanced optical scattering from metal surface [5–7].

But SERS has not been developed to be a powerful detection technique so far, which is attributed to three main reasons [8–11]. First, the phenomenon-surface enhanced Raman scattering is obvious in only silver, gold, and copper, so the research and application of SERS ares limited. Second, there are limited kinds of processes that can achieve SERS substrate with high sensitivity successfully, and the practical SERS active substrate has not been realized. Third, the mechanism of surface enhanced Raman scattering is not totally clear.

Among so many theories about SERS that are proposed to explain the experimental characteristics of surface enhanced Raman scattering, it is widely accepted that the SERS effect mainly contributed to electromagnetic (EM) enhancement and chemical (CM) enhancement [12–19]. The electromagnetic (EM) enhancement is caused by local surface plasmon resonannce (LSPR) and depends on the material, shape, and size of the nanoscale substrate. There is a resonant frequency corresponding to a specific substrate, and the local surface plasmon resonance (LSPR) of substrate is driven into collective oscillator and generates the largest electromagnetic field when the light is incident with the resonant frequency. Electromagnetic enhancement relies on the location of molecules confined within the electromagnetic field and contributes an enhancement of more than 10^5. While, the chemical enhancement is caused by a resonance Raman-like process associated with charge transfer between the analyte molecule and the metal surface and always contributes an enhancement factor of 100. The charge transfer state increase the possibility of Raman transition by providing a pathway for resonant excitation. So chemical enhancement is dependent on site and analyte and happens just when the molecule is adsorbed directly on the metal roughened surface. Surface enhanced Raman scattering is generated from the combination of electromagnetic enhancement and chemical enhancement. However, SERS involves the complicated interactions and coupling between molecule and metal surface. So the theoretical work about SERS still attracts so much attention from scholars in different fields.

The research about surface enhanced Raman scattering revived in 1977, with the reported progress that Kneipp, Nie, and their coworkers brought SERS into the level of single molecule detection [20–23]. It is found that most SERS active substrate is nonuniform [24–28]. For instance, it is realized in silver colloid that nanoparticle with different sizes generates Raman signal with different enhancement factors [20]. From theoretical perspective, the highly active spots in SERS substrate are called "hotspots". The estimated SERS enhancement at hotspots reaches up to 14 orders, much higher than the enhancement factor achieved before; the result pushed research on SERS into a new era.

2. The Nanofabrication of SERS Substrate

The SERS substrate is the most important factor that affects the enhancement in SERS experiment, so the fabrication of SERS substrate always attracts most attention. Surface enhanced Raman scattering depends on the excitation of local surface plasmon resonance (LSPR), so it is critical to control the factors affecting LSPR such as the size, shape, material, interparticle spacing, and dielectric environment of substrate to improve its sensitivity and reproducibility [29]. By the efforts from many researchers, various kinds of SERS substrates are invented.

2.1. The Electrochemical Fabrication Method. Oxidation-reduction cycle electrode [1] is the first found SERS electrode with reproducible enhancement factor of 10^6, and it is widely applied in research on catalytic reactions and other electrochemically active systems [30]. Because of its low enhancement factor among the main kinds of SERS substrates, oxidation-reduction cycle electrode was not thought to be a very promising SERS substrate. But recently new process make roughen electrode more attractive. Duan et al. reported a simple, effective, and inexpensive method to fabricate an ordered Au particle array with hierarchical surface roughness on an indium tin oxide substrate based on an ordered alumina through-pore template [31]. The alumina template is produced by solution dipping on colloidal monolayer through an electrochemical deposition strategy. The array consists of periodically arranged and isolated Au microparticles, which show nanoscaled surface roughness. Essentially, this hierarchically rough particle array exhibits strong surface enhanced Raman scattering effect using Rhodamine 6G (R6G) as probe molecules, associated with its surface geometry. Aqueous solution of 10^{-6} M R6G can be detected by such substrate. Huang et al. fabricated a kind of SERS substrate named Ag-capped Au nanopillar arrays by electrochemical deposition strategy [32]. Arrays of Au nanopillars with tube-like structures in the root parts were electrodeposited into the nanochannels of a porous AAO templates. After the AAO template is removed, the Ag is sputtered to adjust the space of Au nanopillar's tip. The Au nanopillar sputtered by Ag is proved to have capability of detecting 10^{-6} M R6G.

2.2. The Fabrication Method by Electron Beam Deposition. The LSPR wavelength of metal island field can be tuned by changing the film's thickness and confluence, but the enhancement factor of metal island film is lower than other SERS substrate. Glancing angle deposition (GLAD) [33, 34] is a promising method to improve the morphology and SERS property of metal film. Zhou et al. fabricated aligned, single crystalline Ag nanorods on planar Si subatrates by GLAD technique, with sample substrate cooled by liquid nitrogen in e-beam deposition system [35]. It is successful to detect aqueous solution of 1×10^{-12} mol$\cdot$L^{-1} Rhodamine 6G by the porous Ag film with nanorods, the morphology and SERS performance of Ag film with aligned nanorods is shown in Figure 1. Then it is found that bends between two Ag arms with different orientations has better SERS sensitivity as hotspots, so the Ag film with nanorods of zigzag structure is an ideal way to obtain better performance [36].

Zhou et al. deposited Ag, Al, Si, and Ti thin films with thickness d = 25 nm, 100 nm, and 400 nm, respectively, to achieve layers with different reflectivity. Then Ag nanorods was deposited on the thin film layer. The SERS intensity of the Ag nanorods grown on Ag thin film are higher than others, and the SERS intensity of the Ag nanorods on Al film is larger than that on Ti film, and the Ag nanorods on Si film show the smallest SERS intensity. It is inferred that the larger the underlayer reflectivity, the larger the SERS performance of substrate. So the predeposition of Ag layer under Ag nanorods can be an effective way to promote the SERS performance of Ag nanorods [37].

It is realized that most SERS-active substrates are nonuniform before, and it is of specific interest to research hotspots where some spots and sites exhibit extrodinary SERS sensitivity and improve SERS substrate from randomly rough surface to highly ordered nanostructures [38]. Zhang et al. made Ag nanorods in film grow into periodic patterns at a micro-nano scale, and it is shown that the Ag nanorods film with periodic patterns exhibits better SERS performance than Ag film with nanorods arranged randomly as before [39]. The Ag film with periodic patterns is fabricated by the help of electron beam lithography. Hexagonal lattices (200 μm $\times$ 200 μm) of silicon patterns (~1 μm in diameter)

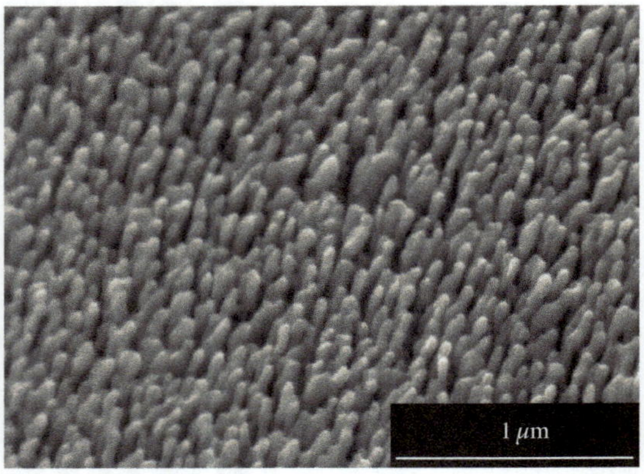

(a)

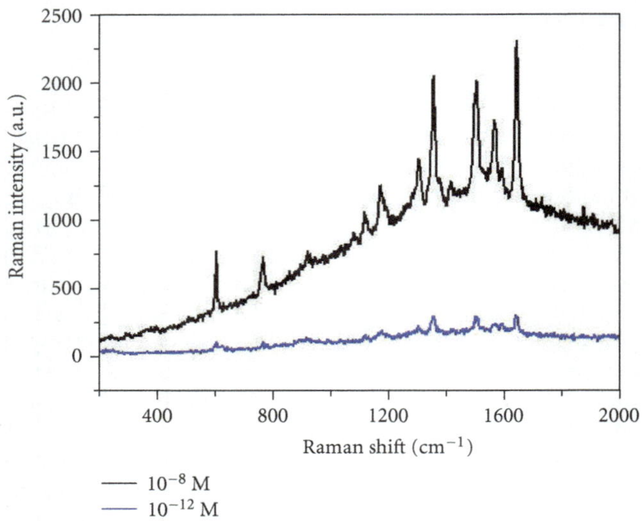

(b)

FIGURE 1: (a) The SEM image showing the morphology of porous Ag film on silicon substrate. (b) The Raman spectra of 10^{-8} M and 10^{-12} M Rhodamine 6G (R6G) in porous Ag film examined by Raman spectrometer.

were fabricated on the substrates by electron beam lithography, where the separation distance of the patterns was controlled to be 0 (closely-packed), 50 nm, 100 nm, 200 nm, 300 nm, 400 nm, 500 nm, and 600 nm, respectively. Then Ag film with hexagonal patterns is deposited on silicon substrate with hexagonal patterns in a high vacuum e-beam deposition system, The morphology of Ag film with periodic patterns of different separation is shown in Figure 2. Characterized by Rhodamine 6G (R6G), the intensity of peak 612 cm^{-1} and 1508 cm^{-1} in Raman spectrum of R6G from Ag film with periodic patterns is shown in Figure 3. In comparison with signals from the planar area, it is noticed that when the patterns are closely packed (0 nm), or separated at ~200 nm, the signals are further enhanced by ~20%. When the diameter of periodic pattern in Ag film is 400 nm (as shown in Figure 4), different regulation of SERS performance is shown in Figure 5. In comparison with the unpatterned area, we notice that when the patterns are separated at ~200 nm, the signals are further enhanced by ~5 times compared with the unpatterned substrates. And the dependence behavior of the two peaks on the separation distance for pattern diameter of both 1 μm and 400 nm is proved to be reproducible. So the SERS performance of Ag film with patterns' seperation of 200 nm is the best both among the Ag film with other separation, when the diameter of patterns is 1 μm or 400 nm. So it is demonstrated that the diameter and separation of patterns in Ag film both have effects on the SERS performance of Ag film, and the phenomenon suggests possibly a coupling effect at nano-micro scales on the localized electrical field. Furthermore, Chu et al. fabricated a gold disk array on a continuous gold film with silica spacer and demonstrated strong coupling between LSPs and surface plasmon polaritons (SPPs) [40]. It is predicted by simulations that the enhancement factor of each disk in array above a gold film is twice as large as that of a single disk above a gold film.

2.3. Other Fabrication Method. Colloidal nanoparticle substrate is easy to synthesize without sophisticated instrumentation and realizes single molecule detection initiatively. But colloidal nanoparticle substrate is unstable, so inmobilized colloidal nanoparticle is developed. Tong et al. synthesized self-assembly colloidal gold nanoparticles, and investigated the electromagnetic enhancement coupling effect appearing within the self-assembly nanoparticles [41].

Electron beam lithography substrates have been used to investigate the magnitude of the enhancement factor while the interparticle separation is varied, enhancement factors as large as 10^8 have been achieved [42]. A soft technique known as nanosphere lithography has been used to fabricate SERS substrates to research atomic clusters, biomolecules, and redox couples in ultrahigh vacuum (UHV), electrochemically controlled, and ambient environments [43–45]. Van Duyne et al. used nanosphere lithography to fabricate surface confined Ag nanoparticles. Nanosphere lithography is an excellent approach to fabrication nanoparticle arrays with shape, size, and interparticle spacing precisely controlled. Meanwhile, metal film over nanosphere surfaces provides extremely SERS sensitivity for applications, and it is found that nanosphere lithography derived nanoparticles exhibit strong UV-visible extinction bands that are different from those of bulk metal. The LSPR spectra of nanosphere lithography derived nanoparticles can easily be tuned from the near-UV to the mid-IR by changing the size or shape of the nanoparticles [46]. Researchers continue to develop new SERS substrates to improve the sensitivity and reproducibility of its performance.

3. The Application of SERS Substrate

SERS has enormous potential for the detection of chemical and biological agents. The well resolved bands from Raman scattering allow the identification of composites, and low strength of signal from water is beneficial for the detection of biological agents [38].

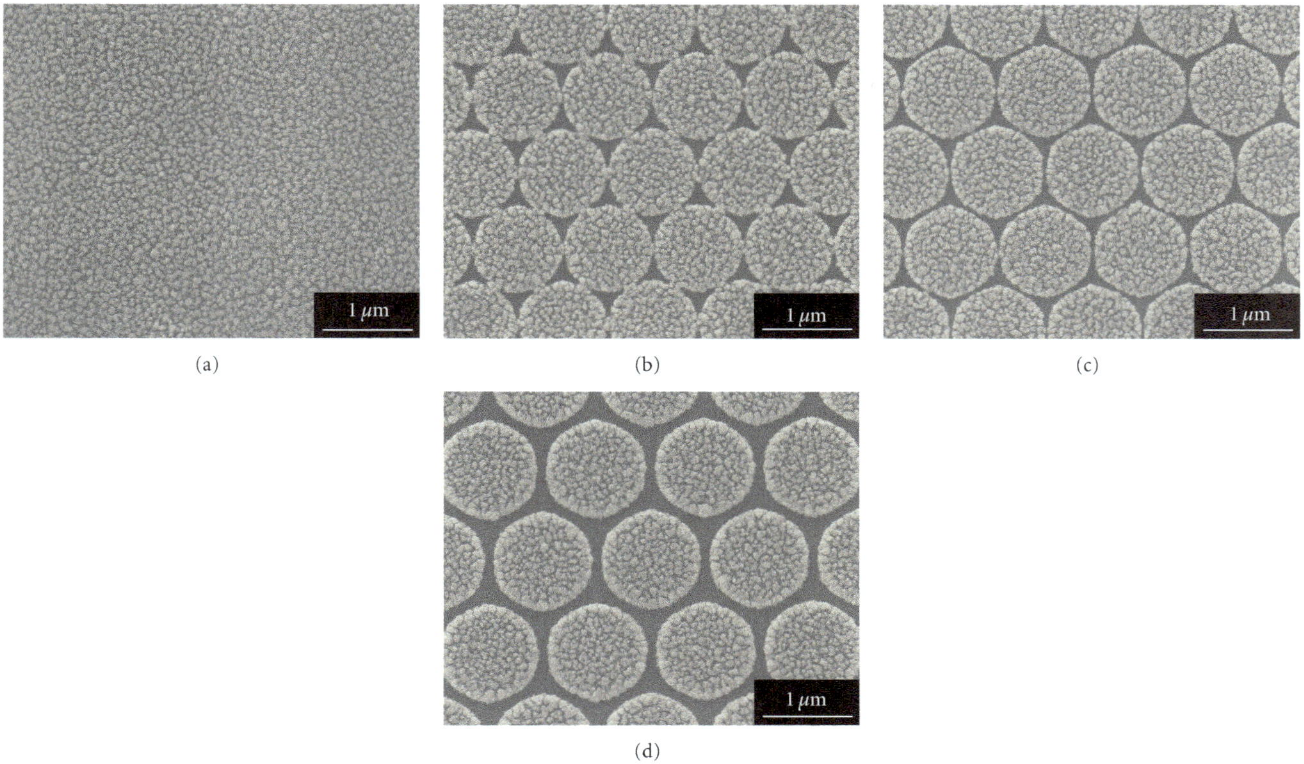

FIGURE 2: The SEM image of porous Ag film with different arrangement of nanorods (diameter of 1 μm). (a) Random arrangement; (b) periodical clusters packed closely; (c) periodical clusters with 50 nm separation; (d) periodical clusters with 200 nm separation.

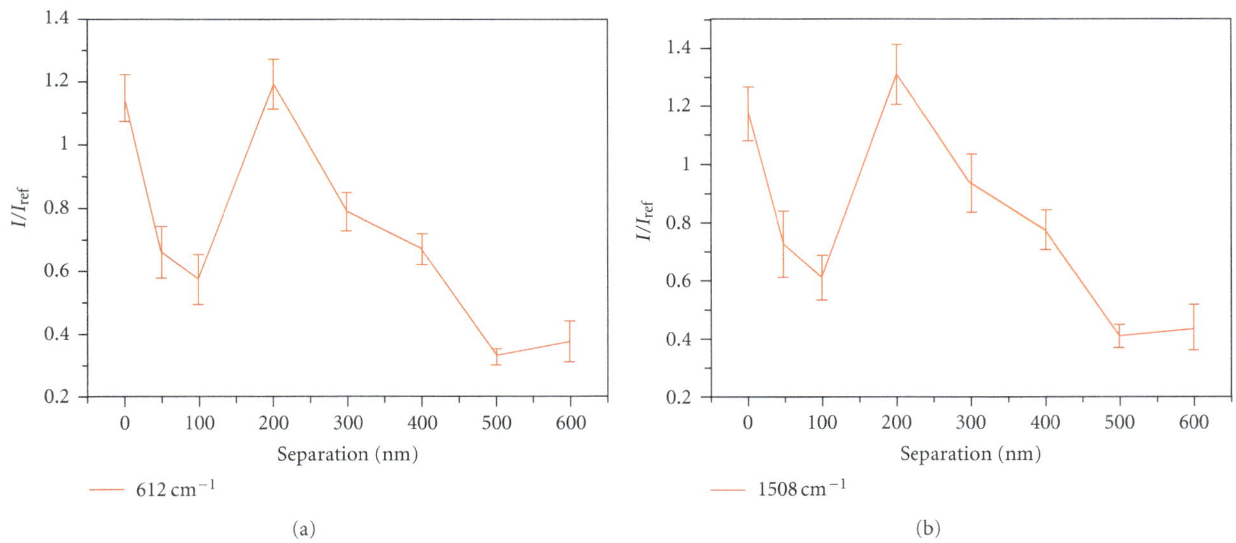

FIGURE 3: The intensity of the two characteristic peaks of 10^{-6} M R6G on hexagonal lattices (with diameter of 1 μm) of Ag nanorods. (a) 612 cm^{-1} and (b) 1508 cm^{-1} that are normalized to the intensity obtained from the planar substrate.

It is a tendency that SERS substrate is made into sensor to facilitate the detection course. Van Duyne et al. has successfully made SERS substrate into a implantable glucose sensor, and the SERS substrate-silver surface is modified with a hydrophilic partition layer to realize the preconcentration of glucose within the range of electromagnetic field. Glucose was detected and quantified with the accuracy approaching the requirements for a biomedical device. And SERS has also

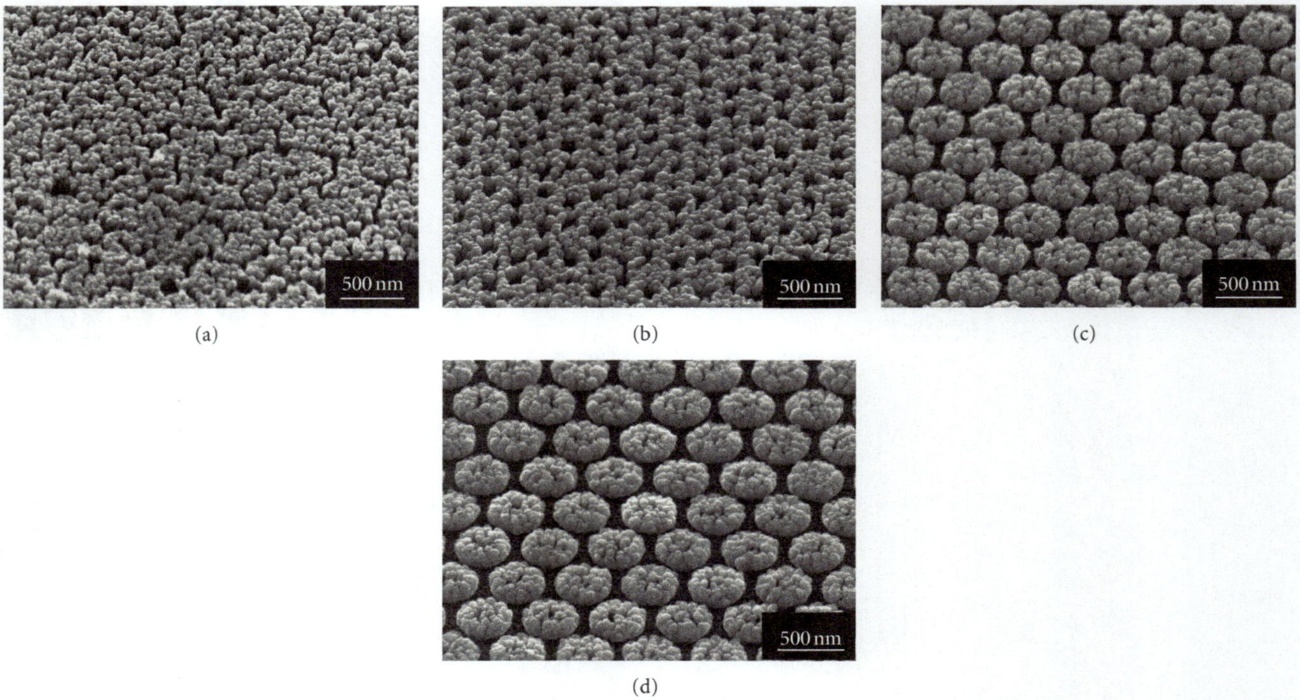

FIGURE 4: Oblique-view SEM images of vertically aligned Ag nanorods deposited on (a) unpatterned silicon; and on hexagonal lattices of silicon patterns that are separated by (b) 0 nm (closely-packed); (c) 50 nm; (d) 100 nm, with patterns diameter of 400 nm.

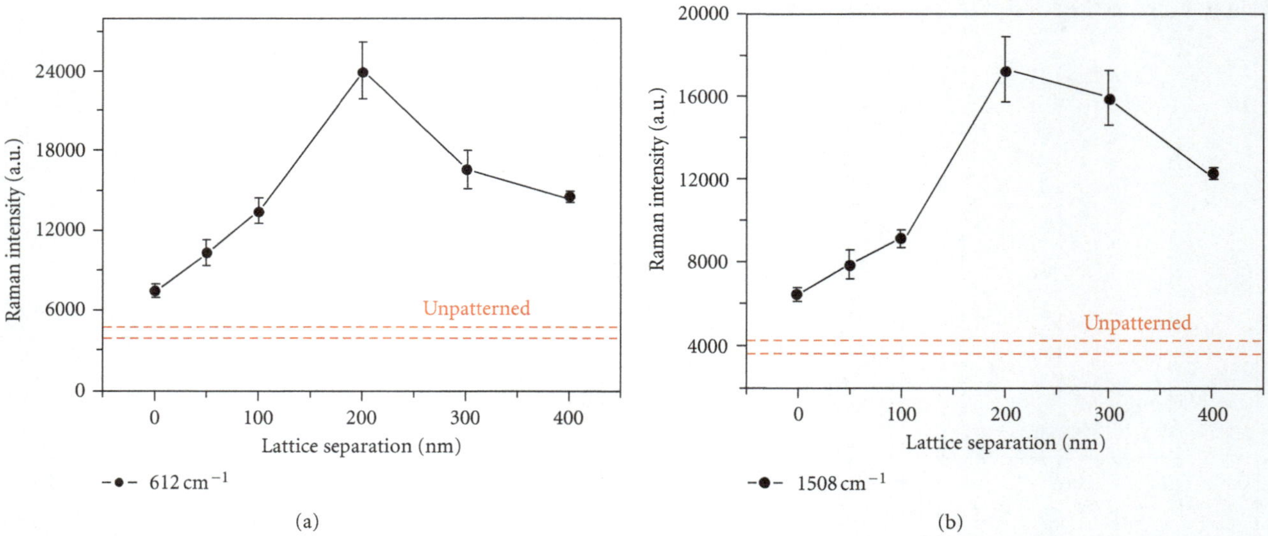

FIGURE 5: The intensity of the two characteristic peaks of 10^{-6} M R6G on hexagonal lattices of Ag nanorods (with patterns diameter of 400 nm): (a) 612 cm^{-1} and (b) 1508 cm^{-1}.

been applied in military [47]. Silver nanowires were used to detect chemical warfare agents at a sensitivity of trace level [48].

Because of the excellent dielectric properties, polychlorinated biphenyls (PCBs) were used widely in transformers, heat transfer, capacitors, and so forth and had caused severe environmental pollution around the world [49]. It was found recently that PCBs are harmful to human body when the density exceeds the critical dose and PCBs can be accumulated through food chain from trace amount. While the most popular detection technique for PCBs is the combination of high-resolution gas chromatography and mass spectrometry, the method requires sophisticated devices, standard samples, complicated pretreatments of samples, proper experimental environment, and so forth. So new method that can detect PCBs at trace amount rapidly is necessary, and surface enhanced Raman scattering has potential to realize the goal.

Zhou et al. detected and distinguished the isomers of monochlorobiphenyls, tetrachlorobiphenyls, and 2,3,3′,4,4′-pentachlorinated biphenyls successfully [50–53]. To facilitate the distinguishing of polychlorobiphenyls, Zhou et al. simulated the vibration mode of polychlorobiphenyls by density functional theory, and the results are beneficial for distinguishing the agents after achieving the Raman spectrum.

Sometimes, the detection is carried out indirectly in biomolecules. For example, Cao et al. used ssDNA with a gold nanoparticle and SERS label to detect specific DNA molecule. If the probe is exposed to the target DNA, the hybridization of DNA occurred with SERS detection of the Raman label. The method detects the target DNA not by detection of DNA directly but by detection of Raman label's signal indirectly [54].

4. Conclusion

Recently, the development of fabrication and application of substrates for surface enhanced Raman scattering is driven by nanotechnology. More and more SERS substrates with high sensitivity and reproducibility are invented by electrochemical deposition, physical vapor deposition of metal film, metal nanoparticle colloids, and so forth and applied into various fields, such as detection of pollutants at trace level, surface analysis, biomolecule, bacteria detection, and so forth. With the development of SERS substrate, Raman spectrometer, and tip enhanced Raman scattering (as the combination of surface enhanced Raman scattering and atomic force microscopy), surface enhanced Raman scattering will be more and more applicable as a detection tool.

Acknowledgments

The authors are grateful to the financial support by the National Natural Science Foundation of China (50931002) and the National Basic Research Program of China (973 Program, 2007CB936601).

References

[1] M. Fleischmann, P. J. Hendra, and A. J. McQuillan, "Raman spectra of pyridine adsorbed at a silver electrode," *Chemical Physics Letters*, vol. 26, no. 2, pp. 163–166, 1974.

[2] M. Moskovits, "Surface-enhanced spectroscopy," *Reviews of Modern Physics*, vol. 57, no. 3, pp. 783–826, 1985.

[3] A. Kudelski, "Analytical applications of Raman spectroscopy," *Talanta*, vol. 76, no. 1, pp. 1–8, 2008.

[4] D. L. Jeanmaire and R. P. Van Duyne, "Surface Raman spectroelectrochemistry Part I. Heterocyclic, aromatic, and aliphatic amines adsorbed on the anodized silver electrode," *Journal of Electroanalytical Chemistry*, vol. 84, no. 1, pp. 1–20, 1977.

[5] R. K. Chang, "Surface enhanced Raman-scattering at electrodes—a status-report," *Physical Chemistry Chemical Physics*, vol. 91, no. 4, pp. 296–305, 1987.

[6] R. L. Garrell, "Surface-enhanced Raman spectroscopy," *Analytical Chemistry*, vol. 61, no. 6, pp. 401A–411A, 1989.

[7] M. Moskovits, "The dependence of the metal-molecule vibrational frequency on the mass of the adsorbate and its relevance to the role of adatoms in surface-enhanced Raman scattering," *Chemical Physics Letters*, vol. 98, no. 5, pp. 498–502, 1983.

[8] A. G. Brolo, M. Odziemkowski, and D. E. Irish, "An in situ Raman spectroscopic study of electrochemical processes in mercury-solution interphases," *Journal of Raman Spectroscopy*, vol. 29, no. 8, pp. 713–719, 1998.

[9] B. H. Loo, "Surface enhanced Raman scattering from pyridine adsorbed on cadmium," *The Journal of Chemical Physics*, vol. 75, no. 12, pp. 5955–5956, 1981.

[10] T. López-Ríos and Y. Gao, "Modification by pd adsorbates of the surface enhanced Raman scattering at Ag surfaces," *Surface Science*, vol. 205, no. 3, pp. 569–590, 1988.

[11] Y. Gao and T. Lopez-Rios, "Raman scattering of pyridine coadsorbed with Al on quenched Ag films: evidence of Raman enhancement in the pores," *Surface Science*, vol. 198, no. 3, pp. 509–523, 1988.

[12] M. Moskovits, "Surface-enhancedspectroscopy," *Reviews of Modern Physics*, vol. 57, no. 3, pp. 783–826, 1985.

[13] M. Moskovits, "Surface selection rules," *Journal of Chemical Physics*, vol. 77, no. 9, pp. 4408–4416, 1982.

[14] M. Moskovits, L. L. Tay, J. Yang, and T. Haslett, "SERS and the single molecule," in *Topics in Applied Physics*, pp. 215–226, Springer, Berlin, Germany, 2002.

[15] A. Otto, I. Mrozek, H. Grabhorn, and W. Akemann, "Surface-enhanced Raman scattering," *Journal of Physics*, vol. 4, no. 5, pp. 1143–1212, 1992.

[16] M. Kerker, "Electromagnetic model for surface-enhanced Raman scattering (SERS) on metal colloids," *Accounts of Chemical Research*, vol. 17, no. 8, pp. 271–277, 1984.

[17] H. Metiu and P. Das, "The electromagnetic theory of surface enhanced spectroscopy," *Annual Review of Physical Chemistry*, vol. 35, pp. 507–536, 1984.

[18] G. C. Schatz, "Theoretical studies of surface enhanced Raman scattering," *Accounts of Chemical Research*, vol. 17, no. 10, pp. 370–376, 1984.

[19] J. R. Lombardi, R. L. Birke, T. H. Lu, and J. Xu, "Charge-transfer theory of surface enhanced Raman spectroscopy: Herzberg-Teller contributions," *Journal of Chemical Physics*, vol. 84, no. 8, pp. 4174–4180, 1986.

[20] S. Nie and S. R. Emory, "Probing single molecules and single nanoparticles by surface-enhanced Raman scattering," *Science*, vol. 275, no. 5303, pp. 1102–1106, 1997.

[21] K. Kneipp, Y. Wang, H. Kneipp et al., "Single molecule detection using surface-enhanced Raman scattering (SERS)," *Physical Review Letters*, vol. 78, no. 9, pp. 1667–1670, 1997.

[22] X. M. Qian and S. M. Nie, "Single-molecule and single-nanoparticle SERS: from fundamental mechanisms to biomedical applications," *Chemical Society Reviews*, vol. 37, no. 5, pp. 912–920, 2008.

[23] K. Kneipp, H. Kneipp, I. Itzkan, R. R. Dasari, and M. S. Feld, "Ultrasensitive chemical analysis by Raman spectroscopy," *Chemical Reviews*, vol. 99, no. 10, pp. 2957–2976, 1999.

[24] T. E. Furtak, G. Trott, and B. H. Loo, "Enhanced light scattering from the metal/solution interface: chemical origins," *Surface Science*, vol. 101, no. 1–3, pp. 374–380, 1980.

[25] I. Pockrand and A. Otto, "Surface enhanced Raman scattering (SERS): annealing the silver substrate," *Solid State Communications*, vol. 38, no. 12, pp. 1159–1163, 1981.

[26] T. Watanabe, N. Yanagihara, K. Honda, B. Pettinger, and L. Moerl, "Effects of underpotentially deposited TI and Pb submonolayers on the surface-enhanced Raman scattering (SERS) from pyridine at Ag electrodes," *Chemical Physics Letters*, vol. 96, no. 6, pp. 649–655, 1983.

[27] D. P. Tsai, J. Kovacs, Z. H. Wang et al., "Photon scanning tunneling microscopy images of optical excitations of fractal metal colloid clusters," *Physical Review Letters*, vol. 72, no. 26, pp. 4149–4152, 1994.

[28] P. Zhang, T. L. Haslett, C. Douketis, and M. Moskovits, "Mode localization in self-affine fractal interfaces observed by near-field microscopy," *Physical Review B*, vol. 57, no. 24, pp. 15513–15518, 1998.

[29] C. L. Haynes and R. P. Van Duyne, "Nanosphere lithography: a versatile nanofabrication tool for studies of size-dependent nanoparticle optics," *Journal of Physical Chemistry B*, vol. 105, no. 24, pp. 5599–5611, 2001.

[30] M. J. Weaver, "Surface-enhanced Raman spectroscopy as a versatile in situ probe of chemisorption in catalytic electrochemical and gaseous environments," *Journal of Raman Spectroscopy*, vol. 33, no. 5, pp. 309–317, 2002.

[31] G. T. Duan, W. P. Cai, Y. Y. Luo, Y. Li, and Y. Lei, "Hierarchical surface rough ordered Au particle arrays and their surface enhanced Raman scattering," *Applied Physics Letters*, vol. 89, no. 18, Article ID 181918, 2006.

[32] Z. L. Huang, G. W. Meng, Q. Huang, Y. J. Yang, C. H. Zhu, and C. L. Tang, "Improved SERS performance from Au nanopillar arrays by abridging the pillar tip spacing by Ag sputtering," *Advanced Materials*, vol. 22, no. 37, pp. 4136–4139, 2010.

[33] M. Malac, R. F. Egerton, M. J. Brett, and B. Dick, "Fabrication of submicrometer regular arrays of pillars and helices," *Journal of Vacuum Science and Technology B*, vol. 17, no. 6, pp. 2671–2674, 1999.

[34] B. Dick, M. J. Brett, and T. Smy, "Controlled growth of periodic pillars by glancing angle deposition," *Journal of Vacuum Science and Technology B*, vol. 21, no. 1, pp. 23–28, 2003.

[35] Q. Zhou, Z. C. Li, Y. Yang, and Z. J. Zhang, "Arrays of aligned, single crystalline silver nanorods for trace amount detection," *Journal of Physics D*, vol. 41, no. 15, Article ID 152007, 2008.

[36] Q. Zhou, Y. He, J. Abell, Z. Zhang, and Y. Zhao, "Optical properties and surface enhanced raman scattering of L-shaped silver nanorod arrays," *Journal of Physical Chemistry C*, vol. 115, no. 29, pp. 14131–14140, 2011.

[37] Q. Zhou, Y. J. Liu, Y. P. He, Z. J. Zhang, and Y. P. Zhao, "The effect of underlayer thin films on the surface-enhanced Raman scattering response of Ag nanorod substrates," *Applied Physics Letters*, vol. 97, no. 12, Article ID 121902, 2010.

[38] Z. Q. Tian, B. Ren, and D. Y. Wu, "Surface-enhanced Raman scattering: from noble to transition metals and from rough surfaces to ordered nanostructures," *Journal of Physical Chemistry B*, vol. 106, no. 37, pp. 9463–9483, 2002.

[39] X. Zhang, Q. Zhou, J. Ni, Z. C. Li, and Z. J. Zhang, "Surface-enhanced Raman scattering from a hexagonal lattice of micropatterns of vertically aligned Ag nanorods," *Physica E*, vol. 44, no. 2, pp. 460–463, 2011.

[40] Y. Z. Chu, M. G. Banaee, and K. B. Crozier, "Double-resonance plasmon substrates for surface-enhanced Raman scattering with enhancement at excitation and stokes frequencies," *ACS Nano*, vol. 4, no. 5, pp. 2804–2810, 2010.

[41] L. M. Tong, T. Zhu, and Z. F. Liu, "Approaching the electromagnetic mechanism of surface-enhanced Raman scattering: from self-assembled arrays to individual gold nanoparticles," *Chemical Society Reviews*, vol. 40, no. 3, pp. 1296–1304, 2011.

[42] L. Gunnarsson, E. J. Bjerneld, H. Xu, S. Petronis, B. Kasemo, and M. Käll, "Interparticle coupling effects in nanofabricated substrates for surface-enhanced Raman scattering," *Applied Physics Letters*, vol. 78, no. 6, pp. 802–804, 2001.

[43] L. A. Dick, A. J. Haes, and R. P. Van Duyne, "Distance and orientation dependence of heterogeneous electron transfer: a surface-enhanced resonance Raman scattering study of cytochrome c bound to carboxylic acid terminated alkanethiols adsorbed on silver electrodes," *Journal of Physical Chemistry B*, vol. 104, no. 49, pp. 11752–11762, 2000.

[44] L. A. Dick, A. D. McFarland, C. L. Haynes, and R. P. Van Duyne, "Metal film over nanosphere (MFON) electrodes for surface-enhanced Raman spectroscopy (SERS): improvements in surface nanostructure stability and suppression of irreversible loss," *Journal of Physical Chemistry B*, vol. 106, no. 4, pp. 853–860, 2002.

[45] M. Litorja, C. L. Haynes, A. J. Haes, T. R. Jensen, and R. P. Van Duyne, "Surface-enhanced Raman scattering detected temperature programmed desorption: optical properties, nanostructure, and stability of silver film over SiO_2 nanosphere surfaces," *Journal of Physical Chemistry B*, vol. 105, no. 29, pp. 6907–6915, 2001.

[46] J. A. Dieringer, A. D. McFarland, N. C. Shah et al., "Surface enhanced Raman spectroscopy: new materials, concepts, characterization tools, and applications," *Faraday Discussions*, vol. 132, pp. 9–26, 2006.

[47] C. R. Yonzon, C. L. Haynes, X. Y. Zhang, J. T. Walsh, and R. P. Van Duyne, "A glucose biosensor based on surface-enhanced Raman scattering: improved partition layer, temporal stability, reversibility, and resistance to serum protein interference," *Analytical Chemistry*, vol. 76, no. 1, pp. 78–85, 2004.

[48] A. Tao, F. Kim, C. Hess et al., "Langmuir-Blodgett silver nanowire monolayers for molecular sensing using surface-enhanced Raman spectroscopy," *Nano Letters*, vol. 3, no. 9, pp. 1229–1233, 2003.

[49] G. Ross, "The public health implications of polychlorinated biphenyls (PCBs) in the environment," *Ecotoxicology and Environmental Safety*, vol. 59, no. 3, pp. 275–291, 2004.

[50] Q. Zhou, Y. Yang, J. Ni, Z. C. Li, and Z. J. Zhang, "Rapid detection of 2, 3, 3′, 4, 4′-pentachlorinated biphenyls by silver nanorods-enhanced Raman spectroscopy," *Physica E*, vol. 42, no. 5, pp. 1717–1720, 2010.

[51] Q. Zhou, X. Zhang, Y. Huang, Z. C. Li, and Z. J. Zhang, "Rapid detection of polychlorinated biphenyls at trace levels in real environmental samples by surface-enhanced Raman scattering," *Sensors*, vol. 11, no. 11, pp. 10851–10858, 2011.

[52] Q. Zhou, Y. Yang, J. E. Ni, Z. C. Li, and Z. J. Zhang, "Rapid recognition of isomers of monochlorobiphenyls at trace levels by surface-enhanced Raman scattering using Ag nanorods as a substrate," *Nano Research*, vol. 3, no. 6, pp. 423–428, 2010.

[53] X. Zhang, Q. Zhou, Y. Huang, Z. C. Li, and Z. J. Zhang, "Contrastive analysis of the Raman spectra of polychlorinated benzene: hexachlorobenzene and benzene," *Sensors*, vol. 11, no. 12, pp. 11510–11515, 2011.

[54] Y. Cao, R. C. Jin, and C. A. Mirkin, "Nanoparticles with Raman spectroscopic fingerprints for DNA and RNA detection," *Science*, vol. 297, no. 5586, pp. 1536–1540, 2002.

18

A Study of the Deformation, Network, and Aging of Polyethylene Oxide Films by Infrared Spectroscopy and Calorimetric Measurements

Carl Bergeron,[1] Etienne Perrier,[2] Aymeric Potier,[2] and Geneviève Delmas[1]

[1] Départment de Chimie, Université du Québec à Montréal, CP.888, Suc. Centre-Ville, Montreal, QC, Canada H3C 3P8
[2] IUT, University of Auvergne, 63001Clermont-Ferrand, France

Correspondence should be addressed to Carl Bergeron, carl.karine@videotron.ca

Academic Editor: Karol Jackowski

The calorimetric and infrared (IR) spectroscopy measurements of polyethylene oxide (PEO) are used to evaluate the deformation and relaxation that films experience during a temperature cycle (30°C–90°C–30°C). After melting, the intensity of some bands decreases by 10 to 70%. During the temperature cycle, the C–O band in the 1100 cm^{-1} region and the C–C–O deformation bands at 650 and 500 cm^{-1} show some new features. A network of cooperative oxygen-hydrogen interactions between the PEO chains form in films with special history, namely, in thermally treated films, in thin films prepared from gel forming solutions, and in thick films after aging. The interchain interaction network is suggested from the IR absorption bands in the 1200 and 900 cm^{-1} region and also from small bands at 1144 and 956 cm^{-1}. The network seems absent or reduced in thin films. IR spectroscopy appears a sensitive technique to study chain conformations in PEO films and in other materials where order, disorder, and the formation of intermolecular interactions coexist.

1. Introduction

The physico-chemical properties of pure polyethylenoxide (PEO) such as crystallinity [1–3], thermal properties [2, 3], and morphology [4, 5] have been investigated. The polar and nonpolar parts of the PEO sequence give rise to a wide range of properties unknown to polyolefins. An intriguing property of PEO in dilute and very dilute solutions is the formation of stable clusters [6, 7]. Nonequilibrium effects related to the polymer-solvent interactions have been studied [8]. Associations have been investigated by several techniques in order to understand the thermodynamics of their origin and obtain some insight into protein-water interactions [9]. The effect of shear on solutions and resulting degradation has been analysed [10].

The vibrational spectra of oriented PEO films have been measured [11, 12] and the positions of the bands were compared to those calculated. Structural models and comparison with the IR spectra of small molecules have suggested the assignments of the bands between 3000 and 400 cm^{-1} observed for the solid and for the melt. Ionic conductivity, effect of surface on conformations, and crystallinity have been investigated in PEO systems in fundamental and applied FTIR studies [13–16]. FTIR spectroscopy is used to study the specific interactions between the PEO chains and small molecules [13]. Chain-chain interactions in composites and chain-ion interactions leave a trace on the vibrational spectrum. They increase the parallel portion of the vibrations at the expanse of the perpendicular ones, especially in the CO band at 1115–1109 cm^{-1}. Thin films of pure PEO and composites have been investigated by different techniques including IR spectroscopy. The effect of the adsorption on a solid surface like KBr or silicon substrates on the chain mobility was investigated through the intensity of bands specific to the crystal and the melt. Mobility is found to be reduced by the interaction of the PEO chains with the surface with consequence on the crystallization kinetics [14]. PEO crystallizes in a monoclinic cell, with the chain adopting a 7/2 helix but in a triclinic cell with a planar zigzag conformation when the sample is distorted.

The interesting and wide-range properties of this polymer have been exploited in biological and pharmaceutical research. A variety of copolymers of PEO and polyolefins have been synthesized. PEO chains can be chemically linked to a variety of molecules (peptides, proteins, antibody fragments, and small molecules) and have been used for drug delivery [15, 16]. Polymer partitioning is a key factor in the formation of biphasic systems and is of importance in the transport of active agents linked to PEO across cell membranes [15].

Stable associations and micelle formations are found in *block* copolymers $(A\text{-}A)_m\text{-}(B\text{-}B)_n$ with different polarities of the A-A and B-B groups. They have been observed surprisingly in *random* polyethylene-polypropylene copolymers, where the $-CH_2-CH_2-$segments and the $-CH(CH_3)-CH_2-$ segments are not polar [17]. A noncrystalline organization can grow in the absence of hydrogen bonds or other strong interactions to minimize free energy. Unexpected variation of the PEO IR spectra with film thickness and solvent history are reported below and in unpublished work in our group suggesting a certain similarity between the spontaneous organization of the PEO chains and those of the PE-PP polymer.

A standard strategy to investigate the phase content in polymers is to observe the effect of sample history (deformation, solvent, or thermal treatment) on the vibrational spectra or on the calorimetric trace. Distortion is usually achieved by an exterior mechanical force. In our laboratory, a T-ramp was used to deform the sample, events which can be followed by calorimetric measurements or by IR spectroscopy. A slow T-ramp is required to achieve and record sample deformation and to study the signals associated with the changes. The calorimetric trace displays not only the main peak of fusion, but also other signals which are found in the melt above T_m. This is a general feature [18–30] found on several polyolefins, for amylose and PEO (Appendix). The IR spectra reveal important changes due to the treatment, such as band displacement and change of intensity or position of the base-line related to the sample. For polyolefins, PET [31] and amylose [32, 33], the integrated absorbance permits to determine (with some assumptions), the ordered, the disordered, and the strained fractions in a sample. For amylose pastes [32, 33], the combination of FTIR and calorimetric analyses illustrates conformation modifications with thermal treatment. Bands were tentatively assigned to simple and double helices, whose respective content was treatment dependent [32, 33]. The sensitivity of some bands of the IR spectra to sample history has been investigated in mixtures of polymers [34]. Films with specific properties (such as high drawing ability and high resistance [22]) have been obtained, using as strategy of morphology modification, a gel whose chains are entangled. Indicators of network organization were found on the calorimetric signals and on the IR spectra [18, 30].

The possibility of finding the same concepts at work for PEO as for PE and amylose, through special thermal treatments and techniques like FTIR and calorimetry, seemed worthwhile and has been investigated [35–40]. Development of strain during aging or through phase changes initiated by treatment or by interactions with other molecules has been followed by ATR and FTIR spectroscopy. Results on pure PEO are likely to give useful information and trends in the mixtures prepared at different temperatures for biochemical and pharmaceutical research and applications. Due to its sensitivity to strain [11, 12], IR spectroscopy is a useful technique to follow deformation of PEO samples during the T-ramp.

In the present work, two strategies were used to change the sample morphology and follow their effect on the IR spectrum. A film prepared in standard conditions was submitted to a T-ramp covering a wide T interval and other films were obtained from entanglement-enriched solutions and analysed at RT as a function of aging time. This controlled preparation and treatment and the knowledge of spontaneous growth of a physical network can be of help for investigating the effect of PEO processing and aging on its in-use properties such as resistance to wear, brittleness, and permeability.

2. Experimental

2.1. Materials. PEO is from the Aldrich Company. The sample is characterized by $MW = 10^4$. In order to prepare gels and network-rich films, solutions are made with a higher MW (Aldrich, $MW = 10^6$) or a higher concentration of the lower MW sample. The polymers used as received are kept in desiccators. The solvents are of chromatographic grade from Aldrich.

2.2. Calorimetry. Standard DSC measurements were performed on a Perkin Elmer DSC-7, calibrated with Indium. Slow calorimetry was done on a Setaram C80 calorimeter, as in our previous works [18–30, 32, 33]. Table 1 gives the characteristics of the calorimetric traces obtained with standard conditions and with slow calorimetry [36, 39]. A calorimetric trace using a slow T-ramp (3 K/h) is given in the Appendix.

2.3. FTIR. The FTIR instrument is a Bomen spectrometer (MB series, Bomen Inc, Canada) used in transmission mode. The spectra were recorded after 50 scans at a resolution of $1\,cm^{-1}$. The IR chamber is flushed by dry nitrogen during the measurements.

2.3.1. Spectra as a Function of Temperature. The IR spectra were taken using a home-made oven [28, 30, 32, 33]. The sample and a thin thermocouple are placed between two KBr disks and maintained with Teflon tape in a well isolated oven. The heating wires of the oven and the thermocouple are plugged into a precise temperature controller. The temperature cycle is 30°C–90°C–30°C. The heating ramp includes a step of 6 K at 0.5 K/min, followed by a one-minute isotherm and repeated until the maximum temperature. The cooling rate is 1.7–0.7 K/min between 90 and 60°C and 0.2 K/min around T_c. An IR spectrum was taken every 13 min.

Table 1: Characteristics of the melting traces of the PEO sample in different conditions.

	Ramp (K/min)	ΔH^a (J/g)	T_m (°C)	T_c (°C)	Events in the melt[b,c]	
					ΔH_{relax} (J/g)	T_{relax} (°C)
Standard	5	147	61	46		
Nonstandard	0.05	130	68	44	50–100	80–110

[a]The value of αc is found to be 0.72 from ΔH_0 = 204 J/g.
[b] The value of ΔH_{relax} and the range of T_{relax} are sensitive to the conditions [22–30, 36, 37].
[c]The signal in the melt is endothermic for a deformation and exothermic for a relaxation.

2.3.2. Baseline Subtraction. The wires emit IR signals, which vary with the heating power. The signal from the cell has been measured in a blank experiment at the same rate of heating and cooling as in the experiment. It was subtracted from the data obtained with the sample at every temperature [29, 32, 36, 39].

2.3.3. Sample Preparation. The film submitted to the T-ramp is prepared as follows: the dissolution of the polymer in toluene with a 0.5% concentration is assisted by sonication during 20 min. The container is surrounded by iced water to avoid a rise of temperature. The power of the sonicator is carefully adjusted by trial and error [36, 39] to avoid sample degradation. The constancy of the sample MW is checked by measuring the solution viscosity. After solvent evaporation at RT, the film, left to dry during two days, is placed between two KBr plates held by Teflon tape. The treatment of this film called thermally treated or (Therm-T) consists of a T-ramp, starting at 30°C up to the maximum temperature (90°C) and cooling to RT. The film has a thickness of 15 μm. There is no trace of oxidative degradation on the IR spectra of Therm-T film after the T-cycle.

Free standing films: for comparison with Therm-T, free standing films of different history are prepared and analyzed at RT. The film (Reference) is prepared from a 1% solution in toluene. Dissolution at 40°C is assisted by a magnetic stirrer during two hours. In order to minimize the strain, evaporation is performed at 4°C. Reference is a film, whose IR spectrum seems to have reached equilibrium (at the time of the analysis). The other free-standing films are obtained from solutions enriched in entanglements. Solvent evaporation is made at RT.

2.3.4. Aging. The films are analyzed usually two days after the solvent has evaporated. The thicker films are analyzed again three days after taking the first spectrum and in some cases after seven or more days.

2.3.5. Change in Chain Orientation and Band Intensity Variation at T_m. Since PEO repetitive unit contains a polar bond, chain reorganization is likely to affect the vibrational energies of the bonds. The average orientation of the bonds in the chains from the crystal to the melt will be modified and consequently, the IR band position and its maximum intensity will be affected. The terms parallel or perpendicular represent the direction of the transition dipole moments, which are parallel and perpendicular to the helical axes of PEO. These terms are satisfactory to describe changes in the solid submitted to different histories or those prepared in different solvents [35]. The perpendicular bands are also called helical bands and the parallel bands are those associated with a transplanar zigzag conformation. The latter is present in deformed samples or when the chains interact with others [34] or with ions [13]. An improved *intramolecular* interaction between the CH_2 segments and the oxygen atoms was, according to P. J. Flory earlier, at the origin of the helical conformation of the PEO chains. However, the *intermolecular* interaction between several CH_2 groups and several oxygen atoms of two different chains is more feasible for chains in the zigzag conformation. The conformation will change from helix to zigzag to minimize the free energy engaged in interchain association.

2.3.6. Change in the Band Position. For PE, reduced cohesion between the chains in the melt shifts some bands towards lower energies [29, 30]. The doublet of the CH_2 rocking is at 720–730 cm^{-1}. A new band is observed at 712–715 cm^{-1} in the nascent materials where the chains are strained. In suitable conditions, the deformed crystals change from the orthorhombic to monoclinic modification. In PEO, the change of the position of the maximum intensity is due mainly to changes of orientation. New bands are formed in M, so that there is an unavoidable ambiguity in comparing small bands in C and M.

The polarized IR spectra of strongly oriented PEO samples [11] enriched in parallel bands by deformation are informative. In Table 2, the comparison of the different bands is made for a qualitative evaluation, using ε for short [11, 12]. The values of the extinction coefficients are not given explicitly, only the bands and the maximum absorbance with different light orientations are shown. For the doublets (1359–1342, 1242–1235, and 963–947 cm^{-1}) and for the 1280, 1260 and 844 cm^{-1} bands, ε is reported to change greatly with change of light polarization in the deformed samples (by a factor of 2 and 3). On the other hand, the parallel and perpendicular bands present, in the $-CH_2$ stretching region and the CO stretching region, have similar values of ε, so one can expect minimum change at T_m. Assuming that the trends observed for the highly stretched sample remain valid for Therm-T, one can expect that the change of absorbance, Δ(Abs), at T_m will depend on the region of the spectrum. The data of the last row of Table 3 which lists the changes of

TABLE 2: Relative values of the extinction coefficient, ε, for orientation-sensitive bands of PEO [11, 12].

CH stretching 3000–2650	Overlap‖ and ⊥ vibrations	$\varepsilon\| \approx \varepsilon \perp$
CH bending 1500-1400	1470: ‖ 1463: ⊥	$\varepsilon\| \approx \varepsilon \perp$
CH_2 wagging/twisting	1343: ‖ 1359:⊥	$\varepsilon\| > \varepsilon \perp$
CH_2 wagging/twisting	1280 ⊥	$\varepsilon\| < \varepsilon \perp$
CH_2, CO rocking/stretching	1241: ‖ 1235 ⊥	$\varepsilon\| > \varepsilon \perp$
CH_2, CO rocking/stretching	Overlap ‖ and ⊥ 1200–990	$\varepsilon\| \approx \varepsilon \perp$
CH_2 rocking CO stretching	1060 ⊥	$\varepsilon\| < \varepsilon \perp$
CH_2 rocking CO stretching	963: ‖ 947: ⊥	$\varepsilon\| > \varepsilon \perp$
CO/CH_2 rocking/stretching	844: ⊥	$\varepsilon\| < \varepsilon \perp$

TABLE 3: Variation of the standard integrated absorbance between C and M[a] for two large regions and for a doublet[c].

	3000–2640 (cm^{-1})			1500–770(cm^{-1})			1000–900 (cm^{-1})		
	Std Abs[a]	Max Abs[b]	Band position (cm^{-1})	Std Abs	Max Abs[b]	Band position (cm^{-1})	Std. Abs[c]	Max Abs.	Band position (cm^{-1})
C	78.3	0.92	2884	139	1.55	1114	7.7–7.0	0.46–0.40	963–947
M	72.6	0.72	2861	131	1.05	1109	3.5	0.25	991–948
%[c]	−9	−25		−6	−40		−70	−(10–45)	

[a]Using the standard method.
[b]From the spectra at 66°C on heating, M, and 30°C on cooling, C.
[c]Bands are displaced from C to M. Using nonstandard absorbance including NDA, the change is smaller.

integrated absorbance at T_m for different regions are consistent with this assumption.

2.3.7. Integrated Absorbance. The maximum absorbance of a band (at ν_i) is widely used as an analytical tool to measure the concentration of a molecule present in a material containing a bond vibrating at a specific energy (ν_i). Most of the studies have used relative intensities rather than absolute intensities because the intensities depend on many factors, some of which are difficult to control. However, it was found that in PE films prepared in controlled conditions, absorbance could be directly related to the sample thickness. A reliable extinction coefficient, ε, was obtained through the analysis of films of different thickness but the same history [26, 29, 35–40]. The value of ε is dependant on the film history. When the morphology of the sample and not its chemical composition is of interest, the integrated absorbance is to be preferred to the maximum absorbance, because it takes into account the change of band-shape and the possible displacement of bands associated with the sample history. The extinction coefficient, ε, calculated by dividing the integrated absorbance by the film thickness has the unit of ($cm^{-1}\,\mu m^{-1}$).

2.3.8. Absorbance, Standard, and Nonstandard: Nondifferentiated Absorbance (NDA). The IR spectra of polymers and gels present a feature which does not exist for small molecules. There is a nondifferentiated absorbance (NDA) which is situated between the base line and the spectrum. It corresponds to the different conformations which exist in the sample, the bonds of which have about the same frequency, the same energy and the same probability of occurrence. NDA is found mostly between 1200 and 900 cm^{-1} in the PEO spectra. The size of NDA gives some information on the amorphous or gel content of the sample. NDA has not received much attention in the IR literature, although it is sizeable in samples with a low crystallinity.

The evaluation of the *standard* integrated absorbance between ν_1 and ν_2 is straightforward, when one uses as base line a straight line between the absorbance A(ν_1) and A(ν_2) on the trace. To evaluate the *nonstandard* integrated absorbance, one chooses as base line, a horizontal line starting at for example 1200 cm^{-1} and going to the bond frequency of interest for instance 900 cm^{-1}. The difference between A (900 cm^{-1}) and the point on the base line at 900 cm^{-1} gives a qualitative information on the NDA at 900 cm^{-1}. Alternatively, the difference of integrated absorbance between the nonstandard absorbance using two points A (1200 cm^{-1}) and Baseline (900 cm^{-1}) and the standard value is a measurement of the NDA in that region.

In PEO samples, the standard absorbance does not vary much at T_m when measured between two distant values of the wavenumber as reported in Table 3. It is the proportion of nondifferentiated absorbance (NDA) over the total absorbance which changes at T_m. The value of Δ(Abs) between the solid and the melt is unambiguous for the –CH_2 stretching band, because this band has no NDA but for other bands and doublets with a smaller intensity, the value of Δ(Abs) at T_m or T_c depends on the limits of the integration chosen.

Free-standing samples have been analyzed at different light polarization. The spectra of the equilibrium films were not different with the light orientation at 90°, 0°, 45°, but the maxima of the bands of thick films in the process of aging is displaced when changing the light polarization.

2.3.9. Film Thickness. The film thickness is measured with a calliper. In a series of films in equilibrium of different

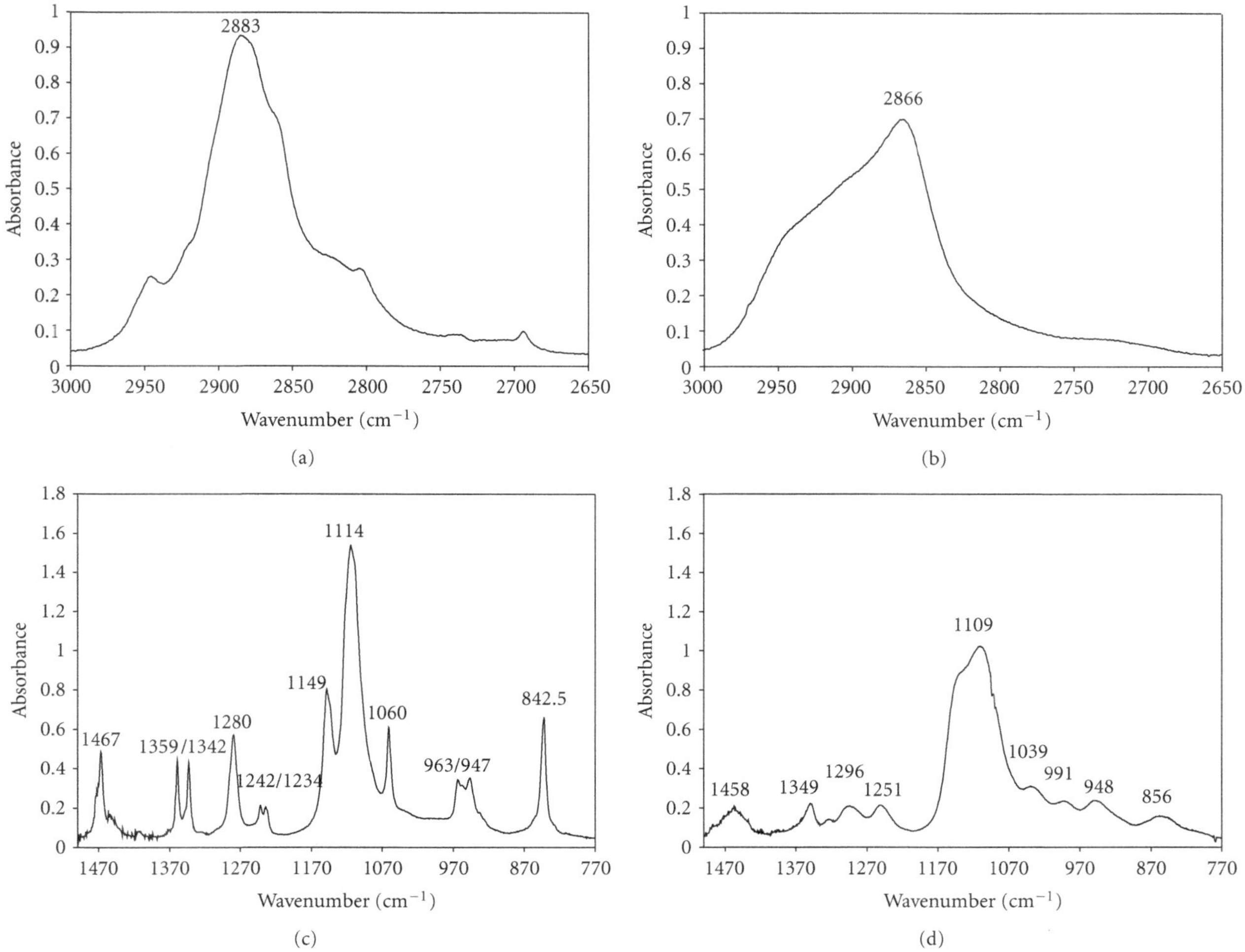

FIGURE 1: Spectrum between 3000 and 2660 cm^{-1}. (a) Sample in the solid state at 30°C on cooling at the end the T-cycle; (b) sample in the liquid state. Spectrum between 1500 and 770 cm^{-1}; (c) sample in the solid sample at 30°C on cooling at the end the T-cycle; (d) sample in the liquid state.

thickness but the same history, the integrated absorbance of the $-CH_2$ stretching group is found to vary regularly with thickness so that a value of $\varepsilon = 5$ ($cm^{-1}\,\mu m^{-1}$) [35] was obtained. For very thin films whose thickness is not precise when measured with a calliper, their thickness is measured through ε and the integrated absorbance of the CH_2 stretching of their spectrum. In films in the process of aging, in which this region is not stable, another crystalline band relatively insensitive to conformations is used. The two films which will be compared in detail are Therm-T (15.0 μm) and Reference (11.5 μm).

2.3.10. IR and Network Tracer. The presence of a network in a material is revealed by its IR spectrum if the junctions are made of interacting hydrogen bonds. The frequency of the OH bond is lower than in a material with free OH bonds. In systems where the $-CH_2-O-CH_2$ segment is involved in a cooperative interaction with segments of another chain, one observes also a shift in the main peak of the CO. However, the difference of bond energy between the oxygen atoms which are part of the interaction and outside it is likely to be small. Consequently, only small differences of band position are expected and overlapping of bands composing the CO vibrations is likely to take place.

3. Results and Discussion

The vibrational spectra presented in this work focus on two subjects. The first is related to the changes in the IR spectra at melting or just before melting (Figures 1 and 2) and the observation of indicators of tension inside the sample during the ramp cycle, especially in the 1100–1090 cm^{-1} (Figure 3) and the 650–500 cm^{-1} region (Figures 4-5). The second part will focus on the observation of indicators of network organization at RT. Network-enriched films have been prepared and the results are compared to those observed in non-enriched samples such as a thin sample made from a dilute solution named Reference (Figures 6 and 7). While looking at the spectra of other films which would be similar to Reference albeit thicker, it was found that the thicker films cannot be

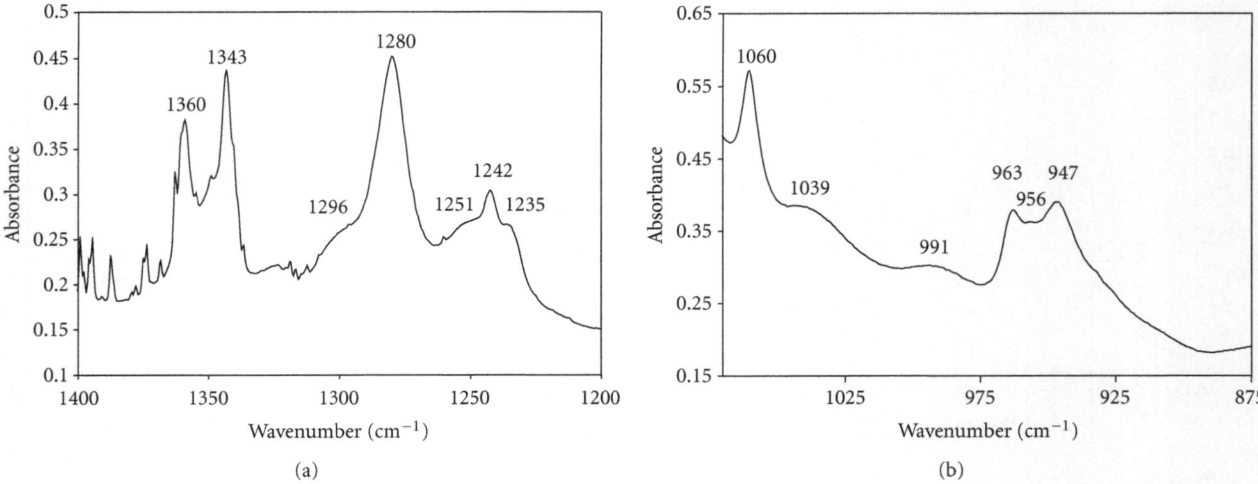

FIGURE 2: Transition spectra at 60°C: The two regions show more bands than those observed in the solid and the melt (Figure 1(b)). (a) The 1400–1200 cm^{-1} interval (with eight bands). (b) The 1070–875 cm^{-1} interval (with six bands). Note the small band at 956 cm^{-1}.

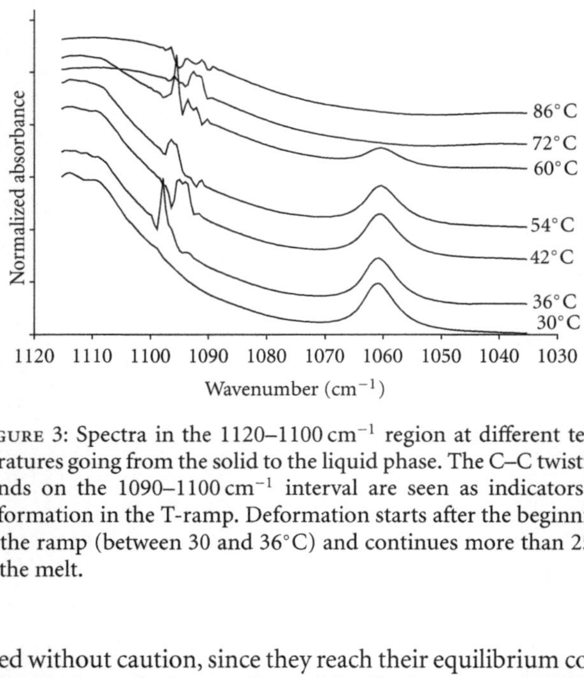

FIGURE 3: Spectra in the 1120–1100 cm^{-1} region at different temperatures going from the solid to the liquid phase. The C–C twisting bands on the 1090–1100 cm^{-1} interval are seen as indicators of deformation in the T-ramp. Deformation starts after the beginning of the ramp (between 30 and 36°C) and continues more than 25 K in the melt.

used without caution, since they reach their equilibrium conformation much slower. Figure 8(a–d) gives an example of nonequilibrium thick films.

3.1. Melting. Melting changes considerably the absorption in the 1500–770 cm^{-1} region as reported earlier [12]. The bands associated with crystallinity, namely, the $-CH_2$ bending (1467 cm^{-1}), the $-CH_2$–CO rocking/stretching (843 cm^{-1}) and the band at 1060 cm^{-1} are no longer present or are reduced and displaced. The fine structure of several of the doublets is replaced by a large band and the CO band is wider and its maximum displaced (1109 cm^{-1}).

In the $-CH_2$ stretching and the CO regions (Table 3), the positions of the maximum intensity are seen to be shifted upon melting from 2883 cm^{-1} to 2861 cm^{-1} and from 1114 to 1109 cm^{-1}. The integrated intensity is reduced (columns 1 and 4) by less than 10% for the two vibrations. For these two groups of vibration modes, the value of $\varepsilon\|$ is similar to that of $\varepsilon_\perp$. The band maximum intensity is reduced but the band is wider, keeping the integrated intensity at about the same value. The increased width indicates a larger range of conformations. Note that the values given for the 1500–770 cm^{-1} region cover also the $-CH_2$ bending, twisting and wagging vibration modes. The change for the CO peak, not reported in Table 3, is the same (−5%) as the (−6%) given for the larger interval because the contribution of the small peaks is small (and $\varepsilon\|$ and $\varepsilon_\perp$ have different magnitudes). The intensity of the doublet (at 963–947 cm^{-1}) associated with the CH_2 rocking/CO stretching vibration is reduced by 70%. The two peaks at 963 and 947 cm^{-1} are replaced by flattened bands at 991 and 948 cm^{-1} in the melt.

The moderate change of intensity of the two large bands of PEO at T_m compared to the larger variation observed in PE is likely a consequence of the rather more disordered morphology of solid PEO. The density of a solid is usually a good indicator of the sample order. The density of PE is reduced by 20% at T_m, while the density of PEO is not very affected by melting.

The gradual change of the sample organization can be followed due to the slow T-ramp, better in some regions than in others. Melting is a slow process in the ramp conditions. For example, the maximum intensity of the $-CH_2$ stretching band is at 2883 cm^{-1} at 56–60°C, 2875 cm^{-1} at 66°C, and at 2866 cm^{-1} at 72°C. Figure 2 shows the spectrum in two different regions around T_m. Due to the slow motion of the chains, bands specific to C and M can be observed in this transition spectrum at 60°C. In Figure 2(a), the doublet at 1360–1343 cm^{-1} and the band at 1280 cm^{-1} of C are still very present, while the parallel bands at 1296, 1251 cm^{-1} of M have started to grow. In Figure 2(b), the crystalline band at 1060 cm^{-1} and the doublet at 963–947 cm^{-1} coexist with the M bands at 1039 cm^{-1} and 991 cm^{-1}. The two spectra in

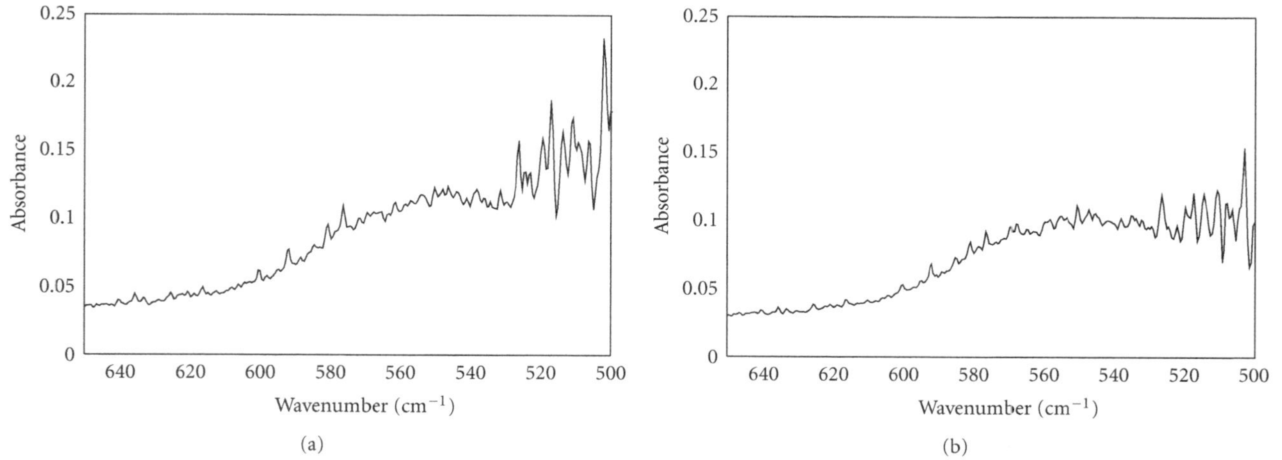

FIGURE 4: Spectra of Therm-T in the melt between 650 and 500 cm^{-1}, (a) in the ascending ramp at 90°C and (b) in the descending ramp before crystallisation.

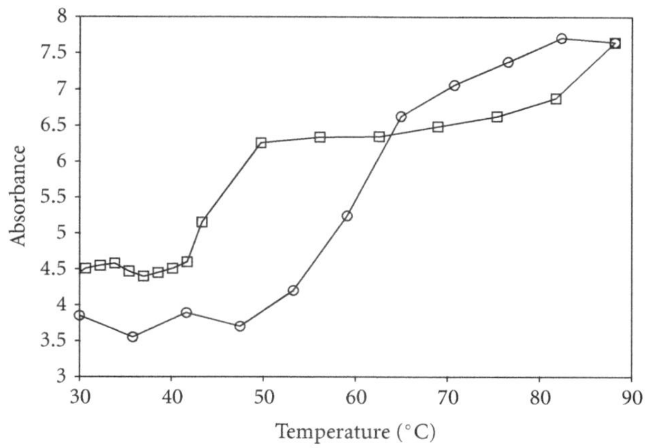

FIGURE 5: Spectra of Therm-T: variation as a function of T of the integrated nonstandard absorbance in the C–C–O bending region (650–500 cm^{-1}). Heating (○) and cooling (□) ramp. The integrated absorbance at 30°C for the nontreated sample (Reference) is not shown on Figure 5 but is found lower (2.5) than that observed in Therm-T at the beginning of the T-ramp (3.8).

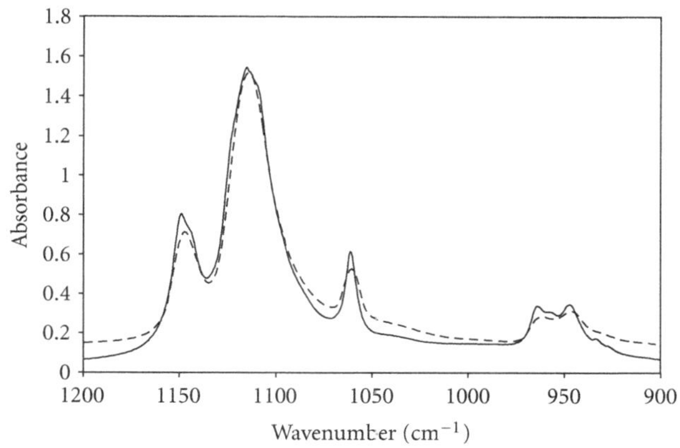

FIGURE 6: Tracers of network organization at 1144 cm^{-1} and 956 cm^{-1} and sample history, seen in Therm-T (solid line) but not apparent in Reference (dotted line).

Figure 2 suggest that the value of the free energy of the conformations associated with these bands is rather similar over a range of temperature. Around T_m, the presence of polar groups along the chains in PEO maintains energetically favourable interchain interactions both in the solid and the melt phases. The relaxation time of the long chains has been investigated by different techniques in polystyrene [41]. The movement of the chains followed by slow calorimetry (Appendix) in the melt of PE and PEO was better understood through the results found on amorphous polystyrene.

3.2. Strain Building. Features in two regions of the IR spectra during the T-ramp investigation can be associated with deformation. These are seen in the CO stretching band (1200–1000 cm^{-1}) and in the C–C–O bending region (650–500 cm^{-1}). They will be discussed below.

3.3. CO Stretching Vibration at 1190–1090 cm^{-1}. The low frequency side of the CO band is presented in Figure 3. The different traces shown between 1120 and 1030 cm^{-1} are a selection of IR spectra at different temperatures in the *ascending* ramp. The spectrum at 30°C shows only the perpendicular band at 1060 cm^{-1}. Above 60°C, this band disappears but returns as expected on crystallization. The spectra above 30°C show new signals between 1090 and 1100 cm^{-1}, which vary in position, shape, and intensity. This is a characteristic associated with the superposition of several events, which can be interpreted as follows. When the T-ramp starts, the polymer is submitted to a strain, which continues until the temperature reaches its maximum at 90°C. The events in the calorimetric trace (Appendix) show that other movements occur at the same time. Relaxation of parts of the stressed

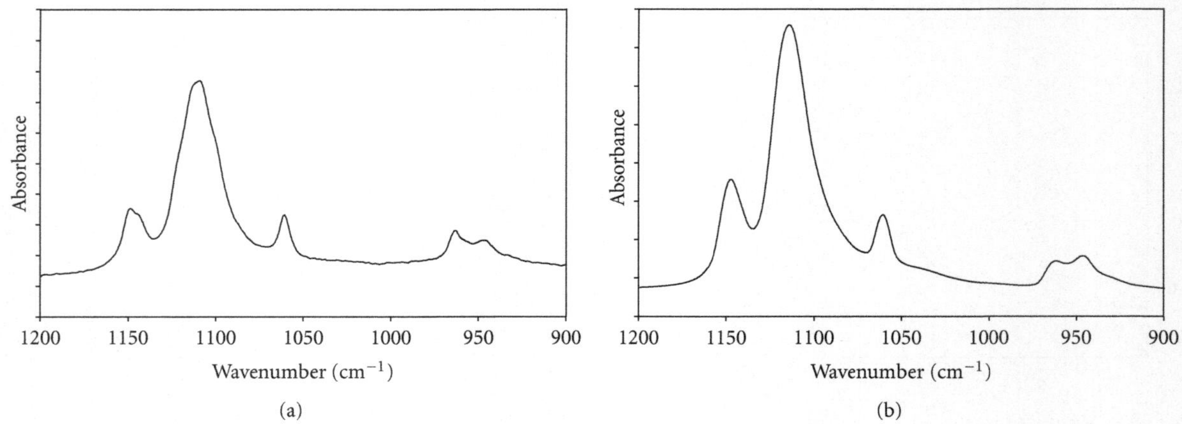

Figure 7: Tracers of network organization at 1144 cm^{-1} and 956 cm^{-1} and sample history: (a) thin film prepared from a gel (2.3 μm); (b) therm-T, the transition spectrum at 60°C (15 μm).

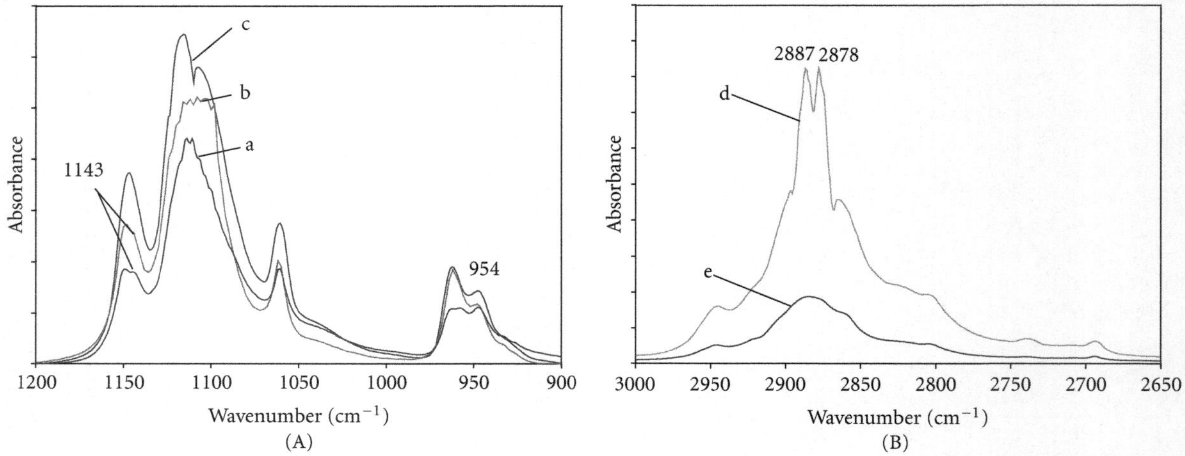

Figure 8: The CO and CH_2 stretching bands in thick films in the process of aging Trace (a) Multipeak CO band in a 27 μm film showing tracers of network at 1144 cm^{-1} and 956 cm^{-1} Trace (b) Multipeak CO band in a 30 μm film with the same tracers at 1144 cm^{-1} and 956 cm^{-1}. Trace (c) Double peak CO band in a 45 μm film. The 1144 cm^{-1} and 956 cm^{-1} bands are not apparent at that stage. Trace (d) Multipeak in the CH_2 stretching band (56 μm). The films are aged three days except b, aged one week. For comparison, the CH_2 stretching band of a 15 μm film after the temperature treatment (Therm-T) is given in the trace (e).

regions takes place concomitantly with the formation of stress in other regions. Relaxation is registered around the frequency of the vibration of the CO stretching, but with a different pattern, which overlaps the frequency due to the strain. One cannot separate the contributions of tension and relaxation on the IR spectra. Consequently, the vibrational spectra are complex and their overall shape changes with temperature, with alternation of sharp bands and wider bands. The sharper absorptions seen at slightly different positions at 36 and 60°C in Figure 3 are not unique to these temperatures, neither are the wider peaks observed at 42, 54, 72°C. The same spectra measured at 48, 68, 78 and 84°C (not shown) display also a combination of sharp and wider bands. The new bands are still present above 61°C, the value of T_m observed in standard calorimetry. At 84°C, well above T_m and 6 K below the maximum temperature, the intensity of the signal is reduced as is the rate of the temperature increase programmed by the temperature control. The IR spectrum at $T = 60$°C will be further discussed below, in relation to the presence of indicators of interchain interactions.

3.4. The C–C–O Bending Region between 650–500 cm^{-1} as a Function of T. Figures 4(a) and 4(b) give the trace in the 650–500 cm^{-1} region. The trace a is in the melt at the maximum of deformation (90°C), trace b is in the descending T-ramp. The two traces are constituted of a large flat peak between 600 and 510 cm^{-1}and of a region between 510 and 500 cm^{-1} which is different for the two temperatures. In highly deformed and crystalline samples [12], a doublet is observed at 529–508 cm^{-1} instead of the flat peak observed in Therm-T. The doublet intensity changes greatly with orientation [12]. It is attributed to a C–C–O bending mode. In the present nonoriented sample, the doublet appears as a group of peaks whose maximum intensity change with temperature and tension on the sample. There is not a unique way to trace the variation of this region with temperature. The non-standard integrated absorbance has been chosen with limits between 650 and 500 cm^{-1}. The integration is between the trace and a horizontal base line with an ordinate at 0.04. The values of the integrated absorbance as a function of temperature are reported in Figure 5.

The integrated absorbance increases from 4 to 7.5 cm^{-1} as the temperature is raised from 30°C to 90°C. This increase is associated with an accumulation of deformations of the C–C–O bonds during the rise of T (○). When T decreases (□), the intensity of the vibration mode is reduced as displayed in the figure. There are interesting differences of response to the deformation between the C-C twisting band absorption (Figure 3) and the C–C–O bending regions (Figure 4). The new bands at 1096 cm^{-1} in Figure 3 appear at the beginning of the ramp (36°C), while the intensity of the C–C–O bending mode in Figure 5 is still small at 54°C. The increase of the ordinate of Figure 5 in the T-ramp is 4.2. Part of the deformation takes place below and at T_m (2.5) and part in the melt (1.7).

The nature of the atoms involved in the two types of vibration is probably at the origin of the difference in the onset of the signal. In the C–C twisting bands, the change which takes place between 1100 and 1090 cm^{-1} involves two (or possibly three) atoms in the chain. The T-ramp induces an immediate response such as a change of angle between adjacent bonds due to rotation. A different movement associated with relaxation following strain takes place a few K higher and leads to a shift of the signal (less than 5 cm^{-1}). The successive events do not seem to be additive. On the other hand, the signal of the C–C–O bending mode (650 and 500 cm^{-1}) which is spread over 150 cm^{-1} involves three or more atoms in the chains. The IR spectroscopic response is delayed by 30 K (or 60 min), until there is enough mobility in the material for the chains to move cooperatively. The changes which are additive become larger as the maximum temperature are reached. The deformation in the C–C–O bending mode is reversible (□) in the conditions of the T-ramp. The intensity decreases on cooling and reaches a value of 4.5 cm^{-1} which is somewhat higher than that at the beginning of the ramp (3.8 cm^{-1}). Suggesting a memory of the treatment. The integrated absorbance data reported in Figure 5 include deformation and relaxation.

The exact values of the integrated absorbance reported in Figure 5 and their variation with temperature depend on the position of the base line. Others integration while being less direct than the one explained above give similar trends.

3.5. Other Events in the Melt. The intensity changes only slightly but consistently in the melt between T_m and 90°C. The intensity of the $-CH_2$ wagging and twisting vibrations between 1400 and 1200 cm^{-1} decreases by 5% in the ascending ramp but regains on cooling exactly the same value. The intensity of the $-CH_2$ rocking band decreases and regains 7% in the same T ramp. These variations parallel those of the heat flow observed by calorimetry in the melt during a slow T-ramp (Appendix).

3.6. Films at RT: Differences between Therm-T and Reference. A comparison was made between Therm-T (after the temperature cycle) and a solution-grown film called Reference, in order to locate a possible remnant at RT of the effect of the ramp on the morphology. The reference has been prepared in conditions chosen to minimize the stress as reported in the experimental part. The absorbance of the doublets at 1359/1343, 1242/1234, and 963/947 indicate that the films have the same fraction of $\parallel$ and $\perp$ orientations.

3.7. Thin Films. Some differences are seen in the 1200–900 cm^{-1} region for these two thin films. The non-differenciated absorbance (NDA) is larger in Therm-T than in Reference. The difference between the positions of the baseline at 1200 cm^{-1} and 970 cm^{-1} is larger for Therm-T (solid line) than for Reference (dotted line). In the CO band, Therm-T has a large shoulder at 1109 cm^{-1} close to the band at 1114 cm^{-1}. Reference has a maximum at 1111.8 cm^{-1} and no apparent shoulder is observed at lower frequency. A parallel band has been calculated and found experimentally at 1103 cm^{-1} [11, 12]. The increase of the parallel band which displaces the band towards low frequency in Therm-T is assigned to chain-chain interactions.

This shift from helical to transplanar zigzag conformation takes place when chains interact, as observed repeatedly in PEO ions systems and in composites [34]. Small bands at 1144 cm^{-1} and at 956 cm^{-1} (between the two peaks at 963 and 947 cm^{-1}) are well seen in Therm-T (Figure 6), but it is not obvious in Reference (Figure 6). These bands are of interest as potential indicators of a specific network organization as discussed below.

3.8. Indicators of the Network Organization and Effect of Aging. In order to confirm the assignment of the 1144 cm^{-1} and the 956 cm^{-1} band to a network organization, films were prepared in conditions where the network fraction was likely to be either important or small. The strategies to prepare a network-rich film are diverse: using a solution of a high MW, or a semiconcentrated solution, or stirring vigorously during dissolution, or annealing the melt in specific conditions. Thin and thick films have been prepared in network-forming conditions, such as films obtained from gels, films obtained from melts left standing at 90°C, films made from 3%, and 4% solution and films prepared from a high molecular weight sample. The 1144 cm^{-1} and 956 cm^{-1} bands are seen in all these films. The IR spectra of two of them are shown in Figures 7(a) and 7(b). The film whose spectrum is shown in Figure 7(a) is 2.3 μm thick and was prepared from a gel. The two bands at 1144 and 956 cm^{-1} are present. Figure 7(b) is Therm-T at 60°C (presented already in Figures 2(a) and 2(b) on a narrow interval of frequencies). The spectrum at that temperature has interesting features in relation to the network bands. The crystalline band at 1149 cm^{-1} (at 60°C) is reduced, so that it forms with the network band at 1144 cm^{-1} almost a flat peak. This is also the case for the 956 cm^{-1} band (between the bands at 963 and 946 cm^{-1}) which has disappeared. The deformation peak at 1096 cm^{-1}, seen in Figure 3, is quite intense at that temperature. The presence of the network bands in Therm-T (15 μm) leads to the reconsideration of the limit of thickness. Therm-T is probably at the limit between thin and thick films.

3.9. Decomposition of the Bands. Decomposition of the bands was done on the 963–947 cm^{-1} doublet for Reference (Figure 6) and the film from the gel (Figure 7(a)). In the film not enriched in network, the fraction at 956 cm^{-1} of the total peak is found to be 0.14, while it is 0.33 for the film made from a gel. The film from the gel has a higher fraction of parallel band at 963 cm^{-1} seen on the spectra and in the decomposition. However, this does not seem to displace the network band, since in other films prepared in network-enriching conditions and where the parallel and perpendicular fraction have the same magnitude, the decomposition leads also to a higher fraction of the 956 cm^{-1} band.

3.10. Thicker Films. The analysis of thicker films leads to the finding that their spectra change with time and that even those prepared in dilute solution reach, after some time (days and weeks), the profile of thin films prepared in network-forming conditions.

Four spectra are given as examples of the effect of thickness and aging on the CO and the CH_2 stretching band.

3.10.1. CO Band. The film of 27 μm whose spectrum is shown in Figure 8(a) has been prepared by cooling a melt annealed at 90°C and is aged for three days. The bands associated with a network are seen at 1144 cm^{-1} and 956 cm^{-1}. The CO peak is multipeak (1114 cm^{-1}, 1110 cm^{-1}) with small shoulders at 1106 cm^{-1} and at a lower wavenumber. The spectrum of the 30 μm film of Figure 8(b) has been aged during one week and shows also the 1144 cm^{-1} and 956 cm^{-1} bands. The spectrum of the 45 μm film of Figure 8(c) has been aged during three days. The 1144 cm^{-1} and 956 cm^{-1} bands are not apparent at that stage. The maximum absorbance of the CO peak is higher than after one day of aging, two maxima are seen at 1115.6 cm^{-1} and 1107.5 cm^{-1}. After a longer aging, the spectrum of that film (or of others with the same history) will develop towards that of Therm-T.

3.10.2. CH_2 Stretching Band. The trace in Figure 8(d) is that of a 56 μm film, aged three days. The maximum absorbance of the CH_2 stretching region is high in relation with its thickness with two thin peaks at 2887 and 2878 cm^{-1}. The maximum absorbance diminish after a longer aging and the two peaks are replaced by one situated in between the previous peaks, at 2883 cm^{-1}, with a shoulder at 2866 cm^{-1} (Figure 1(a)).

It is worth noting that a relatively small difference in thickness (from 5 to 15 μm for instance) increases sufficiently, in thicker films, the time during which the chains have enough mobility to reach an organization different from that of thin films prepared in the same conditions. When chains must move to let favourable interactions between the oxygen and hydrogen atoms to form, it is not unexpected that equilibrium be slow after the initial temporary organization which takes place during drying. In thin films prepared in network-forming conditions, the network is already in place in the solution so that the time in the drying phase is not determining. Slow movements of the chains have been observed in polystyrene by different techniques [41].

The importance of film thickness and time of measurement on the shape and height of the complex CO peak explains the difficulty in finding a reliable extinction coefficient of this region by measuring films of different thickness at about the same time of aging [35]. The crystalline bands of medium intensity such as that at 1280, 1060, 842 cm^{-1} have been reported to be little sensitive to conformational order [34] so they change little during aging in films of different thickness.

The C–C bonds are somewhat deformed to form the cooperative association between the oxygen and hydrogen atoms. Deformation of chain entangled in a network will be different from that of less entangled chains. In the 650–500 cm^{-1} region, deformation as measured from the integrated absorbance increases with T but is significant at RT. The difference between the value of the integrated absorbance of Therm-T in Figure 6 at RT (4.5) and that of Reference (2.8) can count also as an indicator of network. The spectra of Therm-T display irregularities while that of Reference is smoother. Both have a peak at 530 cm^{-1}.

Solutions have been prepared with the sample with MW = 10^6. The dissolution is made at 40°C with a magnetic stirrer and evaporation takes place at RT. The spectra of the films made with a higher MW are somewhat more turbid than those prepared with the MW = 10^5 sample. The tracers of network, the bands at 1144 and 956 cm^{-1} are well seen on these films. As the spectra were comparable to those made from concentrated solutions of MW = 10^5, the spectra of these films have not been presented in the present work.

3.11. PEO and PE Extinction Coefficients. The measure of ε through the analyses of films of different thickness is not as straightforward for the large bands as it is in PE films due to the slow maturing of thicker films as remarked above. One can however make some comparison for thin films prepared in dilute solution which have reached in equilibrium. The ratio ε(PE)/ε(PEO) has been measured [29, 30, 35–40] for different regions. The two extremes of this ratio are the CH_2 stretching (2.2) and gauche (0.076). The well-ordered chains of the orthorhombic crystals of PE lead to a high value of ratio ε(PE) and of the ratio ε(PE)/ε(PEO). On the other hand, due to the helical conformation of the PEO chains and the zigzag morphology of PE, $\varepsilon_{\text{gauche (PEO)}}$ is high and the ratio ratio ε(PE)/ε(PEO) is low. The ratio for the rocking and bending are intermediary (0.71) and (1.40). The rocking and bending are composites with different composition of CH_2 and CO vibrations which explain the variability of this ratio.

4. Conclusion

Spectra of PEO placed between KBr plaques have been obtained during melting and crystallization in a slow T-ramp. Finding IR indicators of the change of morphology between the solid and the melt, of strain building and of network organization, were the aims of this investigation. The spectra of free standing films with different history are compared to those of the treated film.

In PE, melting leads to the disappearance of the doublets whose origin was the dense orthorhombic crystals. In PEO, the doublets associated with perpendicular and parallel bands, are replaced above T_m by smaller bands, other crystalline bands have their intensity reduced. The integrated absorbance of the largest bands, that of the CH_2 stretching and that of the complex CO band on the other hand, remains at the same value in the melt while their position and width change. The change of order at melting as seen by the IR spectra is large for PE and small in part for PEO because solid PEO seems already quite disordered. At T_m, the density of PE is reduced by 20% while it diminishes by only a few percent in PEO.

Several indicators of tension are found on the spectra taken at different temperatures. New peaks at 1090–1096 cm^{-1} in the vicinity of the maximum of the complex CO develop as soon as the T-ramp starts. They persist in the melt until the maximum temperature reached. Stretching and relaxation occur simultaneously or in turn during the ramp as seen on the calorimetric trace (Appendix). Another region sensitive to strain is the C–C–O bending region (650 and 500 cm^{-1}). The variation of the integrated absorbance parallels the T-ramp. It increases in the solid and the melt in the ascending ramp and decreases in the descending ramp.

Films prepared using network enriching conditions (solutions of high MW or of a higher concentration) were compared with other films. Two small bands at 1144 cm^{-1} and 956 cm^{-1}, found in films prepared from network enriched solutions and in Therm-T, are associated with chains in a network. In thin films prepared from dilute solutions these indicators of network are less apparent. This suggests that with enough time, cooperative associations between the oxygen and hydrogen atoms lead to the formation, during the drying and aging process, of a stable network. In thin films which solidify fast, the network is more fragmented and incomplete. A parallel band at low wavenumber in the complex CO peak increases when a chain interacts with another chain in a composite [34] or with an ion. A similar displacement takes place for autoassociation between PEO chains.

The spectra of thick films (20 μm) have been followed during their maturing. The complex CO band and the CH_2 stretching regions for instance are seen to change greatly in band intensity and position before reaching equilibrium where the indicators of network are found.

An example of a trace of melting in a slow T-ramp is given in the appendix. Besides the main endotherm of melting, other events associated with deformation/relaxation are observed.

The trace shows the main endotherm peak as expected when the crystals of PEO melt. Between RT and 60°C, an endothermic flat peak is interpreted as the absorption of heat when the sample is stretched in the T-ramp. Other endothermic peaks between 95 and 105 and 118 and 122°C have the same origin. The exothermic peak between 105 and 115°C is associated with the relaxation of stretched chains. The sequence of deformation/relaxation is dependant on the conditions of the T-ramp. The observation of these events in a standard DSC calorimeter using a 1 K/min ramp was possible in specific conditions, namely, a small mass (0.2 mg) and an open cell to avoid restraint on the sample [27].

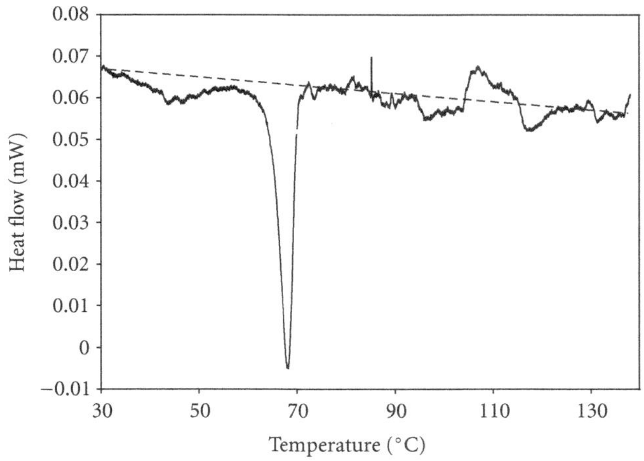

Figure 9: Melting trace of PEO in a 0.05 K/min ramp between 28 and 140°C. Endotherms and exotherms associated with chain deformation and relaxation accompany the main peak of fusion.

Appendix

Calorimetric Trace in a Slow T-Ramp (0.05 K/Min). The trace of melting shown as Figure 9 has special features compared to the standard melting trace. T_m is displaced at higher T (68°C instead of 61°C) and with ΔH_m = 130 J/g. Endotherms and exotherms are seen on the trace. The first endotherm between 30°C and 58°C corresponds to the deformation of the sample (33 J/g). Above T_m, between 72°C and 104°C, deformation and relaxation overlap (27 J/g). A peak of relaxation follows with a maximum at 107°C (−36 J/g). Another peak of deformation has a maximum at about 115°C (14 J/g). In the slow ramp, a fraction of the entangled chains in a crystalline organization is subject to a continuous stress. The melting of the strained crystals does not take place but at a higher T. Consequently, the value of ΔH_m of the main peak is reduced of about 10% (130 J/g) compared to the standard DSC value (147 J/g).

Conditions: 0.05 K/min between 28 and 140°C, m = 2.38 mg. Sample placed on Hg in a glass tube. The glass is sealed under vacuum.

The Growth of Tension in a T-Ramp: from Calorimetry to FTIR. The heat of dissolution ΔH_{diss} of UHMWPE in different nonpolar solvents was measured in order to understand gel formation. Because of the slow kinetics of dissolution, the rate of heating ν was reduced until a homogenous gel was obtained in the revolving calorimeter. ΔH_{diss} was found to vary with the solvent; an unexpected result since in nonpolar solvents, ΔH_{diss} is due almost entirely to the heat of fusion of orthorhombic crystals. With $v = 0.05$ K/min or less, a second endotherm, ΔH_{high-T} was observed. An experiment in the same condition but without solvent lead to traces showing the same calorimetric features as those found to study the gels.

In the slow T-ramp, deformation of the sample is followed, at higher T, by relaxation. The time interval between the successive movements of the chains in the noncrystalline regions and in the melt is large, of the order of hours rather than minutes. The slow T-ramp of the calorimeter is suited to follow these events. Such slow motions have been observed recently by neutron scattering and imaging on polystyrene over a range of temperature [41]. One of the conclusions of the investigation was that orientation at the length scale of the entire chain decays considerably more slowly than at the smaller entanglement length. The findings of the PS were instrumental in interpreting the events in the melt PE taking place during a slow T-ramp. Thermal events in the melt were found a general occurrence in polyolefins and other polymers when melting took place in a low T-ramp using condition similar to those in the C80 calorimeter. The relaxation with a rapid kinetics is observed easily while deformation is continuous and sometimes overlaps the base line or the beginning of melting.

By changing dramatically the conditions of melting, events different from the melting of long-range order could be observed [27]. Measurements were made with a standard DSC ($v = 1$ K/min) with a small mass to increase heat transfer (0.5–1.5 mg) and an open cell to avoid restraint of chain motion. In these DSC experiments, the large unambiguous values of ΔH_{low-T} and ΔH_{high-T} were too large to sustain completely our first interpretation [18–22] of the events which stood like this: the slow ramp generates strain in crystals whose melting point is above the value of T_m of non-strained crystals. The two effects described above complete each other.

Looking for IR indicators of deformation/relaxation in a slow T-ramp was the obvious step to follow after the calorimetric study. The changing shape of the bands between 1090 and 1100 cm^{-1} and between 650–400 cm^{-1} found in this work are seen as indicators of slow deformation and slow relaxation of PEO chains. These events can take place in succession or simultaneously.

Acknowledgments

The support of the NSERC National Science and Engineering Research Council of Canada is appreciated. The contributions of Selenis Canada (PTT Poly Canada) for C. Bergeron and of the University of Clermont-Ferrand for A. Potier and E. Perrier and other students have been helpful.

References

[1] P. Lightfoot, M. A. Mehta, and P. G. Bruce, "Crystal structure of the polymer electrolyte poly(ethylene oxide)$_3$:$LiCF_3SO_3$," *Science*, vol. 262, no. 5135, pp. 883–885, 1993.

[2] B. Wunderlich, *Macromolecular Physics*, vol. 3, Academic Press, London, UK, 1980.

[3] G. Strobl, *The Physics of Polymers*, Springer, Berlin, Germany, 1997.

[4] J. M. Schultz and M. J. Miles, "AFM study of morphological development during the melt-crystallization of poly(ethylene oxide)," *Journal of Polymer Science B*, vol. 36, no. 13, pp. 2311–2325, 1998.

[5] K. Dalnoki-Veress, J. A. Forrest, M. V. Massa, A. Pratt, and A. Williams, "Crystal growth rate in ultrathin films of poly(ethylene oxide)," *Journal of Polymer Science B*, vol. 39, no. 21, pp. 2615–2621, 2001.

[6] M. Polverari and T. G. M. Van De Ven, "Dilute aqueous poly (ethylene oxide) solutions: clusters and single molecules in thermodynamic equilibrium," *Journal of Physical Chemistry*, vol. 100, no. 32, pp. 13687–13695, 1996.

[7] D. L. Ho, B. Hammouda, and S. R. Kline, "Clustering of poly (ethylene oxide) in water revisited," *Journal of Polymer Science B*, vol. 41, no. 1, pp. 135–138, 2002.

[8] A. K. Khasanova and B. A. Wolf, "PEO/$CHCl_3$. Crystallinity of the polymer and vapor pressure of the solvent. Equilibrium and nonequilibrium phenomena," *Macromolecules*, vol. 36, no. 17, pp. 6645–6652, 2003.

[9] M. Duval, "Monitoring of cluster formation and elimination in PEO solutions," *Macromolecules*, vol. 33, no. 21, pp. 7862–7867, 2000.

[10] M. Duval and F. Boué, "Dilute polyethylene oxide aqueous solutions in a turbulent flow," *Macromolecules*, vol. 40, no. 23, pp. 8384–8388, 2007.

[11] T. Miyazawa, K. Fukushima, and Y. Ideguchi, "Molecular vibrations and structure of high polymers. III. Polarized infrared spectra, normal vibrations, and helical conformation of polyethylene glycol," *The Journal of Chemical Physics*, vol. 37, no. 12, pp. 2764–2776, 1962.

[12] H. Matsuura and T. Miyazawa, "Vibrational analysis of molten poly (ethylene glycol)," *Journal of Polymer Science Part A-2*, vol. 7, pp. 1735–1744, 1969.

[13] R. Frech, S. Chintapalli, P. G. Bruce, and C. A. Vincent, "Crystalline and amorphous phases in the poly(ethylene oxide)-$LiCF_3SO_3$ system," *Macromolecules*, vol. 32, no. 3, pp. 808–813, 1999.

[14] Y. Geng, G. Wang, Y. Cong, L. Bai, L. Li, and C. Yang, "Surface adsorption-induced conformational ordering and crystallization of polyethylene oxide," *Journal of Polymer Science B*, vol. 48, no. 2, pp. 106–112, 2010.

[15] M. Spitzer, E. Sabadini, and W. Loh, "Entropically driven partitioning of ethylene oxide oligomers and polymers in aqueous/organic biphasic systems," *Journal of Physical Chemistry B*, vol. 106, no. 48, pp. 12448–12452, 2002.

[16] J. M. Harris, *Polyethylene Glycol Chemistry, Biotechnical and Biomedical Applications*, Plenum Press, New York, NY, USA, 1992.

[17] F. Gouanvé, H. Phuong-Nguyen, Z. F. Hamida, and G. Delmas, "A new technique to characterize mono-molecular micelles in random ethylene-propylene copolymers," *Colloid and Polymer Science*, vol. 283, no. 9, pp. 994–1006, 2005.

[18] H. Phuong-Nguyen, *Melting Behaviour of Strained Crystals of Ultra High Molecular Weight Polyethylene with and without Solvent*, Ph.D. thesis, McGill University, Montreal, Canada, 1990.

[19] H. Phuong-Nguyen and G. Delmas, "Arrested melting due to strain in ultrahigh molecular weight polyethylene," *Macromolecules*, vol. 25, no. 1, pp. 408–413, 1992.

[20] H. Phuong-Nguyen and G. Delmas, "Thermal analysis and model of ultrahigh molecular weight polyethylene gels," *Macromolecules*, vol. 25, no. 1, pp. 414–421, 1992.
[21] G. Delmas, "Irreversible formation of a network during melting/dissolution of nascent PE," *Journal of Polymer Science B*, vol. 31, no. 13, pp. 2011–2018, 1993.
[22] H. Phuong-Nguyen and G. Delmas, "Information given by slow melting on phase content and maximum drawability of high molecular weight polyethylene films," *Journal of Materials Science*, vol. 29, no. 14, pp. 3612–3620, 1994.
[23] X. Zhang, H. Phuong-Nguyen, P. Bernazzani, I. Lapes, and G. Delmas, "Information on the noncrystalline phase of nascent iPP given by slow calorimetry," *Canadian Journal of Chemistry*, vol. 75, no. 10, pp. 1354–1362, 1997.
[24] H. Phuong-Nguyen and G. Delmas, "Crystallization of Isotactic Poly (propylene) in Solution as Followed by Slow Calorimetry," *Collection of Czechoslovak Chemical Communications*, vol. 60, pp. 1905–1924, 1995.
[25] H. Phuong-Nguyen, G. Charlet, and G. Delmas, "Slow calorimetry and heat of fusion of poly(4-methyl pentene-1)," *Journal of Thermal Analysis*, vol. 46, no. 3-4, pp. 809–832, 1996.
[26] Z. Ferhat-Hamida, H. Phuong-Nguyen, P. Bernazzani, A. Haine, and G. Delmas, "LLDPE's grown with metallocene and Ziegler-Natta catalysts: events in the melt and FTIR analysis," *Journal of Materials Science*, vol. 42, no. 9, pp. 3138–3154, 2007.
[27] I. Jedidi, Z. Ferhat-Hamida, and G. Delmas, "Calorimetry in nonstandard conditions: the noncrystalline phases of linear polyethylene," *Journal of Polymer Science B*, vol. 45, no. 15, pp. 1932–1949, 2007.
[28] P. Bernazzani, *Contrôle de la morphologie des systèmes amylose-eau et caractérisation par DSC lent et FTIR*, Ph.D. thesis, University of Quebec at Trois-Rivières, Trois-Rivières, Canada, 2000.
[29] P. Bernazzani, V. T. Bich, H. Phuong-Nguyen et al., "FTIR analysis of the phase content in lowdensity polyethylene," *Canadian Journal of Chemistry*, vol. 76, no. 11, pp. 1674–1687, 1998.
[30] P. Bernazzani, *Analyse de la composition phasique du polyéthylène par FTIR: Mise en évidence d'une phase tendue*, M.S. thesis, University of Quebec, Montreal, Canada, 1994.
[31] A. Ajji, J. Guèvremont, K. C. Cole, and M. M. Dumoulin, "Orientation and structure of drawn poly(ethylene terephthalate)," *Polymer*, vol. 37, no. 16, pp. 3707–3714, 1996.
[32] P. Bernazzani, C. Chapados, and G. Delmas, "Double-helical network in amylose as seen by slow calorimetry and FTIR," *Journal of Polymer Science B*, vol. 38, no. 12, pp. 1662–1677, 2000.
[33] P. Bernazzani, C. Chapados, and G. Delmas, "Phase change in amylose-water mixtures as seen by Fourier transform infrared," *Biopolymers*, vol. 58, no. 3, pp. 305–318, 2001.
[34] J. M. Marentette and G. R. Brown, "The crystallization of poly(ethylene oxide) in blends with neat and plasticized poly(vinyl chloride)," *Polymer*, vol. 39, no. 6-7, pp. 1415–1427, 1998.
[35] C. Bergeron, A. Potier, E. Perrier, and G. Delmas, Part 2 of this work, to be published.
[36] A. Potier, "Report 2002," University of Quebec at Montreal and IUT- University of Auvergne 2002.
[37] A. Pages, "Report 2007," University of Quebec at Montreal and IUT- University of Auvergne 2007.
[38] J. Morales, "Report 2005," University of Quebec at Montreal and IUT- University of Auvergne.
[39] E. Perrier, "Report 2003," University of Quebec at Montreal and IUT- University of Auvergne.
[40] A. Cohade, "Report 2003," University of Quebec at Montreal and IUT- University of Auvergne.
[41] J. Bent, L. R. Hutchings, R. W. Richards et al., "Neutron-mapping polymer flow: scattering, flow visualization, and molecular theory," *Science*, vol. 301, no. 5640, pp. 1691–1695, 2003.

19

Protein Interactions Investigated by the Raman Spectroscopy for Biosensor Applications

R. P. Kengne-Momo,[1,2] Ph. Daniel,[1] F. Lagarde,[1] Y. L. Jeyachandran,[1] J. F. Pilard,[3] M. J. Durand-Thouand,[4] and G. Thouand[4]

[1] *Laboratoire PEC, UMR CNRS 6087, Université du Maine, A.O. Messiaen, 72085 Le Mans, France*
[2] *Laboratoire de Chimie Organique, Université de Yaoundé I, B.P 812, Yaoundé, Cameroon*
[3] *Laboratoire UCO2M, UMR CNRS 6011, Université du Maine, A.O. Messiaen, 72085 Le Mans, France*
[4] *Laboratoire CBAC, UMR CNRS GEPEA 6144, IUT Université de Nantes, 85035 La Roche Sur Yon, France*

Correspondence should be addressed to Ph. Daniel, philippe.daniel@univ-lemans.fr

Academic Editor: Jaan Laane

Interaction and surface binding characteristics of staphylococcal protein A (SpA) and an anti-*Escherichia coli* immunoglobulin G (IgG) were studied using the Raman spectroscopy. The tyrosine amino acid residues present in the α-helix structure of SpA were found to be involved in interaction with IgG. In bulk interaction condition the native structure of proteins was almost preserved where interaction-related changes were observed in the overall secondary structure (α-helix) of SpA. In the adsorbed state, the protein structure was largely modified, which allowed the identification of tyrosine amino acids involved in SpA and IgG interaction. This study constitutes a direct Raman spectroscopic investigation of SpA and IgG (receptor-antibody) interaction mechanism in the goal of a future biosensor application for detection of pathogenic microorganisms.

1. Introduction

Creating protein-protein and protein-surface interactions is the primary step of biological processes in bioimplants, biofouling, immunoassays, and biosensors [1]. Therefore, in the past few years, a great deal for research has been focussed on the understanding of the fundamental structure and physicochemical properties involved in these interactions [2, 3], because the applications are strongly connected in the good knowledge of these mechanisms. Structure and orientation changes are often a part of proteins interaction due to various factors such as specific/nonspecific attachment, hydration/dehydration, steric hindrance, competitive adsorption/desorption processes, and experimental conditions [4, 5].

However it appears that the protein interactions and their resulting structural conformations are very complicated processes. Facing protein interaction complexity, various analytical tools have been developed to characterize the exhibited different phenomena, physico-chemical behaviors, and interaction structures [4, 6–8]. Among these tools, the vibrational spectroscopic techniques, such as Raman spectroscopy, are considered as extremely useful tools to chemically and structurally investigate all material compounds, of which protein analysis now constitute a possible and interesting field of applications, more particularly, to determine secondary conformations and interaction structures [9–11]. Actually the Raman scattering technique is a vibrational molecular spectroscopy which, derives from an inelastic light scattering process. With the Raman spectroscopy, a laser photon is scattered by a sample molecule and loses (or gains) energy during the process. The amount of energy lost is seen as a change in energy (wavelength) of the irradiating photon. This energy loss is characteristic for a particular bond in the molecule. The Raman signal is producing a precise spectral sample fingerprint, unique to each atom, group of atoms or individual molecule. In this respect it is similar to the more commonly found FT-IR spectroscopy. However, unlike FT-IR, there are a distinct number of advantages when using the Raman: for instance (i) Raman can be used to analyse aqueous solutions, (ii) the intensity of spectral features is directly proportional to the concentration of the particular species, (iii) Raman requires little or no sample preparation, and (iv) the use of a Raman microscope provides very high level of spatial resolution and depth discrimination. These advantages and its highly specific nature mean that Raman has become a very powerful tool for analysis and chemical monitoring.

Indeed, the recent technological improvements in Raman spectroscopy, especially in lasers, optics, detectors, and data analysis, make this type of investigation easier. More precisely the sensitivity has increased several orders of magnitude because of advances in optical filters and phonon detectors [12]. For instance single-channel detectors were replaced by CCD detectors whose main advantage is to obtain the spectral information in a single recording. Additionally fluorescence is now minimized by using red/IR excitation in the 630–850 nm excitation, which became possible by the new photon detectors with high efficiency in this region [13]. Finally problems with interpreting Raman data have been reduced with the avaiability of powerful software which are able to discriminate throughout statistical methods [14]. Raman has the main advantage of being a nondestructive and noninvasive method which will not disturb the studied system, which is a major asset. Finally this optical technique can provide rapid, objective, and reproducible measurements on biological samples. Moreover, the Raman spectroscopy is particularly suitable for detection and analysis of the changes in molecular polarizability of biomolecules, changes involved in the modifications of proteins conformation in binding or interaction mechanisms [9–11, 15, 16]. For instance, the well-resolved amide III band in the Raman spectrum has been used to determine precisely the secondary conformation of protein [17]. The spectral changes associated with aromatic amino acids such as tyrosine, phenylalanine, and tryptophan residues were also used to track some interactions induced by structural modifications [15, 16].

SpA and IgG systems were studied because of their importance in affinity biosensor applications. By binding to specific proteins, the fragment crystallizable regions ensure that each antibody generates an appropriate immune response for a given antigen. In the present work, the Raman spectroscopy was used to study the native and interaction structures of a staphylococcal protein A (SpA) and an immunoglobulin G (IgG) and their complex on pure gold (p-Au) and self-assembled monolayer (SAM) functionalized gold (SAM-Au) surfaces. This study is of major importance in the goal of a future bacteria immobilization and detection in the frame of a biosensor design using Raman spectroscopy. Actually one possible biosensing method can be the functionalization of dedicated surfaces in order to specifically detect and fix microorganisms like pathogenic bacteria. The biological elements could be then controlled by a fingerprint technique of which Raman or infrared spectroscopies are possible tools. However, the efficiency of biomolecules detection depends strongly on the success of surface functionalization and on the nature of biomolecules immobilization. The sequence of functionalization includes generally 3 steps: (i) the first step is to functionalize a selected surface (could be gold as in our case) by a chemical way for instance [18], (ii) the second step is bind linker or spacer onto the functionalized polymer surface, and (iii) the final step is then to covalently attach a bioactive compound to the functionalized surface. The second step is actually the more complex one because it includes the fixation of proteins by a covalent way before the immobilization of antibody which is then specific of bioelements that we wish to detect. The full knowledge of protein-antibody interaction is then very important, especially with investigation of the Raman spectroscopy since it is one possible fingerprint technique used in alternative biosensing equipment.

The protein A (SpA) and antibody (IgG) are well known for a long time from a structural point of view, and their interactions were intensively studied as it could be seen in the abundant literature (see for instance the book chapters 19, 20, 21 and some selected examples of articles 22–27). However there are only few Raman spectroscopic investigation published at our knowledge concerning protein A (SpA) and IgG (see for instance [19–22]), and consequently it appears original to understand from a spectroscopic point of view, by a Raman study, the interaction of these both molecules since the potential applications for future biosensors are important.

2. Experimental Details

2.1. Materials. Gold films of thickness and surface roughness around 100 nm and 15 nm, respectively, sputter coated on (100) silicon wafer, were used as substrates. SpA (Cat. no. p6031) purchased from Sigma and an anti-*E. coli* IgG (Cat. no. GTX40640) purchased from Interchim were used as received without further purification. Sterilised phosphate-buffered saline (PBS) of pH 7.42 was used as protein solvent. Millipore water (18.2 MΩ·cm) was used throughout the experiment.

2.2. Substrate Preparation. SAM-Au surfaces were prepared using thiourea and glutaraldehyde. The gold film substrates were first washed in water and ethanol for 20 min each, cleaned in 1% aqua regia for 5 s, and rinsed well in water. The cleaned substrates were immerged in 250 mM of aqueous thiourea solution for 24 hours. Then the substrates were activated in 5% (v/v) of glutaraldehyde in PBS for 20 min. Finally the substrates were rinsed in PBS in order to eliminate nonreacted material and then be used for protein experiments.

As reference surfaces, p-Au surfaces were cleaned in absolute ethanol (Sigma) for 20 min and washed thoroughly in water before experiments.

2.3. Protein Immobilization. PBS (pH 7.4) was used to prepare SpA solution and reconstitute the original IgG solution (4 mg/mL, 0.1% Sodium Azide, 0.01 M PBS, pH 7.2) in identical concentrations. The final concentrations of both SpA and IgG solutions were 0.15 mg/mL.

To study the interaction structure of the proteins on SAM-Au surfaces, two distinct experiments were run: (i) in one experiment SpA and IgG were consecutively adsorbed (contact presence of 2 hours each); (ii) in the second one, the SpA and IgG solutions were previously mixed (SpA + IgG) in one solution, kept stable for 1 h at 20°C, and only then adsorbed on the surface (contact presence 2 h). On p-Au surface (reference), SpA and IgG were consecutively adsorbed for an incubation time of 16 h each.

After adsorption times, all the samples were washed in PBS (10 min.), rinsed in water, and immediately used for Raman measurements. Additionally, the mixed (SpA + IgG) solution was studied in drop coated dry state on p-Au surface to analyse the bulk protein-protein-interaction-related structural changes. The Raman spectra were also recorded for the as-received SpA powder and IgG solution (4 mg/ml, drop coated in dried state) on p-Au surfaces as reference of their native structure.

2.4. Raman Measurements. Raman spectrometer XploRA (model from Horiba) equipped with a confocal microscope was used. The Raman signals were recorded in a spectral range of 400–3100 cm^{-1} with an integration time of 300 s using a 785 nm red laser excitation, at 10 mW power, in combination with a 100x objective magnification (NA = 0.9) of an Olympus BX41 transmission and reflection illumination microscope (Olympus, France) for focus and collection of Raman-scattered light. The confocal hole was set to the minimum value. The dispersive system is comprised of a single grating with 600 groves/mm, and the Raman signal is collected thanks to a front illuminated CCD detector cooled by thermoelectric Peltier air effect (produced by Horiba, 1024×256 chip size). A filter rejects the Rayleigh contribution. For each sample, 3 series of independent measurements were carried out and for each set, the Raman experiment was repeated 6 times to check the reproducibility of the measurement and to minimize the possible artifacts and parasitic signals. Each considered spectrum was preprocessed with a 9-point Savitzky-Golay function corresponding to a denoising routine (smoothing filter), background corrected with a polynomial function, and vector normalized using the Labspec software. A more detailed description of spectral acquisition and treatment is presented elsewhere [23].

3. Results and Discussion

SpA and IgG are globular proteins with average molecular weights ~42 kDa and ~150 kDa, respectively. The secondary conformation of SpA molecules is rich in α-helix structure (54%) and also contains 10–20% of β-structure and a large part of coil structure [24]. IgG molecules are predominantly made of β-sheet (47%), 7% of α-helices, and remaining, percentage of turns and coils [25]. First of all the reference Raman spectra of both SpA and IgG proteins were recorded; as evidenced in Figure 1, the as-received SpA and IgG proteins clearly revealed their characteristic native secondary conformations. The amide I band around 1659 cm^{-1} and a relatively higher intense band at 1336 cm^{-1} in the amide III region were observed in SpA spectrum which are related to its α-helix secondary structure characteristics [17]. SpA spectrum also featured a vibration band around

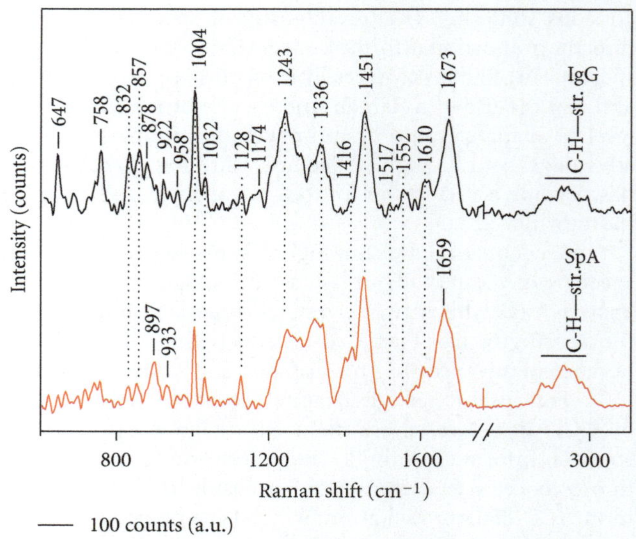

FIGURE 1: The Raman spectra recorded for the as-received SpA and IgG proteins. Conditions of spectra recording: time acquisition 300 s, wavelength 785 nm, nominal power laser 10 mW, optical microscope objective 100x.

933 cm^{-1} associated with rocking vibration of $CH_3(\delta CH_3)$ terminal groups of α-helix structure. The predominant β-sheet structure in IgG was identified by the characteristically amide I band around 1673 cm^{-1} and a higher intense band at 1243 cm^{-1} in the amide III region [17]. The other vibration bands generally associated with protein structures are assigned as follows from the literature [26]. The C–H stretching (νC–H) vibration bands of aliphatic side-chains were observed around 2860 cm^{-1} to 2940 cm^{-1}. The CH_2 deformation (ρCH_2) band was observed around 1451 cm^{-1}. Vibration bands associated with phenylalanine amino acid residue were observed around 1610 cm^{-1} and 1004 cm^{-1}. The bands associated with tyrosine were observed around 857 cm^{-1} (buried), 832 cm^{-1} (exposed), 758 cm^{-1}, and 647 cm^{-1} (skeletal). Tryptophan (indole rings) residue bands were observed around 1552 cm^{-1} and 1416 cm^{-1}. The bands in the region 920 cm^{-1} to 880 cm^{-1} could be assigned to ρCH_2 and the ones around 958 cm^{-1} to ρCH_3 vibrations. Backbone skeletal νC–C vibration bands were also observed in the region of 1170 cm^{-1} to 1032 cm^{-1}. Table 1 summarizes the suggested assignment.

The Raman spectrum measured for the drop coated layer of previously mixed SpA + IgG solution (0.15 mg/mL) on p-Au surface is shown in Figure 2. Interesting and new spectral features, assumed to be produced by SpA-IgG interaction, were observed. The amide I band was observed around 1672 cm^{-1} corresponding to the β-sheet structure, characteristically considered coming from IgG. However, the amide III region (1240 cm^{-1} to 1350 cm^{-1}) showed characteristics of α-helix structure with a relatively higher intensity band around 1338 cm^{-1} associated with bending δC_α–H vibration in main-chain of α-helix. Moreover, the band at 932 cm^{-1} associated with α-helix CH_3 groups increased in intensity compared with the reference SpA spectrum (see Figure 1).

TABLE 1: Interpretation of the reference Raman spectra of proteins SpA and IgG. The bands characteristics of SpA or IgG, respectively, are indicated in bold letters for characterization of the intensity level: vw: very weak, w: weak, m: medium, s: strong, vs: very strong.

Frequency (cm^{-1})	Compound	Assignment	Intensity
647	SpA/IgG	Tyrosine	vw/m
758	SpA/IgG	Tyrosine	vw/m
832	SpA/IgG	Tyrosine	w/m
857	SpA/IgG	Tyrosine	w/m
880–920	SpA/IgG	ρCH_2	m/m
933	**SpA**	**δCH_3**	**w**
958	SpA/IgG	ρCH_3	w/w
1004	SpA/IgG	Phenylalanine	s/vs
1032–1170	SpA/IgG	νC–C or alkyl C–N	w/w
1243	**IgG**	**Amide III**	**vs**
1336	**SpA**	**Amide III**	**vs**
1416	SpA/IgG	Tryptophan	s/vs
1451	SpA/IgG	ρCH_2	vs/vs
1552	SpA/IgG	Tryptophan	w/w
1610	SpA/IgG	Phenylalanine	m/m
1659	**SpA**	**Amide I**	**vs**
1673	**IgG**	**Amide I**	**vs**
2860	SpA/IgG	νC–H	m/w
2940	SpA/IgG	νC–H	m/w

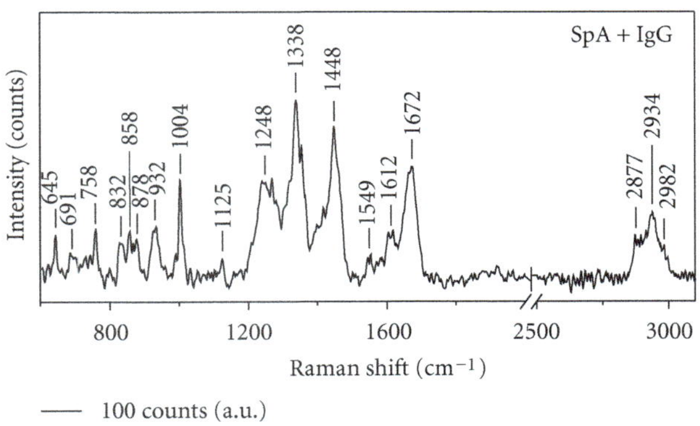

FIGURE 2: The Raman spectra of drop-coated SpA + IgG mixture in dry state. Conditions of spectra recording: time acquisition 300 s, wavelength 785 nm, nominal power laser 10 mW, optical microscope objective 100x.

These observations show that the α-helix main-chain of SpA is predominantly involved in interaction with IgG. The other vibration bands observed in SpA-IgG interacting spectrum correspond to the bands observed in native SpA and IgG spectra (see Figure 1) with small shifts. These shifts can be attributed to the modification of the local structure of each entity due to the interaction. Note that the stretching band around 1032 cm^{-1} associated with cis conformation of skeletal νC–C that is identified in the native spectra was not observed after interaction.

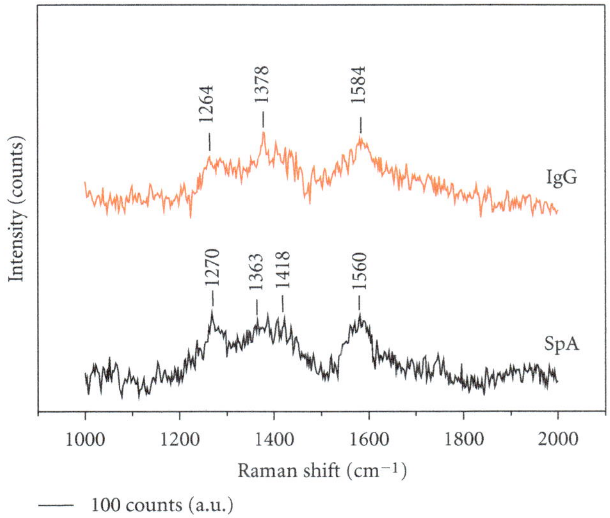

FIGURE 3: The Raman spectra of drop-coated SpA + IgG mixture in dry state. Conditions of spectra recording: time acquisition 300 s, wavelength 785 nm, nominal power laser 10 mW, optical microscope objective 100x.

As introduced, our main goal is the study of the protein-surface interaction by adsorbing the proteins in p-Au and also SAM-Au surfaces. The Raman spectra of SpA and IgG consecutively adsorbed on p-Au surface are shown in Figure 3. Adsorption of SpA and IgG on p-Au surface was found to be very low to produce any significant spectral features other than the overlapped broad vibration bands in the region of amide (I and III) and aliphatic side-chain groups. However note that it is not possible to avoid conformational changes to the native structure that may be induced by studying the proteins on the p-Au surface.

On the contrary, SAM-Au surfaces showed strong binding with proteins. Additionally the results are evidently more pertinent because the SAM layer has the potential to protect the proteins from interacting with and denaturing on the Au surface. Figure 4(a) shows the Raman spectra of SpA and IgG proteins consecutively adsorbed on SAM-Au surface, and Figure 4(b) shows the spectrum of molecules adsorbed from SpA + IgG interacting solution on SAM-Au surface. Despite the incubation time of 2 h, strong and well-resolved Raman signals from proteins were identified. The protein molecules adsorbed on SAM-Au surfaces were found to be largely modified in amide main chain and aliphatic side-chain structure as compared to their native structure due to strong surface binding. The proposed Raman assignments are then done in the next part according to the reference assignment shown in Table 1. However, the Raman signals associated with amino acid residues such as tyrosine (637–800 cm^{-1}) and phenylalanine were found to be enhanced. The Raman spectra of Figure 4 featured some of the common vibration bands of proteins such as amide I around 1656 cm^{-1} and 1633 cm^{-1}, aliphatic side chains from 1420 cm^{-1} to 1466 cm^{-1}, amide III from 1311 cm^{-1} to 1278 cm^{-1}, and alkyl C–N or skeletal νC–C between

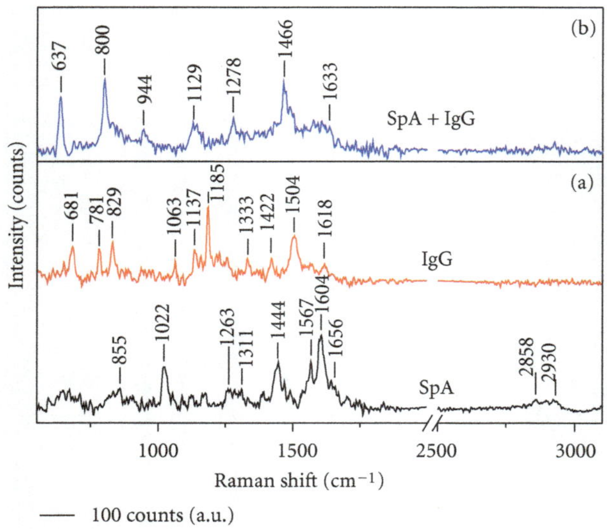

Figure 4: The Raman spectra of (a) consecutively adsorbed SpA and IgG on SAM-Au surface and (b) SpA + IgG interacting molecules adsorbed from SpA and IgG mixed solution on SAM-Au surface. Conditions of spectra recording: time acquisition 300 s, wavelength 785 nm, nominal power laser 10 mW, optical microscope objective 100x.

1130 cm^{-1} and 1050 cm^{-1}. The spectra also featured some enhancement in band intensity or new bands produced due to strong surface binding of proteins with SAM-Au.

The Raman bands observed in the SpA spectrum (Figure 4(a)) around 1604 cm^{-1}, 1567 cm^{-1}, and 855 cm^{-1} could be assigned to phenylalanine, tryptophan, and tyrosine amino acid residues, respectively. The spectrum of adsorbed IgG (Figure 4(a)) on SpA adsorbed SAM-Au layer featured most of the Raman signals associated with tyrosine amino acid residue. The bands observed around 1618 cm^{-1}, 1504 cm^{-1}, 829 cm^{-1}, 781 cm^{-1}, and 681 cm^{-1} could be associated to tyrosine residue. However the origin of an intense band at 1185 cm^{-1} in the IgG spectrum is not clear. This signal may be produced from δCCH vibration in IgG or from the conjugated biotin molecule. In Figure 4(b), the spectrum of (SpA + IgG) solution also showed strong Raman signals associated with tyrosine residue around 800 cm^{-1} and 637 cm^{-1} with a shift and a small signal around 944 cm^{-1} associated with CH_3 groups in α-helix. An important observation from the IgG and SpA + IgG spectra is the predominant presence of tyrosine bands. Tyrosine residues are known to be involved in receptor (SpA)-antibody (IgG) binding [27]. Our observation is a spectroscopic evidence for the involvement of tyrosine residue in SpA-IgG binding at adsorbed state. However, in the bulk state binding of SpA and IgG (Figure 2), the involvement of tyrosine was not clearly revealed, but changes in α-helix main chain of SpA were observed. These two different kinds of information explain that the SpA and IgG binding is mediated by tyrosine amino acid residues present in the α-helix main chain of SpA molecule. This may be one of the important reasons that make SpA a good receptor molecule for IgG binding in affinity biosensors.

Considering the enhancement of some bands (tyrosine and phenylalanine), one open question could be suggested: why are these bands enhanced because a classical dispersive Raman spectroscopy was used in this work? Actually, even if the process is not well understood, the Au substrate could increase the Raman signal of SAMs and the molecules adsorbed on it if the Au surface is constituted of nanometers-scale domains [28], which is the case here. It means that we can suggest that this enhancement is not probably due to denaturation of the surface.

The Raman signal enhancement is a mechanism which can be usually observed in SERS method which generally associate silver or gold colloids or rough metal surfaces to investigated molecules. Considering metal nanoclusters (e.g., gold or silver nanoparticles), they provide a strong enhancement of the Raman signals of molecules attached to them (see for instance [29–31]). In particular, gold nanoparticles may significantly reduce the adsorption of molecules with fluorescent backgrounds when they are used as Raman substrates. It has been shown that SERS detection method with gold nanoparticles probes generally greatly increase specificity and detection. Specific recognition of proteins by conjugates of gold nanoparticles and antibodies followed by an enhancement step based on the electroless deposition of silver onto the gold particles has been employed for protein identification in histochemistry using dot and blot assays [32, 33]. Using the same detection principle, a nanoparticle-based Raman spectroscopic encoding approach for DNA, RNA, or protein detection was developed (see for instance [34]) some years ago. Then the SERS method is clearly a good process for the Raman protein signal increasing. The SERS method does not need any additional Raman equipment but, if the enhancement can be very strong (up to a 10^6 factor), it can be strongly dependent on the aggregations of the particles [35, 36] and then gives selective enhancement of bands. We suppose that this phenomenon could explain and could be connected to our results of specific bands enhancements.

4. Conclusions

The native structure and interaction structure in bulk and adsorbed state of SpA and an IgG were studied by the Raman spectroscopy. SpA and IgG interaction in bulk state showed significant structural changes in the α-helix secondary structure of SpA; however, the adsorbed state on SAM-Au surfaces clearly revealed the involvement of tyrosine amino acid residues in the interaction mechanism. It is concluded that the tyrosine residues present in the α-helix of SpA mediate the interaction with IgG, and this characteristic makes SpA a unique receptor molecule for IgG in biosensors applications. Furthermore, the different tyrosine bands observed in the adsorbed state spectra of proteins are associated with different forms of tyrosine such as buried, exposed, free, and skeletal. This information could be used for advance understanding of SpA-IgG binding on surfaces. This work will be very useful considering the potential applications in biosensors for detection of pathogens. Actually one possible alternative way for the detection of pathogenic bacteria is to imagine a specific functionalized surface (SAM-Au surface for instance) which will be able to fix microorganisms and could be detected by spectroscopy. However the understanding of this SpA-IgG interaction mechanism appears to be very important previously to any fixation of biological element.

Acknowledgments

The present work was supported by the Grant RMB-BIORAM booked by the Région Pays de la Loire, Le Conseil Général de la Vendée, La Ville de la Roche sur Yon.

References

[1] R. A. Latour Jr., "Protein-surface interactions," in *Biomaterials: Protein-Surface Interactions: Encyclopedia of Biomaterials and Biomedical Engineering*, G. E. Wnek and G. L. Bowlin, Eds., pp. 270–284, Informa Healthcare, New York, NY, USA, 2008.

[2] I. S. Moreira, P. A. Fernandes, and M. J. Ramos, "Hot spots—a review of the protein-protein interface determinant amino-acid residues," *Proteins*, vol. 68, no. 4, pp. 803–812, 2007.

[3] J. A. Mielczarski, J. Dong, and E. Mielczarski, "Real time evaluation of composition and structure of concanavalin a adsorbed on a polystyrene surface," *Journal of Physical Chemistry B*, vol. 112, no. 16, pp. 5228–5237, 2008.

[4] Y. L. Jeyachandran, E. Mielczarski, B. Rai, and J. A. Mielczarski, "Quantitative and qualitative evaluation of adsorption/desorption of bovine serum albumin on hydrophilic and hydrophobic surfaces," *Langmuir*, vol. 25, no. 19, pp. 11614–11620, 2009.

[5] M. Rankl, T. Ruckstuhl, M. Rabe, G. R. J. Artus, A. Walser, and S. Seeger, "Conformational reorientation of immunoglobulin G during nonspecific interaction with surfaces," *ChemPhysChem*, vol. 7, no. 4, pp. 837–846, 2006.

[6] J. Pande, C. Pande, D. Gilg, M. Vašák, R. Callender, and J. H. R. Kägi, "Raman, infrared, and circular dichroism spectroscopic studies on metallothionein: a predominantly "turn"-containing protein," *Biochemistry*, vol. 25, no. 19, pp. 5526–5532, 1986.

[7] Z. Q. Wen, L. Hecht, and L. D. Barron, "β-Sheet and associated turn signatures in vibrational Raman optical activity spectra of proteins," *Protein Science*, vol. 3, no. 3, pp. 435–439, 1994.

[8] M. Malmsten, "Ellipsometry studies of fibronectin adsorption," *Colloids and Surfaces B*, vol. 3, no. 6, pp. 371–381, 1995.

[9] G. J. Thomas Jr., "New structural insights from Raman spectroscopy of proteins and their assemblies," *Biopolymers*, vol. 67, no. 4-5, pp. 214–225, 2002.

[10] R. Tuma, "Raman spectroscopy of proteins: from peptides to large assemblies," *Journal of Raman Spectroscopy*, vol. 36, no. 4, pp. 307–319, 2005.

[11] J. Filik and N. Stone, "Drop coating deposition Raman spectroscopy of protein mixtures," *Analyst*, vol. 132, no. 6, pp. 544–550, 2007.

[12] X. Peng, J. Jonas, and J. L. Silva, "Molten-globule conformation of Arc repressor monomers determined by high- pressure 1H NMR spectroscopy," *Proceedings of the National Academy of Sciences of the United States of America*, vol. 90, no. 5, pp. 1776–1780, 1993.

[13] J. Dong, D. Dinakarpandian, and P. R. Carey, "Extending the raman analysis of biological samples to the 100 micromolar concentration range," *Applied Spectroscopy*, vol. 52, no. 8, pp. 1117–1122, 1998.

[14] R. Goodacre, E. M. Timmins, R. Burton et al., "Rapid identification of urinary tract infection bacteria using hyperspectral whole-organism fingerprinting and artificial neural networks," *Microbiology*, vol. 144, no. 5, pp. 1157–1170, 1998.

[15] A. M. Ahern and R. L. Garrell, "Protein-metal interactions in protein-colloid conjugates probed by surface-enhanced raman spectroscopy," *Langmuir*, vol. 7, no. 2, pp. 254–261, 1991.

[16] S. F. El-Mashtoly, S. Yamauchi, M. Kumauchi, N. Hamada, F. Tokunaga, and M. Unno, "Structural changes during the photocycle of photoactive yellow protein monitored by ultraviolet resonance Raman spectra of tyrosine and tryptophan," *Journal of Physical Chemistry B*, vol. 109, no. 49, pp. 23666–23673, 2005.

[17] S. Cai and B. R. Singh, "A distinct utility of the amide III infrared band for secondary structure estimation of aqueous protein solutions using partial least squares methods," *Biochemistry*, vol. 43, no. 9, pp. 2541–2549, 2004.

[18] R. P. Kengne-Momo, Y. L. Jeyachandran, A. Assaf et al., "A simple method of surface functionalisation for immunospecific immobilisation of proteins," *Analytical and Bioanalytical Chemistry*, vol. 398, no. 3, pp. 1249–1255, 2010.

[19] Z. A. Combs, S. Chang, T. Clark, S. Singamaneni, K. D. Anderson, and V. V. Tsukruk, "Label-free raman mapping of surface distribution of protein A and IgG biomolecules," *Langmuir*, vol. 27, no. 6, pp. 3198–3205, 2011.

[20] J. Duarte, M. T. Pacheco, R. Z. Machado, L. Silveira, R. A. Zangaro, and A. B. Villaverd, "Use of near-infrared raman spectroscopy to detect IgG and IgM antibodies against Toxoplasma gondii in serum samples of domestic cats," *Cellular and Molecular Biology*, vol. 48, no. 5, pp. 585–589, 2002.

[21] R. Kumar, M. M. Bajaj, and B. Swaroop, "Conformational changes in the IgG molecule of lepromatous sera using laser Raman spectroscopy," *Indian Journal of Leprosy*, vol. 60, no. 3, pp. 363–373, 1988.

[22] P. C. Painter and L. E. Mosher, "The low-frequency Raman spectrum of an antibody molecule: bovine IgG," *Biopolymers*, vol. 18, no. 12, pp. 3121–3123, 1979.

[23] P. Daniel, P. Picart, L. Bendriaa et al., "Effects of toxic organotin compounds on bacteria investigated by micro-Raman spectroscopy," *Spectroscopy Letters*, vol. 41, no. 1, pp. 19–28, 2008.

[24] I. Sjoholm, "Protein A from Staphylococcus aureus. Spectropolarimetric and spectrophotometric studies," *European Journal of Biochemistry*, vol. 51, no. 1, pp. 55–61, 1975.

[25] M. Marquart, J. Deisenhofer, and R. Huber, "Crystallographic refinement and atomic models of the intact immunoglobulin molecule Kol and its antigen-binding fragment at 3.0 θ and 1.9 θ resolution," *Journal of Molecular Biology*, vol. 141, no. 4, pp. 369–391, 1980.

[26] D. Naumann, "Ft-infrared and Ft-raman spectroscopy in biomedical research," *Applied Spectroscopy Reviews*, vol. 36, no. 2-3, pp. 239–298, 2001.

[27] G. C. Stone, U. Sjobring, L. Bjorck, J. Sjoquist, C. V. Barber, and F. A. Nardella, "The Fc binding site for streptococcal protein G is in the Cγ2-Cγ3 interface region of IgG and is related to the sites that bind staphylococcal protein A and human rheumatoid factors," *Journal of Immunology*, vol. 143, no. 2, pp. 565–570, 1989.

[28] T. Wadayama and M. Oishi, "Surface-enhanced Raman spectral study of Au nano-particles/alkanethiol self-assembled

monolayers/Au(1 1 1) heterostructures," *Surface Science*, vol. 600, no. 18, pp. 4352–4356, 2006.

[29] S. Nie and S. R. Emory, "Probing single molecules and single nanoparticles by surface-enhanced Raman scattering," *Science*, vol. 275, no. 5303, pp. 1102–1106, 1997.

[30] R. A. Reynolds III, C. A. Mirkin, and R. L. Letsinger, "Homogeneous, nanoparticle-based quantitative colorimetric detection of oligonucleotides," *Journal of the American Chemical Society*, vol. 122, no. 15, pp. 3795–3796, 2000.

[31] T. A. Taton, G. Lu, and C. A. Mirkin, "Two-color labeling of oligonucleotide arrays via size-selective scattering of nanoparticle probes," *Journal of the American Chemical Society*, vol. 123, no. 21, pp. 5164–5165, 2001.

[32] C. S. Holgate, P. Jackson, P. N. Cowen, and C. C. Bird, "Immunogold-silver staining: new method of immunostaining with enhanced sensitivity," *Journal of Histochemistry and Cytochemistry*, vol. 31, no. 7, pp. 938–944, 1983.

[33] M. Moeremans, G. Daneels, and A. Van Dijck, "Sensitive visualization of antigen-antibody reactions in dot and blot immune overlay assays with immunogold and immunogold/silver staining," *Journal of Immunological Methods*, vol. 74, no. 2, pp. 353–360, 1984.

[34] Y. C. Cao, R. Jin, and C. A. Mirkin, "Nanoparticles with Raman spectroscopic fingerprints for DNA and RNA detection," *Science*, vol. 297, no. 5586, pp. 1536–1540, 2002.

[35] S. Sánchez-Cortés, J. V. García-Ramos, G. Morcillo, and A. Tinti, "Morphological study of silver colloids employed in surface-enhanced raman spectroscopy: activation when exciting in visible and near-infrared regions," *Journal of Colloid And Interface Science*, vol. 175, no. 2, pp. 358–368, 1995.

[36] J. Neng, M. H. Harpster, H. Zhang, J. O. Mecham, W. C. Wilson, and P. A. Johnson, "A versatile SERS-based immunoassay for immunoglobulin detection using antigen-coated gold nanoparticles and malachite green-conjugated protein A/G," *Biosensors and Bioelectronics*, vol. 26, no. 3, pp. 1009–1015, 2010.

Permissions

The contributors of this book come from diverse backgrounds, making this book a truly international effort. This book will bring forth new frontiers with its revolutionizing research information and detailed analysis of the nascent developments around the world.

We would like to thank all the contributing authors for lending their expertise to make the book truly unique. They have played a crucial role in the development of this book. Without their invaluable contributions this book wouldn't have been possible. They have made vital efforts to compile up to date information on the varied aspects of this subject to make this book a valuable addition to the collection of many professionals and students.

This book was conceptualized with the vision of imparting up-to-date information and advanced data in this field. To ensure the same, a matchless editorial board was set up. Every individual on the board went through rigorous rounds of assessment to prove their worth. After which they invested a large part of their time researching and compiling the most relevant data for our readers. Conferences and sessions were held from time to time between the editorial board and the contributing authors to present the data in the most comprehensible form. The editorial team has worked tirelessly to provide valuable and valid information to help people across the globe.

Every chapter published in this book has been scrutinized by our experts. Their significance has been extensively debated. The topics covered herein carry significant findings which will fuel the growth of the discipline. They may even be implemented as practical applications or may be referred to as a beginning point for another development. Chapters in this book were first published by Hindawi Publishing Corporation; hereby published with permission under the Creative Commons Attribution License or equivalent.

The editorial board has been involved in producing this book since its inception. They have spent rigorous hours researching and exploring the diverse topics which have resulted in the successful publishing of this book. They have passed on their knowledge of decades through this book. To expedite this challenging task, the publisher supported the team at every step. A small team of assistant editors was also appointed to further simplify the editing procedure and attain best results for the readers.

Our editorial team has been hand-picked from every corner of the world. Their multi-ethnicity adds dynamic inputs to the discussions which result in innovative outcomes. These outcomes are then further discussed with the researchers and contributors who give their valuable feedback and opinion regarding the same. The feedback is then collaborated with the researches and they are edited in a comprehensive manner to aid the understanding of the subject.

Apart from the editorial board, the designing team has also invested a significant amount of their time in understanding the subject and creating the most relevant covers. They scrutinized every image to scout for the most suitable representation of the subject and create an appropriate cover for the book.

The publishing team has been involved in this book since its early stages. They were actively engaged in every process, be it collecting the data, connecting with the contributors or procuring relevant information. The team has been an ardent support to the editorial, designing and production team. Their endless efforts to recruit the best for this project, has resulted in the accomplishment of this book. They are a veteran in the field of academics and their pool of knowledge is as vast as their experience in printing. Their expertise and guidance has proved useful at every step. Their uncompromising quality standards have made this book an exceptional effort. Their encouragement from time to time has been an inspiration for everyone.

The publisher and the editorial board hope that this book will prove to be a valuable piece of knowledge for researchers, students, practitioners and scholars across the globe.

List of Contributors

Victoria V. Roshchina
Laboratory of Microspectral Analysis of Cells and Cellular Systems and Optical Department Russian Academy of Sciences, Institute of Cell Biophysics, RAS, Institutskaya Str. 3, Pushchino, Moscow Region 142290, Russia

Anneli Ehlerding, Ida Johansson, Sara Wallin, and Henric Östmark
Department of Energetic Materials, Swedish Defense Research Agency (FOI), 147 25 Tumba, Sweden

Alexander Soin, Tatiana Maryutina, Natalya Musina and Andrey Soin
United Research and Development Centre, Leninsky prospect 55/1 Building 2, Moscow 119333, Russia

Tatiana Maryutina and Andrey Soin
Vernadsky Institute of Geochemistry and Analytical Chemistry, Russian Academy of Sciences, Kosygin Street 19, Moscow 119991, Russia

Atanu Sengupta, Chetan Shende, Stuart Farquharson and Frank Inscore
Real-Time Analyzers Inc., Unit 8, 362 Industrial Park Road, Middletown, CT 06457, USA

Evdokia Kastanos
Department of Life and Health Sciences, University of Nicosia, 1700 Nicosia, Cyprus

Alexandros Kyriakides, Katerina Hadjigeorgiou and Costas Pitris
Department of Electrical and Computer Engineering, University of Cyprus, 1678 Nicosia, Cyprus

Keisuke Niwase
Hyogo University of Teacher Education, Hyogo 673-1494, Kato, Japan

Briauna Hawthorne and Haiyan Fan-Hagenstein
Chemistry Department, Claflin University, Orangeburg, SC 29115, USA

Elizabeth Wood, Jessica Smith and Timothy Hanks
Chemistry Department, Furman University, Greenville, SC 29613, USA

Marek Tulej, Andreas Riedo, Maria Iakovleva and PeterWurz
Space Research & Planetary Sciences, Institute of Physics, University Bern, Sidlerstrasse 5, 3012 Bern, Switzerland

Mohamed A. El Hamd
Department of Pharmaceutical Analytical Chemistry, Faculty of Pharmacy, Al-Azhar University, Assiut 71524, Egypt

Sayed M. Derayea
Department of Pharmaceutical Analytical Chemistry, Faculty of Pharmacy, Minia University, Minia 61519, Egypt

Osama H. Abdelmageed
Department of Pharmaceutical Chemistry, Faculty of Pharmacy, King Abdulaziz University, Jeddah 21589, Saudi Arabia

Hassan F. Askal
Department of Pharmaceutical Analytical Chemistry, Faculty of Pharmacy, Assiut University, Assiut 71526, Egypt

Ana Rei, Michael Belsley, M. Isabel C. Ferreira and Peter Schellenberg
Centro de F´ısica, Universidade do Minho, Campus de Gualtar, 4710-057 Braga, Portugal

Graham Hungerford
HORIBA Jobin Yvon IBH Ltd., Skypark 5, 45 Finnieston Street, Glasgow G3 8JU, UK

Kunal B. Modi
Department of Physics, Saurashtra University, Rajkot, Gujarat 360005, India

Nimish H. Vasoya
Om Shanti Engineering College, Morbi Road, Rajkot, Gujarat 360030, India

Vinay K. Lakhani
Department of Physics, Bahauddin Science College, Junagadh, Gujarat 362001, India

Tushar K. Pathak
Government Engineering College, Kalawad Road, Rajkot, Gujarat 360005, India

P. M. G. Nambissan
Applied Nuclear Physics Division, Saha Institute of Nuclear Physics, Kolkata,West Bengal 700064, India

Kirill O. Bugaev, Anastasia A. Zelenina and Vladimir A. Volodin
A.V. Rzhanov Institute of Semiconductor Physics, Russian Academy of Sciences, Lavrentieva Avenue 13, Novosibirsk 630090, Russia
Physical Faculty, Novosibirsk State University, Pirogova Street 2, Novosibirsk 630090, Russia

Qiaoli Yue, Tongfei Shen, Chaohui Gao and Jifeng Liu
Shandong Provincial Key Laboratory of Chemical Energy Storage and Novel Cell Technology, Department of Chemistry, Liaocheng University, Liaocheng 252059, Shandong, China

Changna Wang
Department of Chemistry and Biochemistry, Brigham Young University, Provo, UT 84602, USA

Timothy J. Johnson, Yin-Fong Su, Kristin H. Jarman, Brenda M. Kunkel, Jerome C. Birnbaum, Alan G. Joly, Eric G. Stephan, Russell G. Tonkyn and Robert G. Ewing
Pacific Northwest National Laboratory, P.O. Box 999, Richland, WA 99354, USA

Glen C. Dunham
Dunham Research, 3251 Kemptown Church Road, Monrovia, MD 21770, USA

Zeev Karpas, Gabi Cohen, Elias Atweh, Geoff Barnard and Moshe Golan
3QBD Ltd., 8 Haplada Street, Arad 89106, Israel

Jarmo T. Alander, Vladimir Bochko, Birgitta Martinkauppi and Timo Mantere
Department of Electrical and Energy Engineering, University of Vaasa, 65101 Vaasa, Finland

Sirinnapa Saranwong
Bruker Optics KK, 1-4-1 Shinkawa, Chuo-ku, Tokyo 104-0033, Japan

Xian Zhang, Qin Zhou, Yu Huang, Zhengcao Li and Zhengjun Zhang
Advanced Materials Laboratory, Department of Materials Science and Engineering, Tsinghua University, Beijing 100084, China

Qin Zhou
Institute of Nuclear and New Energy Technology, Tsinghua University, Beijing 100084, China

Carl Bergeron and Geneviéve Delmas
Départment de Chimie, Université du Québec à Montréal, CP.888, Suc. Centre-Ville, Montreal, QC, Canada H3C 3P8

Etienne Perrier and Aymeric Potier
IUT, University of Auvergne, 63001Clermont-Ferrand, France

R. P. Kengne-Momo, Ph. Daniel, F. Lagarde and Y. L. Jeyachandran
Laboratoire PEC, UMR CNRS 6087, Université du Maine, A.O. Messiaen, 72085 Le Mans, France

R. P. Kengne-Momo
Laboratoire de Chimie Organique, Université de Yaoundé I, B.P 812, Yaoundé, Cameroon

J. F. Pilard
Laboratoire UCO2M, UMR CNRS 6011, Université du Maine, A.O. Messiaen, 72085 Le Mans, France

M. J. Durand-Thouand and G. Thouand
Laboratoire CBAC, UMR CNRS GEPEA 6144, IUT Université de Nantes, 85035 La Roche Sur Yon, France